INSTRUMENTAL METHODS OF CHEMICAL ANALYSIS

Instrumental Methods of Chemical Analysis

Third Edition

GALEN W. EWING

Professor of Chemistry
Seton Hall University

McGRAW-HILL BOOK COMPANY

New York   St. Louis   San Francisco
Toronto   London   Sydney

INSTRUMENTAL METHODS OF
CHEMICAL ANALYSIS

Library of Congress Catalog Card Number 68-25651
19851

34567890MAMM7543210

Preface

As in previous editions, the general objective in this book is to survey modern analytical instruments and techniques and to present sufficient theory for their comprehension. Emphasis is placed on the possibilities and limitations inherent in the various methods.

The text is planned for use in upper-level undergraduate or first-year graduate classes. To be taught most effectively, this course should follow work in elementary quantitative analysis and a year of physics; it may follow or run concurrently with physical chemistry.

It is always a difficult matter to decide what to include and what to omit. The words "analytical" and "instrumental" are not amenable to objective definition. With respect to the former, H. A. Laitinen has written: "The vital point here is that if the research is aimed at methods of solution of a measurement problem, it is properly classified as analytical chemistry, whereas the interpretation of the results of the measurements infringes upon other fields of chemistry." (Editorial, *Anal. Chem.*, **38**, 1441, 1966.) I have attempted to include just enough interpretive material to suggest the areas in which a method can be useful.

With respect to the term "instrumental," I have tried to be led more by usefulness to the chemistry student than by a strict definition of the term. Thus such an important instrument as the analytical balance is not discussed, whereas paper chromatography is.

The principal changes in the third edition are a reduction in space devoted to the theory of potentiometric and conductometric titrations, to refractometry, and to classical emission spectrography. Much more attention is given to gas chromatography and to recent modifications of polarography and related techniques. A unified treatment is attempted in the analytical applications of flames. Similarly many of the techniques of separation are grouped together to emphasize their basic similarities. Optical rotatory dispersion, circular dichroism, microwave absorption, and chronopotentiometry are subjects newly introduced.

The chapter on electronics is expanded and updated, with increased emphasis on solid-state devices and particularly on the unique and valuable properties of operational amplifiers. This chapter is self-supporting, and some instructors may wish to use it to introduce the course.

Stylistically, it seems appropriate to let *cuvette* follow *pipette* and *burette* in dropping the final two letters. *Millimeter-of-mercury* has given way to *torr*, *cycle-per-second* to *hertz*, and *millimicron* to *nanometer* (which, along with *kilometer*, should be pronounced more like *centimeter* than like *thermometer*).

Mention of the products of individual manufacturers does not necessarily imply that I consider them superior to competing items. The aim is to describe instruments typical of their class or possessing some special features of interest, not to write a complete catalog of analytical apparatus.

I wish to express my sincere appreciation to my colleagues and students, past and present, who have offered advice and pointed out errors. Especial thanks go to Professors J. M. Fitzgerald and R. F. Hirsch, who have read carefully and critically many chapters of manuscript. My thanks also to the personnel of instrument companies and distributors, too numerous to list, without whose cooperation the book could not be a success.

Galen W. Ewing

Contents

INSTRUMENTAL METHODS OF CHEMICAL ANALYSIS

1
Introduction

Analytical chemistry may be defined as the science and art of determining the composition of materials in terms of the elements or compounds which they contain. Historically, the development of analytical methods has followed closely the introduction of new measuring instruments. The first quantitative analyses were gravimetric, made possible by the invention of a precise balance. It was soon found that carefully calibrated glassware made possible considerable saving of time through the volumetric measurement of gravimetrically standardized solutions.

In the closing decades of the nineteenth century, the invention of the spectroscope brought with it an analytical approach which proved to be extremely fruitful. At first it could be applied only qualitatively; gravimetric and volumetric methods remained for many years the only quantitative procedures available for nearly all analyses. Gradually a few colorimetric and nephelometric methods were introduced, principally for substances for which other techniques were unknown or unreliable. Then it was found that electrical measurements could detect end points in titrations. In the years since about 1930, the rapid development of the vacuum-tube amplifier and the photoelectric tube, and more recently, of transistors and other semiconductor devices, has resulted in the establishment of many analytical methods based upon them. Today the chemist, whether he calls himself an analytical specialist or not, must have a working knowledge of a dozen or so instruments which were virtually unknown a generation ago.

Nearly any physical property characteristic of a particular element or compound can be made the basis of a method for its analysis. Thus the absorption of light, the conductivity of a solution, or the ionizability of a gas can each serve as an analytical tool. A whole series of related techniques depends upon the varying electrical properties of different elements, as evidenced by their redox potentials. The phenomena of artificial radioactivity have led to several analytical methods of extremely great significance. It is the purpose of this book to investigate the possibilities of many of these modern instrumental methods of analysis.

PHYSICAL PROPERTIES USEFUL IN ANALYSIS

The following is a list of physical properties which have been found applicable to chemical analysis. The list is not exhaustive, but it certainly includes all those properties which have been extensively investigated, as well as some not yet fully exploited.

EXTENSIVE PROPERTIES

1. Mass
2. Volume (of a liquid or a gas)

MECHANICAL PROPERTIES

1. Specific gravity (or density)
2. Surface tension
3. Viscosity
4. Velocity of sound

PROPERTIES INVOLVING INTERACTION WITH RADIANT ENERGY

1. Absorption of radiation
2. Scattering of radiation
3. Raman effect
4. Emission of radiation
5. Refractive index and refractive dispersion
6. Rotation of the plane of polarized light and rotatory dispersion
7. Circular dichroism
8. Fluorescence and phosphorescence
9. Diffraction phenomena
10. Nuclear and electron magnetic resonance

ELECTRICAL PROPERTIES

1. Half-cell potentials
2. Current-voltage characteristics
3. Electric conductivity
4. Dielectric constant
5. Magnetic susceptibility

THERMAL PROPERTIES

1. Transition temperatures
2. Heats of reaction
3. Thermal conductivity (of a gas)

NUCLEAR PROPERTIES

1. Radioactivity
2. Isotopic mass

METHODS OF SEPARATION PRIOR TO ANALYSIS

It would be desirable to discover analytical methods which are *specific* for each element or radical or class of compounds. Unfortunately only a few methods are completely specific,* and it is therefore frequently necessary to perform quantitative separations with the objective either of isolating the desired constituent in a measurable form or of removing interfering substances. Some methods of separation are the following:

1. Precipitation
2. Electrodeposition
3. Formation of complexes
4. Distillation
5. Solvent extraction or sublation
6. Partition chromatography
7. Adsorption chromatography
8. Ion exchange
9. Electrophoresis
10. Dialysis

FUNDAMENTAL AND DERIVED PHYSICAL QUANTITIES

Fundamental physical quantities that can be measured directly are surprisingly few. Most of the measurements which we make in the laboratory consist essentially in the observation of linear or angular displace-

* An example of an analysis which is specific is the resonance absorption of radiation by atoms of the same element giving rise to the radiation (atomic absorption).

ment, by comparison with some kind of scale.   In using the analytical balance, we actually note the displacement of a pointer or its equivalent, and adjust the weights to bring the displacement to zero.   The buret is read by observation of the linear displacement of the meniscus from its initial to its final position.   Electrical measurements are made through the angular displacement of meter needles or of potentiometer dials, and so on.   Many other quantities, such as the intensities of light or sound, must serve only as null indicators unless a device is available to convert the quantity to a form which can be read on a meter.   It is the function of the instrument to translate chemical composition into information directly observable by the operator.   In nearly all cases, the instrument acts either directly or indirectly as a *comparator*, in that the unknown is evaluated relative to a standard.

Most of the analytical methods to be described rest on sound mathematical theory.   Occasionally there is reported an experimental procedure which is mostly empirical, with little theoretical background.   Such a method may be usable for analytical purposes, but it must be proved valid by exhaustive study and independent checking of data so that the analyst may have certain knowledge of what he is actually measuring.

TITRATION

Titration is defined as the measurement of an unknown constituent by establishment of the exactly equivalent amount of some standard reagent.   Physical measurements are involved in two ways: in the detection of the equivalence point and in the measurement of the quantity of reagent consumed.   Usually, and unless otherwise specified, the quantity of reagent is measured volumetrically with a buret.   The chief exception is the *coulometric titration*, where the reagent is generated electrolytically on the spot as required, and its quantity determined by electrical measurements; *photochemical* generation has recently been demonstrated as suitable for titrations.

BIBLIOGRAPHY

The student who wishes to pursue in greater depth any of the topics mentioned in this book has many avenues to which to turn.   There are of course the general sources, such as *Chemical Abstracts*, applicable to all branches of chemistry.

In the analytical field there is a great proliferation of journals of primary interest.   *Analytical Chemistry, Analytica Chimica Acta, Talanta, The Analyst* (which includes *Analytical Abstracts*), and the *Zeitschrift für analytische Chemie* attempt general analytical coverage. In specific fields are the *Journal of Electroanalytical Chemistry,* the *Journal of Chromatography,* the *Journal of Gas Chromatography, Spectrochimica Acta, Analytical Biochemistry,* and many others. With emphasis on instruments per se, one finds the *Review of Scientific Instruments,* the *Journal of Scientific Instruments, Instrumentation Technology* (formerly the *Instrument Society of America Journal*), and the *Instrument Society of America Transactions.* The *Journal of Chemical Education* runs a monthly column on topics in chemical instrumentation, in addition to many articles of analytical interest.

On the theoretical side, the *Treatise on Analytical Chemistry,* edited by I. M. Kolthoff and P. J. Elving (Interscience Publishers, Division of John Wiley & Sons, Inc.), is invaluable, especially Part I. Also not to be overlooked is the series *Advances in Analytical Chemistry and Instrumentation,* edited by C. N. Reilley (and, for vol. 5, F. W. McLafferty), and published by the same house.

The *Annual Reviews* issue of *Analytical Chemistry,* published each April, contains critical reviews in all fields of analysis; in even years the reviews are classed by the analytical principles involved, and in odd years by fields of application. The *Treatise on Analytical Chemistry* and the *Annual Reviews,* taken together, provide the best entry into a new field.

An immense amount of useful information, with succinct reviews of theoretical principles, has been collected under the editorship of L. Meites in the *Handbook of Analytical Chemistry,* published by McGraw-Hill Book Company in 1963.

2
Introduction to Optical Methods

A major class of analytical methods is based on the interaction of radiant energy with matter. In the present chapter we shall review some pertinent properties, both of radiation and of matter, and then discuss those features of optical instrumentation which apply to all or several spectral regions in common. In subsequent chapters each major spectral range (visible, ultraviolet, infrared, x-ray, microwave) will be considered separately, with respect to instrumentation and chemical application.

THE NATURE OF RADIANT ENERGY

An investigation into the properties of radiant energy reveals an essential duality in our understanding of its nature. In some respects its properties are those of a wave, while in others it is apparent that the radiation consists of a series of discrete packets of energy (*photons*). The photon concept is almost always required in the rigorous treatment of the interactions of radiation with matter, although the wave picture may be used to give approximately correct results when large numbers of photons are involved.

Radiant energy can be described in terms of a number of properties or parameters. The *frequency* ν is the number of oscillations per second described by the electromagnetic wave; the units of frequency are the *hertz* (1 Hz = 1 cycle-sec^{-1}) and the *fresnel* (10^{12} Hz). The *velocity c* of propagation is very nearly 2.998×10^8 m-sec^{-1} for radiation traveling through a vacuum, and somewhat less for passage through various transparent media.

The *wavelength* λ is the distance between adjacent crests of the wave in a beam of radiation. It is given by the ratio of the velocity to the frequency. The units of wavelength are *angstroms* (1 Å = 10^{-10} m), *microns* (1 μ = 10^{-6} m), or *nanometers* (1 nm = 10^{-9} m = $10^{-3} \mu$ = 10 Å). The nanometer is also designated the millimicron (mμ); the term nanometer follows the recommendations of the National Bureau of Standards in 1963.[10] Another quantity which is convenient in some circumstances is the *wave number* λ^{-1} which is the number of waves per centimeter.* The unit for wave number is the *reciprocal centimeter* (cm^{-1}), for which the name *kaiser* (K) has been suggested.

The velocity, wavelength, frequency, and wave number are related by the expressions

$$c = \nu\lambda = \frac{\nu}{\lambda^{-1}} \tag{2-1}$$

The energy content E of a photon is directly proportional to the frequency

$$E = h\nu = h\frac{c}{\lambda} = hc\lambda^{-1} \tag{2-2}$$

where h is Planck's universal constant, very close to 6.6256×10^{-27} erg-sec.[10] Thus there is an inverse relationship between energy content and wavelength, but a direct relation between energy and frequency or wave number. It is for this reason that the presentation of spectra in terms of frequency or wave number rather than wavelength is gaining favor.

It is convenient, particularly with nuclear radiations and x-rays, to characterize the radiation by the energy content of its photons in *electron volts* (eV); 1 eV = 1.602×10^{-12} erg, and corresponds to fre-

* It is unfortunate that the symbol $\bar{\nu}$ is often used to indicate the wave number, because of possible confusion with ν for frequency; in certain areas of physics it is usual to interchange these symbols. In this book we will use the slightly cumbersome but unambiguous symbol λ^{-1}. There can be no justification for expressions such as "a frequency of 1600 cm^{-1}," which is all too often found in the literature; frequency *must* have the dimensions of reciprocal time, *never* reciprocal distance.

quency $\nu = 2.4186 \times 10^{14}$ Hz or to the (in vacuo) wavelength

$$\lambda = 1.2395 \times 10^{-6} \text{ m}$$

The multiples keV and MeV are frequently encountered.

A beam of radiation consists of energy which is emitted from a source and propagated through a medium or series of media to a receptor where it is absorbed. On its way from source to ultimate absorber, the beam may suffer partial absorption by the media through which it passes, it may be changed in direction by reflection, refraction, or diffraction, and it may become partially or wholly polarized.

Since energy per unit time is power, it is correct to speak of the *radiant power* of the beam, a quantity often loosely referred to as intensity. *Intensity* more correctly refers to the power emitted by the source per unit solid angle in a particular direction. A photoelectric cell gives a response related to the total *power* incident upon its sensitive surface. A photographic plate, on the other hand, integrates the power over the period of time of exposure to the beam, and hence its response (silver deposit) is a function of the total incident *energy* (rather than power) per unit area. In both photoelectric cells and photographic plates, as well as in the human eye, the sensitivity is a more or less complicated function of the wavelength, and this must be taken into consideration in their use.

SPECTRAL REGIONS

The spectrum of radiant energy is conveniently broken down into several regions, as shown in Table 2-1. The limits of these regions are deter-

Table 2-1   Regions of the electromagnetic spectrum*

Designation	Wavelength limits		Frequency limits, Hz†	Wave number limits, cm⁻¹†
	Usual units	Meters		
X-rays	10^{-2}–10^2 Å	10^{-12}–10^{-8}	10^{20}–10^{16}	
Far ultraviolet	10–200 nm	10^{-8}–2×10^{-7}	10^{16}–10^{15}	
Near ultraviolet	200–400 nm	2×10^{-7}–4.0×10^{-7}	10^{15}–7.5×10^{14}	
Visible	400–750 nm	4.0×10^{-7}–7.5×10^{-7}	7.5×10^{14}–4.0×10^{14}	25,000–13,000
Near infrared‡	0.75–2.5 μ	7.5×10^{-7}–2.5×10^{-6}	4.0×10^{14}–1.2×10^{14}	13,000–4000
Mid infrared‡	2.5–50 μ	2.5×10^{-6}–5.0×10^{-5}	1.2×10^{14}–6.0×10^{12}	4000–200
Far infrared‡	50–1000 μ	5.0×10^{-5}–1×10^{-3}	6×10^{12}–10^{11}	200–10
Microwaves	0.1–100 cm	1×10^{-3}–1	10^{11}–10^8	10–10^{-2}
Radio waves	1–1000 m	1–10^3	10^8–10^5	

* Where a numerical factor is omitted, it is because the precision of delineation of the region does not warrant a greater number of significant figures.
† Calculated from $\nu = c/\lambda$, where $c = 3.0 \times 10^8$ m/sec.
‡ The limits for the subdivisions of the infrared follow the recommendations of the Triple Commission for Spectroscopy; *J. Opt. Soc. Am.*, **52**:476 (1962).

mined by the practical limits of appropriate experimental methods of production and detection of radiations. The figures quoted in the table are not in themselves especially significant, and should be considered only as rough boundaries.

The differentiation of spectral regions has additional significance to the chemist in that the physical interactions follow different mechanisms and provide different kinds of information. The most important atomic or molecular transitions pertinent to the successive regions are:

X-ray...................... K- and L-shell electrons
Far ultraviolet.............. Middle-shell electrons
Near ultraviolet and visible.... Valence electrons
Near and mid-infrared .... Molecular vibrations
Far infrared .... Molecular rotations and low-lying
                                                            vibrations
Microwave .... Molecular rotations

These will now be considered in detail.

INTERACTIONS WITH MATTER: ATOMIC SPECTRA

Electromagnetic radiation originates in the deceleration of electrically charged particles and can be absorbed by the reverse process, contributing its energy to produce acceleration. Hence an understanding of the interactions between matter and radiation can only be built upon a knowledge of the structure of atoms and molecules. Figures 2-1 and 2-2 show typical energy-level diagrams for an atom and a molecule, respectively.

Consider first the energy levels of an atom, which are represented by horizontal lines in Fig. 2-1. The vertical lines in this figure indicate permitted electronic transitions between the various levels. The greater the vertical distance between two levels, the greater the energy difference, and the greater energy a photon must have to be emitted or absorbed in such a transition.

In a normal (i.e., unexcited) atom, the electrons occupy as many levels as needed, starting with the lowest ($1s$) and proceeding upward according to the well-known quantum rules. Sodium, for example, has 11 electrons, designated $1s^2$, $2s^2$, $2p^6$, $3s^1$. The $3s$ electron is the least strongly held and hence can easily be pushed upward from the $3s$ to the $3p$ level, which is an example of electronic excitation. This can be accomplished by providing energy in any one of a number of forms. The excited electron has a strong tendency to return to its normal state, the $3s$ level, and in doing so emits a quantum of radiation (a photon). This

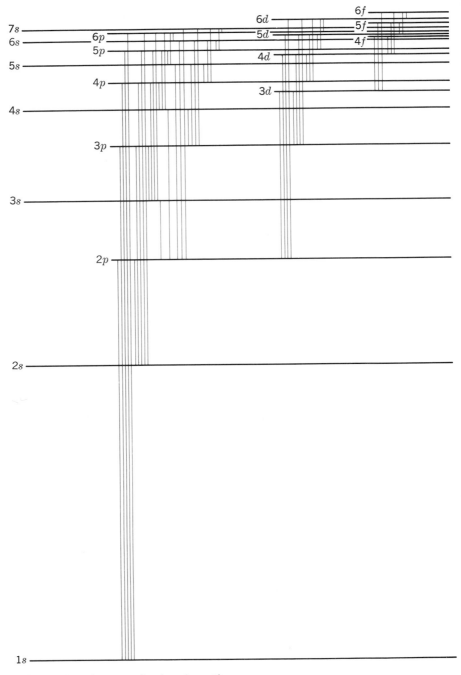

Fig. 2-1 Atomic energy levels, schematic.

emitted photon possesses a very definite and uniform amount of energy, dictated by the spacing of the energy levels. In the present example, it constitutes the familiar yellow light characteristic of flames or lamps containing sodium in the vaporized state. This simple case, where the outer electron is raised by one energy level and then returns, is known as *resonance radiation*. The important analytical technique known as *atomic absorption* is based on this phenomenon.

If the electron is given more than enough energy to produce resonance, it may become more highly excited, raised to some level higher than $3p$, say $4p$. Then it may not drop back to $3s$ by a single process, but may pause at intermediate levels, like a ball rolling down steps. This situation no longer fits the definition of resonance radiation, but is more complex. For one thing, not all conceivable transitions are actually possible—some are "forbidden" by the selection rules of quantum mechanics. (Thus the $3s$ electron of sodium cannot be raised to the $4s$ state.)

With a highly potent source of energy, many electrons (not only the outermost) in any element can be excited to varying degrees, and the resulting radiation may contain as many as several thousand discrete and reproducible wavelengths, mostly in the ultraviolet and visible regions. This is the basis of the analytical method of *emission spectroscopy*.

If the source of excitation is extremely energetic, an inner electron can be torn entirely away from its atom. An electron from some higher level will then drop in to fill the vacancy. Since the energy change corresponding to this transition is much greater than in the case of excited outer electrons, the photons radiated will be of much greater frequency and correspondingly shorter wavelength. This describes the emission of x-rays from atoms subjected to bombardment, for example, by a beam of fast-moving electrons.

MOLECULAR SPECTRA

In a typical molecule, as contrasted with an atom, a few energy levels might show relations such as those of Fig. 2-2. The molecule to which this diagram applies has a singlet ground state designated S_0, which represents its normal, unexcited condition. Two series of excited states can exist, the *singlet* series, $S_1, S_2, \ldots S_n$, and the *triplet* series, $T_1, T_2, \ldots T_m$. A triplet level generally has less energy than the corresponding singlet. These two series refer to a difference in the net electronic spin of atoms in the various levels.

It is difficult to effect a change in electron spin, so the absorption

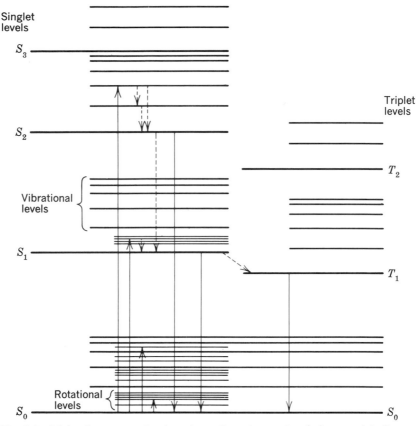

Fig. 2-2   Molecular energy levels, schematic.   Arrows headed upward indicate absorption of radiation; solid arrows downward, emission of radiation; broken arrows, nonradiative degradation.   Rotational sublevels are associated with all vibrational levels, though only a few are shown.

of radiant energy is restricted to raising an atom from S_0 to a higher S level or sublevel.   Triplet states can be reached only by a roundabout process.   Similarly it is difficult for a molecule in a triplet state to drop back to the ground level.

Each electronic level (S or T) has associated with it a series of *vibrational sublevels,* which correspond to the energy required to excite various modes of vibration within the molecule.   The sublevels related to each vibrational level correspond to the energy of rotation of atoms or groups of atoms within the molecule, and so are called *rotational sublevels.*

Excellent discussions of this whole field can be found in the books by Barrow,[1] by Jaffé and Orchin,[6] and others.   A convenient summary has been given by Jaffé and Miller[5] in the *Journal of Chemical Education*.

ABSORPTION SPECTRA

Transitions within molecules are usually studied by the selective *absorption* of radiation passing through them, and less commonly by emission processes such as fluorescence and phosphorescence.   Transitions between electronic levels are found in the ultraviolet and visible regions; those between vibrational levels, within the same electronic level, are in the near- and mid-infrared; and those between neighboring rotational levels, in the far-infrared and microwave regions.

Electronic transitions involve jumps to and from the various sublevels, so that ultraviolet absorption spectra always consist of *bands* because of the spread of energies of photons required to excite molecules from all the many vibrational and rotational states of the ground level to a similarly large number of vibrational and rotational states of the excited levels.

Ultraviolet spectra frequently show *fine structure* corresponding to these various possible transitions.   The fine structure is sharpened or increased in detail by making observations in the gas phase rather than liquid, and also by cooling the sample, as with liquid nitrogen.   These specialized techniques are not usually required in analytical work.

Absorption spectra are readily measured in each spectral region and are of great utility in analytical studies, as will become amply evident in subsequent chapters.

FLUORESCENCE

The energy gained by a molecule which absorbs a photon does not remain in that molecule, but is lost or degraded by any of several mechanisms. It may be emitted as radiation of the same wavelength as that absorbed (*resonant fluorescence*).   Of greater importance in solution chemistry is the case where a part of the energy is degraded to heat, lowering the net energy of the molecule to the lowest vibrational and rotational level within the same electronic (singlet) level.   The remainder of the energy is then radiated, to return the molecule to its ground state.   This is the phenomenon of *fluorescence*.   The emitted radiation has less energy per photon than the exciting radiation, and hence a longer wavelength.

Many organic and some inorganic compounds, when irradiated with ultraviolet, fluoresce in the visible spectrum.   Fluorescence is also important in the x-ray field, where irradiation of a sample with high-energy x-rays is a convenient method of exciting x-ray spectra of lower frequency.

PHOSPHORESCENCE

In some molecules it is possible for a nonradiative transition to take place from an excited singlet state to the corresponding triplet level, from which the remaining energy is radiated as the molecule reverts to its ground state. The triplet state, however, is metastable, which means that the probability of transition back to the singlet ground state is small. As a result, phosphorescence may last for a measurable time interval after the exciting radiation is turned off, contrasting with fluorescence which has no measurable persistence.

RAMAN SPECTRA

A phenomenon which bears some relation to fluorescence is the *Raman effect*. Here, also, radiation is emitted from the sample with a change in wavelength from the exciting incident radiation. But whereas to excite fluorescence the primary radiation must be absorbed by the sample, the incident radiation in order to produce the Raman effect must *not* be appreciably absorbed. The shift in wavelength in the Raman effect is caused by the extraction of energy from the quanta of incident radiation to raise molecules to higher vibrational states. The emergent quanta can thus be thought of as the same ones which entered, but with less energy.

Since the vibrational levels are subject to quantum rules, the energy change in the Raman effect is also quantized, and discrete wavelength shifts are observed. Occasionally there is a Raman shift toward higher energies. This effect arises with molecules which are so easily excited that a relatively large proportion already possess vibrational energy in excess of their ground states. The excess energy can be lost to the radiation in the reverse of the more common process. The lines so produced in the spectrum are known as *anti-Stokes* lines in contrast with the *Stokes* lines, which are of longer wavelength than the exciting source.

Vibrational transitions can be observed both in infrared absorption and in the Raman effect, but not all possible transitions are observable in both. It can be shown by quantum-mechanical theory that infrared absorption will result only from those vibrations which are accompanied by a change in the *dipole moment* of the molecule, that is, where the centers of positive and negative charge are displaced from each other in varying degree as the molecule vibrates. On the other hand, only vibrational modes which result in a change in *polarizability* will be visible in Raman spectroscopy. Polarizability may be defined as the ability of a molecule to be deformed by an electric field, separating temporarily the centers of positive and negative charge.

Thus for a study of molecular structure, the Raman and infrared techniques supplement each other; for analytical purposes, the choice may depend upon such factors as convenience and the availability of equipment.

REFRACTION

We now turn from atomic and molecular phenomena to "bulk" phenomena concerning matter in its interaction with radiation.

The *index of refraction* is an important bulk property of matter. It is defined as the ratio of the velocity of radiation of a particular frequency in a vacuum to that in a medium. The variation of the refractive index of a substance with wavelength is called its *refractive dispersion*, or simply its *dispersion*. The dispersion of a substance throughout the electromagnetic spectrum is intimately related to the degree to which

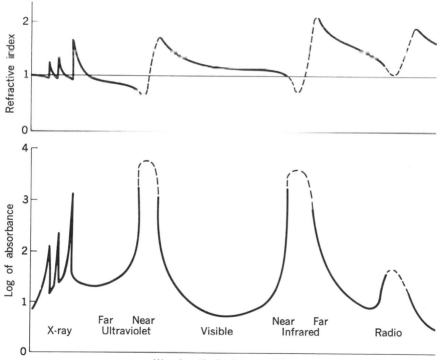

Fig. 2-3 Refractive index and absorbance as functions of wavelength over the whole electromagnetic spectrum. Schematic for a hypothetical substance. (*Upper section after Jenkins and White.*[7])

radiation is absorbed. In regions of high transparency, the refractive index decreases with increasing wavelength (not linearly); in regions of high absorbance, the index is usually difficult to measure precisely, but must show an abrupt rise with increasing wavelength. Figure 2-3 shows schematically the absorption spectrum and the dispersion curve* of a substance which is transparent to visible radiations. The shape of the dispersion curve in regions of transparency is an important property, particularly with solids, because it is this curve which dictates the design of lenses and prisms.

POLARIZATION AND OPTICAL ACTIVITY

Yet another property sometimes shown by matter is its ability to polarize light. A beam of normal radiation can be thought of as a bundle of waves with their vibratory motions distributed over a family of planes, all of which include the line of propagation. Figure 2-4a shows a cross section of such a ray which is proceeding in a direction perpendicular to the plane of the paper. If this beam of light is passed through a component called a *polarizer*, each separate wave of the bundle, for example, that vibrating along the vector **AOA'** (Fig. 2-4b), is resolved into its components **BOB'** and **COC'** in the directions of the orthogonal X and Y axes characteristic of the polarizer. The polarizing material has the property of eliminating one of these component vibrations (say **COC'**)

* The existence of indices of refraction less than unity in portions of this curve suggests velocities of radiation greater than the in vacuo value, in violation of relativity theory. Actually this is due to a difference in the definition of velocity. See Jenkins and White,[7] p. 477.

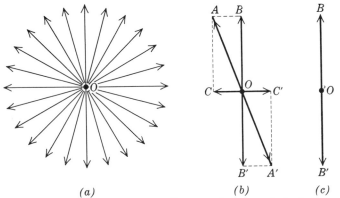

(a)	*(b)*	*(c)*

Fig. 2-4 Vibration vectors in ordinary and plane-polarized electromagnetic radiation.

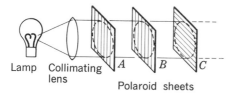

Lamp Collimating A B C
        lens
        Polaroid sheets

Fig. 2-5 Plane-polarization of radiant energy.

and passing the other (**BOB′**). Thus the emergent beam consists of vibrations in one plane only (Fig. 2-4c) and is said to be *plane-polarized*.

A second polarizer (called the *analyzer*) placed in the beam will similarly pass only that component of the light vibrating parallel to its axis. Since the beam is already polarized, that means that in one position essentially all the radiation will come through, but turning the analyzer through a 90° angle will reduce the power to zero. This is illustrated in Fig. 2-5; radiation from a lamp, rendered parallel by a collimating lens, passes through polarizer A, which has its axis vertical. The analyzer B, also with a vertical axis, has no further effect on the beam, but C, with its axis oriented horizontally, cuts the light to zero. If C is rotated in its own plane, the power of the radiation transmitted varies as the sine of the angle. Two polarizers placed in series are said to be *crossed* if their axes are mutually perpendicular. A beam of radiation may possess any degree of plane polarization from zero (complete symmetry) to 100 percent (complete polarization).

Polarization is of importance in chemistry because of the ability exhibited by some crystals and liquids to rotate the plane of polarized light passed through them. This property is known as *optical activity*. Its variation with wavelength is called *rotatory dispersion* and is related to regions of absorption in much the same manner as the refractive dispersion.

A number of transparent crystalline materials show a phenomenon known as *double refraction* or *birefringence*, which is evidenced by the fact that a beam of light passing into the crystal is split into two beams of equal power which diverge from each other at a small angle. The two beams are found to be plane-polarized at right angles to each other. This effect is of great value in the identification and study of crystals; it is also important in that it permits design of devices useful in the production and measurement of polarized light.

A number of optical components can serve as polarizers. One class consists of a variety of prisms made of birefringent crystals, particularly quartz or calcite, cut with particular reference to their optical axes, and mounted in pairs, either in contact or with an air space. Each type of dual prism is able to resolve a beam of unpolarized radiation into X- and Y-polarized components. Those prisms in which we are interested are so designed that they pass one component nearly undeviated while they direct the other away from the optical axis of the system. Such prism assemblies usually go by the names of their inventors; best known is the Nicol, but somewhat superior for instrument use are the Glan-Thompson and the Rochon. For details see any optics text.

Another kind of polarizer depends on the combined effects of myriad submicroscopic crystals embedded in a film of plastic material. A stress is applied during manufacture so that all the tiny crystals line up with their axes parallel. *Polaroid* is the best known example; it is much less expensive than a crystal prism, especially where a large area is required, but cannot be expected to be so perfect optically.

The concepts of circularly and elliptically polarized radiation are considered in Chap. 10.

PRACTICAL SOURCES OF RADIATION

From a purely physical standpoint it is convenient to classify sources according to whether they produce continuous or discontinuous spectra.

Continuous sources (sometimes called "white" sources) emit radiations over a wide band of wavelengths. They are utilized in the study of absorption spectra, and as illuminants in such other fields as microscopy and turbidimetry.

The most familiar sources of continuous radiation are based on incandescence. Any substance at a temperature above absolute zero emits radiation. The theory of this thermal emission has been rather thoroughly worked out, in terms of an ideal emitter, called a *black body*. Figure 2-6 shows the manner in which black-body radiation is distributed as a function of wavelength for various temperatures.[7] Note that the wavelength corresponding to maximum energy moves both toward higher energies and toward shorter wavelengths as the temperature is raised. This means that incandescent sources are very practical in the infrared and visible, but must be operated at inconveniently high temperatures for appreciable ultraviolet coverage.

Actual materials may vary to a considerable degree from the black-body curves, in that they may give less emission in some wavelength regions for a given temperature. These are sometimes called "gray bodies."

In the part of the infrared region which is of greatest analytical importance, the *Nernst glower* is the most widely used continuous source. It consists of a rod or hollow tube about 2 cm long by 1 mm in diameter, made by sintering together oxides of such elements as cerium, zirconium, thorium, and yttrium. It is maintained at a high temperature by electrical heating and can be operated in air, since it is not subject to oxida-

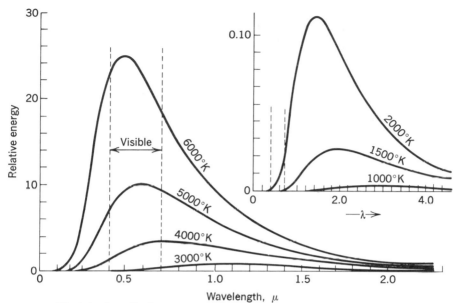

Fig. 2-6 Black-body radiation as a function of temperature. (*Adapted from Jenkins and White,[7] p. 432.*)

tion. Another source, the *Globar*, is a rod of silicon carbide of somewhat larger dimensions than the Nernst glower. Although the Globar is operated at lower temperatures (to avoid oxidation), it must be water-cooled because of the tendency to overheat at the terminals; it has greater emissivity than the glower at wavelengths longer than about 30 μ. A simple coil of *Nichrome* wire can be used as an infrared source if the required wavelength range and intensity are not too great.

For the near infrared and visible regions, a *tungsten* filament lamp in a glass or silica bulb is invariably used. Its life at high temperatures can be extended by the inclusion of a small pressure of iodine vapor, the *quartz-iodine* lamp. The iodine apparently reacts with vaporized tungsten to form a volatile compound which pyrolyzes when it comes into contact with the hot filament, redepositing the tungsten atoms on the filament rather than allowing them to accumulate on the cooler walls of the bulb.

For a continuous source in the ultraviolet, recourse must be taken to a relatively high-pressure *gas discharge*. An electric discharge through a gas typically produces a line spectrum. At low pressures each "line" approaches a single wavelength, but as the pressure is increased, the lines broaden in proportion. At sufficiently high pressures, neighboring lines coalesce, and a continuous spectrum results. The pressure required for a given degree of broadening depends in a complex manner on the molecular weight of the gas.   Continuous spectra can be obtained from discharges in various gases, for example, in *xenon* at several atmospheres and in *mercury* vapor at pressures which may go higher than 100 atm. The very useful continuous discharges in *hydrogen* and *deuterium* in the neighborhood of 10-torr pressure arise because of the complete dissociation of these gases at the voltages applied.   Electrons are continually dropping from varying "infinite" distances to any of the atomic or molecular energy levels, resulting in a continuous spectrum with a definite cutoff limit toward longer wavelengths, and with considerable superimposed fine structure.[11]

All of these gaseous sources are usable through the ultraviolet (from as low as 160 nm for a specially processed hydrogen lamp) into the near infrared (about 3.5 μ for xenon), though no one lamp will cover this wide a range.   One of the most intense sources is a high-pressure quartz-capillary mercury lamp, but it is seldom selected for analytical instruments, because it requires water cooling, which is inconvenient.   A less powerful mercury lamp, which does not require forced cooling, is useful over nearly as wide a range.

The xenon lamp is the next most intense and covers the widest spectral range; it requires a special power supply (as does the mercury lamp also) and is relatively costly.   The hydrogen arc lamp is convenient to use and relatively inexpensive, but its output continuum does not extend to wavelengths higher than about 375 nm.   The deuterium lamp has about the same range as the hydrogen, produces somewhat greater intensity, has a longer life expectancy, and about twice the cost.

In the x-ray region, radiation results from the bombardment of a target, usually by an electron beam.   The interaction between the projectile electrons and the atomic electrons of the target produces a continuous spectrum because the incoming electrons are slowed down gradually by successive interactions.   The deceleration of these charged particles results in radiation characterized by a broad spectral band with a definite minimum wavelength corresponding to the maximum energy of the electrons in the beam.   Continuous x-ray spectra always have a number of narrow emission lines superimposed upon them, the result of excited atomic electrons falling back into the ground-state levels.

At the other end of the spectrum, in the microwave and radio wave regions, there is no convenient mechanism for producing a continuum of radiations simultaneously.   This shortcoming is more than outweighed by the convenience of "scanning" the wavelength region by means of a variable-frequency electronic oscillator.   The techniques are so different from those encountered in other parts of the electromagnetic spectrum that they will not be discussed further in this introductory chapter.

LINE SOURCES

Sources producing discrete wavelengths are required for some instrumental applications. In the ultraviolet and visible regions a line spectrum can be obtained easily from an arc discharge in a gas containing excited monatomic neutral or ionic species such as a metallic vapor or a noble gas. Excitation may be thermal, as in a flame, as well as electric, in an arc or spark. Elements can be found which give lines spaced throughout most of the ultraviolet and visible regions.

Characteristic infrared emission lines can be obtained from heated polyatomic gases. X-ray line spectra have already been mentioned.

A special type of line source in the visible and ultraviolet is the *hollow-cathode lamp*. This device is filled with a noble gas at a pressure of a few torrs, to sustain an arc discharge. The cathode is in the form of a hollow cylinder, closed at one end, out of which the radiation is obtained. The anode is a straight wire at one side. The cathode is made of (or lined with) the metal whose line spectrum is desired. The energy of the arc causes metallic atoms to be ejected from the surface by the process of "sputtering." It is these ejected atoms which become excited and emit their characteristic spectra. If the potential is kept low enough, only the resonance radiation is emitted with appreciable intensity.

LASERS

A laser is a highly specialized source of monochromatic radiation principally in the red and infrared regions. The first example, reported in 1960 and still one of the best, consists of a carefully machined rod of ruby (Al_2O_3 with Cr_2O_3 as a minor constituent) with precisely parallel ends. One end is silvered so that all light approaching from the interior of the crystal is reflected back. The other end is coated with a thin layer of silver so that only a fraction (typically 80 to 90 percent) of the incident light is reflected back. When the rod is subjected to an intense flash of light, as from a xenon discharge lamp (Fig. 2-7), nearly all the chromium atoms become excited, and most of them immediately drop into a metastable energy level (cf. discussion of phosphorescence above). Then the first electrons to return to the ground state from the metastable level radiate photons of the corresponding wavelength, 694.3 nm. Some of this light is directed parallel to the axis of the rod and is reflected back and forth many times. Laser action results because the presence of this radiant energy at exactly the required frequency *stimulates* emission from the remainder of the metastable chromium atoms, so that the radiant flux builds up rapidly. At every reflection from the partially silvered

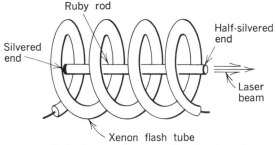

Fig. 2-7 Ruby laser, schematic; the simplest of many types.

end some light escapes, forming the output of the device. The action is so efficient that a large pulse of monochromatic light is emitted within a period of 0.5 msec. The power in each pulse may reach the megawatt level.

Lasers can be made with other solid active materials, notably neodymium oxide, and with certain liquids and gases. Gas lasers are energized by a high-voltage electric discharge within the gas itself, eliminating the need for an external light source. They can be made to emit light continuously as well as in pulses; this light is even more nearly monochromatic, but with less power, than a ruby laser.

Light from a laser has several unique properties. It is highly *monochromatic*, though only a relatively small number of discrete wavelengths can so far be produced. The light emitted is *coherent*, which means that the waves originating from all the atoms of the emitting substance are in phase with each other (not true of conventional light sources). Partly as a consequence of the coherency, the collimated beam of laser radiation has very little tendency to spread out (lose collimation) as it propagates. This permits a large amount of energy to be concentrated on a small target, even at a considerable distance.

The importance of lasers for analytical purposes lies in the high degree of monochromaticity and the high power levels which can be achieved. They find application as a source of localized heating, as an exciter for Raman spectroscopy, as an illuminant for precise interferometry, etc.

WAVELENGTH SELECTION

In the study of absorption spectra and absorptiometric analysis, it is usually necessary to utilize a rather narrow band of wavelengths. In some instances a line source provides the narrow band required, but more often it is desirable to start with radiation from a continuous source, and select a band of wavelengths from it. This results in greater flexibility than a line source would provide, as the chosen band can be taken at any desired location within the range covered by the source.

There are two basic methods of wavelength selection: (1) the use of filters and (2) geometrical dispersion by means of a prism or diffraction grating.

FILTERS

A filter is a device which will transmit radiations of some wavelengths but absorb wholly or partially other wavelengths. Filters employed in the visible region are usually of colored glass. Dyed gelatin or similar materials can be used, but though cheaper, are much less permanent in nature. A great number of glass filters are available, more or less evenly spaced through the visible region. Figure 2-8 shows the transmission curves of a series of filters manufactured by the Corning Glass Works.

Filters are also made which function on the interference principle. Figure 2-9 represents a section through such an *interference filter*. This device consists of a layer of transparent material such as magnesium fluoride, which is coated on each side with a thin film of silver. Each silver film reflects about half and transmits the other half of any radiation which strikes it. Part of the incident radiation is reflected repeatedly by the silver layers, but at each reflection some is transmitted outward. The several emergent rays to the right reinforce each other for those wavelengths which are exactly *even* multiples of the distance separating

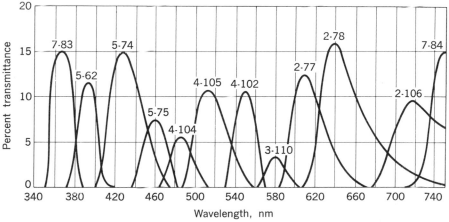

Fig. 2-8 Transmission spectra of some glass filters. (*Corning Glass Works, Corning, N.Y.*)

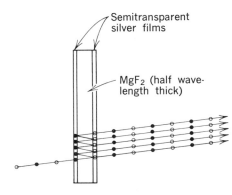

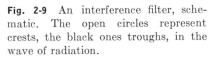

Fig. 2-9 An interference filter, schematic. The open circles represent crests, the black ones troughs, in the wave of radiation.

the silver films. For other wavelengths, the beams interfere destructively, so that essentially no energy is passed. In a commercial interference filter, the layers shown in the figure are sandwiched between two transparent plates. Transmission curves for a few interference filters are shown in Fig. 2-10. The wavelength bands isolated are much narrower and the peak transmittances much greater than is the case with colored glass filters.

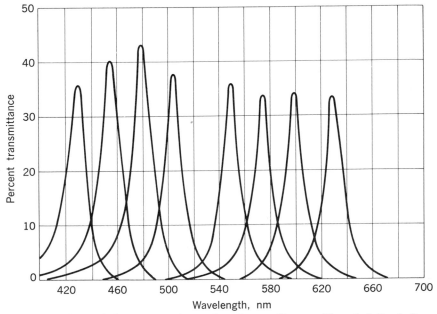

Fig. 2-10 Transmission spectra of some interference filters. (*Bausch & Lomb, Inc., Rochester, N.Y.*)

Filters of one type or another are available which, alone or in combination, permit selection of wavelength bands at almost any region of the spectrum, from x-rays through the infrared.

MONOCHROMATORS

A monochromator is a device which will isolate a band of wavelengths, usually much narrower than is obtainable by a filter. The most significant distinction between the two is that the location of the passed band can be varied over a comparatively wide range of the spectrum with a monochromator, whereas a filter permits only slight, if any, spectral variation.

The monochromator consists of a dispersing element (a prism or a diffraction grating), together with two narrow slits which serve as entrance and exit ports for the radiation. The entrance slit defines a narrow beam of radiation which falls on the dispersing element. The action of this component is to deflect the beam through an angle which depends on wavelength, and thus "fans out" the beam, as shown for the simplest cases in the visible region in Fig. 2-11. An exit slit can be positioned so as to pass a narrow band of wavelengths at any point in the spectrum. (A practical monochromator generally requires lenses or mirrors, or both, incidental to its major function.)

The narrower the exit slit and the more distant from the prism or grating it is located, the better will be the resolution of wavelengths. Even with a very narrow slit, approaching zero width, the emergent beam will have a finite width, and, in addition, a series of lesser maxima on each side, produced by diffraction. This makes it necessary to establish some arbitrary criterion by which to measure the separation of two neighboring wavelengths. This was done many years ago by Lord Rayleigh. According to Rayleigh's criterion, two wavelengths differing by $\Delta\lambda$ are said to be resolved when the central maximum of one coincides

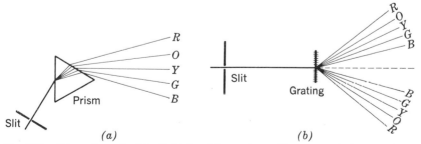

Fig. 2-11 Dispersion of white light by (a) a prism, (b) a transmission grating.

with the first minimum of the other.   The *resolving power* of the monochromator is then defined as

$$R = \lambda/\Delta\lambda$$

where λ is the average of the two wavelengths.

A *spectrograph* is an instrument similar to a monochromator with
the exit slit omitted.   A photographic film or plate is mounted so that
successive wavelengths are focused at successive points.   Thus an entire
spectral region can be photographed simultaneously.

A *spectrophotometer* is an instrument consisting of a source of continuous radiation, a monochromator, and a detector, such as a photoelectric cell, suitable for observing and measuring an absorption spectrum.

Except for the source and the camera or photoelectric detector, the
design problems are the same for monochromator, spectrograph, and
spectrophotometer, so it is convenient to discuss them together.

DISPERSION BY PRISMS

Prisms are suitable dispersing elements from the near-ultraviolet through
the mid-infrared regions, but are not generally applicable elsewhere.   In
principle, any transparent medium may be used to fashion a prism, but its
usefulness is determined by its *dispersion*, that is, the rate of change of
its index of refraction with wavelength.

In the ultraviolet, the only generally useful solid materials with
adequate transparency and dispersion are silica and alumina.*   *Silica*
can be used either as quartz or in the vitreous form, often called "fused
quartz."   *Alumina* is employed in the form of artificial sapphire.   All of
these transmit freely from somewhat below 200 nm in the ultraviolet
up to about 4 μ in the infrared.   Most vitreous silica is inferior in optical
quality to the best quartz or sapphire, but is considerably less expensive.
In recent years progress has been made in improving the transmission
of silica either into the far ultraviolet or into the infrared, though apparently both objectives have not been met in the same samples.

In the visible region silica and sapphire are inferior to optical glass
with respect to dispersion.   For infrared work beyond about 3 μ, prisms
are usually made of alkali halide salts, such as NaCl, KBr, or CsBr.
(Other infrared-transmitting materials will be considered in Chap. 5.)

The prism monochromator which is conceptually simplest is based
on a 60° prism with two lenses, as in Fig. 2-12.   Radiation enters through

---

* Several ionic crystals, such as NaCl, also show these properties in the ultraviolet,
but are seldom used in this region.   One disadvantage is that intense ultraviolet
irradiation tends to produce color centers in the salt crystal, thereby reducing its
transparency with time.

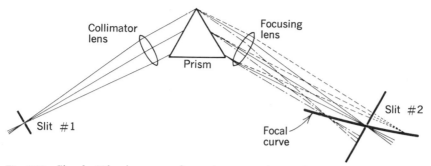

Fig. 2-12 Simple 60° prism monochromator or spectrograph.

slit No. 1, is rendered parallel by a collimating lens, and falls at an oblique angle on one face of the prism. The dispersed radiation emerging from the prism is focused by a second lens, so that the desired wavelength is centered on the exit slit. Because the lens is not achromatic, successive wavelengths are focused at different distances. The locus of foci is called the *focal curve*. To change the wavelength, the slit could be moved along the focal curve, but usually the focusing lens and exit slit are mounted on an arm which turns as a unit around a pivot beneath the center of the prism. If this instrument is to be used as a spectrograph, the photographic plate must lie along the focal curve.

A complication arises when this type of optical system is designed for the ultraviolet region, due to the peculiar nature of crystalline quartz.* Quartz is not only doubly refracting, it is also optically active. This means that a beam of light, in passing through quartz, is split into two beams, a situation intolerable in such precision optical elements as lenses and prisms.

It is possible to overcome this difficulty by judicious use of the dextro- and levorotatory forms of quartz so that rotatory effects of the two forms cancel each other. Historically, Cornu solved the problem by means of a 60° prism composed of two 30° halves, one of each type of quartz. More significant, as it turned out, was the invention of the Littrow mount for a prism and mirror combination. In the Littrow design for a spectrograph (Fig. 2-13), the radiation entering through the slit is collimated by a quartz lens, then dispersed by a 30° quartz prism with a mirrored back. The dispersed rays are focused by the same quartz lens on a photographic plate which is tilted and slightly bent to follow the focal curve. Passage of radiation through the same quartz optics, first in one direction and then in the other, eliminates all difficulties due to polarization.

* Similar, but not identical, considerations apply to sapphire, which is highly birefringent, but which is seldom used in complex optical systems.

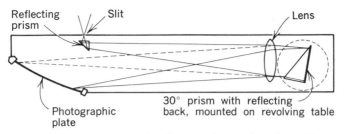

Fig. 2-13 Littrow-mounted prism spectrograph, schematic. The prism is mounted on a table which can be turned to adjust the wavelength range on the photographic plate.

Littrow prism spectrographs are seldom built now, though many are still in use, but the contribution of this design has been great, because the same mounting, or slight variations of it, have been found extremely useful, not only with quartz, but also with isotropic materials, such as the sodium chloride prisms used in the infrared. The Littrow mounting is also an excellent design for use with a plane reflection grating, to be discussed in the following section.

Various other types of prisms are employed occasionally when particular features are desired, such as constant deviation, zero deviation, and autocollimation. Some of these are very ingenious. For details, the student is referred to any of the textbooks on spectroscopy.

DISPERSION BY GRATINGS

A beam of monochromatic radiation, in passing through a transparent plate which has a large number of very fine parallel lines ruled on it, is found to be split into a number of beams. One of these proceeds straight forward, as though the plate were unruled. The other beams are deviated from this forward direction, as in Fig. 2-11(b), through angles which depend upon the spacing of the ruled lines and upon the wavelength of the radiation. This can be explained by the assumption that each clear portion between the lines, when illuminated from behind, acts as though it were itself a source of the radiation which emanates from it in all forward directions (Huygens' principle). However, the rays coming from these numerous secondary sources will be destroyed by interference in most directions. Only at those angles where the geometry is just right will the beams reinforce each other. Figure 2-14 shows one of the possible deviated beams. The angle of deviation is θ, the difference in the length of the path taken by beamlets from successive transparent areas is a, and the distance between the centers of adjacent lines (the grating space) is d. Thus $a = d \sin \theta$. The many beamlets can only reinforce each other when the difference in path length is equal to an integral number of wavelengths of the radiation. This gives the fundamental relation called the *grating equation:*

$$n\lambda = d \sin \theta \qquad (2\text{-}3)$$

where n is any integer, 0, 1, 2, 3, . . . , called the *order*, and λ is the wavelength.

It follows from Eq. (2-3) that, if a beam of polychromatic radiation is passed through the grating, it will be fanned out into a series of spectra located symmetrically on each side of the normal to the grating. On each side there will be a spectrum corresponding to each of the first few values of n. The equation further shows that for a particular angle θ there will be several wavelengths for which the value of $n\lambda$ is the same. For example, a grating with 2000 lines per cm (grating space $d = \frac{1}{2000} = 5 \times 10^{-4}$ cm) will deflect through an angle $\theta = 6.00°$ radiation of those wavelengths given by

$$\lambda = \frac{d \sin \theta}{n} = \frac{(5 \times 10^{-4})(\sin 6.00°)}{n}$$

$$= \frac{(5 \times 10^{-4})(0.1045)}{n} = \frac{0.5225 \times 10^{-4}}{n} \text{ cm} = \frac{522.5}{n} \text{ nm}$$

The actual wavelengths corresponding to successive orders at this angle will be

Order n	1	2	3	4	. . .
Wavelength λ, nm	522.5	261.2	174.2	130.6	. . .

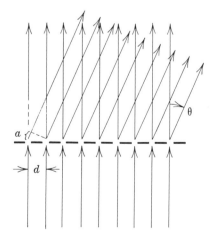

Fig. 2-14 Diffraction at a plane grating.

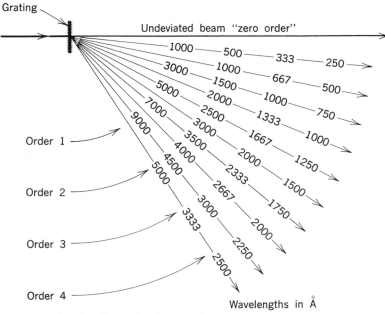

Fig. 2-15   Overlapping orders in a grating spectrum.

This relation is shown diagrammatically in Fig. 2-15, which gives selected wavelengths for the first four orders on one side of the normal.

The fact that successive orders of spectra overlap might be thought to be a great drawback, but in practice it gives little trouble. If the spectrum is to be observed visually, the question will not arise, since the visible regions (400 to 750 nm) of the various orders do not overlap. If the spectrum is to be recorded photographically, the spectral sensitivity of the plate will limit the degree of overlap, but some may still be encountered. Overlapping may be reduced or eliminated by placing ahead of the grating an auxiliary prism of small deviation, called a *foreprism* or *Order-Sorter*, or by the use of absorbing filters which remove one region of the spectrum while passing another.

The grating discussed above is of the type known as a plane *transmission grating*. In practical instruments other than small, hand-held ones, *reflection gratings* are more common. In these the lines are engraved on the surface of a mirror, which may be either a polished metal slab or a glass plate on which has been deposited a thin, metallic film.

It is possible to rule a grating in such a way as to throw a maximum fraction of the radiant energy into those wavelengths which are diffracted at a selected angle. This is accomplished by ruling with a specially shaped diamond point held at a specified angle. The resulting grating is called an *echelette,* and is said to have been given a *blaze* at a particular angle. Figure 2-16 shows the geometry of a portion of an echelette reflection grating. The wider faces of the grooves make an angle ϕ with the surface of the grating. A ray incident at angle α will be reflected from the groove face at angle β such that $\alpha + \phi = \beta - \phi$. The rays reflected from successive grooves then undergo interference as already described. Because of the efficiency of specular reflection at the metal surface, much more energy will be diffracted at this angle (β) than at any other, for a given value of α. The energy will be only slightly less at angles close to β, so the grating can be used to advantage for a considerable wavelength span in a given order. A grating blazed for a particular wavelength in the first order will also be blazed for half that wavelength in the second order, one-third of it in the third order, etc. Very little energy will be found at the symmetric position on the other side of the normal to the grating, and very little in the "zero order."

In general, best results are obtained with a grating in which the spacing between lines is of the same order of magnitude as the wavelength region to be dispersed. For special purposes, other grating spaces may be found useful. An *echelle,* for example, is a grating with step-shaped rulings a few hundred times wider than the average wavelength to be studied. It must be used at an order n of 100 or so, which produces difficult problems with overlapping of orders, but it is capable of tremendous dispersion.

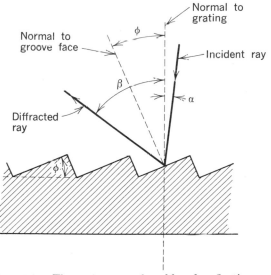

Fig. 2-16 The geometry of a blazed reflection grating.

The manufacture of precision gratings is a very exacting piece of work. It is performed with an extremely precise and delicate machine, called a *ruling engine,* which scribes the fine parallel lines with a diamond point. Only the largest precision spectrographs use original gratings, because of their expense. Less costly spectrographs and practically all grating spectrophotometers use *replica gratings,* made by casting a plastic material on an original grating, then stripping it off and mounting it on a rigid support. The art of replication has reached the point where gratings made this way are of very nearly as good quality as the originals.

There are many ways in which a diffraction grating can be mounted in designing a monochromator or spectrograph. The two most used today both require a plane reflection grating. The first of these is the Littrow mounting, already mentioned. A diagram of it would be identical with Fig. 2-13, with the prism replaced by the grating, on the same rotatable table. A concave mirror is more often used than a lens for collimation and focusing because it is equally effective at all (optical) wavelengths, which is not true of a lens.

The other mounting for a plane grating was invented by *Ebert* in 1889, but was little used until resurrected and improved by Fastie[4] (1952). In this design (Fig. 2-17) a single, large, spherical mirror serves for both collimation and focusing, with symmetrically placed slits. Wavelength selection is effected by rotating the grating. *Czerny and Turner*[2] (1930) suggested using two small, spherical mirrors, mounted symmetrically, to save the expense of the large Ebert mirror, much of which was unused, and most current instruments with this geometry incorporate the best features of the Czerny-Turner and Fastie designs.

There is little choice between the modified Littrow and the modified Ebert mountings. The Littrow is slightly more compact, and saves one mirror, but the two slits must be close together, usually one above the other, which cramps the design somewhat.

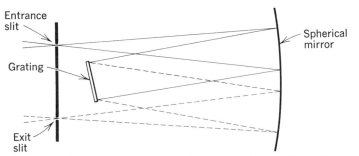

Fig. 2-17 Ebert mounting for a plane reflection grating. The wavelength is selected by turning the grating around a vertical axis at its center.

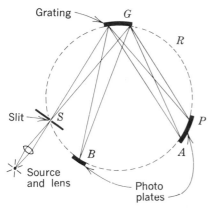

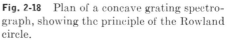

Fig. 2-18 Plan of a concave grating spectrograph, showing the principle of the Rowland circle.

Another class of instruments makes use of a reflection grating in which the lines are ruled on a concave spherical surface of rather long radius of curvature. Rowland in 1882 discovered the following important principle of design which bears his name: If a circle (the *Rowland circle*) be drawn tangent to the concave grating at its center, but with its diameter equal to the radius of curvature of the grating, then the diffracted image of the entrance slit will lie on the circle if the slit itself is on the circle. This will apply to all wavelengths of all orders of diffraction, and is illustrated in Fig. 2-18. Several mechanical designs have been devised to make use of this principle, ranging from very large spectrographs (35 ft or more in diameter) down to small, portable spectrophotometers.

Instruments based on the Rowland circle suffer from the inherent defect of *astigmatism*. This means that, although the image of the entrance slit is very sharply focused along the circle (i.e., its wavelength can be measured with precision), its height is not sharply defined. This makes quantitative measurements difficult, as some light from the entrance slit is masked off by the top and bottom of the exit slit.

DISPERSION CURVES

The dispersion of a spectrograph is usually defined as the derivative $d\lambda/dx$, where x is distance measured along the focal curve, i.e., on the surface of the developed photoplate. It may be specified in angstroms per millimeter or other similar units. In a monochromator or spectrophotometer the corresponding quantity is the *effective band width*, in angstroms (or microns, etc.) per millimeter of slit width. This is convenient, especially in instruments in which the slit width can be varied.

A grating instrument produces a *normal* spectrum, that is, one which

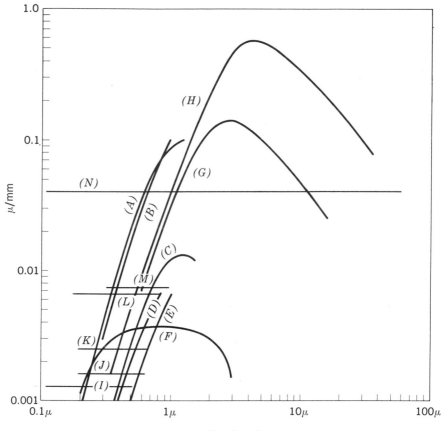

Fig. 2-19 Effective band-width curves and useful ranges for several spectrophotometers and spectrographs. The curves refer to the following: (A) Beckman Model DU, quartz Littrow; curve extends to 0.00083 μ/mm at 0.2 μ; (B) Beckman Model B, Féry glass prism; (C) Model LEP of Laboratoire d'Électronique et de Physique Appliquées, Paris, four quartz prisms in series; curve extends to 0.00058 μ/mm at 0.3 μ; (D) Bausch & Lomb large quartz Littrow spectrograph; curve extends to 0.0001 μ/mm at 0.2 μ; (E) same, with glass optics; curve extends to 0.0003 μ/mm at 0.35 μ; (F) Cary Model 14, combining a quartz Littrow prism and 600 line per millimeter grating monochromators; (G) Beckman IR-4 with NaCl prism, extended; (H) same, with CsBr prism, extended; curves (G) and (H) were prepared from Beckman's literature for the regions to the right of the maxima and extended to shorter wavelengths by normalized comparison with data from Harshaw Chemical Co.; this extended region does not correspond to any commercial instrument; (I) McPherson Model 218 vacuum monochromator with 2400 line per millimeter grating; (J) Bausch & Lomb Spectronic-505 with 1200 line per millimeter grating; (K) Durram Stopped-Flow spectrophotometer, 1180 line per millimeter grating; (L) Phoenix Precision dual-wavelength spectrophotometer, 600 line per millimeter grating; (M) Bausch & Lomb Spectronic-20, 600 line per millimeter grating; (N) McPherson, same as (I), with 75 line per millimeter grating.

is spread out uniformly on a wavelength scale. The dispersion or band width is then uniform over the whole spectrum. A prism, on the other hand, gives an unequally spaced spectrum in which the longer wavelengths are crowded together as compared with the shorter. The band width for a given width of slit is no longer constant, but differs from one wavelength to another and from one design of instrument to another.

Figure 2-19 shows the band-width curves for a number of commercial monochromators and spectrographs. Notice that, other variables being equal, the spectral purity obtainable is better the smaller the ratio of band width to slit, that is, the lower down on the graph one can operate. Note also that for grating instruments, the curve is merely a horizontal line. Relative position on this graph tells only part of the story, of course. A grating instrument with a band width of 0.002 μ/mm may give an *actual* band width of much less than 0.002 μ, if the intensity of the lamp, the sensitivity of the detector, and other variables are such as to permit a slit narrower than 1 mm.

DETECTORS OF RADIATION

The power of electromagnetic radiations can be detected and measured by means of chemical effects, the production of heat, the production of electronic changes in matter, and (in the microwave region) by direct electromagnetic induction.

PHOTOCHEMICAL DETECTION

Photochemical detectors are *integrating* devices, in that they give a cumulative response to all incident radiation over the period of time during which they are exposed, without regard to variations in flux over short time spans.

PHOTOGRAPHY. Sufficiently energetic radiation is capable of dissociating a variety of chemical compounds, including the silver halides. We can write, for example,

$$AgBr \xrightarrow{(h\nu)} Ag + Br$$

where $(h\nu)$ on the arrow denotes the required radiant energy. This is the basis for the whole field of silver photography.

In the usual form, the silver salt is prepared as tiny grains imbedded in a matrix of gelatin. This mixture, called the *photographic emulsion*, is coated as a thin layer on glass plates or cellulose acetate film. The primary photochemical process indicated above produces only invisibly small centers of metallic silver, the *latent image*, on the surface of those grains which were exposed to radiation. To obtain measurable quantities of silver, the exposed plate is subjected to the action of a chemical reducing agent, the *developer*. The silver of the latent image acts catalytically so that the quantity of silver chemically deposited is reproducibly related to the amount initially liberated by the radiation.

The amount of metallic silver at any point on a developed plate is measured by an instrument called a *densitometer*, the principle of which is shown in Fig. 2-20. The radiation from an incandescent lamp is focused on the photo plate and again on the sensitive surface of a photoelectric cell or photomultiplier tube. The output of the photocell is then a measure of the amount of radiation transmitted through any selected point on the plate. If the power transmitted through a clear portion of the plate is designated by P_0 and that through the silver deposit by P, then the *optical density** is defined as

$$D = \log \frac{P_0}{P} \tag{2-4}$$

Photographic detection is subject to a serious limitation in the nonlinearity of the photographic emulsion with respect to the quantity of radiant energy received. The relationship is shown by the *H and D curve* (Hurter and Driffield), in which the optical density is plotted as a function of the logarithm of the exposure (Fig. 2-21). *Exposure* is the photographic term designating the total amount of radiant energy received, the product of radiant power by time. As shown by the figure, the photographic response follows the logarithmic relation very

---

* This resembles the definition of absorbance to be introduced in the next chapter; the distinction is that D as here used refers not to a chemical sample, but to a photographic plate.

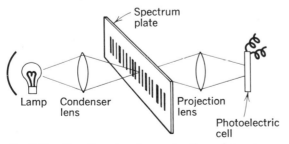

Fig. 2-20 Densitometer for evaluation of spectrum plates.

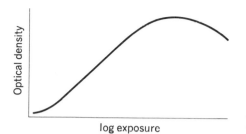

Optical density

log exposure

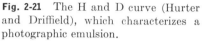

Fig. 2-21 The H and D curve (Hurter and Driffield), which characterizes a photographic emulsion.

closely over a considerable range (the linear portion of the curve) but flattens off toward both extremes. This means that the photographic method of measuring radiation cannot be used for very faint radiation (except by unduly prolonged exposure time) nor for too powerful radiation. Indeed, if the radiation is much more powerful or the exposure time much longer, *reversal* will result, with partial destruction of the latent image.

The definition of exposure as the product of radiant power and time implies that the response of the photographic plate is proportional to this product, a relation known as the *reciprocity law*. Unfortunately this is not always strictly true. Longer times than the law would predict are often required for either very low or very high illumination, even within the nearly linear part of the H and D curve. This effect becomes particularly important when a plate is being used to integrate the energy in a pulsating beam of radiation.

As in any application of photography, the exposed plate or film must be bathed successively in developer and fixer, with adequate rinsing after each of these steps. For photometric purposes a developer which gives high contrast is preferred. The recommendations of the manufacturer of the plates should be followed for best results. Care must be taken to provide adequate agitation; otherwise uneven development may occur, leading to error in the results of a quantitative measurement. In any analysis where two silver deposits are to be compared quantitatively, the processing must be identical. The effects of differences of temperature, time of development, aging of solutions, etc., are all important. There may also be differences in the sensitivity of plates, especially if taken from different packages, due to differences in age, temperature of storage, etc.

Photographic detection is most useful in the visible, ultraviolet, and x-ray regions. Plates and films are available with a variety of characteristics, including spectral response, sensitivity (usually called "speed"), contrast (slope of the linear portion of the H and D curve), and graininess. Ordinary photographic materials lose sensitivity rapidly beyond about 250 nm in the ultraviolet, where gelatin ceases to be transparent, then regain sensitivity in the x-ray region.

In the ultraviolet beyond 250 nm, plates can be sensitized by coating them with a substance which fluoresces under ultraviolet radiation; the useful range can be extended by this means to wavelengths as short as 50 nm. Following exposure, but preceding development, the plate must be bathed in a suitable solvent to remove the fluorescent material. Special plates, called *Schumann plates*, can also be used in the far ultraviolet; these utilize a very thin emulsion with a minimal amount of gelatin, and have been found sensitive as far down the scale as 1 nm. Schumann plates require special handling to avoid abrading the delicate emulsion.

Photographic film is used extensively in the x-ray region. For this application it is commonly coated on both sides, to increase the degree of absorption of the penetrating radiation.

The upper wavelength limit of ordinary emulsions is about 450 to 500 nm. This can be extended to about 650 nm ("pan" films) and as far as about 1200 nm ("hypersensitized" infrared film) by incorporating certain dyes into the emulsion.

For more detail about scientific photography, the reader is referred to publications of the Eastman Kodak Company,[3] or to textbooks on photography.

IONIZATION OF A GAS. Another effect of energetic radiation which may be classed as photochemical is the production of ion pairs in a gas. This is highly useful for detection and measurement in the x-ray region and is also applicable to the observation of nuclear radiations, both particulate and electromagnetic. The devices employed in such measurements are ionization chambers, proportional counters, and Geiger counters, which differ from each other mostly in the electrical system for collecting the ions. The details of these methods will be postponed to Chap. 16.

Various other photochemical processes have been utilized for measuring radiation. An example is the *actinometer* in which oxalic acid undergoes decomposition catalyzed by photoactivated uranyl ion, UO_2^{++}. The absorbed radiant energy can be determined indirectly by titration of the residual oxalic acid.[8]

VACUUM PHOTOTUBES

According to the classical photoelectric effect, photons with greater than some critical energy content, when incident on a metallic surface, cause the ejection of electrons. This principle is used in vacuum and gas-filled photoemissive tubes. Such a tube consists of two electrodes encased in a transparent envelope. The cathode is a sheet of metal arranged to receive the radiation to be measured. It is coated with a material which has a great tendency to lose electrons under the triggering influence of radiant energy. The anode is simply a wire which serves to collect the electrons.

The wavelength response of vacuum phototubes depends upon the material of the cathode coating. The alkali metals are particularly suitable. The pure metal is deposited by sublimation in a very thin layer over a subsurface which also has an effect on the sensitivity. Lothian[9] gives the following approximate wavelength limits of various cathode materials:

Cathode	Wavelength range, nm
Sodium (thick layer)	300–500
Potassium (thick layer)	400–500
Silver-cesium oxide-cesium	250–1200
Silver-potassium	200–700
Antimony-cesium	200–660
Bismuth-oxygen-silver-cesium	200–750

The chief advantage of photoemissive tubes, as compared with most of the devices to be mentioned subsequently, lies in their ultraviolet sensitivity.

As an electric circuit component, the vacuum phototube can be considered a variable resistor which changes from perhaps 250 MΩ in the dark to a much lower value of a few megohms in bright light. If operated with a constant applied voltage, the current which flows will be proportional to the *power* of the incident beam of radiation.

Gas phototubes are similar in construction but contain a noble gas at a pressure of a few tenths torr. Ionization of the atoms of the gas by photoelectrons results in perhaps tenfold amplification. Gas tubes are widely used for sound reproduction and for relay applications, but are seldom selected for photometry because they are somewhat less stable and reproducible than their vacuum counterparts.

PHOTOMULTIPLIER TUBES

The photomultiplier is a type of vacuum phototube so designed that an amplification of several millionfold is achieved within one tube. This is

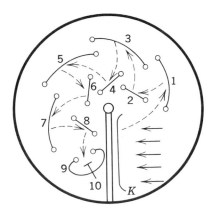

Fig. 2-22 Photomultiplier tube; schematic cross section of one variety.

accomplished by means of *secondary emission*. Electrons are caused to leave the cathode by the action of light, just as in the simple phototube; but in the multiplier these electrons are made to strike a second sensitive surface, a *dynode*, at a more positive potential. Here each electron by its impact causes the release of several secondary electrons. These are in turn accelerated and strike another dynode, where the number of electrons is again increased by a similar factor. This process may be repeated as many as 10 or 15 times. The multiplication per step depends on the applied voltage and is typically 2 or 3, giving (for 10 steps) an amplification of 2^{10} to 3^{10}.

A cross-sectional view of one design of photomultiplier tube is shown in Fig. 2-22. The degree of amplification is very sensitive to changes in the overall voltage. The rectifier which supplies this voltage must

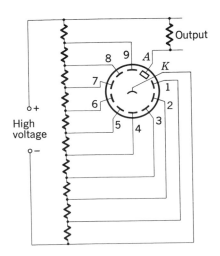

Fig. 2-23 Photomultiplier tube; electric connections.

therefore be stabilized so as to eliminate any fluctuations. The appropriate voltages for successive dynodes are most conveniently obtained by a series string of dropping resistors, as in Fig. 2-23, where K is the photocathode, A the anode, and the numbered elements the dynodes.

While an ordinary phototube may be rated at several microamperes output per lumen of radiation, the multiplier tube may bear a rating in microamperes per *micro*lumen. More than a few milliamperes will damage the tube, so it can be utilized only at low light intensities, such as are frequently encountered in spectrophotometers and fluorimeters.

SEMICONDUCTOR PHOTOCELLS

A *semiconductor* is a crystalline substance which is intermediate between metallic conductors on one hand and nonconducting insulators on the other. The energy with which bonding electrons are held in fixed orbitals in semiconductor crystals is such that they can be dislodged with comparative ease by incident radiation. They are then free to move about within the lattice. The "hole" from which an electron has been taken acts as though it were an entity itself, bearing a positive charge. Under the influence of an applied electric field, both electrons and holes move, in opposite directions, thus constituting an electric current.

A unit designed to take advantage of this effect is a *photoconductive cell*. Some materials of which such cells can be made are the following:

Material	Long wavelength limit, μ
Cadmium sulfide	0.60
Cadmium selenide	0.95
Lead sulfide	4.00
Silicon	1.50
Selenium	0.80

Toward shorter wavelengths, sensitivity falls off gradually through the visible region. A typical cadmium sulfide cell has a resistance of about 25 MΩ or greater in the dark, and perhaps 500 Ω in normal room light. The logarithm of the resistance is nearly linear with the logarithm of the illumination.

It is possible to prepare a contact interface between a semiconductor and a metal electrode in such a way that a rectifying action results. Electrons are able to pass easily from the semiconductor to the metal, but only with great difficulty in the opposite direction. Since illumination provides an increased concentration of electrons in the semiconductor,

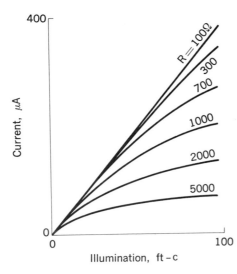

Fig. 2-24 Characteristics of a typical photovoltaic cell.

a net flow results which manifests itself as a negative potential on the metal electrode relative to a nonrectifying metallic junction elsewhere. A photocell making use of this effect is called a *barrier-layer* or *photovoltaic cell.* Typically a layer of crystalline selenium or silicon is deposited on a metal base, then covered with a film of silver or other noble metal so thin as to be transparent to incident radiation.

Photovoltaic cells are very extensively used in photometric applications. They produce a sufficient current to operate a microammeter without electronic amplification. The spectral sensitivity covers the visible region, with a maximum in the green. The sensitivity curve closely resembles that of the human eye.

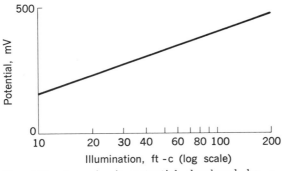

Fig. 2-25 Open-circuit potential developed by a photovoltaic cell.

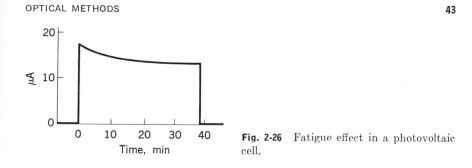

Fig. 2-26 Fatigue effect in a photovoltaic cell.

The characteristic curves of photovoltaic cells are usually plotted as output current as a function of illumination (Fig. 2-24). The output is greatly affected by the resistance of the external circuit, which should be kept as low as possible if maximum sensitivity and linearity are desired. The open-circuit voltage, as measured with a potentiometer or electronic voltmeter, is a nearly logarithmic function of the illumination (Fig. 2-25). A related device, the photosensitive junction diode, is discussed in Chap. 26.

When a photovoltaic cell is suddenly exposed to a bright light, the electric output will abruptly assume a value somewhat above its true equilibrium value, and then drop exponentially to its final level (Fig. 2-26). This effect, known as *fatigue*, can cause large errors if its presence is not realized. It can be minimized by careful selection of the optimum level of illumination, resistance of the measuring circuit, etc. It is largely eliminated in properly designed two-cell circuits.

The *phototransistor* is a unit which combines the photosensitivity of a semiconductor with the amplifying ability of a transistor. It is used primarily in automatic control devices and data processors.

There is a considerable variation with temperature in most semiconductor devices, which must never be overlooked.

DETECTION THROUGH FLUORESCENCE

Radiation may sometimes be detected indirectly by a process of wavelength conversion. The two common applications are in photography, where ultraviolet sensitization by fluorescence has already been described, and in the x-ray region. A widely useful method for detecting x-rays (also particles from nuclear decay) is by the counting of scintillations. The rays are absorbed in a material which will fluoresce, producing a tiny flash of visible light (a *scintillation*) for each absorbed x-ray photon. The flashes are observed and counted by a photomultiplier tube with its associated electronics. A widely used fluorescent material for this application is crystalline sodium iodide which has been activated by admixture of a trace of thallous iodide. This detection system is discussed in more detail in Chap. 16.

THERMAL DETECTION

In principle, radiant energy in any region can be measured by conversion to heat, followed by a measurement of temperature rise. In fact this is the one method which is capable of quantitative theoretical interpretation, and so is used in absolute determinations and as a reference detector for calibration purposes. It is seldom chosen as a working detector, except in the infrared region, because other devices are more sensitive and convenient.

Thermal detectors have a relatively long time constant, since each time they receive a signal and are warmed up by it, they must have time to cool off again to be ready for the next signal. Hence radiation cannot be chopped at a frequency greater than about 15 to 20 Hz. In the mid- and far-infrared ranges no more convenient wide-band detector is known.

For a practical thermal detector, the heat capacity must be kept to a minimum so that a very small amount of heat will produce a measurable temperature rise. It is usual to let the radiation fall on a tiny strip of blackened metal foil (perhaps 0.5 by 5 mm), to the back of which is cemented the detector element itself. If this detector consists of a number of thermocouple junctions, it is called a *thermopile;* if it is a thermistor or platinum resistance thermometer, it is known as a *bolometer.* The sensitivities of the two types are about equal. An interesting account of the fabrication of these tiny detectors has been given by Strong.[12]

The measurement of power in the microwave region can be carried out with an absorber of heat called a *microwave calorimeter.* Again, so far as chemical applications are concerned, this is a reference standard for calibrating practical detectors of other types.

PHOTOMETRY

There are a number of arrangements of optical and electrical components with which the relative power in a beam of radiation can be measured. The absolute determination of power is quite difficult and is seldom required in chemical applications. Generally we are interested in the ratio of the powers of two related beams. An important example is found in the detector of a spectrophotometer. Consider Fig. 2-27: P_s, the power of the beam coming from the source and incident on the cuvet,*

---

* *Cuvet* is the name given to the transparent container for a liquid specimen to be inserted into a photometer, particularly in the ultraviolet and visible regions.

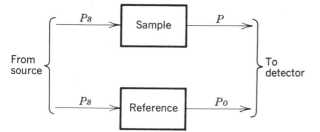

Fig. 2-27 Power relations in a double-beam photometer.
P_s, the power from the source, is assumed equal on the two
cells; P_0 is passed by the reference, P by the sample
solution.

is necessarily greater than P, the power of the transmitted radiation
proceeding to the detector.

If we are interested in determining the power absorbed by the
sample, it is not sufficient to relate P to P_s, for power will be lost from
the beam through other mechanisms than absorption by the sample.
In order to eliminate these extraneous losses, it is essential to compare the
power emerging from the sample cuvet with that from the same or an
identical cuvet containing a reference material (Fig. 2-27). We can
assume that the radiant powers incident on the two cuvets are equal,
for means must be provided in any spectrophotometer whereby they
can be equalized. The contents of the reference cuvet should be as
nearly as possible identical to the sample with respect to absorbing
impurities, refractive index, and if measurements are to extend to more
than a narrow wavelength band, to dispersion as well. The refractive
index is important in that it determines the amount of reflective losses
at the inner surfaces of the cuvet.

In practice, we need not measure P_s, but only P_0 and P. The ratio
of these quantities is the *transmittance* T where

$$T = \frac{P}{P_0} \tag{2-5}$$

With detectors which give an output proportional to the incident
power, T can be determined easily by successive or simultaneous measurements of P_0 and P. Detectors which meet this requirement include the
photovoltaic cell when measured with a low-resistance meter, and the
vacuum phototube and photomultiplier operated at a constant applied
voltage. It is convenient to set a sensitivity control so that $P_0 = 100$;
P will then measure percent transmittance (percent T) directly.

On the other hand, those detectors in which the response is a
logarithmic function of the radiant power give more conveniently the

quantity $\log T = k(P_0 - P)$, where k is a proportionality constant. Photoconductive cells and open-circuit photovoltaic cells fall in this class, as does also a photomultiplier operated at constant current. It will be seen in the next chapter that there is a linear relation between the logarithm of T and the number of absorbing molecules in the path of the radiation, which suggests the desirability of logarithmic detectors.

The photometric method wherein successive measurements are made of P_0 and P is called a *single-beam* system. It is inherently liable to error arising from fluctuations in the intensity of the source. To be preferred, though more expensive, is the *dual-beam* system of photometry, of which there are several modifications. In each, the beam of radiation is split into two branches. One branch passes through the reference sample (the blank), while the other traverses the sample in an identical container. The beam-splitting device may be a rotating mirror or some equivalent, which sends the beam alternately through one branch and the other, at a frequency of perhaps 10 to 50 Hz. Figure 2-28 shows a representative dual-beam photometer. The signals from the two phototubes are compared electrically in such a way that the meter or recorder responds to the ratio of the two powers, but not to fluctuations in the line voltage, which affect both beams equally.

In other modifications of the dual-beam principle, the two beams are recombined after passage through the two samples and fall on a single photocell. These and other modifications are detailed later in the descriptions of specific instruments.

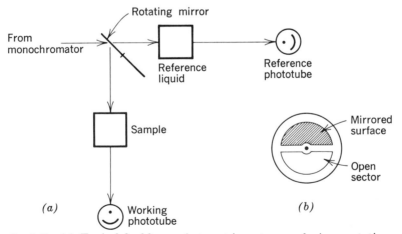

Fig. 2-28 (*a*) Typical dual-beam photometric system employing a rotating mirrored sector and two phototubes. (*b*) Detail of the rotating sector.

PROBLEMS

2-1 The legal definition of the meter in the United States[10] is 1,650,763.73 wavelengths in vacuo of the $2p_{10}-5d_5$ transition in isotopically pure ^{86}Kr. Compute the wave number in reciprocal centimeters; the wavelength in angstroms, in nanometers, and in microns; the frequency in hertz and fresnels; and the energy per photon in electron volts, for this radiation. Take the velocity of light in vacuo as $(2.997925 \pm 0.000003) \times 10^{10}$ cm-sec^{-1} and Planck's constant as $(6.6256 \pm 0.0005) \times 10^{-27}$ erg-sec. Give each answer with the appropriate uncertainty limits. (If you do not have access to a calculator which can handle enough digits, round off the numbers as necessary and explain the extent to which this degrades your calculations.)

2-2 A transmission grating for a spectrograph has 15,000 lines per in. Calculate the wavelengths which will be diffracted at an angle θ of 25° for the first two orders. At what angles would radiation of 2537 Å appear in each of the first two orders?

REFERENCES

1. Barrow, G. M.: "Introduction to Molecular Spectroscopy," McGraw-Hill Book Company, New York, 1962.
2. Czerny, M., and A. F. Turner: Z. Physik, **61**:792 (1930).
3. Kodak Plates and Films for Science and Industry, Eastman Kodak Company, Rochester, N.Y.
4. Fastie, W. G.: J. Opt. Soc. Am., **42**:641 (1952).
5. Jaffe, H. H., and A. L. Miller. J. Chem. Educ., **43**:400 (1966); G. W. Ewing and J. M. Fitzgerald, J. Chem. Educ., **44**:622 (1967).
6. Jaffé, H. H., and M. Orchin: "Theory and Applications of Ultraviolet Spectroscopy," John Wiley & Sons, Inc., New York, 1962.
7. Jenkins, F. A., and H. E. White: "Fundamentals of Optics," 3d ed., McGraw-Hill Book Company, New York, 1957.
8. Leighton, W. G., and G. S. Forbes: J. Am. Chem. Soc., **52**:3139 (1930); G. W. Castellan, "Physical Chemistry," p. 667, Addison-Wesley Publishing Company, Inc., Reading, Mass., 1964.
9. Lothian, G. F.: "Absorption Spectrophotometry," 2d ed., Hilger and Watts, Ltd., London, 1958.
10. Natl. Bur. Std. Tech. News Bull., February and October, 1963.
11. Penner, S. S.: "Quantitative Molecular Spectroscopy and Gas Emissivities," chap. 3, Addison-Wesley Publishing Company, Inc., Reading, Mass., 1959.
12. Strong, J.: "Procedures in Experimental Physics," chap. 8, Prentice-Hall, Inc., Englewood Cliffs, N.J., 1938.

3

The Absorption of Radiation: Ultraviolet and Visible

If a beam of white light is allowed to pass through a glass cuvet filled with liquid, the emergent radiation will be less powerful than that entering. The diminution in power may be approximately equal over the whole wavelength range, or it may be of different extent for different colors. The loss, as we have seen previously, is due in part to reflections at the surfaces and in part to scattering by any suspended particles present; but it is primarily accounted for by the *absorption* of radiant energy by the liquid. In clear solutions scattering may be reduced to a vanishingly small amount by ordinary care. In the study of colloidal suspensions the scattered fraction may itself be important as an analytical tool (Chap. 6).

The extent to which energy is absorbed by the liquid is generally greater for some of the colors making up the white light than for others, with the result that the emergent beam is colored. Table 3-1 gives the colors of radiation of successive wavelength ranges, together with their complements. These ranges are approximate only, as different observers may give widely varying figures. The apparent color of the solution is always the *complement* of the color which is absorbed. Thus a solution which absorbs in the blue region (465 to 480 nm) will appear yellow, one which absorbs green will appear purple, etc.

Table 3-1 Colors of visible radiation

Approximate wavelength range, nm	Color	Complement
400–465	Violet	Yellow-green
465–482	Blue	Yellow
482–487	Greenish blue	Orange
487–493	Blue-green	Red-orange
493–498	Bluish green	Red
498–530	Green	Red-purple
530–559	Yellowish green	Reddish purple
559–571	Yellow-green	Purple
571–576	Greenish yellow	Violet
576–580	Yellow	Blue
580–587	Yellowish orange	Blue
587–597	Orange	Greenish blue
597–617	Reddish orange	Blue-green
617–780	Red	Blue-green

SOURCE: D. B. Judd, in M. G. Mellon (ed.), "Analytical Absorption Spectroscopy," p. 525, John Wiley & Sons, Inc., New York, 1950.

In referring to color, we are of course restricting the discussion to the visible region of the spectrum, but many of the concepts and analytical methods will carry over with no change in principle into both the ultraviolet and infrared ranges.

To the analytical chemist, the importance of colored solutions lies in the fact that the radiation which is absorbed is characteristic of the material doing the absorbing. A solution containing the hydrated cupric ion is blue, because this ion absorbs yellow and is transparent to other colors. Thus a solution of a copper salt may be analyzed by measuring the degree of absorption of yellow light under standardized conditions. Any soluble colored material may be determined quantitatively in this way. In addition, many substances which are colorless or only faintly colored may be analyzed by adding a substance which will react with them to form an intensely colored compound. Thus the addition of ammonia to the copper solution produces a much more intense color than that of the aqueous cupric ion itself and, therefore, provides a much more sensitive analytical method.

The general term for chemical analysis through measurement of absorption of radiation is *absorptiometry*. *Colorimetry* should be applied only in relation to the visible spectral region. *Spectrophotometry* is a division of absorptiometry which refers particularly to the use of the spectrophotometer. The term *photometry* is too general to be very useful in the present connection; it may be interpreted to include methods of emission spectroscopy as well as all of absorptiometry.

SELECTIVE ABSORPTION

The absorption of radiant energy in the visible and ultraviolet spectral regions depends primarily upon the number and arrangement of the electrons in the absorbing molecules or ions.

Among inorganic substances, selective absorption may be expected whenever an unfilled electronic energy level is covered or protected by a completed energy level, usually formed by means of coordinate covalences with other atoms.

Consider copper as an example. The simple ion Cu^{++} is never found in aqueous solution (though often written as such), because it has a great tendency to form coordinate bonds with any available molecules or ions which carry unshared pairs of electrons. Such unshared pairs are present in water, ammonia, cyanide ion, chloride ion, and many other entities. The structure of the cupric ion which is coordinated with any of these Lewis bases has only 17 electrons in the third (M) major energy level, whereas the fourth (N) level contains a stable octet. The several ions do not have identical colors, as the nature of the ligand has an effect on the energies of the electrons.

Selective absorption among organic compounds is again related to a deficiency of electrons in the molecule. Completely saturated compounds show no selective absorption throughout the visible and ultraviolet regions. Compounds which contain a double bond absorb strongly in the far ultraviolet (195 nm for ethylene). Conjugated double bonds (i.e., alternating single and double) produce absorption at longer wavelengths. The more extensive the conjugated system, the longer will be the wavelengths at which absorption is observed.

If the system is extended far enough, the absorption enters the visible region, and color results. Thus β-carotene, with 11 conjugated double bonds, absorbs strongly in the region 420 to 480 nm and hence is yellow-green in appearance. The complete conjugated system in a compound is called its *chromophore*.

The wavelengths of the absorption maxima of a compound provide a means for identifying the chromophore which it contains. The spectra are in general modified by the presence of various atomic groups when these are substituted for hydrogen atoms on the carbons of the chromophore system. Such substituents usually have the effect of shifting the absorption bands toward longer wavelengths and changing their absorbance values. Substituents which produce these effects are known as *auxochromes*.

In Table 3-2 are listed a number of organic compounds containing representative chromophores, together with their characteristic wavelengths of maximum absorption and molar absorptivity values.*

Table 3-2   Representative chromophores*

Compound	Chromophore	Solvent	λ_{max}, nm	Log ϵ †
Octene-3	C=C	Hexane	185	3.9
			230	0.3
Acetylene	C≡C	(Vapor)	173	3.8
Acetone	C=O	Hexane	188	2.9
			279	1.2
Diazoethyl acetate	N=N	Ethanol	252	3.9
			371	1.1
Butadiene	C=C—C=C	Hexane	217	4.3
Crotonaldehyde	C=C—C=O	Ethanol	217	4.2
			321	1.3
Dimethylglyoxime	N=C—C=N	Ethanol	226	4.2
Octatrienol	C=C—C=C—C=C	Ethanol	265	4.7
Decatetraenol	[—C=C—]₄	Ethanol	300	4.8
Vitamin A	[—C=C—]₅	Ethanol	328	3.7
Benzene		Hexane	198	3.9
			255	2.4
1,4-Benzoquinone		Hexane	245	5.2
	O=⟨ ⟩=O		285	2.7
			435	1.2
Naphthalene		Ethanol	220	5.0
			275	3.7
			314	2.5
Diphenyl		Hexane	246	4.3

* Data collected from various sources; to be taken as illustrative only.
† Defined in Table 3-4.

In aromatic compounds, the benzene ring is the simplest chromophore. Two or more rings in conjugation, as in either naphthalene or diphenyl, again shift the absorption toward the visible. Table 3-3 shows the effects of some auxochromes on the absorption of benzene.

* This quantity will be defined later.

Table 3-3   Effect of auxochromes on the benzene chromophore[36]

Compound	Solvent	Ethylenic band		Benzenoid band	
		λ_{max}, nm	Log ϵ	λ_{max}, nm	Log ϵ
Benzene	Cyclohexane	198	3.90	255	2.36
Anilinium ion	Aq. acid	203	3.88	254	2.20
Chlorobenzene	Ethanol	210	3.88	257	2.23
Thiophenol	Hexane	236	4.00	269	2.85
Phenol	Water	210.5	3.79	270	3.16
Aniline	Water	230	3.93	280	3.16
Phenolate ion	Aq. base	235	3.97	287	3.42

The quinoid ring, such as that of benzoquinone, $O=C \begin{smallmatrix} C=C \\ \\ C=C \end{smallmatrix} C=O$

is much more effective as a chromophore than is the benzene ring.   An example which contrasts the two types is phenolphthalein, which has the following structures in acidic and basic solutions, respectively:

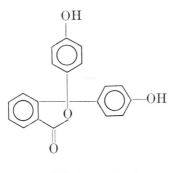

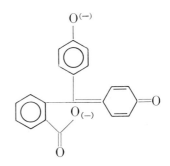

Colorless molecule
(in acid solution)

Red anion (in basic
solution)

In the colorless form, conjugation does not extend outside the individual aromatic rings (except that one ring is conjugated with a carbonyl group). In the red form, however, one ring has been converted to the corresponding quinone, which results in extending the conjugation to include the central carbon and, through it, the other two rings.   So we conclude that the entire anion constitutes a chromophore, whereas in acid solution, the molecule contains three separate and nearly identical lesser chromophores, the benzene rings.

The degree of absorption of a substance is conveniently plotted as a function of wavelength or frequency, and such a plot is referred to as an *absorption spectrum*.

STRUCTURE DETERMINATION

Ultraviolet and visible absorption spectra provide a valuable tool in the identification of unsaturated organic compounds and in the elucidation of their structure. The detailed correlation of molecular structure with absorption bands is too extensive a subject to be adequately covered in this text, but a few examples may be given to illustrate the utility of the method.

Information concerning a compound of unknown structure can sometimes be obtained by a direct comparison of its absorption spectrum with those of model compounds of known structure. For example, in the investigation of the compound cannabidiol[1] (a substance isolated from Minnesota wild hemp), chemical evidence showed its structure to be either *A* or *B* (in Fig. 3-1) but could not decide between these two. Ultraviolet absorption spectra were determined for cannabidiol and for the model compounds, 5-amylresorcinol and 4-amylcatechol (*C* and *D* respectively, Fig. 3-1). It is seen that the spectrum of the unknown resembles *C* very closely, whereas *D* is quite different. This observation constitutes strong evidence in favor of *A* rather than *B* for the structure of cannabidiol.

A particularly profitable correlation[46] is that concerned with unsaturated ketones of the general formula

$$R—\underset{\underset{O}{\|}}{C}—\underset{\underset{x}{|}}{C}=\underset{\underset{y}{|}}{C}—z$$

The wavelength of maximum absorption is found to depend on the number of substituents in the positions marked x, y, and z and on the presence of cyclic structures:

Structure	λ_{max}, nm
Monosubstituted, x or y	225
Disubstituted, x, y, or y, z:	
No exocyclic double bond	235
One exocyclic double bond	240
Trisubstituted, x, y, z:	
No exocyclic double bond	247
One exocyclic double bond	252

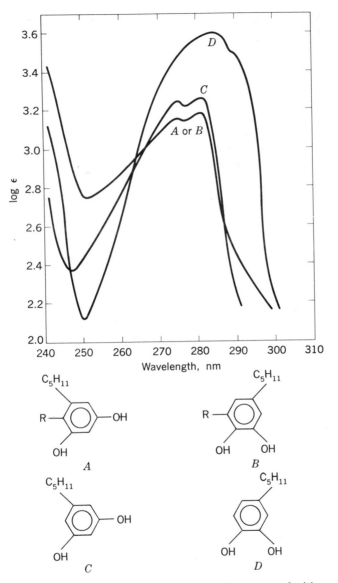

Fig. 3-1 Absorption spectrum of cannabidiol compared with those of certain phenols. (*Journal of the American Chemical Society.*)

A wealth of information and numerous examples along these lines can be found in the literature.

The variation of the absorption spectrum of an acid-base indicator as a function of pH provides an excellent method for determining the pK value of the indicator. In Fig. 3-2 are plotted the absorption curves for phenol red at a series of pH values. It is seen that with increasing pH the absorption at λ 610 nm increases, while the lesser absorption at λ 430 decreases. Note that the several curves cross very nearly at a common point at λ 495; this is called an *isoabsorptive point* or *isosbestic point* and is characteristic of a system consisting of two chromophores which are interconvertible so that the total quantity is constant.

If now we plot the absorbance at λ 615 against pH, an S-shaped curve is obtained (Fig. 3-3). The horizontal portion to the left corresponds to the acidic form of the indicator, while the upper portion to the

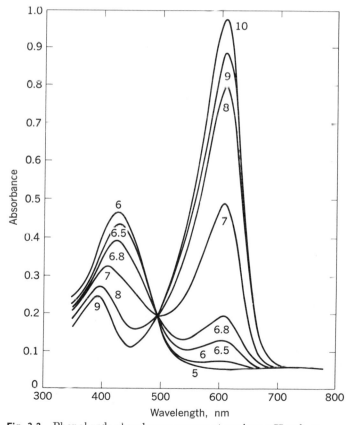

Fig. 3-2 Phenol red; absorbance curves at various pH values.

right corresponds to nearly complete conversion to the basic form.   Since
the pK is defined as the pH value for which one-half of the indicator is in
the basic form, one-half in the acid form, this point is determined by the
intersection of the curve with a horizontal line midway between the left
and right segments.

The dissociation constants of compounds with absorption in the
ultraviolet rather than the visible can often be determined by a similar
procedure.   Examples   are   theobromine[39]   (λ 240   nm),   theophylline[39]
(λ 240), and benzotriazole[12] (λ 274).

It is instructive to plot the data in such cases in three dimensions
(see Fig. 3-4).   In this presentation, which has been called a *stereospectrogram*, the three axes refer to wavelength, pH, and absorbance, respectively.[2]   Note that the isoabsorptive point on graph *F* corresponds to
a straight line parallel to the pH axis on the stereospectrogram.

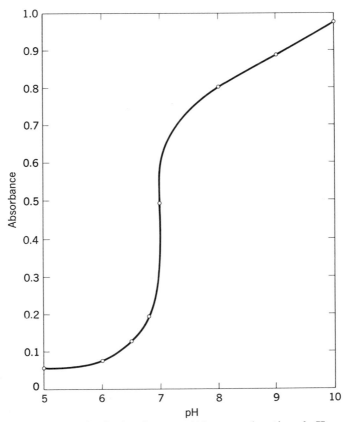

Fig. 3-3   Phenol red; absorbance at 615 nm as a function of pH.

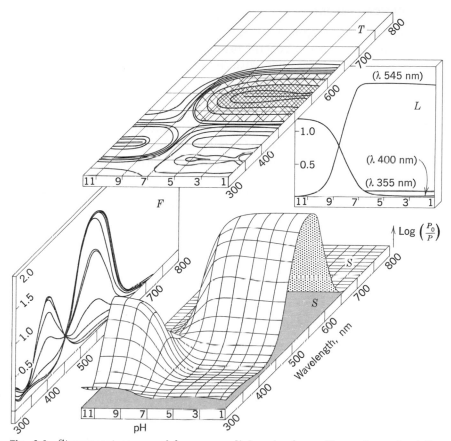

Fig. 3-4 Stereospectrogram of benzeneazodiphenylamine. (*From the work of Dr. H. Jaffé, as quoted by Archibald.*[2]) The stereospectrogram S represents a three-dimensional plot of the absorbance (vertical axis) as a function of pH (left-right axis) and wavelength (oblique axis). F, T, and L are the three two-dimensional modes of representation of the same data. F corresponds to Fig. 3-2, L to Fig. 3-3. In T, each line is an isoabsorbance line, and the graph can be read in the same manner as a topographical map.

The existence of an isoabsorptive point is evidence that a chemical equilibrium exists between two species. If instead of one absorption peak disappearing as another appears (the condition for an isoabsorptive point), the wavelength of a maximum shifts gradually upon change of pH, concentration, or other variable, then one can safely assume that the change is due to a physical interaction between the absorbing substance and its environment.

MATHEMATICAL THEORY

The quantitative treatment of the absorption of radiant energy by matter depends upon the general principle known as *Beer's law.* Consider a transparent container with plane, parallel faces traversed by monochromatic radiation. Losses by reflection at the surfaces and absorption by the material of the container itself will be neglected for the moment. Suppose that the container is filled with an absorbing liquid (either alone or dissolved in a nonabsorbing solvent). It is clear that the power of the radiation will be less the further it penetrates into the liquid and the greater the concentration of absorbing material. More generally stated, the beam of radiation will diminish in power in proportion to the number of absorbing molecules* in the path of the beam. The quantitative statement of this relation is Beer's law:† *Successive increments in the number of identical absorbing molecules in the path of a beam of monochromatic radiation absorb equal fractions of the radiant energy traversing them.* In terms of the calculus, this may be stated‡ as

$$\frac{dP}{dn} = -kP \tag{3-1}$$

where dP is the power absorbed at power level P by an increment dn in the number of absorbing molecules; k is a constant of proportionality. Rearrangement followed by integration between limits gives

$$\int_{P_0}^{P} \frac{dP}{P} = -k \int_0^N dn$$

$$\ln \frac{P}{P_0} = -kN \tag{3-2}$$

where P and P_0 are defined as previously, and N is the number of absorbing molecules traversed, for a beam of 1 cm² cross section. For a beam of cross-sectional area s cm², the right member of Eq. (3-2) must be multiplied by s:

$$\ln \frac{P}{P_0} = -kNs \tag{3-3}$$

---

* Molecules only if the absorbing substance is molecular, ions if it is ionized, etc.

† This relation is sometimes known as the Beer-Lambert or the Bouguer-Beer law, as contributions to its development have been made by Lambert and by Bouguer (as well as by others). For interesting discussion of the law, see papers by Liebhafsky and Pfeiffer[25] and by Hughes.[22]

‡ Beer's law is frequently stated in terms of intensities rather than radiant powers, with the letter I in place of P. We follow the concepts introduced in Chap. 2 and in Ref. 22.

The quantity Ns is a measure of the number of particles which are effective in absorbing radiation. A more useful measure, however, is the product of concentration c and length of path b, so that we shall write

$$\ln \frac{P}{P_0} = -kbc \tag{3-4}$$

For convenience, we shall replace k by another constant, a, which includes the factor for conversion of natural to common logarithms:

$$\log \frac{P_0}{P} = A = abc \tag{3-5}$$

Note that the ratio P_0/P has been inverted to remove the negative sign. The quantity $\log P_0/P$ is so important that it is given a special symbol A and called the *absorbance*. The shortest statement of Beer's law is thus $A = abc$.

Since the transmitted power P can vary between the limits 0 and P_0, the logarithm of the ratio, in theory, can vary from zero (the logarithm of 1) to infinity. In practice, however, absorbances greater than 2 or 3 are seldom usable, and the range which will give adequate analytical precision is even more limited, the exact permissible values being determined in part by the type of measuring instrument employed.

ABSORPTIVITY

The constant a of Eq. (3-5) is called the *absorptivity*. It is characteristic of a particular combination of solute and solvent for a given wavelength. Its units are dependent upon those chosen for b and c (b is customarily in centimeters), and the symbol varies accordingly, as indicated in Table 3-4. Other symbols and names which have been widely prevalent in the past are included for reference. The present notation is that proposed by the Joint Committee on Nomenclature in Applied Spectroscopy, established by the Society for Applied Spectroscopy and the American Society for Testing and Materials, as set forth in its report published in 1952.[22] Compare also papers by Brode[7] and by Gibson.[15]

It must be noted carefully that the *absorptivity* is a property of a substance (an intensive property), while the *absorbance* is a property of a particular sample (an extensive property) and will therefore vary with the concentration and thickness of the container.

Percent transmittance, percent $T = 100P/P_0$, is a convenient quantity if the transmitted radiation is of more interest than the chemical nature of the absorbing material. Color filters for colorimetry or photography are commonly rated in terms of percent transmittance. The

Table 3-4   Units and symbols for use with Beer's law

Accepted symbol	Definition*	Accepted name	Obsolete or alternate	
			Symbol	Name
T	P/P_0	Transmittance	. . . . . .	Transmission
A	$\log P_0/P$	Absorbance	D, E	Optical density, extinction
a	A/bc	Absorptivity	k	Extinction coefficient, absorbancy index
ϵ	AM/bc	Molar absorptivity	a_M	Molar (molecular) extinction coefficient, molar absorbancy index
b	. . . .	Length of path	l, d	

* The definitions of P and P_0 are given in the text (Chap. 2). The units of c are grams per liter; of b, centimeters; M is the molecular weight. A symbol formerly much used is $E_{1cm}^{1\%}$, which may be defined as A/bc', where c' is the concentration in percent by weight, and $b = 1$ cm.

absorbance A or the absorptivity a is useful as a measure of the degree of absorption of radiation by colored materials. The symbol a is used if the nature of the absorbing material, and hence its molecular weight, is not known. The molar absorptivity ϵ is preferable if it is desired to compare quantitatively the absorption of various known substances.

Beer's law indicates that the absorptivity is a constant independent of concentration, length of path, and the intensity of the incident radiation. The law provides no hint of the effect of temperature, wavelength, or the nature of the solvent. In practice, the temperature is found to have only secondary effects, unless varied over an unusually wide range. The concentration of the solution will vary slightly with change of temperature because of the volume change. Also, if the absorbing solute is in a state of equilibrium with its ions, with other compounds, with its tautomer, or with undissolved solid (a saturated solution), more or less variation with temperature is to be expected. On the other hand, some substances show quite different absorption if cooled to liquid-nitrogen temperature. For much practical analytical work, temperature effects may be disregarded, especially when the absorption of an unknown is directly compared with a standard, provided both are at the same temperature. In instances where more precise temperature control is necessary, that fact should be specified in the procedure.

The effect of changing the solvent on the absorption of a given solute cannot be predicted in any general way. The analyst is frequently limited to a particular solvent or class of solvents in which his material is soluble, so that the effect of change of solvent may not arise. A further restriction applies particularly to work in the ultraviolet, where many common solvents are no longer transparent. Water, alcohol, ether, and saturated hydrocarbons are satisfactory, but benzene and its derivatives, chloroform, carbon tetrachloride, carbon disulfide, acetone, and many others are not usable except in the region immediately adjoining the visible. Table 3-5 gives the approximate limits of ultraviolet transmission for a number of useful solvents.

Table 3-5 Ultraviolet transmission limits of common solvents*

180–195 nm	*265–275 nm*
Sulfuric acid (96%)	Carbon tetrachloride
Water	Dimethyl sulfoxide
Acetonitrile	Dimethyl formamide
200–210 nm	Acetic acid
Cyclopentane	*280–290 nm*
n-Hexane	Benzene
Glycerol	Toluene
2,2,4-Trimethylpentane	m-Xylene
Methanol	*Above 300 nm*
210–220 nm	Pyridine
n-Butyl alcohol	Acetone
Isopropyl alcohol	Carbon disulfide
Cyclohexane	
Ethyl ether	
245–260 nm	
Chloroform	
Ethyl acetate	
Methyl formate	

* Transmission limits taken arbitrarily at the point where $A = 0.50$ for $b = 10$ mm; within each group, solvents are arranged in approximate order of increasing wavelength limit. Data supplied by Matheson Coleman & Bell, Cincinnati, Ohio.

Even at constant temperature and in a specified solvent, it is sometimes found that the absorptivity may not be truly constant but may deviate toward either greater or smaller values. If the absorbance A is plotted against the concentration, a straight line through the origin should result, according to the prediction of Eq. (3-5) (curve 1, Fig. 3-5). Deviations from the law are designated as positive or negative, according to whether the observed curve is concave upward or concave downward.

It must be realized that conformity to Beer's law is not necessary in order that an absorbing system be useful for quantitative analysis. Once a curve corresponding to Fig. 3-5 is established for the material under specified conditions, it may be used as a calibration curve. The concentration of an unknown may then be read from the curve as soon as its absorbance is found by observation.

In general, Beer's law may be expected to hold reasonably closely for radiation of any given wavelength, but the absorbance will change as the wavelength is varied. The width of the band of wavelengths employed may also affect the apparent value of the absorptivity. A specific example will help to make this clear. Figure 3-6 shows the absorption spectrum of the permanganate ion in water solution. Reference to Table 3-1 shows that a substance absorbing as this does in the range 480 to 570 nm, approximately, should appear red-purple, which of course we know to be true. If an absorption measurement were made on this solution, using radiation passed by a filter of green glass with transmission limits approximately at the wavelengths marked A and F, the effects of the detailed peaks and valleys of the absorption curve would be averaged out, so that the value of the molar absorptivity so determined might be something like 1700 to 1800. However, if by some means the wavelength range of the light were limited to the region B to E, the value found would be the average of the true values within this space, perhaps 2300. If the width of the wavelength band were still further reduced to the region C to D, the molar absorptivity would approach its true value at this wavelength, 2500.

From this line of reasoning, it follows that, if it is desired to determine true absorption curves, it is necessary to use an instrument capable of isolating very narrow wavelength bands of light. Such an instrument is the spectrophotometer.

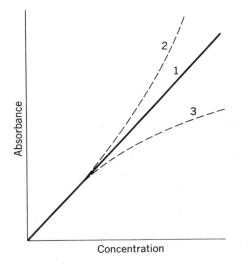

Fig. 3-5 Beer's law (1) obeyed, (2) positive deviation, and (3) negative deviation.

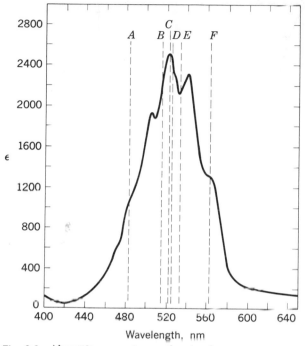

Fig. 3 6   Absorption spectrum of potassium permanganate in water solution.

DEVIATIONS FROM THE ABSORPTION LAW

The shape of an absorption curve may sometimes change with changes in concentration of the solution, and unless precautions are observed, apparent failure of Beer's law will result.   The phenomenon may be due to interaction of the solute molecules with each other or with the solvent. Polymerization, for example, will have such an effect if the monomeric and polymeric molecules have different absorption curves.   An example is benzyl alcohol in carbon tetrachloride, observed in the near infrared region.[13]   This compound exists in a polymeric equilibrium:

$$4C_6H_5CH_2OH \rightleftharpoons (C_6H_5CH_2OH)_4$$

Dissociation of the polymer increases with dilution.   The monomer absorbs at 2.750 to 2.765 μ, whereas the polymer absorbs at 3.000 μ. Hence observation at 2.750 μ will show negative deviation (absorptivity falls off with increasing concentration), whereas at 3.000 μ positive deviation will result.

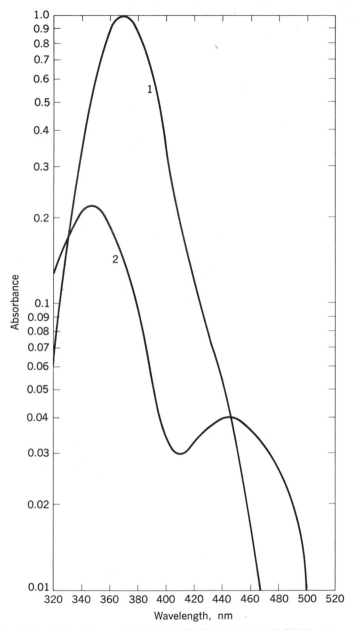

Fig. 3-7 Absorption spectra of (1) K_2CrO_4 in 0.05 N KOH, (2) $K_2Cr_2O_7$ in 3.5 N H_2SO_4. Both correspond to 0.01071 mg of Cr(VI) per ml; path length = 10.0 mm. Data for (1) from National Bureau of Standards, for (2) taken on a Beckman DU spectrophotometer. Absorbance plotted on a logarithmic scale for convenience.

Another, more familiar, example is the change of color of the dichromate ion upon dilution with water. The effect can be predicted from the equilibria

$$Cr_2O_7^{--} + H_2O \rightleftharpoons 2HCrO_4^- \rightleftharpoons 2H^+ + 2CrO_4^{--}$$

(orange) (yellow)

The absorption curves for $K_2Cr_2O_7$ and K_2CrO_4 (equal concentrations in terms of mg of Cr^{VI} per ml) are shown in Fig. 3-7. As a solution of dichromate is progressively diluted with water, the absorbance-concentration curves at the specified wavelengths resemble (qualitatively) the solid curves of Fig. 3-8. The dashed curves in each case show the absorbance which would be found if the dilution were performed at constant acidity or basicity, rather than simply by adding water.

Negative deviation can always be expected when the illumination is not monochromatic. This can be made evident by considering an extreme case. Suppose you are examining a series of solutions of potassium chromate (at constant basic pH). At zero concentration, all colors are transmitted equally. At a high concentration, nearly all the blue will be absorbed, but light of wavelengths above about 500 nm will be freely transmitted as before. At one-half this concentration, one-half the blue light will come through, along with all the higher wavelengths. A moment's thought will show that the total absorbance over all visible wavelengths cannot possibly be linear with concentration. A mathematical proof of this is readily derived.[27]

An interesting experimental study of the effect of changing the slit width, including comparison with a very narrow band of wavelengths from a laser, has recently been published.[20] The authors examined the absorbance of a dye which has a sharp maximum at 635 nm, both against a water reference and differentially against standard solutions. Their

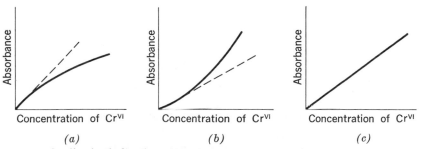

Fig. 3-8 Qualitative indication of Beer's law plots for the chromate-dichromate system upon dilution with water: (a) at λ372 nm; (b) at λ348 nm; and (c) at the isosbestic point, λ446 nm.

spectrophotometer required a slit width of 1.48 mm for a differential analysis with a tungsten lamp, as compared with 0.08 mm with the laser at 632.8 nm. Absorbance-concentration curves were found to be linear as far as they could be followed, about 1.4 absorbance, with the laser, but to show marked negative deviation with the tungsten lamp. If it were not limited to a few discrete wavelengths, the laser would be an ideal spectrophotometric source, combining narrow band width with a relatively high power level.

ADDITIVITY OF ABSORBANCES

In our discussion of Beer's law it was pointed out [Eq. (3-3)] that the absorbance is proportional to the number of particles which are effective

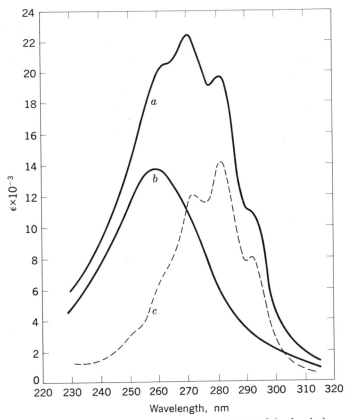

Fig. 3-9 Ultraviolet absorption spectra of (a) 7-dehydrocholes-teryl 4-nitrobenzoate, (b) cyclohexyl 4-nitrobenzoate, and (c) 7-dehydrocholesterol, determined by subtraction; solvent, n-hexane. (*Journal of the American Chemical Society.*)

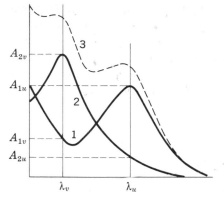

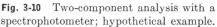

Fig. 3-10 Two-component analysis with a spectrophotometer; hypothetical example.

in absorbing radiation at the specified wavelength. This is easily extended to cover the presence in the same solution of more than one absorbing species. We can write

$$A = \sum_i A_i = b \sum_i a_i c_i \qquad (3\text{-}6)$$

which means that the absorbance is an additive property. This relation presumes, of course, that there is no chemical interaction between solutes.

This additivity can be useful in a number of ways. It permits subtraction from an observed absorbance of the contribution due to solvent or reagents, the familiar use of a "blank." It also enables one to subtract from the spectrum of an unknown the absorbance due to a chromophore known to be contained therein in order to identify a second chromophore. For example, in Fig. 3-9, curve a is the observed absorption spectrum of the 4-nitrobenzoate of the steroid 7-dehydrocholesterol. To identify this ester for what it is, the known spectrum b of another ester of the same acid (cyclohexyl 4-nitrobenzoate) is subtracted point-by-point from curve a. The resulting curve c is found to be essentially identical with the spectrum determined on the free sterol itself.[21]

The additivity of absorbance is also important in multiple analysis, the simultaneous determination of two or more absorbing substances in the same solution. The requirement for multiple analyses is merely that the absorption curves for the individual components do not approach too close to coincidence. Some partial overlapping is permissible, but the greater the overlap, the less the precision of the analysis. The relation can be understood by reference to Fig. 3-10. Curves 1 and 2 are the absorption spectra of the pure components. Curve 3 is the spectrum of a mixture. It is assumed that no other substances present absorb in this region.

Note that the maxima of curve 3 will not in general appear at exactly the same wavelengths as the corresponding maxima of curves 1 and 2. It will be seen that both substances contribute to the absorption at both wavelengths λ_u and λ_v. Let c_1 and c_2 represent the concentrations of the two components in the mixture, and let A_{1u} represent the absorbance of substance 1 at wavelength λ_u, etc., as indicated in the figure. Then the absorbance of curve 3 at λ_v will be given by

$$A_{3v} = A_{2v} + A_{1v} = a_{2v}bc_2 + a_{1v}bc_1$$

and at λ_u by

$$A_{3u} = A_{2u} + A_{1u} = a_{2u}bc_2 + a_{1u}bc_1$$

Since A_{3v} and A_{3u} are determined by experimental observation on the mixture, and the a's are obtained once and for all in advance, then the above equations can be solved simultaneously for c_1 and c_2. For highest precision, a_{1v} and a_{2u} should be as low as possible, while a_{1u} and a_{2v} should be high. An example which shows the possibilities of the method is the simultaneous determination of molybdenum, titanium, and vanadium by means of their colored complexes formed by treatment with hydrogen peroxide.[43] The standard curves are reproduced in Fig. 3-11.

Such multicomponent analysis has achieved its greatest importance in the field of infrared absorption, to be considered in Chap. 5.

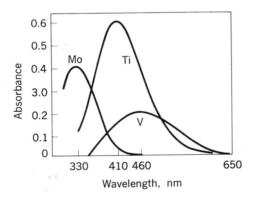

Metal	Absorbances, A		
	330 nm	410 nm	460 nm
Mo	0.416	0.048	0.002
Ti	0.130	0.608	0.410
V	0.000	0.148	0.200

Fig. 3-11 Comparison spectra of the products of reaction of hydrogen peroxide with molybdenum, titanium, and vanadium. Concentration, 4 mg of metal per 100 ml. (*Analytical Chemistry*.)

PHOTOMETRIC ACCURACY

Even for a system which shows no deviation from Beer's law, the concentration range over which photometric analyses are useful is limited at both high and low values. At high concentrations of absorbing material, so little radiant energy will get through that the sensitivity of the photometer becomes inadequate. At low concentrations, on the other hand, the error inherent in reading the galvanometer or other indicating device will become large compared to the quantity which is being measured. In many photoelectric instruments, the deflection of the galvanometer or the setting of a balancing potentiometer is directly proportional to the power of the radiation falling on the photocell. This means that the smallest detectable change in power ΔP will be constant regardless of the absolute value of the power itself. For greatest accuracy in the measurement of absorbance A, however, the increment ΔA, which corresponds to the power change ΔP, must be as small a fraction as possible of the actual absorbance A; in other words, the quantity $\Delta A/A$ must be a minimum. To determine the transmittance for which $\Delta A/A$ is a minimum, it is necessary to differentiate Beer's law twice and to set the second differential equal to zero. It is convenient to rewrite the law in the form

$$A = \log P_0 - \log P \tag{3-7}$$

Then

$$dA = 0 - (\log e) \frac{1}{P} dP$$

from which

$$\frac{1}{A} dA = -\frac{0.4343}{A} \frac{1}{P} dP = -\frac{0.4343}{A} \frac{1}{P_0 10^{-A}} dP$$

Replacing differentials by finite increments,

$$\frac{\Delta A}{A} = -\frac{0.4343 \Delta P}{P_0} \left(\frac{1}{A \; 10^{-A}} \right)$$

Differentiating again (remember that ΔP is a constant),

$$\frac{d(\Delta A/A)}{dA} = -\frac{0.4343 \Delta P}{P_0} \left(\frac{10^A \ln 10}{A} - \frac{10^A}{A^2} \right) \tag{3-8}$$

The condition for the minimum value for $\Delta A/A$ is that the right-hand member of Eq. (3-8) be zero. This means that the factor within the

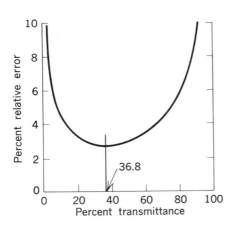

Fig. 3-12 Relative error as a function of transmittance.

parentheses must be zero, or

$$\frac{10^A \ln 10}{A} = \frac{10^A}{A^2}$$

from which

$$A = \frac{1}{\ln 10} = 0.4343 \tag{3-9}$$

This means that the optimum value for the absorbance is 0.4343, which corresponds to a transmittance $T = 36.8$ percent. The relative error in an analysis resulting from a 1 percent error in the photometric measurement for varying transmittance or absorbance is shown graphically in Fig. 3-12. The situation can perhaps be visualized more readily, especially for those who follow the calculus with difficulty, with the aid of Fig. 3-13, in which Beer's law is plotted in the form $P/P_0 = 10^{-abc}$.

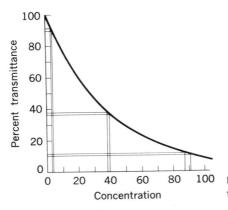

Fig. 3-13 Beer's law, plotted as transmittance versus concentration.

An arbitrary value of $\Delta T = 1$ percent is plotted in three positions, at 10, 37, and 90 percent T. The corresponding uncertainty in percent concentration is largest in absolute value at the 10 percent point, which results in poor precision. At the other extreme, 90 percent T, the uncertainty is much smaller but represents a large fraction of the total concentration, which again gives poor precision. It is evident that there must be some intermediate point where the two tendencies become equal and the error is a minimum. This is the 37 percent point.

It is apparent from Fig. 3-12 that, although the error is least at 37 percent T, it will not be much greater over a range of transmittances about 15 to 65 percent (absorbances 0.8 to 0.2).

The conventional method of plotting a calibration graph for a photometric analysis is either the exponential curve of Fig. 3-13 or the straight line of Fig. 3-5. The latter has the advantage of showing the region over which Beer's law is followed, but it fails to give any indication of the relative precision at various levels of absorbance. Another method of plotting has been suggested,[3, 33]* which gives some added features. Figure 3-14 shows the curve which is obtained by plotting percent transmittance against the logarithm of the concentration. If a sufficient range of concentrations has been covered, an S-shaped curve, sometimes known as a *Ringbom curve*, always results. If the system follows Beer's law, the point of inflection occurs at 37 percent transmittance; if not, the inflection is at some other value but the general form of the curve is the same. The curve generally has a considerable region which is nearly straight. The extent of this straight portion indicates directly the optimum range of concentrations for the particular photometric analysis.

* Both Ringbom and Ayres plot photometric data as *percent absorptancy*, defined as $100 -$ percent T. In the present discussion, this has been recalculated to the more familiar transmittance.

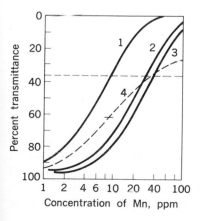

Fig. 3-14 Standard curves for permanganate. The solid curves are determined with a spectrophotometer at wavelengths (1) 526 nm, (2) 480 nm, and (3) 580 nm. The dashed curve (4) is from data taken with a filter photometer with a filter centering at 430 nm. Compare with Fig. 3-6. (*Analytical Chemistry.*)

Furthermore, the precision of the analysis can be estimated from the slope of the curve, since the steeper the curve, the more sensitive the test. It can be shown by a differentiation procedure that, if the absolute photometric error is 1 percent, the percent relative error in the analysis is given by $230/S$, where S is the slope, taken as the transmittance change in percent (read from the ordinate scale) corresponding to a tenfold change in concentration. The relative error in the determination of permanganate by curve 1 of Fig. 3-14 is shown by an application of this relation to be approximately 2.8 percent per 1 percent absolute photometric error. If the error in reading the photometer (reproducibility) is 0.2 percent (a reasonable value with modern instruments), then the relative analysis error is about 0.6 percent. A similar analysis by means of curve 4 would be much less precise. The precision with curves 2 or 3 is about the same as with curve 1, but the range of usable concentrations is shifted to larger values. A detailed comparison of Figs. 3-6 and 3-14 will show the reason for this.

The range of concentrations suitable for analysis with adequate precision, as indicated by the straight portion of a T vs. log c plot, may be too short for application to the unknowns likely to be encountered. There are several methods by which this useful range can be extended in the direction of higher concentrations. The most obvious is simply quantitative dilution of the solution to bring it within the required limits. This approach, if carried too far, may defeat its own purpose, as the cumulative volumetric errors partially offset the gain in photometric precision. Similarly, a more dilute solution can be prepared by taking a smaller sample, the precision being limited only by the sensitivity of the balance available. These methods often suffice to permit the use of sensitive absorptiometric methods with samples rich in the desired constituent.

Reference to Fig. 3-14 shows that analyses may also be carried to higher concentrations by a different choice of wavelength. The useful range for permanganate is about 6 to 60 ppm of manganese at 580 nm, as compared with 2 to 20 ppm at 526 nm.

PRECISION PHOTOMETRY

The standard procedure for absorptiometric analysis, as previously described, requires two preliminary adjustments: (1) the scale must be adjusted to read zero with no light reaching the photocell (lamp turned off or shutter closed), and (2) another adjustment must assure that the scale reads 100 with pure solvent in the beam of radiation. (Here and in what follows we shall assume a linear transmittance scale marked off in 100 divisions. The discussion applies equally to a deflection or null type of photometric circuit.) To complete an analysis with a system known to follow Beer's law, one must make two readings of transmittance, one for a standard, one for the unknown. It is desirable, particularly if some deviation from the law is suspected, to have the concentrations of standard and unknown rather close together.

This standard procedure leaves much to be desired with respect to precision, especially when operating near either end of the scale, i.e., solutions either of very low or very high transmittance relative to the solvent. Consider, for example, an analysis such that the standard solution shows 10 percent transmittance, and the unknown solution, 7 percent. Then both readings which are to be compared must be made utilizing only 10 percent of the instrument's scale. A tenfold increase of precision can be obtained by effectively spreading this first 10 percent to fill the whole scale by the simple expedient of setting the instrument to read 100 with the standard solution in the light beam instead of the solvent alone. This operation is shown schematically in Fig. 3-15, where the horizontal lines represent the scale of a photometer marked off in 100 divisions. The upper scale shows the position of the standard ($T = 10.0$ percent) and unknown ($T = 7.0$ percent) for the conventional method, and the lower scale shows that the transmittance of the unknown becomes 70.0 percent when the photometer is set at 100 percent for the standard.

The calculation of the concentration of the unknown involves no complications over the conventional method, which can be shown as follows. Let us write Beer's law twice, designating the unknown solution by subscript x and the standard by subscript s (subscript 0 still refers to the solvent):

$$\log P_0 - \log P_s = abc_s \qquad\qquad (a)$$

$$\log P_0 - \log P_x = abc_x \qquad\qquad (b)$$

Subtracting Eq. (b) from Eq. (a), we have

$$\log P_x - \log P_s = \log \frac{P_x}{P_s} = ab(c_s - c_x) \qquad\qquad \text{(3-10)}$$

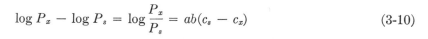

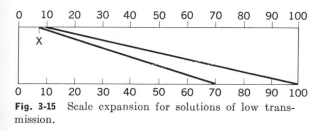

Fig. 3-15 Scale expansion for solutions of low transmission.

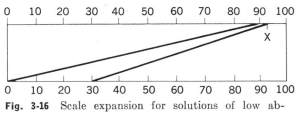

Fig. 3-16 Scale expansion for solutions of low absorption.

It is convenient to designate the quantity $\log P_x/P_s$ as the *relative absorbance*, and P_s/P_x as the *relative transmittance*. Equation (3-10) shows that, as long as Beer's law is followed, the relative absorbance is proportional to the difference in concentration between unknown and standard.

This procedure, sometimes referred to as *differential* or *relative* photometry,[19] has found wide usage. For example, Bastian[6] has obtained excellent results in the determination of copper in brass or bronze by a procedure of this type utilizing the color of the hydrated cupric ion itself. λ 870 nm was chosen because other common ions (Ni^{++}, Co^{++}, Fe^{++}, $Cr_2O_7^{--}$) do not interfere appreciably at this wavelength. Samples running from 1.5 to 1.8 percent were determined to an accuracy of better than 3 parts per 1000 against a 1.500 percent reference standard.

It will be noted that this precision compares favorably with standard gravimetric and volumetric procedures, whereas "ordinary" colorimetric methods are seldom better than 50 parts per 1000.

TRACE ANALYSIS

An analogous situation exists with highly dilute solutions.[29] Figure 3-16 diagrams such a case, for which a standard gives 90 percent transmittance, and the unknown, 93 percent. In conventional operation (upper scale), again only 10 percent of the scale is utilized. In this case, expansion of the scale is accomplished by setting the zero with a standard solution in the light path instead of an opaque shutter. The increase in precision for the example quoted is again tenfold.

Some types of photometers have no provision for offsetting the zero as far as may be required for this method. In general it can be accomplished with an instrument which is equipped with both sensitivity and dark-current controls. The zero is adjusted by the dark-current control (a misnomer, since darkness is now replaced by a finite level of illumination), and the 100 point is set by the use of the sensitivity control. In some instruments this may be rather inconvenient in that setting the full scale point may change the setting for the zero, and vice versa, so that a successive approximation procedure must be applied.

For this method of photometry, it can be shown that the absorbance follows a power series:

$$A = -\log T_2 - \left(\frac{0.4343}{100}\right)\left(\frac{1-T_2}{T_2}\right)R$$
$$+ \left(\frac{0.4343}{2 \times 10^4}\right)\left(\frac{1-T_2}{T_2}\right)^2 R^2 - \cdots \quad (3\text{-}11)$$

where R designates the scale reading corresponding to absorbance A, and T_2 is the transmittance of the reference solution which was used to establish the zero of the scale. This is the equation of a straight line if the term in R^2 is neglected and shows only slight curvature if that term is included. Higher terms are completely negligible.

METHOD OF ULTIMATE PRECISION

The greatest precision of which photometry is capable can be realized by a fourth procedure which combines features of the preceding two methods.[29] Both ends of the scale are established with standard reference solutions—one slightly more concentrated, one slightly more dilute than the unknown. The nearer together the three concentrations are, the higher will be the precision. Figure 3-17 illustrates this case. In this procedure also the absorbance follows a power series:

$$A = -\log T_2 - \left(\frac{0.4343}{100}\right)\left(\frac{T_1 - T_2}{T_2}\right)R$$
$$+ \left(\frac{0.4343}{2 \times 10^4}\right)\left(\frac{T_1 - T_2}{T_2}\right)^2 R^2 - \cdots \quad (3\text{-}12)$$

where T_1 is the transmittance of the reference solution used in setting the 100 point on the scale, and the other symbols retain their previous meanings.

It will be seen that this procedure amounts to close bracketing of the unknown between two standards. If the standards are taken close

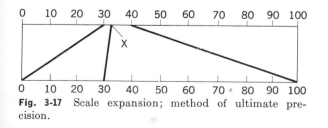

Fig. 3-17 Scale expansion; method of ultimate precision.

enough together, the limiting source of error ceases to be the photometer and becomes the accuracy of preparation of the standard solutions.

To recapitulate, we can distinguish four methods of executing a photometric analysis: case I, the conventional procedure wherein zero is set with the photocell in darkness, 100 with pure solvent; case II, for solutions of high absorbance, where the 100 point is set with reference solution; case III, for solutions of high transmittance where the zero is set with a reference solution; and case IV, where both ends of the scale are set with reference solutions. Cases II and III are both derivable from case IV and can give equivalent precision within their respective domains of usefulness.

With any of these high-precision methods, particular care must be given to potential sources of error which might be negligible in the conventional procedures. This is particularly true of possible mismatching of cuvets.[4] In some photometric instruments, mismatching of photocells must also be considered.

Anyone interested in pursuing further these principles of precision photometry should study the excellent critical article by O'Laughlin and Banks.[28] These authors point out that the use of the spectrophotometer as a null-point device in comparing the sample with a reference solution of known concentration is the most certain method of increasing the relative precision. This is largely due to the cancellation of systematic errors, and applies to any level of concentrations.

APPARATUS

Instruments for the measurement of the selective absorption of radiation by solutions are known as *colorimeters, absorptiometers,* or *spectrophotometers.* The term colorimeter is generally restricted to the simpler visual and photoelectric devices for the visible region. The term absorptiometer includes the class of colorimeters, but can be applied to other spectral regions as well. Spectrophotometers vary from simple absorptiometers only in that a much narrower band of wavelengths is employed, as produced by a monochromator. These classes of instruments vary only in degree, and the distinctions are not sharply drawn.

All forms must have certain features or components in common, as indicated in the following illustrative tabulation. Some of the simpler instruments may omit one or more items, and the sequence in which the radiation passes from one item to another is not always the same.

1. SOURCE OF RADIATION

Incandescent lamp
H_2 or D_2 arc lamp
Daylight

2. INTENSITY CONTROL

Iris diaphragm
Variable slit
Rheostat in lamp circuit

3. WAVELENGTH CONTROL

Color filter
Monochromator

4. SAMPLE HOLDER

Test tube
Cuvet

5. RECEPTOR

Photographic plate
Phototube or photomultiplier
Photocell
Bolometer or thermopile
Eye

6. INDICATOR

Galvanometer
Potentiometer
Pen recorder
Oscilloscope

The classification of absorptiometers as single- or dual-beam instruments has previously been discussed. The double-beam type may have two matched detectors, or the radiation may be flashed alternately over the two paths to a single receptor.

VISUAL COMPARATORS

Before photoelectric instruments were generally available, colorimetric analyses were carried out by simple visual comparison methods. Many of these methods are still prevalent, as the apparatus is less expensive and the precision is adequate for many purposes. An absolute accuracy of ±5 percent may be expected, though it may often be improved by careful attention to details.

The apparatus required for visual comparison methods may be quite simple. One common comparison cell is the *Nessler tube*, a cylindrical glass tube of perhaps 30-cm length, with the lower end closed by a plane window and with an etched mark at a fixed height from the bottom. A series of standard solutions is placed in the Nessler tubes to the exact height of the mark. The unknown, prepared under the same conditions, is placed in another tube and compared with the standards by looking downward through the solutions toward a uniform diffuse light source. The unknown can thus be bracketed between two standards, one slightly deeper and one slightly lighter, and the concentration estimated accordingly.

Another visual procedure involves a variation in the depth of the liquid through which the light must pass to reach the eye, so that the intensity of color will match the standard. The observation may be carried out with a *Duboscq comparator*. In this instrument, the unknown and standard solutions are held in glass-bottomed cells mounted on carriages, so that each can be moved vertically by a rack-and-pinion device equipped with a millimeter scale. Fixed glass plungers with polished ends extend into the cells from above; their upper ends lead to a special optical system and eyepiece. Light from below passes through the bottoms of the cells, through the liquids and plungers, and is combined in the eyepiece in such a way that the field of observation is divided into two parts, each illuminated by light passing through one of the cells. The position of the cells is adjusted until the intensity of the field is uniform. From the settings of the depth scales and the concentration of the standard, that of the unknown can be calculated from the equation

$$c = \frac{c'b'}{b} \tag{3-13}$$

where primes refer to the standard. Note that since both cuvets contain the same substance and are equally illuminated, the values of a, P, and P_0 are the same for both sides, and hence do not enter into the equation. With these restrictions, Eq. (3-13) follows directly from Beer's law.

FILTER PHOTOMETERS USING PHOTOVOLTAIC CELLS

The simplest photoelectric photometer (Fig. 3-18) consists of a small incandescent lamp with a concave reflector, an adjustable diaphragm, a

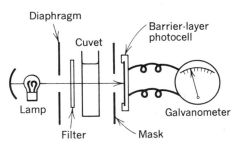

Fig. 3-18 Single-cell photoelectric photometer, schematic.

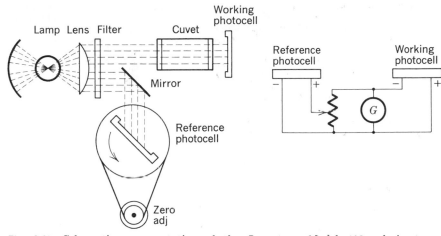

Fig. 3-19 Schematic representation of the Lumetron Model 402 colorimeter. (*Photovolt Corporation, New York.*)

colored-glass filter, and a cuvet—all composing the optical system—a single photocell to receive the radiation, and a directly connected galvanometer. The current output of the cell is directly proportional to the radiant power falling upon it, at a given wavelength. The usual procedure for determining the absorbance of a solution is to adjust the diaphragm so that the meter reads full scale (100) with pure solvent in the cuvet. The sample is inserted without changing the diaphragm; the meter will then read the percent transmittance, from which the absorbance can be calculated.

This simple arrangement possesses the fault that the reading varies with any fluctuation in intensity of the light source. This effect can be avoided by supplying current to the lamp from a storage battery or through a transformer especially designed to maintain a constant voltage. A better remedy is to use two photocells in a double-beam circuit so arranged that the fluctuations, being observed equally by both cells, are canceled out, whereas the effect of the sample is limited to only one of the cells.

There are several ways in which two identical photovoltaic cells may be connected with satisfactory results. Two circuits which are used in commercial photometers are shown in Figs. 3-19 and 3-20. In each, the two photocells are illuminated by a single lamp. The radiation passes through a filter and then is split into two portions: one traverses the cuvet to illuminate the "working" photocell; the other falls directly on the "standard" photocell. The amount reaching the standard cell must be controlled for the zero or comparison setting. In the Lumetron

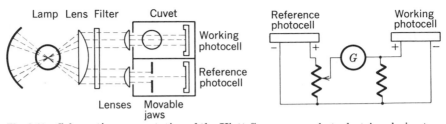

Fig. 3-20   Schematic representation of the Klett-Summerson photoelectric colorimeter. (*Klett Manufacturing Co., New York.*)

instrument, this is accomplished by rotation of the photocell itself about a vertical axis, whereas in the Klett-Summerson, the beam is limited by a pair of movable jaws.   The electrical connections are somewhat different, but in both the operation is the same.   The cuvet is first filled with pure solvent, the voltage-divider knob is set at 100, and the amount of radiant power falling upon the reference cell is adjusted to zero the galvanometer.   The solvent is then replaced by the absorbing solution, and the galvanometer is returned to zero by means of the voltage divider. The dial setting will then read percent transmittance if the scale is linear. Some instruments of this type are provided with a logarithmic scale to read absorbances, and some bear both linear and logarithmic scales engraved on the same dial.

FILTER PHOTOMETERS USING VACUUM PHOTOTUBES

The majority of instruments in this class are intended for special purposes, such as the continuous monitoring of flowing streams of liquid or gas.   An example of such an instrument is the Du Pont 400 Photometric Analyzer.*   This consists of a series of building blocks which can be assembled in various configurations, two of which are shown in Fig. 3-21. In (*a*) the radiation passes through the sample before being split into two beams.   The beams are filtered separately so that the "reference" beam corresponds to wavelengths not absorbed by the material analyzed for, while the "measuring" beam consists of wavelengths which are absorbed by this material.   In the configuration shown at (*b*), which is more conventional, the beam is filtered before splitting.   In either case, the electronic system displays on a meter or recorder the *difference* between the logarithms of the currents passed by the two photocells, which permits calibration in absorbance units or directly in terms of concentration.

Filter photometers for laboratory use are rarely designed to utilize

* E. I. du Pont de Nemours and Co., Inc., Wilmington, Del.

photomultiplier tubes, as their expense together with that of the required high-voltage power supply puts the instrument into a price bracket which usually justifies inclusion of a monochromator. A multipurpose photometer of this type, the Nefluoro-Photometer, was formerly manufactured by Fisher Scientific Company, Pittsburgh, Pennsylvania, and although discontinued, is still in use. It is a double-beam instrument using steady (unchopped) illumination, and an optical null attained by

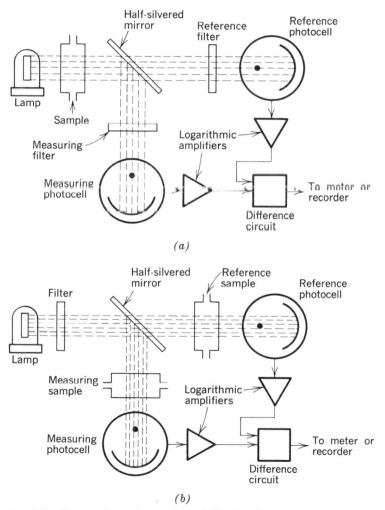

Fig. 3-21 Schematic representation of the Du Pont 400 Photometric Analyzer, shown in two configurations. (*E. I. du Pont de Nemours and Company, Inc., Wilmington, Del.*)

relative rotation of a pair of Polaroid disks. Two matched photomultipliers are required. Interchangeable lamps and optical parts permit its use as a fluorimeter or nephelometer as well as an absorptiometer.

SPECTROPHOTOMETERS

The chief limitation of filter photometers is the breadth of the wavelength band which must be employed. As pointed out previously, there are two disadvantages: (1) true absorption curves cannot be determined, and (2) Beer's law is not followed, which means that the apparent absorptivity will vary, not only with concentration, but also from one photometer to another. To avoid such difficulties, it is necessary to replace the filters with a monochromator capable of isolating a much narrower band of wavelengths.

A very widely used spectrophotometer primarily for the visible range, 340 to 625 nm (but capable of extension to 950 nm by a change of phototube), is the Spectronic-20.* This is a single-beam instrument with a 600 line per millimeter reflection grating (Fig. 3-22). It will be noticed that the plane grating is placed in a convergent rather than collimated light beam. This means that the degree of monochromaticity is less than the maximum obtainable with the same grating and represents a compromise between precision and expense.

The detector is a vacuum phototube, the signal from which is passed through a special amplifier, compensated to reduce the effects of

* Bausch & Lomb, Inc., Rochester, N.Y.

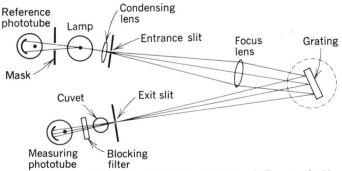

Fig. 3-22 Optical diagram of the Bausch & Lomb Spectronic-20 spectrophotometer. The grating is turned from a panel knob by a cam-and-bar linkage. The blocking filter is used only with the infrared phototube, to eliminate overlapping orders. In one version of the instrument the reference phototube and mask are omitted. (*Bausch & Lomb, Inc., Rochester, N.Y.*)

nonlinearity and drift. Both transmittance and absorbance are indicated on a large panel meter. There are only three controls: wavelength
selection, zero adjustment, and 100-percent adjustment. Spurious deflec-
tions caused by lamp fluctuations are effectively eliminated by a magnetic
or electronic regulator. The latter involves an auxiliary phototube which
monitors the lamp directly. As the Spectronic-20 is a relatively inexpensive and easily operated instrument, it is often used for convenience
where a filter photometer would serve equally well. The monochromator
passes a band of wavelengths 20 nm in width.

ULTRAVIOLET SPECTROPHOTOMETERS

Spectrophotometers for the ultraviolet and visible generally extend down
to somewhere between 165 and 205 nm. The upper limit is never less
than about 650 nm, and many extend to 1000 nm or even farther. Until
about the year 1963, very few instruments were usable below about 210
nm, a limit imposed principally by the absorption of natural quartz.
The advent of instruments with optical parts of specially manufactured
vitreous silica has opened up great possibilities in the observation of
isolated double bonds and certain other less-familiar chromophores in
this region.

The earliest ultraviolet spectrophotometer with photoelectric detec
tion to appear on the American market was the Beckman Model DU.*
This instrument, introduced in 1941, has contributed very greatly to
our knowledge of the ultraviolet region for both theoretical and analytical
purposes. It has been modified somewhat over the years to increase its
sensitivity and the convenience of its operation, but it remains essentially
the same instrument, and is in very wide use today.

The optical system of the Beckman DU is shown in Fig. 3-23. It is
seen to follow the classical Littrow design. There are interchangeable
tungsten and hydrogen or deuterium lamps, and also interchangeable
phototubes, as required to cover the complete range. A photomultiplier
is available as an option, which increases by some tenfold the sensitivity
from the ultraviolet limit up to about 625 nm. The operation of the
DU will be described in some detail because of the important position
of this instrument, and because it is representative of manual spectrophotometers in general.

The DU is provided with three electrical controls (two additional
ones with the photomultiplier attachment). These three are designated
as *sensitivity*, *dark current*, and *transmittance*. Each is to be operated
to bring the galvanometer needle to zero, and does so by adjusting the
potential of the grid of the first amplifier tube so as to offset the drain of

* Beckman Instruments, Inc., Fullerton, Calif.

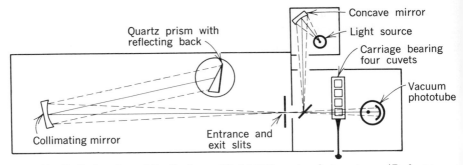

Fig. 3-23   Optical system of the Beckman Model DU spectrophotometer.   (*Beckman Instruments, Inc., Fullerton, Calif.*)

electrons through the phototube under different conditions of illumination, thus maintaining the meter reading constant.   A circuit diagram is given in Fig. 3-24.

Null measurements, as used in the DU, can give greater precision than is obtainable with direct meter indication at comparable cost. There are several reasons for this.   One is that the amplifier need only

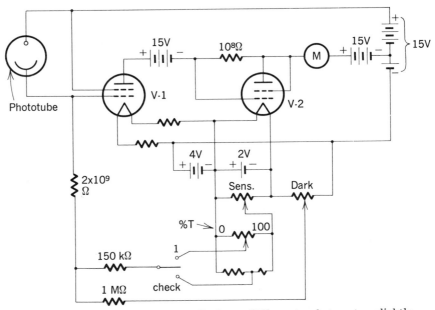

Fig. 3-24   Electronic circuitry of the Beckman DU spectrophotometer, slightly simplified.   The "zero" point on the meter actually corresponds to a small but finite current, the plate current of tube V-2.   (*Beckman Instruments, Inc., Fullerton, Calif.*)

be sensitive, not at the same time linear, as it is always returned to the same signal level when a measurement is taken. Another is that the transmittance scale can be longer, hence can be read with greater precision; in the DU it is about 22 cm in length, twice as long as the meter scale on a comparable direct-reading spectrophotometer.

In the normal procedure for determining the absorbance of a solution, the operator performs the following steps:

1. A solution of the substance in a transparent solvent is placed in one cuvet, a portion of the same solvent in a second identical cuvet.
2. With the phototube in darkness (shutter closed), the galvanometer is zeroed by means of the dark-current control. This is to compensate for the small current which always flows in the phototube circuit, even in the absence of radiation.
3. The wavelength is selected by means of the appropriate dial, which causes rotation of the prism.
4. The proper phototube and source are selected for the wavelength of interest.
5. The solvent cuvet is placed in the measuring position and the shutter opened.
6. The function switch is turned to the "check" position (the electrical equivalent of setting the transmittance dial at 100 percent), and the meter zeroed by adjustment of the sensitivity and slit controls. (The optimum settings of these two controls represent a compromise between high sensitivity with comparatively low precision on one hand, and lower sensitivity with greater photometric precision on the other.)
7. The function switch is advanced to the measuring position, and simultaneously the sample cuvet is moved into the beam of radiation.
8. The needle is now zeroed by turning the transmittance slidewire knob, and the transmittance or absorbance is read directly from the dial and recorded.
9. The above procedure from step 3 on is repeated for every wavelength desired; step 2 is checked occasionally.

An important variation on the conventional manual spectrophotometer, such as the DU, has been introduced by Gilford.*[16,45] No essential changes in the optical or mechanical features are involved, but a different photometric system is employed. The detector is a photomultiplier operated at constant *current*, rather than at constant *voltage*, as in conventional photometers. This is accomplished through a special feedback circuit which electronically adjusts the voltage applied to the photo-

* Gilford Instrument Laboratories, Oberlin, Ohio.

multiplier in an inverse relation to the illumination. Operated in this fashion, the characteristic curve of the photomultiplier, plotted as anode potential against illumination, is very nearly logarithmic, and can be made precisely logarithmic by a simple electronic padding system. The advantage is that the output becomes direct reading in absorbance, and accurately so (within 1 percent or better) up to an absorbance of 3 or even higher, corresponding to 0.1 percent T or less. Furthermore, the stability of the system, both short- and long-term, is unusually good, so that kinetic studies can be followed for hours or days without objectionable zero drift. This device so far has found its greatest usefulness in biochemical work. The principle appears not to have been applied to automatic wavelength-scanning spectrophotometers, though it should be possible to adapt it for such service without too extensive modification.

RECORDING SPECTROPHOTOMETERS

As an example of an excellent recording instrument for the ultraviolet, visible, and near-infrared regions, we will describe the *Cary Model 14*.* The optical system (Fig. 3-25) includes two monochromators, the first with a vitreous silica prism, the second a plane reflection grating (600 lines per millimeter), both in the Czerny-Turner configuration. A combination of cams and bars, driven by a constant-speed motor, ensures that the two monochromators track each other accurately, and provides a scan which is linear in wavelength over nearly the whole range, 186 nm to 2.65 μ.

The dispersion produced by the prism prevents the overlapping of spectral orders from the grating, and the presence of the grating causes the dispersion to be more nearly linear than would be the case with prism alone. The reciprocal dispersion curve, as shown in Fig. 2-19, is intermediate in shape between curves typical of prism and of grating instruments.

The operation of the instrument can be followed in Fig. 3-25. Radiation from the hydrogen lamp A or tungsten lamp c, as selected by the movable mirror a, enters the double monochromator through the entrance slit D. It is dispersed by the 30° Littrow-type prism F and by the grating J; H is an intermediate slit, and L the exit slit. All three slits are variable in width, and operated synchronously by a servo system; slits D and L are always equal in width, while H is slightly wider to make allowance for slight residual inaccuracies in the optical system. Monochromatic radiation from the exit slit L enters the photometer, where it encounters a rotating chopper disk N and a semicircular mirror O, also rotating, which steers it alternately via mirrors R, V, and W, or

* Cary Instruments Division of Varian Associates, Monrovia, Calif.

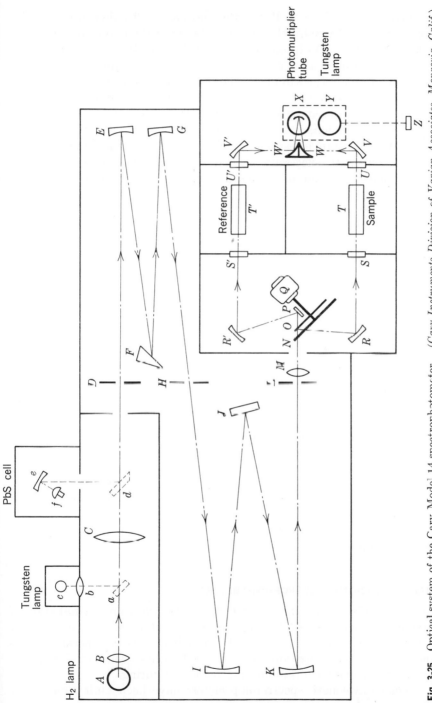

Fig. 3-25 Optical system of the Cary Model 14 spectrophotometer. (*Cary Instruments Division of Varian Associates, Monrovia, Calif.*)

via mirrors P, R', V', and W', to the photomultiplier detector X. The rotating mirror alternates the radiation between the two beams at a frequency of 30 cps. As in any conventional dual-beam photometer, one beam passes through a reference cuvet T', the other through an identical cuvet T containing the sample. The pulses of light from the two paths are out of phase with each other, so that the phototube receives radiation from only one beam at a time. This has the advantage that variations in the sensitivity characteristic of photomultipliers need not be compensated. This feature is especially advantageous in the near infrared, where it avoids the necessity of matching two lead sulfide photoconductive detectors, and of preserving the match through changes of temperature.

Auxiliary photoswitches, operated by the same motor which chops the light beam, are connected electronically in such a way that the output from the photomultiplier is sampled synchronously and split into two electrical signals corresponding to the two optical beams. The signals are fed into a comparator, and the built-in strip-chart recorder plots the ratio, either directly (transmittance) or as its logarithm (absorbance).

For operation in the near infrared, the positions of source and detector are interchanged, so that the radiation traverses the optical system in the opposite direction. The changeover is accomplished by means of plunger Z, which not only moves the photomultiplier out of the way and the tungsten lamp Y into position, but also swings a mirror d so as to deflect the beam to the lead sulfide detector f.

The reason for inversion of the system is that the chopper and other parts in the photometer housing heat up considerably as they are operated and radiate infrared energy on their own. If this energy were allowed to fall directly on the detector without first passing through the monochromator, it would constitute a substantial error. We will see in Chap. 5 that in practically all infrared spectrophotometers, the photometer precedes the monochromator for this reason. In the ultraviolet, the other way is preferable, because intense irradiation with all wavelengths in this region will cause photochemical decomposition in many substances.

DUAL-WAVELENGTH SPECTROPHOTOMETERS

It is possible to design an instrument so that two beams of radiation of different wavelengths pass through a cuvet simultaneously or alternating at the frequency of a chopper. In principle, this can permit simultaneous analysis for two components, and this has been done for special situations involving continuous on-stream monitoring.

Dual-wavelength spectrophotometers have been built along other lines, following the original ideas of B. Chance.[10,32] These are intended primarily for two purposes: (1) the study of kinetics, where the ratio of concentrations of two absorbing species is required, and (2) the measurement of small concentrations of an absorbing substance in the presence of turbidity.[11]

One such is the Aminco-Chance dual-wavelength spectrophotometer.* A schematic of this instrument is shown in Fig. 3-26. The radiation from a common source passes through a Czerny-Turner monochromator equipped with two gratings, separately adjustable, which are so disposed that each grating disperses one half of the beam of radiation

* American Instrument Co., Inc., Silver Spring, Md.

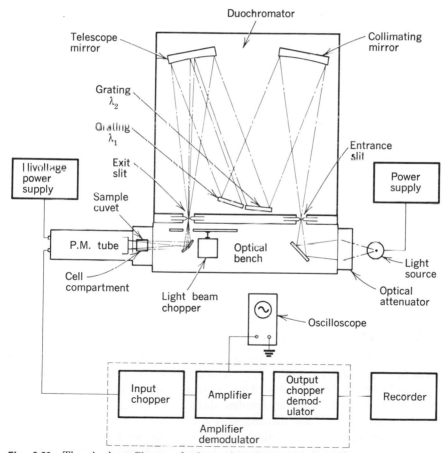

Fig. 3-26 The Aminco-Chance dual-wavelength spectrophotometer. (*American Instrument Co., Inc., Silver Spring, Md.*)

without intercepting the other. (The dual-wavelength monochromator is sometimes called a *duochromator*.) The two half beams are recombined after leaving the monochromator, and pass through a single cuvet to a photomultiplier tube which is mounted close to the cuvet so as to intercept the greatest possible amount of radiation. A mechanical chopper is arranged to pass the two beams alternately, and a synchronized switching device discriminates between the electrical signals corresponding to the two beams. A built-in analog computing element then determines the ratio between the two and displays it on a meter or recorder.

The advantage of this system for use with turbid materials results from the fact that the optical path through the sample is effectively longer than the internal thickness of the cuvet, because the rays suffer multiple reflections at the surfaces of the suspended particles. The effective path length is likely to differ greatly from one sample to another; hence, passing both measuring and reference beams through the identical sample has much merit. The measuring beam will be at the wavelength appropriate to the absorption spectrum of the substance sought. The reference beam should be fixed at a wavelength where no variable component absorbs, but as near to the measuring wavelength as permissible, because the degree of scattering by a suspension is a function of wavelength, a matter we will consider further in Chap. 6.

The usual cancellation of extraneous variables is so effective in this arrangement that, under favorable conditions, a change in absorbance of 0.0002 can be detected in a sample with an apparent absorbance of 2.

For kinetic studies in nonturbid solutions it is often advisable to set the reference at the wavelength of an isosbestic point, since the absorption there will not change during the course of the reaction. If a rapid reaction is being studied, the output can be displayed on an oscillograph or on an oscilloscope equipped with a memory device (storage oscilloscope) for subsequent photography.

We will now consider further practical examples of applied spectrophotometry.

IDENTIFICATION OF A COMPLEX

Many colorimetric analyses, particularly for metals, depend upon the formation of colored complex ions or molecules. It is frequently important to know the molar ratio of metal to reagent in the complex. This can be ascertained from photometric data by three different procedures: (1) the *mole-ratio* method introduced by Yoe and Jones,[47] (2) the method of *continuous variations* attributable to Job and modified by Vosburgh and Cooper,[42] and (3) the *slope-ratio* method of Harvey and Manning.[18]

In the mole-ratio method, the absorbances are measured for a series of solutions which contain varying amounts of one constituent with a constant amount of the other. A plot is prepared of absorbance as a function of the ratio of moles of reagent to moles of metal ion. This is expected to give a straight line from the origin to the point where equivalent amounts of the constituents are present. The curve will then become horizontal, because all of one constituent is used up, and the addition of more of the other constituent can produce no more of the absorbing complex. If the constituent which is in excess itself absorbs at the same wavelength, the curve after the equivalence point will show a slope which is positive but of smaller magnitude than that prior to equivalence. Figure 3-27 shows the results of such an experiment concerning the complex of diphenylcarbazone with mercuric ions.[14]

The method of continuous variations requires a series of solutions of varying mole fractions of the two constituents wherein their *sum* is kept constant. The difference between the measured absorbance and the absorbance calculated for the mixed constituents on the assumption of no reaction between them is plotted against the mole fraction of one of the constituents. The resulting curve will show a maximum (or minimum) at the mole fraction corresponding to that in the complex. An example is shown in Fig. 3-28 for the mercury-diphenylcarbazone complex.[14]

The sharpness of the breaks in the curves of both of these methods of identification of a complex depends on the magnitude of the stability constant. In fact, the degree of curvature often provides a convenient means of measuring the constant.

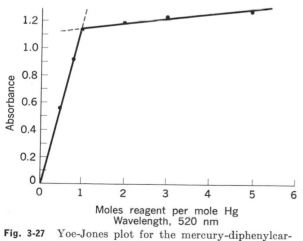

Fig. 3-27 Yoe-Jones plot for the mercury-diphenylcarbazone complex. (*Analytical Chemistry.*)

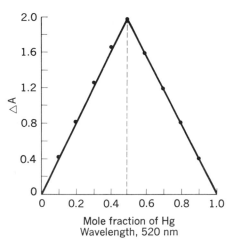

Fig. 3-28 Continuous-variation plot for the mercury-diphenylcarbazone complex. The vertical coordinate ΔA represents the *difference* between the absorbance of the mixed solution and the sum of the absorbances that the reagents would have shown had they not reacted. (*Analytical Chemistry.*)

The Job plot, even when it shows a large degree of curvature in the region of its maximum, nevertheless usually approaches both the zero and unity mole-fraction points in a linear manner. The ratio of the slopes of these two linear portions will equal the ratio of the components in the complex. This is known as the slope-ratio method.

In a system which forms two or more complexes, the slope-ratio method is no longer valid, and the other two methods will only give usable results if at least one of the stability constants is high, and if the several complexes absorb at sufficiently separated wavelengths.

THE CHEMISTRY OF ABSORPTIOMETRIC ANALYSES

A number of analytical procedures are given here to illustrate the variety possible in the application of the photometric methods previously described. They are not intended as laboratory directions, as many operational details are omitted.

For convenience the examples are classified into five groups:

   I. Ultraviolet: self-absorption
  II. Ultraviolet: developed absorption
 III. Visible: self-absorption
 IV. Visible: developed absorption
  V. Visible: indirect methods

In many instances a wide choice is open to the analyst as to which type of apparatus to employ. Thus copper can be determined, with varying degrees of precision, by the ammonia-color method with Nessler tubes, a Duboscq comparator, a filter photometer, or the most elaborate spectrophotometer. In the examples which follow, the photometric instrument is not specified unless a particular method is required for some reason.

CLASS I. DIRECT ULTRAVIOLET ABSORPTION ANALYSIS

A. The percentage of acetone in mixtures with ethers, alcohols, and low-molecular-weight monoolefins can be determined by measurement of the absorption of an aliquot, diluted with isooctane, at 280 nm.[5] A spectrophotometer was used in the reported work, but since the absorption maximum is relatively broad, an ultraviolet filter photometer could undoubtedly be substituted. A precision of ± 0.3 to 0.4 percent based on total volume was obtained.

B. The concentration of ozone in an urban atmosphere (smog) has been estimated by measurement of its absorbance at about 260 nm.[31] A special spectrophotometer was constructed to record the absorption spectrum from 254 to 365 nm automatically at 15-min intervals. Ozone in the range of about 2 to 100 parts in 10^8 was measurable.

C. Benzoic acid can be determined in saponified alkyd resins by measurement of its absorbance at 273 nm following purification by solvent extraction.[38]

D. Twelve of the lanthanide metals have been determined simultaneously with a recording spectrophotometer.[37] The lanthanide ions are characterized by numerous, very sharp absorption bands at isolated wavelengths throughout the visible and ultraviolet regions. Hence it is possible to select wavelengths for a multiple analysis such that very little overlapping is observed, and it is not necessary to undertake a laborious solution of simultaneous equations.

CLASS II. MISCELLANEOUS ULTRAVIOLET METHODS

A. Tellurium can be determined by means of the characteristic absorption of its iodide complex, presumably $[TeI_6]^{--}$, at 335 nm.[23] The sample containing tetravalent tellurium is treated with hydrochloric acid and potassium iodide, and the absorbance determined within 20 min. Bismuth and selenium interfere, but can easily be removed prior to analysis. Fe^{3+}, Cu^{++}, and other substances which oxidize iodide to free iodine must be absent.

B. An indirect method for the determination of calcium in blood serum has been described, based on the oxidation of the equivalent amount of oxalate by excess ceric sulfate and on a spectrophotometric measurement of the residual ceric ion.[44] The calcium for a 1-ml portion of serum is separated as the oxalate, redissolved in sulfuric acid, and treated with dilute ceric sulfate solution which also contains cerous ion (as a catalyst). The excess ceric ion is measured at 315 nm.

CLASS III. DIRECT ABSORPTION ANALYSIS IN THE VISIBLE

For the most part, methods in this class are so obvious that detailed descriptions at this point would serve no purpose. In addition to the salts of those metals which yield colored ions, many organic materials are readily analyzed by their natural colors. The clinical determination of hemoglobin in blood is an example.

CLASS IV. METHODS REQUIRING THE DEVELOPMENT OF VISIBLE COLOR

There are an extremely large number of methods falling in this class. The following are representative:

A. Cobalt is determined as a red complex with nitroso-R salt.[35] The sample is dissolved in acid and treated with an aqueous solution of the reagent. Enough sodium acetate is added to bring the pH close to 5.5. The solution is then diluted to volume, and the absorbance is determined through a green filter.

B. Copper reacts with diphenylthiocarbazone (dithizone), which is green, to give a red-violet product.[35] The sample should contain less than 0.005 mg of copper in a volume of 5 ml of 0.1 N acid. It is shaken with a 0.001 percent solution of dithizone in carbon tetrachloride in a small separatory funnel. The nonaqueous layer will contain a mixture of copper dithizonate and excess dithizone. It is observed in a photoelectric photometer either in the range 500 to 550 or 600 to 650 nm. The reference curve should be constructed from comparable copper solutions treated with dithizone from the same batch, as the reagent does not retain its concentration more than a few weeks. This is known as a *mixed-color* procedure, as the solution contains both the red-violet complex and excess green reagent. If the photometric measurement is made at 500 to 550 nm, which is in the green, the absorbance will be a direct measure of the complex, which absorbs green while the reagent transmits it. If the solution is examined at 600 to 650 nm, the absorbance will measure the excess reagent. Either can be used for the analysis. Dithizone gives similar colors with Mn, Fe, Co, Ni, Cu, Zn, Pd, Ag, Cd, In, Sn, Pt, Au, Hg, Tl, and Pb ions. In spite of this, considerable selectivity can be achieved through close control of the pH at which the carbon tetrachloride (or chloroform) extraction is carried out. Full details are given by Sandell.[35]

C. Chromium and manganese in steel samples can be oxidized to dichromate and permanganate and determined simultaneously as such.[26]

The sample of steel is dissolved in acid, phosphoric acid is added to complex the iron, and a few drops of silver nitrate solution are added as an oxidation catalyst. Potassium persulfate is added to oxidize the chromium and most of the manganese. A small amount of potassium periodate followed by heating ensures complete oxidation of the manganese. The absorbance of the solution is determined in a spectrophotometer at both 440 and 545 nm. The concentrations of manganese and chromium in the steel are then calculated by appropriate simultaneous equations.

CLASS V. MISCELLANEOUS INDIRECT COLORIMETRIC METHODS

A. Arsenic is determined by the *molybdenum-blue* color produced by the reduction of ammonium arsenomolybdate.[35] The arsenic in the sample should not exceed 0.03 g. It is reduced to the trivalent state and distilled as the chloride in order to remove it from nonvolatile substances which would interfere. The distillate is then evaporated to dryness with nitric acid to oxidize the arsenic to As^V. It is then treated with a solution containing hydrazinium sulfate and ammonium molybdate and heated to complete the formation of the blue color. The photometric measurement is made through a red filter. This reaction can be applied to the determination of phosphate, silicate, and germanate, as well as arsenate.

B. Aluminum can be precipitated quantitatively as a complex salt with 8-hydroxyquinoline (oxine) from an alkaline tartrate solution by dropwise addition of the reagent.[35] The filtered precipitate is washed and then dissolved in hydrochloric acid; sulfanilic acid and sodium nitrite are added; and the solution is allowed to stand for a few minutes. Rendering the solution alkaline with sodium hydroxide gives a yellow-red color which can be measured photometrically. Certain other metals give precipitates with oxine and hence may interfere unless the pH is carefully controlled.

PHOTOMETRIC TITRATIONS

In conventional titrimetry the equivalence point in a reaction is detected by visual observation of a change in color, either inherent in one of the reactants (e.g., permanganate) or produced by an indicator. Under favorable conditions, precision within a few tenths of one percent is easily obtainable by operators with normal vision. Good results, however, are difficult or impossible to obtain in cases where the color change is gradual or where the colors of the two forms do not contrast sharply.

Such difficulties can often be overcome by carrying out the titration

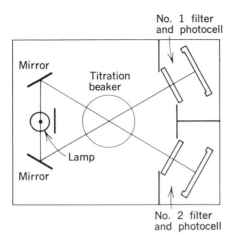

No. 1 filter
and photocell

Mirror

Titration
beaker

Lamp

Mirror

No. 2 filter
and photocell

Fig. 3-29 Suggested photometric titrator.

in a cuvet in a spectrophotometer or filter photometer. An optimum wavelength (or filter) is selected and the zero adjustments made in advance; then a photometric reading is taken following each incremental addition from a buret. Conventional spectrophotometers or colorimeters generally require some structural modification to permit insertion of a tritration vessel of convenient size as well as a buret tip and some type of stirrer. One such instrument, a modified Beckman Model B spectrophotometer, has been described by Goddu and Hume.[17]

A number of automatic or semiautomatic instruments which can carry out all the steps of a titration with a minimum of operator attention are commercially available. In some, the results are plotted by a pen recorder, in others the stopcock of the buret is closed electrically at the end point. Such photometric titrators can be extremely convenient and economical, particularly for repetitive titrations.

A convenient arrangement, suitable for a student-made photometric titrator, is suggested in Fig. 3-29. This device is especially suited for following indicator titrations, in which case the filters shown can be replaced by test tubes containing respectively the same indicator in each of its color forms. By means of mirrors, two beams of radiation from a single lamp are directed through the titration beaker, through the filters, to two photocells. The photocells are to be connected in a balanced circuit. The filters permit one photocell to see only the first colored form of the indicator, the second, the other form. The galvanometer will show a sharp change of deflection at the end point. The theory of such a titrator, called *dichromatic*, can be found in the literature.[34]

The usual photometric titration curve is a plot of absorbance against volume of added reagent. If the absorbing substances (titrant, substance titrated, or both) follow Beer's law, then the titration curve will ideally consist of two straight lines, intersecting at the equivalence point. The intersection is likely to show some degree of curvature, due to incompleteness of the reaction at the equivalence point. This is usually of no great consequence, because the segments of the curve more remote from equivalence are nearly straight, and can be extrapolated to an intersection.

Since the absorbance depends on the concentration, it is necessary to take into account the effect of dilution. The absorbance would change as the result of added solvent even if it contained no reagent. The dilution error can be eliminated in one of two ways, either by the use of a reagent much more concentrated than the solution titrated, so that the added volume can be neglected, or by a simple arithmetic correction factor:

$$A' = A\, \frac{V + v}{V}$$

where A = measured absorbance
   A' = corrected value
   V = original volume of solution
   v = volume of titrant added

An excellent example of a photometric titration is shown in Fig. 3-30, which gives the titration curve for a mixture of m- and p-nitro phenol titrated by sodium hydroxide.[17] The absorbance was measured at 545 mu, a wavelength where the anions of both isomers absorb, but where the corresponding acids do not. The absorptivity of the m-isomer is greater than that of the p-isomer. The p-isomer is neutralized first, because it is the stronger of the two weak acids. The end points corresponding to the two straight-line intersections were in error by slightly over one percent in this particular experiment. It would be impossible to determine these two acids in the presence of each other by visual observation, with or without an indicator, as the color change corresponding to the first equivalence point would be very gradual. For a similar reason, the analysis would be impracticable also by potentiometric (pH-meter) techniques.

This method is excellent for a weak acid which absorbs at a different wavelength from its anion, or if either acid or anion does not absorb. Both the ionization constant and the concentration of the acid must be considered in determining how weak an acid can be titrated. Satisfactory results can be obtained[40] if CK_a, the product of the molar concentration and the acid ionization constant, is greater than about 10^{-12}.

Strong acids cannot be titrated in this way because they are in the ionized state at all times. They may, however, be followed by monitor-

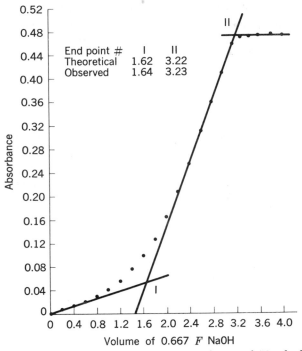

Fig. 3-30 Photometric titration of a mixture of 50 ml of 0.0219 F p-nitrophenol and 50 ml of 0.0213 F m-nitrophenol with 0.667 F sodium hydroxide; wavelength, 545 nm. (*Analytical Chemistry.*)

ing photometrically the absorbance of an added indicator. The indicator is, as usual, a weak acid such that the free acid and its anion absorb at different wavelengths. The indicator is not selected on the same basis as for visual titrations, however. For visual work, it is desirable that the the pK_a of the indicator coincide with the pH of the titration system at its equivalence point. For a photometric system, on the

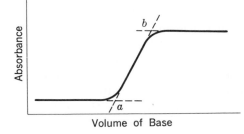

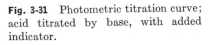

Fig. 3-31 Photometric titration curve; acid titrated by base, with added indicator.

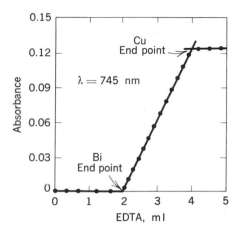

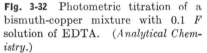

Fig. 3-32 Photometric titration of a bismuth-copper mixture with 0.1 F solution of EDTA. (*Analytical Chemistry.*)

other hand, when measurements are made in the region of absorption of the indicator anion, it is desirable to choose an indicator which has a small enough pK_a value that it does not start to be neutralized until the stronger acid has essentially completely reacted. The equivalence point can then be found by the intersection of two straight-line segments, as at a in Fig. 3-31. Intersection b corresponds to complete titration of the indicator, so this system can be considered to be the titration of a mixture of a strong and a weak acid.

A large variety of photometric titrations have been reported, including not only neutralizations, but also redox and complex-formation reactions.[30] In fact, the titration of various metals by EDTA and similar complexogens is one of the most fruitful applications of photometric titrimetry. Figure 3-32 shows an example wherein bismuth and copper are determined successively in a single titration.[41] Measurements were made at 745 nm, where the Cu-EDTA complex absorbs strongly, but the Bi-EDTA complex does not.

Precipitation titrations represent another important application of photometric end point detection, but discussion of them will be postponed to Chap. 6.

PROBLEMS

3-1 A particular sample of a solution of a colored substance, which is known to follow Beer's law, shows 80 percent transmittance when measured in a cell 1 cm in length. (*a*) Calculate the percent transmittance for a solution of twice the concentration in the same cell. (*b*) What must be the length of cell to give the same transmittance (80 percent) for a solution of twice the original concentration? (*c*) Calculate the percent transmittance of the original solution when contained in a cell 0.5 cm in length. (*d*) If the original concentration was 0.005 percent (weight per volume), what is the value of the absorptivity a?

3-2 In the preparation of a reference curve for an analysis with a photoelectric colorimeter, the following values were obtained:

Concentration mg/liter	P_0	P
0.00	98.0	98.0
1.00	97.0	77.2
2.00	100.0	63.5
3.00	99.5	50.0
4.00	100.0	41.3
5.00	100.0	33.5
6.00	100.0	27.9
7.00	99.0	23.4
8.00	98.2	20.3
9.00	100.0	18.1
10.00	100.0	16.4

(a) Calculate absorbances, and plot against concentration. Do these data indicate a negative or positive deviation from Beer's law, or neither? (b) Plot percent transmittance against the logarithm of the concentration. Using the method of Ringbom, state approximately the range of concentrations which will give adequate precision in this analysis.

3-3 Vitamin D_2 (calciferol), measured at wavelength 264 nm, its maximum, in alcohol as solvent, follows Beer's law over a wide range of concentrations, with a molar absorptivity $\epsilon = 18,200$. (a) If the formula weight is 397, what is the value of a? (b) What range of concentrations, expressed in percent, can be used for analysis if it is desirable to keep the absorbance A between the limits 0.4 to 0.9? Assume $b = 1$ cm.

3-4 A sample of Nichrome wire weighing 0.5000 g was dissolved in acid, and by suitable treatment the chromium in the sample was converted to dichromate ion. The volume at this point was made up to 250 ml. A 5-ml portion was transferred to a cuvet for photometric examination. 1.0000 g of pure $K_2Cr_2O_7$ was dissolved in acidified water to a volume of 250 ml (solution A). Ten milliliters of A diluted with 10 ml water gave solution B. Ten milliliters of A diluted to a volume of 100 ml with water gave solution C. Of each of these, 5 ml was placed in a cuvet for the photometer. The following results were obtained:

Sample	Absorbance
A	0.895
B	0.450
C	0.090
Unknown	0.140

What is the percentage of chromium in the Nichrome?

3-5 A silicate rock is to be analyzed for its chromium content. A sample is ground to a fine powder, and a 0.5000-g portion is weighed out for analysis. By suitable treatment the material is decomposed, the chromium being converted to Na_2CrO_4. The filtered solution is made up to 50.00 ml with 0.2 N H_2SO_4, following the addition of 2 ml of 0.25 percent diphenylcarbazide, a reagent which gives a red-violet color with sexivalent chromium. A standard solution is available which contains 15.00 mg of pure $K_2Cr_2O_7$ per liter. A 5.00-ml aliquot of the standard is treated with 2 ml of the diphenylcarbazide solution and diluted to 50.00 ml with 0.2 N H_2SO_4. The absorbances of the two final solutions are determined in a filter photometer and found to be

Standard $A_s = 0.354$
Unknown $A_x = 0.272$

(a) What is the amount of chromium in the rock, expressed in terms of percent Cr_2O_3? (b) Suppose the same solutions were compared directly with each other in the filter photometer, by the relative method. Describe how this would be done and how the calculation would be performed. Would any additional data be needed, and if so, what? (c) Suppose the same solutions were compared in a Duboscq comparator. Describe how this would be done. Calculate the scale reading for the unknown, provided the standard were set at 20.0 mm.

3-6 For each of the following situations, predict whether Beer's law would show an apparent negative deviation, a positive deviation, or practically none at all. (a) The absorbing substance is the undissociated form of a weak acid. (b) The absorbing entity is the cation in equilibrium with the weak acid. (c) A metal is being determined by means of a color-forming reagent, measured with a photoelectric colorimeter with the appropriate glass filter. (d) In the same system as (c), an insufficient amount of reagent is added to react completely with the three most concentrated samples of the ten examined.

3-7 It is desired to determine the volume of an irregularly shaped water tank. To accomplish this, 1.000 kg of a soluble dye is added, the tank is filled to capacity, and the water is thoroughly mixed by means of a circulating-pump system. A sample is then withdrawn and analyzed for its dye content. A 0.1000-g portion of the original dye was dissolved and diluted to 500 ml (solution A). A portion of this was further diluted with an equal volume of water (solution B). A two-photocell filter photometer, set at zero absorbance with solution B in the cuvet, showed an absorbance of 0.863 for the tank water and 0.750 for solution A. Calculate the capacity of the tank in gallons (1 U.S. gal = 3.785 liters).

3-8 Set up simultaneous equations for the analysis of molybdenum, titanium, and vanadium mixtures by the peroxide method, using the data of Fig. 3-11 and the associated table. In one experiment, a test solution was treated with excess peroxide and perchloric acid and diluted to 50.00 ml. The following absorbances were obtained:

λ, nm	330	410	460
A	0.248	0.857	0.718

Calculate the quantities of the three elements in milligrams present in the sample.

3-9 Caffeine, $C_8H_{10}O_2N_4 \cdot H_2O$ (formula weight = 212.1), has been shown to have an average absorbance $A = 0.510$ for a concentration of 1.000 mg per 100 ml at 272 nm. A sample of 2.500 g of a particular brand of a soluble coffee product was mixed with water to a volume of 500 ml, and a 25-ml aliquot was transferred to an Erlenmeyer flask containing 25 ml of 0.1 N sulfuric acid. This was subjected to the prescribed clarification treatment and made up to 500 ml. A portion of this treated solution showed an absorbance of 0.415 at 272 nm. (a) Calculate the molar absorptivity. (b) Calculate the number of grams of caffeine per pound of soluble coffee (1 lb = 453.6 g). Assume $b = 1$ cm.

3-10 The solubility of barium chromate is to be determined (at 30°C) through the color produced by diphenylcarbazide in a saturated solution. An excess of the solid barium chromate is shaken with water in a constant-temperature bath for a time sufficiently long to ensure the attainment of equilibrium. A 10-ml aliquot of the supernatant liquid is transferred into a 25-ml volumetric flask and treated with 1 ml of 5 N sulfuric acid and 1 ml of a 0.25 percent solution of the reagent and diluted to the mark. The absorbance is then determined through a green filter (540 nm). A standard solution containing 0.800 ppm of chromium (in the sexivalent state) is also measured. The results are: $A_{std} = 0.440$, $A_x = 0.200$. It is reported in the literature that Beer's law is followed with sufficient precision at least as far as 10 ppm. Calculate the solubility of the barium chromate in grams per 100 g of water.

3-11 The following facts are abstracted from a recent article: (a) Both arsenic and antimony can be oxidized from the trivalent to the pentavalent state by bromine, arsenic more readily than antimony. (b) Sb^{III} forms a complex with chloride ion (in 6 N HCl) which absorbs in the ultraviolet at 326 nm, whereas Sb^V, As^{III}, and As^V do not. (c) Potassium bromate and potassium bromide dissolved together in water form a stable solution which will liberate bromine quantitatively, upon being titrated into an acid solution, according to the reaction $BrO_3^- + 5Br^- + 6H^+ \rightarrow 3Br_2 + 3H_2O$. (d) Free bromine, in the presence of excess bromide ion, absorbs strongly in the ultraviolet, including the vicinity of 326 nm, though its maximum is at a shorter wavelength. Bromide alone shows no such absorption.

On the basis of the above statements, show how trivalent arsenic and antimony can be determined in mixed solution in hydrochloric acid by titration with standard bromate-bromide reagent. The absorbance of the solution is to be determined at 326 nm after each addition of reagent. Hint: A curve will result with two breaks corresponding to the two elements sought.

3-12 Water has been determined by spectrophotometric titration in a solvent consisting of anhydrous acetic and sulfuric acids.[8] The reagent is acetic anhydride $(AcO)_2O$. The titration is followed by the absorption of radiation at 257 nm by $(AcO)_2O$. Sketch a titration curve which might result and explain its shape.

3-13 A method has been reported[9] for the spectrophotometric determination of chlorate (ClO_3^-) impurity in ammonium perchlorate (NH_4ClO_4) for use in rocketry. This is based on the reduction of chlorate to free chlorine: $ClO_3^- + 5Cl^- + 6H^+ \rightarrow 3Cl_2 + 3H_2O$. The chlorine then reacts with benzidine (I) to give a colored product (II) with a maximum absorption at 438 nm.

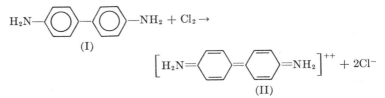

Experiments with standard $KClO_3$ solutions showed the following straight-line relation to hold under prescribed experimental conditions (in 10.0-mm cuvets):

$$A = (1.17 \times 10^3)C - 0.186$$

where C is the formal concentration of $KClO_3$. (a) Explain why (II) gives a colored solution, while (I) does not. (b) What color will the solution of (II) be, and what color of filter would be suitable for its determination in a filter photometer? (c) Does this system obey Beer's law? (d) The term "-0.186" in the equation was ascribed to a reducing impurity in the reagents. Explain why this would be expected to give a subtractive term. (e) A sample of 6.000 grams of commercial NH_4ClO_4 was dissolved in water, treated with HCl and benzidine, and diluted to 100.0 ml. A portion of this was examined in a 10.0-mm cuvet at 438 nm. The absorbance was found to be 0.450. Calculate the concentration of ammonium chlorate in ammonium perchlorate in terms of mole percent.

REFERENCES

1. Adams, R., C. K. Cain, and H. Wolff: *J. Am. Chem. Soc.*, **62**:732 (1940).
2. Archibald, R. M.: *Chem. Eng. News*, **30**:4474 (1952).
3. Ayres, G. H.: *Anal. Chem.*, **21**:652 (1949).
4. Banks, C. V., P. G. Grimes, and R. I. Bystroff: *Anal. Chim. Acta*, **15**:367 (1956).
5. Barthauer, G. L., F. V. Jones, and A. V. Metler: *Ind. Eng. Chem., Anal. Edition*, **18**:354 (1946).
6. Bastian, R.: *Anal. Chem.*, **21**:972 (1949).
7. Brode, W. R.: *J. Opt. Soc. Am.*, **39**:1022 (1949).
8. Bruckenstein, S.: *Anal. Chem.*, **31**:1757 (1959).
9. Burns, E. A.: *Anal. Chem.*, **32**:1800 (1960).
10. Chance, B.: *Rev. Sci. Instr.*, **22**:634 (1951).
11. Cowles, J. C.: *J. Opt. Soc. Am.*, **55**:690 (1965).
12. Fagel, Jr., J. E., and G. W. Ewing: *J. Am. Chem. Soc.*, **73**:4360 (1951).
13. Fox, J. J., and A. E. Martin: *Trans. Faraday Soc.*, **36**:897 (1940).
14. Gerlach, J. L., and R. G. Frazier: *Anal. Chem.*, **30**:1142 (1958).
15. Gibson, K. S.: Spectrophotometry, *Natl. Bur. Std. Circ.* 484, 1949.
16. Gilford, S. R., D. E. Gregg, O. W. Shadle, T. B. Ferguson, and L. A. Marzetta: *Rev. Sci. Instr.*, **24**:696 (1953).
17. Goddu, R. F., and D. N. Hume: *Anal. Chem.*, **26**:1679, 1740 (1954).
18. Harvey, A. E., and D. L. Manning: *J. Am. Chem. Soc.*, **72**:4488 (1950); **74**:4744 (1952).
19. Hiskey, C. F.: *Anal. Chem.*, **21**:1440 (1949).
20. Houle, M. J., and K. Grossaint: *Anal. Chem.*, **38**:768 (1966).
21. Huber, W., G. W. Ewing, and J. Kriger: *J. Am. Chem. Soc.*, **67**:609 (1945).
22. Hughes, H. K., et al.: *Anal. Chem.*, **24**:1349 (1952).
23. Johnson, R. A., and F. P. Kwan: *Anal. Chem.*, **23**:651 (1951).
24. Judd, D. B.: Measurement and Specification of Color, in M. G. Mellon (ed.), "Analytical Absorption Spectroscopy," John Wiley & Sons, Inc., New York, 1950.
25. Liebhafsky, H. A., and H. G. Pfeiffer: *J. Chem. Educ.*, **30**:450 (1953).
26. Lingane, J. J., and J. W. Collat: *Anal. Chem.*, **22**:166 (1950).
27. Meites, L., and H. C. Thomas: "Advanced Analytical Chemistry," p. 255, McGraw-Hill Book Company, New York, 1958.

28. O'Laughlin, J. W., and C. V. Banks: Differential Spectrophotometry, in G. L. Clark (ed.), "The Encyclopedia of Spectroscopy," p. 19, Reinhold Book Corporation, New York, 1960.

29. Reilley, C. N., and C. M. Crawford: *Anal. Chem.*, **27**:716 (1955).

30. Reilley, C. N., R. W. Schmid, and F. S. Sadek: *J. Chem. Educ.*, **36**:555, 619 (1959).

31. Renzetti, N. A.: *Anal. Chem.*, **29**:869 (1957).

32. Rikmenspoel, R.: *Rev. Sci. Instr.*, **36**:497 (1965).

33. Ringbom, A.: *Z. anal. Chem.*, **115**:332 (1939).

34. Ringbom, A., B. Skrifvars, and E. Still: *Anal. Chem.*, **39**:1217 (1967).

35. Sandell, E. B.: "Colorimetric Determination of Traces of Metals," 3d ed., Interscience Publishers (Division of John Wiley & Sons, Inc.), New York, 1959.

36. Silverstein, R. M., and G. C. Bassler: "Spectrometric Identification of Organic Compounds," 2d ed., p. 165, John Wiley & Sons, Inc., New York, 1967.

37. Stewart, D. C., and D. Kato: *Anal. Chem.*, **30**:164 (1958).

38. Swann, M. H., M. L. Adams, and D. J. Weil: *Anal. Chem.*, **28**:72 (1956).

39. Turner, Jr., A., and A. Osol: *J. Am. Pharm. Assoc., Sci. Ed.*, **38**:158 (1949).

40. Underwood, A. L.: Photometric Titrations, in C. N. Reilley (ed.), "Advances in Analytical Chemistry and Instrumentation," vol. 3, pp. 31ff, Interscience Publishers (Division of John Wiley & Sons, Inc.), New York, 1964.

41. Underwood, A. L.: *Anal. Chem.*, **26**:1322 (1954).

42. Vosburgh, W. C., and G. R. Cooper: *J. Am. Chem. Soc.*, **63**:437 (1941).

43. Weissler, A.: *Ind. Eng. Chem., Anal. Edition*, **17**:695 (1945).

44. Weybrew, J. A., G. Matrone, and H. M. Baxley: *Anal. Chem.*, **20**:759 (1948).

45. Wood, W. A., and S. R. Gilford: *Anal. Biochem.*, **2**:589 (1961).

46. Woodward, R. B.: *J. Am. Chem. Soc.*, **63**:1123 (1941); **64**:72, 76 (1942).

47. Yoe, J. H., and A. L. Jones: *Ind. Eng. Chem., Anal. Edition*, **16**:111 (1944).

4
Fluorimetry and Phosphorimetry

It was pointed out in connection with Fig. 2-2 that both fluorescence and phosphorescence constitute possible mechanisms whereby excited molecules can lose energy. During the process of excitation, most of the affected molecules acquire vibrational as well as electronic energy. The greatest tendency is for them to drop to lower vibrational states through collisions. If this radiationless process stops at an excited electronic level, then the molecules will be able to return directly from there to the ground state by the radiation of a quantum of energy (fluorescence); less commonly they may shift to a metastable triplet level before emitting radiation (phosphorescence).

In either case the molecule may end up in any of the vibrational states of the ground level, and so fluorescence and phosphorescence spectra generally consist of many lines, mostly in the visible region. In the presence of a solvent the lines are broadened and fuse together more or less to give a structured spectrum of the same general appearance as an ultraviolet or visible absorption spectrum. The short-wave end of a fluorescence spectrum usually overlaps at least slightly with the long-wave end of the absorption spectrum which gives rise to the excitation, but this overlap may not occur in phosphorescence.

Among organic molecules, fluorescence is to be expected in those species with large, rigid, multicyclic structures. Rigidity is sometimes brought about by complexation with a transition metal, and in such instances fluorescence is likely to provide a highly sensitive and often specific analytical tool for the metal. It is important to guard against photolysis which results from use of too energetic primary radiation (i.e., too short a wavelength of ultraviolet).

Fluorescence is more common and more widely applied in analysis than is phosphorescence, so we will consider it first and in more detail.

FLUORESCENCE

Fluorescent radiation is emitted equally in all directions from the irradiated sample. In some instruments it is observed in the direction opposite to the primary source, in others at an angle, generally 90°. The mathematical treatment is much more complicated than the mathematics of simple absorption leading to Beer's law. For one thing, the amount of primary radiation absorbed varies exponentially throughout the body of the solution, in accordance with the absorption law. For another, the fluorescent light is always subject to at least slight absorption in the solution, and the thickness of solution through which it passes is not constant, because the radiation does not originate at a single point. The complete equations are much too complex to include in the present discussion, but may be found in the literature.

The simplest case is that in which only a single absorbing and fluorescing species is present in a solution. For this case the governing equation is

$$F = P_0 K(1 - 10^{-A}) \tag{4-1}$$

where F is the power of fluorescent light reaching the detector and P_0 is the power of the incident ultraviolet radiation. K, a constant for a given system and instrument, is the factor for conversion of absorbed power to the fraction of unabsorbed fluorescence reaching the detector. A is the absorbance of the solution at the primary wavelength.

This expression can be transformed by application of an exponential series to a more useful form:

$$F = P_0 K \left[2.30A - \frac{(2.30A)^2}{2!} + \frac{(2.30A)^3}{3!} - \cdots \right] \tag{4-2}$$

This means that at low concentrations (A less than about 10^{-2}), where the squared and higher terms become negligible, F is linear with A, and thus with the concentration c:

$$F = 2.30P_0KA = P_0K'abc \tag{4-3}$$

where $K' = 2.30K$
$\qquad a$ = absorptivity
$\qquad b$ = path length in primary direction

At higher concentrations F drops off, giving a curve which resembles negative deviation from Beer's law.

Two important conclusions can be drawn from Eq. (4-3).   1. Since this equation is valid only for small values of A (less than about 0.01, which corresponds to approximately 98 percent transmission), it gives an indication of why fluorescence is most useful at a much lower concentration level than is absorptiometry.   In observing the fluorescence of a dilute solution, the photocell sees the faint light against a dark background, whereas in observing the absorbance of the same solution, it is necessary to measure the removal of a very small fraction of the radiation, which amounts to a precision measurement of the difference between two large numbers.   2. P_0 is greater for primary radiation which is selected by a filter than by a monochromator (for a given source) because of the wider wavelength band passed.   This accounts for the greater sensitivity of a filter fluorimeter than of a spectrofluorimeter.

FLUORIMETERS

Instruments for the measurement of fluorescence are known as *fluorimeters* (sometimes *fluorometers* or *fluophotometers*).   They are comparable to absorptiometers in that the sample is subjected to irradiation, and measurement is made of the power of the radiation leaving the sample.   In the majority of instruments the illumination is at right angles to the direction of observation (Fig. 4-1a).   The 90° construction is not particularly favorable, except for convenience in the layout of parts.   A smaller angle, as in Fig. 4-1b, is advantageous if the solution is turbid or if it absorbs appreciably at the wavelengths of the emitted light (possibly due to other substances present), as the observed fluorescent radiation in this geometry will have originated in the surface layers of the sample, where absorption of the primary ultraviolet is greatest, and will have traversed only a minimal thickness of solution.   The in-line construction of Fig. 4-1c has advantages if theoretical deductions (the quantum yield, for example) are to be drawn from the observations, because the equations involve fewer approximations.[4]

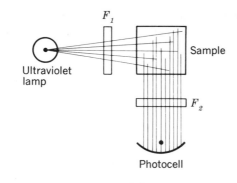

(a)

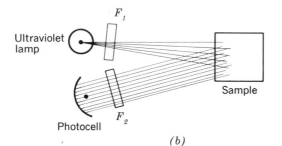

(b)

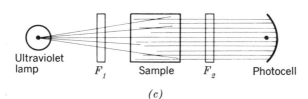

(c)

Fig. 4-1 Alternative fluorimeter geometries: observation (a) at 90°, (b) at a small angle, and (c) in a straight line. F_1 and F_2 are primary and secondary filters, either or both of which may be replaced by monochromators.

FILTER FLUORIMETERS

The basic arrangement for a single-beam, 90°, filter fluorimeter is shown in Fig. 4-2, exemplified by the Farrand Model A-2 Fluorometer.* The primary filter F_1, which will transmit ultraviolet but not visible radiation, is inserted between the lamp and the sample. A secondary filter F_2, which transmits visible and absorbs ultraviolet, goes between the sample and the photomultiplier tube. The relations of filters will be made clear

* Farrand Optical Co., Inc., Bronx, N.Y.

by reference to Fig. 4-5, which illustrates a particular application to be described later.

As in any single-beam photometer, the voltage supplies for both lamp and phototube must be stabilized to preserve constant sensitivity during an analysis. The dual-beam principle can be introduced in either of two ways: the reference photocell may monitor the ultraviolet lamp directly, or it may receive fluorescent light from a standard, sometimes called a *fluorescence generator*. In both arrangements the effect of variations in brilliance of the lamp resulting from line voltage fluctuations is eliminated. The use of the fluorescent standard is to be preferred, however, as it minimizes also the effect of temperature changes and variations in phototube sensitivity and linearity to be expected at widely differing wavelengths.

It is most desirable to employ as reference a standard solution of the substance analyzed, but sufficient purification to remove interfering materials may be a major problem. Acidic solutions of quinine fluoresce

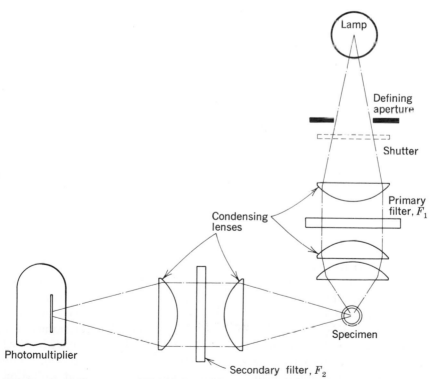

Fig. 4-2 Optical system of Farrand Model A-2 Fluorometer. (*Farrand Optical Co., Inc., Bronx, N.Y.*)

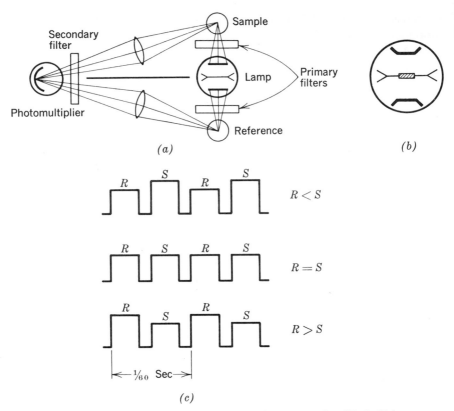

Fig. 4-3   Beckman Ratio Fluorometer: (*a*) layout, somewhat simplified; (*b*) lamp cross section; (*c*) waveforms, where *R* and *S* represent respectively the powers in the reference and sample beams.   (*Beckman Instruments, Inc., Fullerton, Calif.*)

brilliantly and are quite stable, so reference solutions of this substance at various dilutions are frequently used as secondary standards.   Highly stable standards can be made of commercial uranium glass.

An example of a double-beam filter fluorimeter is the Beckman Ratio Fluorometer* (Fig. 4-3).   This instrument makes use of a specially designed mercury-vapor lamp (Fig. 4-3*b*) with two anodes on opposite sides of a central structure which combines a symmetrical cathode with a light shield.   The lamp operates on alternating current so connected that the two anodes receive the discharge, and hence produce radiation, on alternate half-cycles of the exciting voltage.   This has the same effect as a rotating shutter, so that the sample and standard receive identical radiation pulsed at 60 Hz and separated by a short dark interval.   The two beams pass through primary filters, and the fluorescent radiation

* Beckman Instruments, Inc., Fullerton, Calif.

from the two cuvets converge through a common secondary filter onto a single photomultiplier, which therefore sees a series of pulses (Fig. 4-3c). The electronic circuits sort out these pulses and utilize the magnitude of the reference pulses to adjust the dynode voltages on the photomultiplier so as to maintain a constant reference level, not necessarily 100 percent. Then the sample pulses will produce a meter deflection which can be made direct reading in concentration units. Beckman supplies a series of graded uranium-glass standards for use when standard solutions of the substance sought are not available.

SPECTROFLUORIMETERS

These instruments fall into two classes: those which consist of a fluorescence attachment for a spectrophotometer, and those which are self-contained instruments, usually with two monochromators. The attachments provide for illumination of the sample at a 90° angle with ultraviolet from a mercury or xenon lamp via a filter or small, large-aperture, monochromator. The fluorescent light is analyzed by the spectrophotometer which may be scanned either manually or automatically. Even though the spectrophotometer may be of a double-beam type, it can only function in a single-beam mode as a fluorimeter.

A spectrofluorimeter designed as such should be expected to be more efficiently optimized for its intended function than an adapted spectrophotometer. Ideally such an instrument should be double-beam and ratio-recording, it should be capable of scanning either the primary or the secondary radiation (or both) at the choice of the operator, it should have an observation angle (as in Fig. 4-1) adjustable from 180° to as near 0° as practicable, it should have automatic control of the primary beam to maintain constant incident energy on the sample, and provision should be made for regulation of the temperature of both sample and standard. There is no spectrofluorimeter on the market with all these features, though each is available in the products of one or more manufacturers.

Probably the most widely distributed is the Aminco-Bowman Spectrophotofluorometer* (Fig. 4-4), a single-beam instrument incorporating two Czerny-Turner grating monochromators with a 90° sample geometry. Temperature regulation is provided.

Operation of any spectrofluorimeter usually follows some such procedure as this. By rough preliminary observations, a suitable wavelength in the emission spectrum can be chosen, and the second monochromator set at this point. An excitation spectrum can then be determined by scanning the first monochromator. Similarly an emission spectrum can be obtained by scanning the second monochromator with

* American Instrument Co., Inc., Silver Spring, Md.

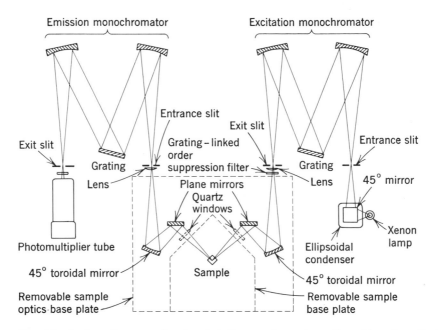

Fig. 4-4 Aminco-Bowman Spectrophotofluorometer, schematic. (*American Instrument Co., Inc., Silver Spring, Md.*)

the first set at a suitable value. With most substances, neither spectrum is found to vary greatly by a change in the wavelength selected for the *other* monochromator, as long as it is at a point of fairly high sensitivity.

Spectrofluorimeters are useful for establishing conditions for analysis and for studying interferences as well as for carrying out actual analyses. For best precision in the latter function, results should be compared with standards run on the same instrument. It is found that supposedly comparable spectra measured on different instruments often give varying results, due primarily to the variations with wavelength of the output of the light sources, the absolute sensitivities of the detectors, and the efficiencies of the monochromators. The observed spectra represent some combination of the spectral properties of the substance itself with artifacts generated by the instrument. Such variations are particularly disturbing when it is desired to measure the quantum efficiency of the fluorescent process.

It is possible to calibrate a spectrofluorimeter by painstaking comparisons involving a standard light source and thermocouple or precalibrated photomultiplier tube. The corrections so derived may be of considerable magnitude—a factor varying from unity to as high as 50 in one reported case.[10] Much of the tedium of such methods may be eliminated by the use of a fluorescent material as a standard of comparison. The aluminum chelate of the dye Pontachrome Blue-black R (the sodium salt of 2,2'-dihydroxy-1,1'-azonaphthalene-4-sulfonic acid) has been found particularly suitable.[1] Application of the calibration techniques described in this reference should go far toward eliminating the reporting of "uncorrected" fluorescence spectra.

An optional accessory is available for the Aminco-Bowman instrument which permits correction for the varying spectral output of the lamp and response of the photomultiplier and other components; it is essentially a special-purpose analog computer, including a calibrated thermocouple which monitors the primary beam as it leaves the first monochromator.

One spectrofluorimeter, the Turner model 210,* incorporates corrections for all these variable factors, through a rather complex and ingenious application of the dual-beam principle.[7]

QUENCHING

This is the name given to any reduction in the intensity of fluorescence due to the specific effects of constituents of the solution itself. Quenching may occur simply as the result of partial absorption of the fluorescent light by some component of the solution. If the fluorescent substance itself is responsible for this absorption, the phenomenon is known as *self-quenching* Quenching may also be due to the possibility of energy transfer by collision of excited molecules of the fluorescent substance with molecules of solvent or other solutes, resulting in a parallel but nonradiative mechanism for return to the ground state.

APPLICATIONS

Visible fluorescence occurs principally in two classes of substances: (1) a large variety of minerals and inorganic "phosphors," and (2) organic and organometallic compounds which have extensive ultraviolet absorption.

In the first of these classes, we will mention only a method for the determination of uranium salts which has been applied extensively in the field of nuclear research.[3] The sample is oxidized by evaporation with nitric acid and then fused with sodium fluoride to a melt containing the fluorides of sodium and uranium. This solidifies to a glass upon cooling. The glass is examined directly in a specially designed fluorime-

* G. K. Turner Associates, Palo Alto, Calif.

ter.   The sensitivity is of the order of 5×10^{-9} g of uranium in a 1-g solid sample.

An example of a fluorimetric analysis in inorganic chemistry is the determination of ruthenium in the presence of other platinum metals.[8] The complex ion of ruthenium(II) with 5-methyl-1,10-phenanthroline fluoresces strongly at pH 6.   Any other element of the platinum group can be present to the extent of at least 30 $\mu g/ml$ without interfering with the determination of ruthenium in the range of 0.3 to 2.0 $\mu g/ml$.   The precision of the determination with the instrument used by the authors was about ± 2 percent.   Palladium forms a precipitate with the reagent, which can be removed by centrifuging.   Iron must not be present, as it forms a complex which, although not fluorescent, absorbs energy too strongly.   Cerium, manganese, silver, and chromate ions also show varying degrees of interference.   The method possesses the great advantage that it is not necessary to separate the ruthenium by distillation of the tetroxide, a requirement of prior methods for this element.

Figure 4-5 shows the excitation and fluorescence spectra of the complex, tris-(5-methyl-1,10-phenanthroline)-Ru(II) ion, together with the transmission spectra of a pair of Corning glass filters which would be appropriate for an angular filter fluorimeter.   Interference filters are available which would permit in-line fluorimetry; for example, Multi-Films No. 90-1-500 as primary and No. 90-2-580 as secondary (Bausch & Lomb, Inc., Rochester, N.Y.).

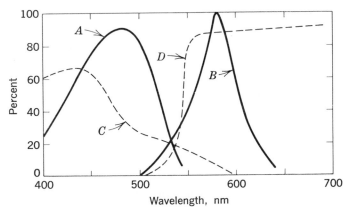

Fig. 4-5   Tris-(5-methyl-1,10-phenanthroline)-Ru(II)   complex ion: (A) excitation spectrum (uncorrected); (B) fluorescent emission spectrum corresponding to excitation at 450 nm; (C) Corning filter No. 5-59; (D) Corning filter No. 3-68.   The vertical scale represents relative radiant power for A and B, percent transmission for C and D.   (*Redrawn from Veening and Brandt*[8] *and from the catalog of Corning Glass Works, Corning, N.Y.*)

As another example, we may mention the determination of small amounts of aluminum in alloys.[9] Aluminum in the range of 0.2 to 25 μg in a volume of 50 ml can be determined with a sensitivity of 1 part in 10^8. The reagent is the dye Pontachrome Blue-black RM, at a pH of 4.8. The method is superior to older techniques in speed, sensitivity, and freedom from interferences.

Among fluorescent compounds, dependence upon the pH of the solution is often marked, and procedures generally call for careful buffering. The amount of fluorescence is also dependent upon the temperature to a greater degree than is usual in absorption spectroscopy. Another hazard in fluorimetry is the ease of contamination with fluorescent substances in amounts undetectable by most other means. Even water and aqueous solutions can extract fluorescent substances from rubber and plastic laboratory apparatus, even from bakelite bottle caps![5]

Thiamine (vitamin B_1) is commonly assayed by the blue fluorescence of its oxidation product, thiochrome.[2] The sample is treated with phosphatase, an enzyme which causes hydrolysis of the phosphate esters of thiamine which are frequently present in food materials. The phosphatase and other insoluble matter is filtered off and the filtrate diluted to a known volume. Two equal aliquots are taken, one for analysis, one for a blank. To the first is added an oxidizing agent (potassium ferricyanide), and to both equal quantities of sodium hydroxide and isobutyl alcohol. After shaking and subsequent removal of the aqueous layer, the alcoholic solution is examined in the fluorimeter. The whole procedure, including a blank, is repeated with a standard thiamine solution.

If other colored or fluorescing substances are present, purification can be effected by passing the solution (following the phosphatase treatment) through a column of adsorbent. Thiamine is preferentially adsorbed, whereas the contaminants pass through. The thiamine is then eluted with an acidified potassium chloride solution, made up to volume, and treated as above.

Riboflavin can also be determined by a fluorescence method.[2] The fluorescent power is dependent in large degree upon the exact conditions and upon the nature and amount of impurities. In order to make certain that any impurities have the same effect upon standard and unknown, the method of standard increment is employed, i.e., the fluorescence of a portion of the standard is measured in the same solution with the unknown. The procedure also takes advantage of the fact that riboflavin can readily be oxidized to a nonfluorescing substance, which is in turn easily reduced to regenerate the vitamin quantitatively.

The sample (a foodstuff) is extracted with acid; the solution is treated to precipitate salts which might interfere and is oxidized with dilute permanganate. The residual fluorescence is determined as a blank.   A slight excess of solid sodium dithionite ($Na_2S_2O_4$) is added as a reducing agent, after which the fluorescence is again determined.   A known volume of standard is then added and the fluorescence measured once again.   The results are calculated according to the following scheme:

Solution	Designation
10 ml oxidized sample + 1 ml water	A
Same + dithionite	B
Same + 1 ml standard	C

Then, since the fluorescent power F is proportional to the concentration of fluorescing material

$$\frac{F_B - F_A}{F_C - F_A} = \frac{m_x}{m_x + m_{std}}$$

where m_x and m_{std} are respectively the masses of riboflavin from sample and standard in the cuvet.

PHOSPHORIMETRY

Few compounds show appreciable phosphorescence at room temperature, but many more display the phenomenon if cooled with liquid nitrogen. The reduction in temperature appears to enhance the probability of transitions from excited singlet states to metastable triplet states, the required condition for phosphorescence, and also to diminish competing mechanisms for nonradiative return to the ground state.

Compounds which phosphoresce are likely to fluoresce as well, and a phosphorimeter must be able to distinguish between the two.   This can be done by means of a rotating shutter (see Fig. 4-6) such that a definite delay is introduced between the times during which the sample is irradiated and observed.   A versatile spectrophosphorimeter has been described by Winefordner and Latz,[11] and a commercial instrument is available.*   Theoretical aspects of optimum shutter design have been discussed in an interesting paper by O'Haver and Winefordner.[6]

Phosphorimetry has not found many practical analytical applications to date, but its potentialities are great.   A large number of organic compounds with conjugated ring systems phosphoresce intensely, and the method should provide excellent opportunities for trace analysis. The subject has been discussed in some detail by Winefordner and Latz,[11]

* American Instrument Co., Inc., Silver Spring, Md.

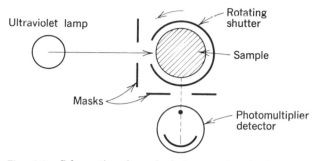

Fig. 4-6 Schematic of a single-beam phosphorimeter. Filters or monochromators may be located in either the primary or secondary beam, or both.

who give references to previous work. They developed a simple, fast procedure for the phosphorimetric determination of aspirin in blood serum and plasma, in which the presence of free salicylate does not interfere.

The best solvents to use in phosphorimetry are those which solidify to glasses at liquid nitrogen temperatures, rather than freezing to a crystalline state. This eliminates the tendency toward segregation of the solute, giving solid solutions instead. A particularly useful solvent is EPA, a mixture of ethyl ether, isopentane, and ethanol in the volume ratio of $5:5:2$.

PROBLEMS

4-1 A 2-g sample of pork is to be analyzed for its vitamin B_1 content by the thiochrome method. It is extracted with hydrochloric acid, treated with phosphatase, and diluted to 100 ml. An aliquot of 15 ml is purified by adsorption and elution, during which process it is diluted to 25 ml. Of this, two 5-ml portions are taken, one of them treated with ferricyanide, and both made up to 10 ml for fluorimetric examination. A standard solution of thiamine containing 0.2 μg per ml is subjected to similar treatment, except that the portion introduced into the adsorption column is made up to its original volume after elution (i.e., it is not diluted). Two 5-ml aliquots are taken, one oxidized, and both are made up to a final volume of 10 ml for measurement of fluorescence. The following observations are recorded:

Solution	Relative fluorescent power
A (standard, oxidized)	62.4
B (standard, blank)	7.0
C (sample, oxidized)	52.0
D (sample, blank)	8.0

Calculate the vitamin B_1 content of the pork in terms of micrograms per gram.

4-2   A 2-g sample of poultry feed is extracted with acid, and the extract is purified and diluted to 100 ml.   A 50-ml aliquot is oxidized with permanganate, the excess of which is removed by reduction with hydrogen peroxide.   The solution is made up to 100 ml, of which 10 ml is transferred to a cuvet, 1 ml water added, and the fluorescence measured (F = 2.2).   A few milligrams of sodium dithionite are added to the cuvet and stirred carefully.   The fluorescence is increased to F = 65.0.   Another 10-ml aliquot of the oxidized solution is introduced into an identical cuvet, and 1 ml of standard riboflavin solution (0.5 μg) is added, followed by dithionite.   The fluorescence observed is F = 86.0.   Calculate the riboflavin content of the feed, in micrograms per gram.

REFERENCES

1. Argauer, R. J., and C. E. White: *Anal. Chem.*, **36**:368 (correction p. 1022) (1964).
2. Association of Vitamin Chemists, Inc., "Methods of Vitamin Assay," 3d ed., Interscience Publishers (Division of John Wiley & Sons, Inc.), New York, 1966.
3. Byrne, J. T.: *Anal. Chem.*, **29**:1408 (1957).
4. Fletcher, M. H.: *Anal. Chem.*, **35**:278, 288 (1963).
5. Kordan, H. A.: *Science*, **149**:1382 (1965).
6. O'Haver, T. C., and J. D. Winefordner: *Anal. Chem.*, **38**:602 (1966).
7. Turner, G. K.: *Science*, **146**:183 (1964).
8. Veening, H., and W. W. Brandt: *Anal. Chem.*, **32**:1426 (1960).
9. Weissler, A., and C. E. White: *Ind. Eng. Chem., Anal. Edition*, **18**:530 (1946).
10. White, C. E., M. Ho, and E. Q. Weimer: *Anal. Chem.*, **32**:438 (1960).
11. Winefordner, J. D., and H. W. Latz: *Anal. Chem.*, **35**:1517 (1963).

GENERAL REFERENCES

Conrad, A. L.: Fluorimetry, in I. M. Kolthoff and P. J. Elving (eds.), "Treatise on Analytical Chemistry," pt. I, vol. 5, chap. 59, Interscience Publishers (Division of John Wiley & Sons, Inc.), New York, 1964.
West, W.: Fluorescence and Phosphorescence, in W. West (ed.), "Chemical Applications of Spectroscopy," chap. 6, being vol. IX of A. Weissberger (ed.), "Technique of Organic Chemistry," Interscience Publishers (Division of John Wiley & Sons, Inc.), New York, 1956.
White, C. E., and A. Weissler: *Anal. Chem.*, **38**:155R (1966).

5
The Absorption of Radiation: Infrared

Whereas the absorption of ultraviolet and visible radiation is conveniently considered as a unit, it is preferable to treat separately the corresponding phenomena in the infrared region. There are two important reasons for this: first, the optical techniques are sufficiently divergent that no spectrophotometers are available to cover without modification both the infrared and the visible-ultraviolet ranges; second, the absorption of infrared rests on a different physical mechanism than does that of visible and ultraviolet radiation.

It was pointed out in Chap. 2 that the absorption of infrared radiation depends on increasing the energy of vibration or rotation associated with a covalent bond, provided that such an increase results in a change in the dipole moment of the molecule. This means that nearly all molecules containing covalent bonds will show some degree of selective absorption in the infrared. The only exceptions are the diatomic elements, such as H_2, N_2, and O_2, because only in these can no mode of vibration or rotation be found which will produce a dipole moment. Even these simple species show slight infrared absorption at high pressures, apparently due to distortions during collisions.

Infrared spectra of polyatomic covalent compounds are often exceedingly complex, consisting of numerous narrow absorption bands, even though the absorbing sample may be a pure liquid. This contrasts strongly with the usual ultraviolet and visible spectra which almost invariably concern samples in the form of dilute liquid solutions. The difference arises in the nature of the interaction between the absorbing molecules and their environment. This interaction has a great effect on electronic transitions occurring within a chromophore, broadening the ultraviolet and visible absorption lines so that they tend to coalesce into wide regions of absorption. Only in the gas phase, and to some extent in nonpolar solvents, is this tendency not seen.

In the infrared, on the other hand, the frequency and absorptivity due to a particular bond usually show only minor alterations with changes in its environment (which includes the rest of its own molecule). The lines are not likely to be broadened so as to coalesce.

Exceptions to this generalization sometimes occur. For example, a long-chain molecule in the liquid phase is free to assume a limitless number of configurations due to free rotation about the many C—C bonds. The spectra of these many forms will be nearly but not quite identical, so that a broadening of absorption bands will appear. Hence it is preferable to examine compounds of this type in the solid phase, if possible.

A typical infrared absorption spectrum, that of isopropylbenzene (without solvent), is reproduced in Fig. 5-1.[1] Figure 5-2 shows the change in appearance of the spectrum of a long-chain compound, stearic acid, as it appears in solution, as a solid film at room temperature, and again at liquid-nitrogen temperature.[17] The concentration of the solution and thickness of the film were adjusted to give convenient traces on the recorder. Notice that by comparison with the room-temperature solid, the solution shows considerable broadening of peaks and loss of detail through much of the range, while a drastic reduction of temperature has the opposite effect.

Infrared spectra are usually plotted as percent transmittance, as in Fig. 5-1, rather than as absorbance. This makes absorption bands appear as dips in the curve rather than as maxima, as conventional in ultraviolet and visible spectra. This method of plotting is not universally employed, however, and Fig. 5-2 shows an inverted format. It is to be regretted that different conventions are followed in the two fields, but as both are deeply entrenched, there seems little prospect of changing either. The independent variable in infrared spectrograms is sometimes the wavelength in microns, and sometimes the wave number in reciprocal centimeters. Some workers in the field feel strongly that the wave-number treatment is preferable, because it is proportional to frequency, and hence more easily correlated with the vibrations within

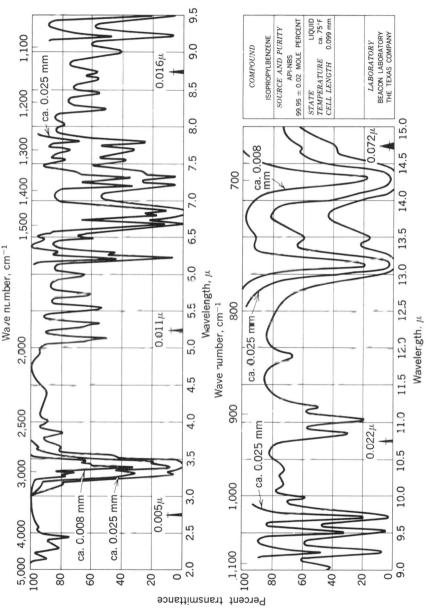

Fig. 5-1 Reproduction of the infrared spectrum of isopropylbenzene, replotted from reference 1; the black triangles indicate spectral band widths at several points.

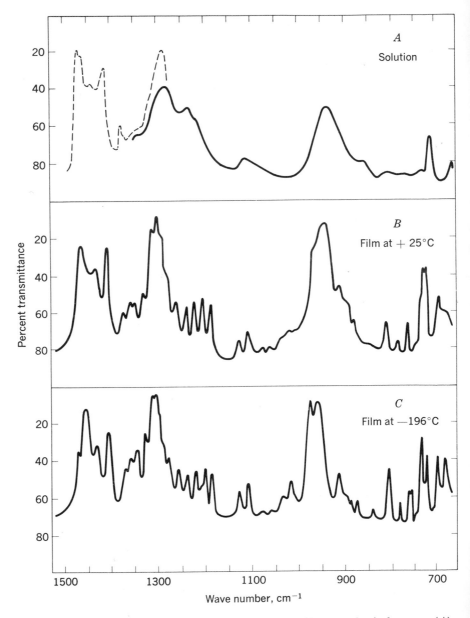

Fig. 5-2 Infrared spectra of stearic acid measured in different physical states: (*A*) in carbon tetrachloride solution (broken curve); in carbon disulfide solution (solid curve); (*B*) film of the β-polymorph at room temperature; and (*C*) at −196°C. (*John Wiley & Sons, Inc., New York.*)

the molecule.   On the other hand, some prefer linear wavelength presentation, because in the NaCl region the highly detailed "fingerprint" portion (about 5 to 15 μ) is spread out conveniently rather than bunched up at one end.   The two presentations are compared in Fig. 5-3.

STRUCTURAL CORRELATIONS

Since most infrared studies are concerned with organic compounds, we will follow suit, only emphasizing that similar principles apply to any substance containing covalent bonds.

It is possible, through a painstaking examination of a large number of spectra of known compounds, to correlate specific vibrational absorption maxima with the atomic groupings responsible for the absorption. Such empirical correlations provide a powerful tool for the identification of a covalent compound.

Some sweeping generalizations can be made.   It is useful to distinguish types of vibration as "stretching," "distortion," "bending," etc.   The shorter infrared wavelengths, from about 0.7 to 4.0 μ, include mostly stretching vibrations of bonds between hydrogen and heavier atoms; this includes the near-infrared region, and is especially useful in the identification of functional groups which contain hydrogen.   The 4.0 to 6.5-μ region contains vibrations of double and triple bonds.   Above this wavelength are found the "skeletal" distortion and bending modes, including C—H bending, etc.

The absorptions in the far-infrared region, beyond about 25 μ, correspond to vibrational modes involving heavy atoms and groups of atoms, including bonds of carbon to phosphorus, silicon, heavy metals, and also heavy metals to oxygen, and many others.   Also found in this region are some low-lying distortional frequencies, such as the ring-puckering mode in four-membered rings, as well as "torsional" vibrations of methyl and other groups, and the majority of purely rotational bands.

Table 5-1 gives an indication of the infrared regions corresponding to frequently occurring bond types.   Much more extensive and detailed correlations, taking account of the intramolecular environment of each bond, are available.   These have been prepared in convenient chart form by Goddu[9] for 1.0 to 3.1 μ and by Colthup[5] for the 2.5 to 25 μ regions.   Bentley[4] has compiled a chart which extends out to 33 μ.   All of these charts are reproduced in Meites' "Handbook of Analytical Chemistry."   Most of the major manufacturers of infrared spectrophotometers publish charts corresponding to the ranges and dispersions of their own instruments.   Beckman Instruments provides such data out as far as 300 μ.   Detailed discussion of such correlations can be found in numerous texts and monographs intended to assist in the determination of the structure of organic compounds.[3, 6, 9, 22, 23]

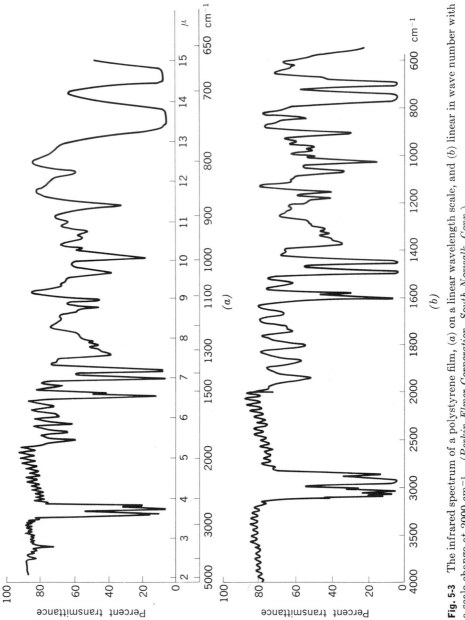

Fig. 5-3 The infrared spectrum of a polystyrene film, (*a*) on a linear wavelength scale, and (*b*) linear in wave number with a scale change at 2000 cm⁻¹.   (*Perkin-Elmer Corporation, South Norwalk, Conn.*)

Table 5-1 Infrared positions of various bond vibrations*

Bond	Mode	Relative strength†	Wavelength, μ	Wave number, cm^{-1}
C—H	Stretch	s	3.0–3.7	2700–3300
C—H	Stretch (2ν)‡	m	1.6–1.8	5600–6300
C—H	Stretch (3ν)‡	w	1.1–1.2	8300–9000
C—H	Stretch (C)§	m	2.0–2.4	4200–5000
C—H	Bend, in-plane	m–s	6.8–7.7	1300–1500
C—H	Bend, out-of-plane	w	12.0–12.5	800–830
C—H	Rocking	w	11.1–16.7	600–900
O—H	Stretch	s	2.7–3.3	3000–3700
O—H	Stretch (2ν)	s	1.4–1.5	6700–7100
O—H	Bending	m–w	6.9–8.3	1200–1500
N—H	Stretch	m	2.7–3.3	3000–3700
N—H	Stretch (2ν)	s	1.4–1.6	6300–7100
N—H	Stretch (3ν)	w	1.0–1.1	9000–10000
N—H	Stretch (C)	m	1.9–2.1	4800–5300
N—H	Bending	s–m	6.1–6.7	1500 1700
N—H	Rocking	s–m	11.1–14.3	700–900
C—C	Stretch	m–w	8.3–12.5	800–1200
C—O	Stretch	m–s	7.7–11.1	900–1300
C—N	Stretch	m–s	7.7–11.1	900–1300
C=C	Stretch	m	5.9–6.3	1600–1700
C=O	Stretch	s	5.4–6.1	1600–1900
C=O	Stretch (2ν)	m	2.8 3.0	3300–3600
C=O	Stretch (3ν)	w	1.9–2.0	5000–5300
C—N	Stretch	m–s	5.9–6.3	1600–1700
C≡C	Stretch	m–w	4.2–4.8	2100–2400
C≡N	Stretch	m	4.2–4.8	2100–2400
C—F		s	7.4–10	1000–1350
C—Cl		s	13–14	710–770
C—Br		s	15–20	500–670
C—I		s	17–21	480–600
Carbonates		s	6.9–7.1	1400–1450
Carbonates		m	11.4–11.6	860–880
Sulfates		s	8.9–9.3	1080–1120
Sulfates		m	14.7–16.4	610–680
Nitrates		s	7.2–7.4	1350–1390
Nitrates		m	11.9–12.3	820–840
Phosphates		w	9.0–10.0	1000–1100
Silicates		. . .	9.0–11.1	900–1100

* Approximate only; fundamentals unless noted; collected from various literature sources.
† s = strong, m = medium, w = weak.
‡ (2ν) means second harmonic or first overtone, etc.
§ (C) means combination frequency.

It must be emphasized that there are many situations where the infrared absorption of a compound is altered more or less extensively by the conditions under which it is observed. These variations in the location of absorption bands make it necessary to utilize the empirical tables and charts with great caution, when attempting to determine the structure of an unknown substance.

The causes of such alterations may lie with the spectrophotometric apparatus—such variables as slit width and scanning rate—and they will be considered later. But the spectrum can be significantly altered by factors concerned directly with the sample. The effects of solvent and temperature have been mentioned previously.

Another type of interaction is exemplified by the effect of hydrogen bonding on the absorption frequency of a carbonyl group. The frequency corresponding to the stretching mode of the C=O bond in a compound dissolved in a nonpolar solvent is lowered by considerable amounts by the formation of hydrogen bonds with some added hydroxylic substance, or by changing to a hydroxylic solvent. The carbonyl absorption is also altered by its environment within the same molecule; the absorption of the C=O bond is significantly different in a carboxylic acid where an intermolecular hydrogen bond (formation of a dimer) can occur, as compared with the ester where there is no possibility of such a bond. The anion is affected by resonance which renders the two oxygen atoms equivalent to each other, so that the double-bond character of the carbonyl is even more profoundly changed. The following tabulation shows representative values for the carbonyl stretching vibrations of a few aliphatic compounds:[6]

Ketone	$5.81–5.85\ \mu$	$1720–1710\ cm^{-1}$
Carboxylic acid monomer	$5.65–5.71$	$1770–1750$
Carboxylic acid dimer	$5.81–5.85$	$1720–1710$
Ester	$5.73–5.80$	$1745–1725$
Salt, asymmetric*	$6.21–6.45$	$1610–1550$
Salt, symmetric*	near 7.14	near 1410

* The anion of a carboxylic acid shows two modes of stretching vibration:

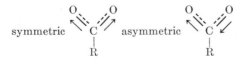

This discussion of the carbonyl absorption is presented as an abbreviated example of the kind of structural considerations which can be of great significance to the organic chemist as well as to the analyst.

One result of the complexity of infrared spectra is that it becomes highly improbable that any two different compounds will have identical curves. Hence an infrared spectrum of a pure compound presents a sure method of identification, provided that the analyst has at hand an extensive compilation or "atlas" of spectra of known compounds. A number of such atlases are available.[1, 19]

Infrared spectra of gaseous samples at relatively low pressures show a great amount of detail, and the best theoretical correlation with the "true" natural frequencies of the unperturbed molecule. They are not extensively utilized in analytical work because of their complexity and because other means of examination are usually more convenient.

We shall now turn our attention to the instrumental aspects of infrared spectroscopy.

MATERIALS OF CONSTRUCTION

There are no solids which are transparent through the entire infrared region of chemical interest. Some of the best are listed in Table 5-2

Table 5-2   Infrared-transmitting solid materials

Material	Limiting wavelengths, μ*		Prisms	
	Short	Long	Beckman[†]	Perkin-Elmer[‡]
Borosilicate glass	0.40	2.6		
Vitreous silica	0.20	4.4		0.17–3.5
Calcite	0.30	5.1		
Sapphire	0.18	6.1		
LiF	0.13	8.5	1.0–6.0	0.5–6.5
CaF$_2$	0.15	12.0	0.5–9.0	0.5–9.5
Si	1.30	13.0		
BaF$_2$	0.16	15.0		
Ge	1.90	23.0		
NaCl	0.22	25.0	1.0–16.0	1.0–15.5
AgCl	0.40	28.0		
KBr	0.24	40.0	10.0–25.0	11.0–25.0
KI	0.35	42.0		
CsBr	0.23	53.0	11.0–35.0	15.0–38.0
CsI	0.26	70.0		

* Maximum and minimum wavelengths are taken from a much longer tabulation given by Holter et al.[15]; they represent the wavelengths at which a sample 2 mm thick has 10 percent transmission.

† Data taken from literature of Beckman Instruments, Inc., referring to the standard interchanges for their IR-4 spectrophotometer.

‡ Data from Perkin-Elmer Corporation, referring to their Model 221, except the data for vitreous silica, which refer to Model 450 with specially selected prism.

with wavelength limits corresponding arbitrarily to the points where a 2-mm thickness will transmit 10 percent of the incident radiation. A substance can still be useful for some purposes beyond this cutoff, for example, as material for a window where a piece thinner than 2 mm can be used, or where relatively high energy levels are encountered. On the other hand, as material for a prism, these figures are too optimistic, for a prism will seldom be less than 20 or 30 mm thick at the base.

Some of the salts must be protected from atmospheric moisture, lest their surfaces become fogged. In addition many are soft and easily scratched. Fluorite, CaF_2, is an exception, as it is hard and not affected by moisture.

In the far infrared, beyond the reach of a cesium bromide prism, the optics must be completely reflective, including reflection gratings rather than prisms. Indeed, even in the mid-infrared region, it is universal practice to substitute concave mirrors for lenses, because of their freedom from chromatic aberration, and because they can be constructed of materials more durable than salts. It is generally necessary to pass the radiation through solid material in two locations, the absorption cell,* and a window at the detector. The latter can be quite thin, as it does not cover a large area. The absorption cell windows are made of salts below their respective cut-off points. In the far infrared, difficulties become apparent. Windows can be made of polyethylene, a polymer with only single C—C and C—H bonds, which is therefore reasonably transparent in the far infrared. Quartz starts to transmit again beyond about 45 μ, and can be used for windows and cells.

SOURCES

Continuous sources in the infrared, as we have seen, are primarily incandescent. The *tungsten* filament lamp is the usual source for near-infrared instruments; the *Globar*, the *Nernst glower*, and *Nichrome* heater are common through the midrange. In the far infrared a high-pressure *mercury arc* is to be preferred. Beckman has applied the quartz mercury lamp in a unique manner in their far-infrared spectrophotometer (IR-11). At the shorter wavelengths, where quartz absorbs, the heated quartz envelope supplies the radiation, while at longer wavelengths, the quartz transmits the energy from the mercury plasma itself. The crossover point falls somewhere between 50 and 100 μ.

DETECTORS

The ultimate in sensitivity and speed of response (which determines the permissible chopping rate) in infrared detection appears to be certain photoconductive devices,[8] but they suffer from practical disadvantages.

* In the infrared region a cuvet is generally called an absorption cell.

If operated at room temperature, they have a very restricted range, usually limited to the near infrared. The range is broadened by drastic cooling. A specially treated germanium unit has been found useful through the entire NaCl region if cooled to 30°K. Obviously such a detector will be applied only in unusually elaborate installations.

In commercial spectrophotometers, lead sulfide photoconductors are regularly employed in instruments limited to the near infrared. Other detectors depend on the rise in temperature of a physically small, blackened, receptor. The temperature rise in a few designs is measured by a bolometer, but more commonly either by a thermocouple or a pneumatic system operating via changes in the pressure of a gas. The thermocouple has previously been discussed.

Pneumatic detectors can be selective or general. A selective detector is filled with an absorbing gas (CO_2, H_2O, hydrocarbons, etc.) which sensitizes it to the analysis of the corresponding gas in mixtures (of which more later). To be nonselective, the gas with which the detector is filled is not intended to absorb radiation directly, but to be warmed by an absorbing solid receptor immersed in the gas.

The chief example of a nonselective pneumatic detector is the *Golay cell*,* which is built into several commercial spectrophotometers. The radiation to be measured is absorbed by a blackened film located in the center of a small gas chamber. The resulting pressure increase in the gas causes a thin, flexible mirror to bulge outward. The convex mirror forms part of an optical system by which light from a small incandescent bulb is focused on a phototube. The signal seen by the phototube is thus modulated in accordance with the power of the radiant beam incident on the gas cell.

The Golay detector is best suited to measurements of radiation chopped at a frequency of 10 to 15 Hz. It has about the same sensitivity as a thermocouple detector in the mid-infrared region, but has a wider range of usefulness. It has been found uniformly sensitive from the ultraviolet, through the visible and infrared, at least as far as a wavelength of 7.5 mm in microwaves, provided that appropriate window materials are selected. It is more expensive and bulky, and somewhat less convenient than many other detectors.

NONDISPERSIVE PHOTOMETERS

It is not practicable to attempt analysis in the infrared by means of a simple filter photometer, as is done so widely in the visible, because the absorption bands of most substances are much narrower and closer

* Invented by Dr. Marcel J. E. Golay, and manufactured by The Eppley Laboratory, Inc., Newport, R.I.

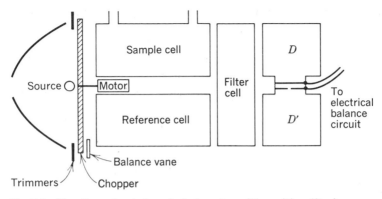

Fig. 5-4   Nondispersive infrared photometer with positive filtering.

together than can be distinguished by available filters. The nearest equivalent to a filter photometer is the *infrared gas analyzer*, which depends on the properties of selective pneumatic detectors. Figure 5-4 shows one possible arrangement.[11] D and D' are two identical vessels, each containing a sample of the gas being determined, usually diluted with argon to reduce the specific heat. The vessels are separated by two electrically insulated diaphragms, one pierced by a hole. The intact diaphragm is free to bend slightly in response to any variation in pressure on its two faces, thus changing the electrical capacitance between it and its pierced counterpart. The pressure in D and D' depends on the temperature, which in turn depends on the amount of radiation absorbed. Hence the pressure in D will vary inversely with the amount of the substance sought which is present in the sample cell. The reference cell is ordinarily filled with dry nitrogen and sealed off. In many simple applications the "filter cell" is not required, but it may be used if necessary to desensitize the instrument to any gas component which has absorption bands overlapping those of the substance sought. A motor-driven chopper is required, because the differential detector is a dynamic device; i.e., it responds to rapid changes more effectively than to gradual drifts. The chopper is designed to interrupt the two beams simultaneously. The capacitor formed by the two diaphragms is incorporated in a high-frequency electronic circuit which energizes a small servomotor to drive a balancing vane across the reference beam to the point where both beams deliver the same power to the detector. The amount of such compensation is indicated on a meter or recorded on a moving chart.

   Photometers of this type are widely used industrially, connected directly to monitor a gas stream. Sensitivities of the order of tenths or hundredths mole percent are representative.

   A similar photometer with a single beam but an ingenious arrangement of two selective detectors in tandem, designed originally for measuring CO and CO_2 in submarines, has been described by Beebe and Liston.[2]

SPECTROPHOTOMETERS

Early infrared spectrophotometers were manual instruments, not unlike the Beckman DU at shorter wavelengths. Manual spectrophotometers in the infrared are seldom used today, and only for special purposes, such as monitoring the effluent from a chromatograph for a constituent of known absorption characteristics.

Automatically scanning spectrophotometers are available in confusing abundance. There are at least 27 models made in the United States, and six actively imported. They can be divided into two main classes: low-cost models for routine identification work, and research instruments equipped with various refinements, such as higher precision, higher resolution, greater wavelength span, more versatility with respect to mode of presentation of the spectrum, or with respect to types of sample holders. Most of these utilize the double-beam principle.

A surprising number of infrared spectrophotometers of various manufacturers and in both price classes have nearly identical optical layouts. The differences arise in the precision tolerances of the many optical and mechanical components, and are reflected primarily in the specifications of wavelength and photometric accuracy and of resolving power, and secondarily in the variety of automated features for the convenience of the operator.

One of the most common optical systems is shown in Fig. 5-5a. Energy from the source N, which may be a Nernst glower or a Nichrome wire, is reflected by two sets of mirrors to form two symmetrical beams, one of which passes through the sample cell, the other a blank or reference cell. The two beams are reflected by a further array of mirrors so that they are both focused on the slit S_1. C is a motor-driven, circular, chopping disk, half reflecting, half cut away (cf. Fig. 2-28b), which permits the two beams to reach the slit during alternate time periods. Slit S_1 is the entrance slit to a Littrow monochromator with a 60° prism and mirror. The selected wavelength band passes out through the exit slit S_2 and is brought to a focus on the thermocouple detector D.

The detector sees a square wave corresponding in amplitude to the difference in powers in the two beams. This ac signal, after amplification, drives a servomotor to control the position of a comb attenuator A, which adjusts the power of the reference beam to equal that of the sample beam, an *optical null*. The motor simultaneously positions the pen on the built-in recorder.

The same optical train may be modified to use a pair of diffraction

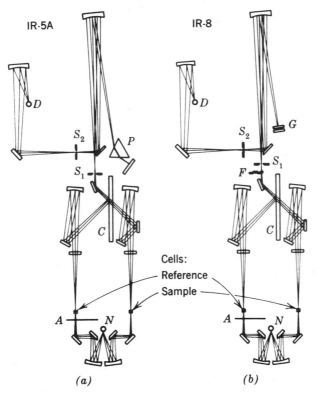

Fig. 5-5 Optical systems of two Beckman infrared spectrophotometers: (*a*) a prism instrument (IR-5A); (*b*) a grating model (IR-8). (*Beckman Instruments, Inc., Fullerton, Calif.*)

gratings mounted back-to-back (Fig. 5-5*b*).   The first grating has 300 lines per millimeter, blazed at 3 μ, and is in position, slowly rotating, while the spectrophotometer is scanning from 2 to about 5 μ.   When 5 μ is reached, the grating table abruptly turns to bring the second grating (100 lines per millimeter, blazed at 7.5 μ) into position, and the scan resumes, covering the range from 5 to 16 μ.   Both gratings are operated in the first order.   Four filters F are required to eliminate higher orders; they are positioned sequentially by the drive mechanism at the required wavelengths, behind the entrance slit.   The spectra produced by some of the less expensive grating instruments show a gap between the regions scanned by the two gratings, a slightly distracting feature which does not appear in spectra from prism instruments nor from more elaborate grating spectrophotometers where the motion of the paper or pen carriage is halted during the shift from one grating to the other.

Either prism or grating spectra may be linear in either wavelength or wave number, according to the design of the cams and bars making up the mechanical linkage between the scanning motor and the prism or grating table. This is, of course, simplest for the case of linear wavelength with grating dispersion. From the standpoint of the operator, however, one is as convenient to use as another.

Table 5-3 gives comparative sensitivity and accuracy data for a few representative infrared spectrophotometers of various prices as an indication of what may be expected. The data, taken from manufacturers' literature, have in some instances been transposed from one mode of presentation to another to facilitate comparison. The figures for resolution are approximations which refer to about the middle of the range for each instrument. (For a thorough discussion of resolution in the infrared, its significance, and how it is measured, see the paper by Sloane and Cavenah.)[21] Prices have been rounded off, and are for illustration only.

It is evident from this tabulation that increased price may be accounted for by either increased range, increased accuracy, increased resolution, or combinations of these.

FAR-INFRARED SPECTROPHOTOMETERS

Instruments for the region beyond about 50 μ are much fewer in number and more varied in design. Beckman's Model IR-11 covers the range from 12.5 to 300 μ, the Perkin-Elmer Model 301 covers from 15 to 400 μ, and the FIR-64 of Acton* extends from 40 to 600 μ. All these utilize a high-pressure mercury lamp source, at least for the longer wavelengths, and a Golay pneumatic detector. Both the Beckman and Perkin-Elmer instruments are double-beam, but the Acton is single. All are provided with cases which can be purged with dry air to minimize atmospheric absorption; the Acton is installed in an evacuable chamber.

Several gratings must be interchanged at successive wavelength intervals; for the longest wavelengths a grating may have as few as two or three lines per millimeter. To cut out background radiation and unwanted orders, filters must be present. There are two types of filters suitable for far-infrared applications, both depending on selective reflection rather than transmission. One of these utilizes the phenomenon of "Reststrahlen," the narrow bands of high reflection in crystalline substances which correspond to the refractive index maxima associated with areas of high absorption.[15] The other type consists of a *scatter plate*, which may be a metallic film deposited on a rough-ground glass plate or a grating with its rulings horizontal rather than parallel to the

* Acton Laboratories, Inc., Acton, Mass.

Table 5-3 A few infrared spectrophotometers: partial specifications

Make, model	List price	Type and range	Wavelength accuracy	Wavelength-reproducibility	Maximum stray light	Transmission reproducibility	Resolution
Beckman Microspec	$ 2,900	Filter wedge 2.5–14.5 μ	1%	≯0.5%	4.0%	2.0%	<1.8%
Perkin-Elmer 137B	5,300	NaCl prism 2.5–15 μ	0.03 μ	0.01 μ	0.1%	0.5%	
Beckman IR-8	6,500	Dual gratings 2.5–16 μ	0.008 μ (<5 μ) 0.015 μ (>5 μ)	0.005 μ (<5 μ) 0.01 μ (>5 μ)	1.5%	1.0%	0.2%
Perkin-Elmer 21	11,000	NaCl prism 0.7–15.5 μ	0.015 μ	0.005 μ	2.0%	0.5%	1.7%
Beckman IR-4	13,500	Dual prisms (NaCl) 1–16 μ	0.015 μ	0.008 μ	0.1%	0.2%	0.10%
Perkin-Elmer 421	18,000	Dual gratings 2.5–18 μ	0.010 μ (at 10 μ)	0.005 μ (at 10 μ)	0.1%	0.5%	0.13%
Beckman IR-11	35,000	Four gratings 12.5–300 μ	1 μ (at 100 μ)	0.5 μ (at 100 μ)	0.4% (at 50 μ) 4.0% (at 150 μ)	1.0%	1.5–0.25%

vertical slits. The scattering is very effective at reducing the amount of higher frequency radiation passing through the optical system but has practically no deleterious effect on the long waves of interest.

It is particularly important in the far infrared to chop the radiation prior to its passage through the sample and monochromator, because spontaneous thermal radiation from the sample and optical elements may be far from negligible. Chopping, together with a tuned amplifier, enables the detector to respond only to radiation from the source itself.

It has been found practicable to determine far infrared spectra by means of special adaptations of *interferometry*. The principal advantage lies in the fact that no slits are needed, which means that the detector can observe a more powerful beam. This permits either a shorter time for obtaining a given spectrum, or greater sensitivity in the same time interval. Such an instrument, covering the extremely wide range of 20 to 1000 μ, is available from Beckman (Model FS-620). Block* makes a spectrometer similar in principle, limited to the near and mid-infrared. These instruments are too complex to describe here, particularly as they are so new that they have not yet established a place for themselves in analytical work. Descriptions can be found in the literature.

CALIBRATION AND STANDARDIZATION

The calibration of a spectrophotometer with respect to both wavelength and transmittance may change gradually with continued use, as the result of mechanical wear, fogging of optical surfaces, or aging of components, so that periodic checks are highly advisable.

The wavelength or wave-number scale can in theory be calibrated by the dispersion geometry of a grating of known spacing or a prism of known refractive index, but this is impracticable as a routine procedure. The most common check in the sodium chloride region is to run the spectrum of a thin sheet of polystyrene as a secondary standard. Infrared manufacturers usually supply mounted samples for this purpose, along with a standard spectrum for comparison. The spectra (see Fig. 5-3) show easily recognized absorption bands distributed throughout the range.

Another convenient wavelength check is performed by switching to single-beam operation and scanning the spectrum with no sample present. Under this condition the absorption of atmospheric water vapor and carbon dioxide will be clearly seen (Fig. 5-6), and as their wavelengths are known with precision, the scale can readily be checked at a number of points. This figure makes evident one of the major advantages of a dual-beam spectrophotometer—the cancellation of atmospheric absorption.

* Block Engineering Co., Inc., Cambridge, Mass.

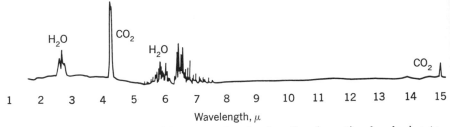

1　　2　　3　　4　　5　　6　　7　　8　　9　　10　　11　　12　　13　　14　　15

Wavelength, μ

Fig. 5-6　Single-beam trace without sample, showing the absorption bands due to atmospheric water and carbon dioxide. （*Plenum Publishing Corp., New York.*）

If the spectrophotometer is to be used for quantitative measurements, the linearity of its photometric scale must also be verified occasionally. This can be done roughly by measuring the polystyrene spectrum, or the apparent transmittance of a series of thicknesses of a liquid such as benzene, at selected wavelengths. A more precise (and more expensive) method is by the use of calibration wheels. These are opaque disks with cutaway sectors of known angle, designed to be mounted in the sample position and rotated at a speed great in comparison with that of the beam chopper. The sectors are precalibrated in terms of percent transmittance (or absorbance).

QUANTITATIVE ANALYSIS

Beer's law, as presented in Chap. 3, applies equally to the infrared region of the spectrum:

$$A = \log \frac{P_0}{P} = abc \tag{5-1}$$

In the near infrared, 1- to 10-mm cuvets and dilute solutions are the rule, and no special difficulties arise in the use of this relation. Beyond this region, however, the path length b is usually much less, and the concentration greater, because of the lack of suitable solvents for the infrared, a matter which will be considered later. The greater concentration is apt to cause deviations from the law, due to molecular interactions. Furthermore, it is often difficult to make an accurate photometric measurement because of the overlap of absorption bands. We will consider each of these features in turn.

The path length can be measured by a number of methods. In some cell designs it can be done with a micrometer microscope. In a cell which has plane parallel walls, interference fringes resulting from multiple internal reflections can easily be recorded by running the spectrophotometer so as to indicate the apparent transmission through the empty cell against air in the reference path.[6] The resulting trace will resemble Fig. 5-7. The value of b can then be computed from the equation

$$b = \frac{n}{2(\lambda_1^{-1} - \lambda_2^{-1})} \tag{5-2}$$

where n is the number of fringes between wave numbers λ_1^{-1} and λ_2^{-1}.

Some cells do not have walls of sufficient flatness to give sharp interference fringes. To determine the effective or average path length in such cells, it is necessary to use Beer's law "backwards," by measuring the absorbance of a solution of known absorptivity and concentration.

Deviations from Beer's law due to high concentration can be handled only by careful calibration with a series of solutions of known concentrations. The plot of absorbance against concentration may not only be curved, it may intersect the axis of zero concentration at a finite value, corresponding to *background absorption*. This quantity can be found by extrapolation and subtracted from all values of apparent absorbance, if a true value for the absorptivity is desired.

The experimental determination of absorbance often presents difficulties on account of the background absorption and overlapping bands of other materials present. Suppose it is desired to find the absorbance corresponding to the first band shown in Fig. 5-8. A *base line* is drawn across the shoulders of the band; the quantities P_0 and P can then be measured as shown, and the absorbance calculated. However, with an absorption such as the second in the same figure, the proper location of the base line is open to much doubt; a few possibilities are suggested. In this situation, which frequently occurs, the only pro-

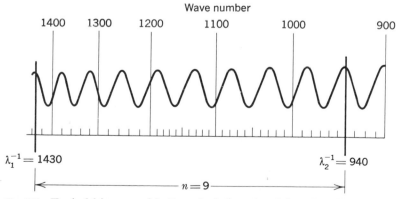

Fig. 5-7 Typical fringes used in the calculation of path length. (*Allyn and Bacon, Inc., Boston.*)

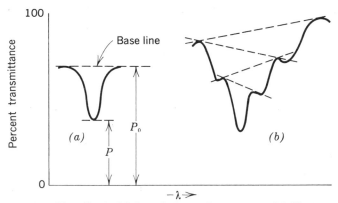

Fig. 5-8 Hypothetical infrared absorption spectra: (*a*) illustrating the use of the base-line measurement technique; (*b*) showing uncertainty as to where to draw the base line.

cedure is to standardize on one particular way to draw the line. The certainty of one's results could be improved by analyzing several absorption bands in the same spectrum; consistent results would go far to prove the adequacy of the method.

The present and future importance of infrared spectrophotometric analysis lies more and more with the qualitative identification of substances, pure or in mixtures, and as a tool in the establishment of structure. The quantitative aspects have diminished in importance partly because of their inherent difficulties, and largely because quicker and more convenient methods of quantitation have emerged, especially gas chromatography. A typical procedure for the examination of a sufficiently volatile mixture is separation of the components on a gas chromatograph, which tells one the number and relative quantities of each, followed by qualitative identification of each component by infrared spectrophotometry.

RAPID–SCAN SPECTROPHOTOMETERS

Conventional spectrophotometers which are intended to bring out all the detail of a complex spectrum must be operated at slow scanning speeds because of the limited response rate of the detector and recorder. Also, the slits must be kept as narrow as possible to give good resolution.

If it is desired to obtain a spectrum at high speed (a few seconds), one must sacrifice resolution. An instrument in this class is the Beckman IR-102, which is specifically designed to accept the effluent from a gas chromatograph without prior isolation and to identify substances as they are eluted. Figure 5-9 shows spectra of the same compound

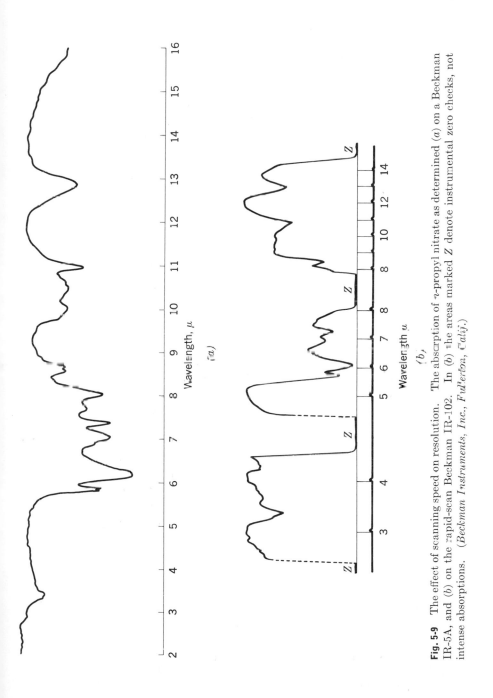

Fig. 5-9 The effect of scanning speed on resolution. The absorption of n-propyl nitrate as determined (a) on a Beckman IR-5A, and (b) on the rapid-scan Beckman IR-102. In (b) the areas marked Z denote instrumental zero checks, not intense absorptions. (*Beckman Instruments, Inc., Fullerton, Calif.*)

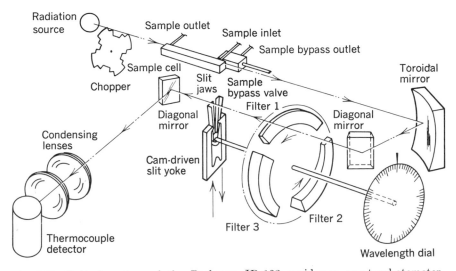

Fig. 5-10 Optical system of the Beckman IR-102 rapid-scan spectrophotometer. Filters 1, 2, and 3 are specially designed interference wedges which act as wavelength selectors. (*Beckman Instruments, Inc., Fullerton, Calif.*)

taken on the IR-102 and for comparison on the IR-5A (the instrument of Fig. 5-5).   The IR-102 is a single-beam instrument; hence, background absorption is superimposed on the desired spectrum.   The response is linear in wavelength in three segments with nonsignificant spaces between. The wavelengths are selected by means of a series of three interference wedge filters sequentially mounted on a rotating wheel; each wedge isolates the wavelengths corresponding to one of the regions in the recorded spectrogram.   This arrangement takes the place of conventional prisms or gratings.   Figure 5-10 shows the schematic diagram.

The spectra produced by the IR-102 are not so accurate as those produced by the slower scan instruments, and so a file of spectra of known compounds determined on the same instrument must be consulted for identification.

PREPARATION OF SAMPLES

Gaseous samples can be examined in an infrared spectrophotometer with no prior preparation other than removal of water vapor.   Long path lengths may be required.   The IR-102 is provided with cells of 10-, 20-, and 30-cm length.   Some other models can accept multiple-pass cells in which the beam is reflected back and forth up to a total length of perhaps 10 m.

Liquids are frequently handled pure, i.e., without solvent, in thin layers, largely because there are no liquids available for solvents which are entirely free of absorption on their own account. Carbon tetrachloride is satisfactory over a considerable range, but dissolves only a limited number of substances. Chloroform, cyclohexane, and other liquids can be used in restricted wavelength regions and in very thin layers. Figure 5-11 shows in chart form the spectral regions where various solvents can be used. The degree of absorption which can be tolerated in the solvent will, of course, depend upon the sensitivity of the spectrophotometer. It can be canceled out, at least to a good approximation, by the usual means of placing a blank in the reference beam.

A variety of absorption cells for liquids is available, ranging from inexpensive ones designed to be discarded as soon as they become at all fouled, to cells which can be disassembled for cleaning, and cells provided with a screw thread to allow adjustment of thickness. The *cavity cell* is a convenient form, which is made by machining a parallel-sided hole in a single block of sodium chloride or other salt. The inner surfaces cannot be polished, so the thickness cannot be measured by the interference-fringe method, but rather by measuring the absorbance of a known liquid such as benzene.

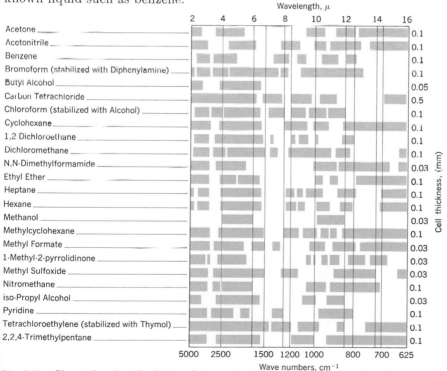

Fig. 5-11 Chart showing the infrared transmission regions for a number of solvents. The rectangles designate the areas of transmission. (*Eastman Kodak Co., Rochester, N.Y.*)

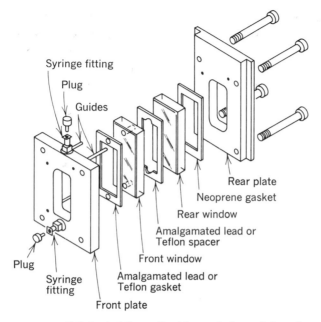

Syringe fitting
Plug
Guides
Plug
Syringe fitting
Front plate
Amalgamated lead or Teflon gasket
Front window
Amalgamated lead or Teflon spacer
Rear window
Neoprene gasket
Rear plate

Fig. 5-12   Cell for holding a liquid sample in an infrared spectrophotometer.   (*Beckman Instruments, Inc., Fullerton, Calif.*)

A demountable cell is shown in Fig. 5-12.   It consists of a pair of salt plates separated by a shim or gasket made of metal or sometimes of Teflon, the whole held together as a sandwich by metal clamps.   Two holes are drilled through the metal frame and one salt plate, for filling and flushing.   A cell can generally be used many times before it becomes necessary to disassemble it and clean or repolish the plates.   The cells are usually filled, emptied, and rinsed with the aid of a hypodermic syringe (without needle), the nib of which fits the holes in the cell frame.

Liquids of high viscosity are often simply sandwiched as a layer between two salt plates, since it is not easy to introduce viscous liquids into preassembled cells.

Solid samples can be prepared for analysis by incorporating them into a pressed plate or pellet of potassium bromide or (less commonly) potassium iodide or cesium bromide.   A weighed portion of the powdered sample is thoroughly mixed (as in a ball mill) with a weighed quantity of highly purified and desiccated salt powder.   Then the mixture is submitted to a pressure of several tons in an evacuated die, to produce a highly transparent plate or disk which can be inserted into a special holder for the spectrophotometer.   The disk typically is about 1 cm in diameter and perhaps 0.5 mm in thickness. The exact thickness is required for quantitative purposes; sometimes it can be determined from the dimensions of the die, or it can be measured with a micrometer caliper.

There are hazards in the use of the potassium bromide pellet technique. If the particle size is not sufficiently small, excessive scattering results. Some heat-sensitive materials (steroids, for example) may show signs of partial decomposition, presumably arising from the heat generated in crushing crystals of salt. The technique has been considered in critical detail in a paper by Hannah.[12]

Solid samples can also be examined in the form of a thin layer deposited by sublimation or solvent evaporation on the surface of a salt plate. Another procedure with much to recommend it is called *mulling*. The powdered sample is mixed to form a paste with a little heavy paraffin oil (medicinal grade Nujol is often used). The oil has only a few isolated absorption bands, specifically at about 3.5, 6.9, and 7.2 μ. If these bands interfere, the mull may be made with Fluorolube, a fluocarbon material which has no absorption at wavelengths shorter than about 7.7 μ. The mull is sandwiched between salt plates for measurement.

An advantage of either the mull or the pressed disk is that scattering of radiation is reduced to a minimum; this is a source of trouble when a powdered sample is run as such.

ATTENUATED TOTAL REFLECTION

It is well known that when a beam of radiation encounters an interface between two media, approaching it from the side of higher refractive index, total reflection occurs if the angle of incidence is greater than some critical angle, the value of which depends upon the two refractive indices. Not so generally realized, though predicted by electromagnetic theory, is the fact that in total reflection some portion of the energy of the radiation actually crosses the boundary and returns. This is suggested by the oversimplified sketch of Fig. 5-13. If the less dense medium absorbs at the wavelength of the radiation, the reflected beam will contain less energy than the incident, and a wavelength scan will produce an absorption spectrum.

In principle, this will apply to any spectral region, but it has been

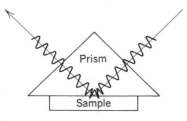

Fig. 5-13 Schematic representation of the total internal reflection of a beam of radiation, showing some degree of penetration into the substrate. (*Wilks Scientific Corporation, South Norwalk, Conn.*)

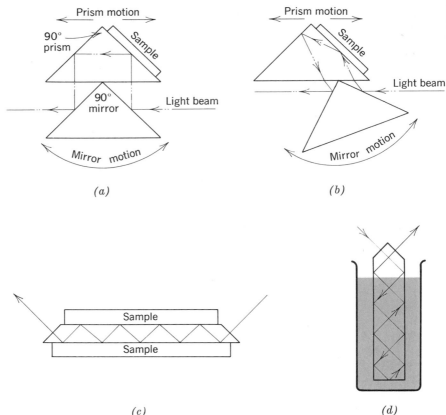

Fig. 5-14   ATR apparatus: (*a* and *b*) an arrangement whereby the entering and leaving beams are in the same straight line; the lower prism and the left face of the upper must be silvered; motion of the parts as indicated will permit the angle of reflection at the sample to be varied; (*c*) a device for multiple reflections; (*d*) an immersible plunger to permit multiple reflections with a bulk liquid sample.   (*a* to *c*, *Wilks Scientific Corporation, South Norwalk, Conn.; d, Analytical Chemistry.*)

found most useful in the infrared, where very thin layers of sample are often required.   The distance to which the radiation appears to penetrate in internal reflection depends on the wavelength, but is of the order of 5 μ or less in the sodium chloride region.   The phenomenon goes under a variety of names, including attenuated total reflectance (ATR), frustrated internal reflectance (FIR), and when several reflectances are utilized, multiple internal reflectance (MIR), or frustrated multiple internal reflectance (FMIR); we will use the abbreviation ATR.

A number of sample holders have been devised to make use of ATR for infrared analysis, mostly designed to fit the sample compartment of conventional spectrophotometers; a few are shown in Fig. 5-14.   At (*a*) is an arrangement for a single reflection at the sample surface, at a variable angle. By moving the lower prism (b), which has silvered surfaces and hence reflects at any angle, the angle of incidence at the sample can be varied, still maintaining the entering and exiting beams on the same straight line. The drawing at (c) shows a linear multiple reflection device[13] especially convenient for solid samples, which are clamped in optical contact with the two surfaces of a slab of transparent solid of high index, several centimeters long. Figure 5-14(d) represents an elongated prism with plane parallel sides, so arranged that the radiation enters and leaves at the same end;[14] this type can be surrounded by a liquid sample, as shown, or can be clamped between solids.

ATR spectra are not identical to those obtained conventionally, but are quite similar. The distortion becomes greater as the angle of incidence approaches the critical angle. The further from the critical angle, however, the less is the absorption, and so a greater number of reflections must be introduced.

ATR has been found most useful with opaque materials which must be observed in the solid state. Applications include studies of rubber and other polymeric materials, adsorbed surface films, and coatings such as paints.

MICROWAVE ABSORPTION

Absorption in the microwave region[10] can well be considered an extension of the far infrared, as the phenomena giving rise to the absorption spectra are primarily rotational transitions in molecules possessing a permanent dipole moment. Absorptions can also be observed under appropriate conditions in molecules possessing magnetic moments, such as O_2, NO, NO_2, ClO_2, and in free radicals. The microwave spectral region most fruitful for chemical spectroscopy lies between approximately 8 and 40 GHz. [1 GHz (gigahertz) = 10^9 Hz.] As we will see in a later chapter, this includes the frequencies useful in electron-spin magnetic resonance (ESR), so that the two methods share some common features, though not based on the same molecular mechanisms.

There are no microwave absorptions which are characteristic of specific bond types or functional groups, such as we often find in optical and infrared spectra. Qualitative analysis can only be applied by means of comparison with known spectra. The microwave region is excellent as a complement to the infrared in the identification of relatively small covalent compounds by the "fingerprint" approach. This is partly due to the very great number of frequencies available. If absorption bands can be resolved at a separation of 200 kHz, a not unreasonable figure, this gives a total of 160,000 spaces or "channels" in the 8- to

40-GHz region [(40 − 8) × $10^9/2$ × 10^5 = 1.6 × 10^5]. By comparison, if we assume that an average resolution of 1 cm^{-1} is attainable over the infrared region from the visible up to 50 μ, only 10,000 spaces are available. Of course there are many frequencies which do not appear in any known spectra, so these figures may be somewhat misleading; nevertheless, an extremely large number of compounds can in principle be examined in a microwave absorptiometer with little probability of overlapping. Only gaseous samples can be studied, though the vapor pressure need not be high; 10^{-1} to 10^{-3} torr is the usual pressure range.[7,10]

The chief limitation arises from interactions between the rotational energy levels associated with various bonds within the molecule. If there are more than three or four rotors present, so many interactions will appear that the spectrum will become a mass of close lines, few if any of which are sufficiently intense to be useful. Hence large molecules cannot be studied to advantage unless cyclic structures prevent rotation around some of the bonds.

The quantitative absorption law (Lambert's law) can be expressed in the form

$$P = P_0 \cdot 10^{-\alpha b} \text{ or } \log \frac{P_0}{P} = \alpha b \qquad (5\text{-}3)$$

where P_0 and P represent respectively the radiant (microwave) power incident upon and passing through the absorption cell, which has an effective length b. The absorption coefficient α corresponds to the absorbance A, as usually employed in the optical region, taken per unit length of the absorption cell. Ideally it is a function only of the number of absorbing molecules per unit path length, and hence should be related to the partial pressure of the substance in a mixture of gases.

The degree to which such a relation is valid depends on the way in which the absorption coefficient is measured and utilized. At low pressures and power levels, where saturation effects are not evident, the pressure is proportional to α_{max}, the height of the absorption maximum. However at higher pressures, the value of α_{max} becomes constant, and an increase in pressure results in the broadening of the line as measured at half-height. Above this point it can be shown that the integrated absorption is very nearly proportional to the pressure:

$$\int \alpha(\nu) \, d\nu = Kp \qquad (5\text{-}4)$$

where ν = frequency
$\quad$ p = pressure
$\quad$ K = a proportionality constant.

Figure 5-15 shows the absorption curves of a typical band at a series of pressures.

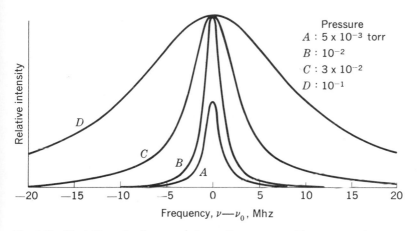

Fig. 5-15 Variation of microwave absorption curves with pressure change. Note that the peak intensity remains constant over a wide pressure range. (*John Wiley & Sons, Inc., New York.*)

If a nonreacting gas which does not absorb at the same frequency is added to the sample, the value of the integrated absorption is found to be proportional to the mole fraction of the absorbing gas. Hence this integral (the area beneath the curve) is a valid quantitative analytical tool. It is necessary to be certain that the power level is not too high, so that neither the absorbing sample nor the crystal rectifier detector becomes saturated. The temperature also must be controlled. With due care standard deviations of the order of 3 to 5 percent are to be expected (but remember that the sample may be only a few micromoles).

In this connection it must be emphasized that both theory and techniques in microwave absorptiometry are still young. There is reason to believe that the signal-to-noise ratio can be increased drastically over the best now available, so that the precision as an analytical tool should be improved considerably.

MICROWAVE INSTRUMENTATION

The details of theory and construction of apparatus cannot be discussed here for lack of space; it has been reviewed recently.[7] Spectrophotometers have been constructed which are comparable to single- and double-beam instruments in the more familiar spectral ranges. The source is either a klystron or a backward-wave oscillator, which can be swept through a considerable frequency range, generating essentially monochromatic radiation at each point. The beam chopper is replaced by an electronic system of modulation based on the Stark or (less commonly) the Zeeman effect. A crystal rectifier serves as detector of radiation.

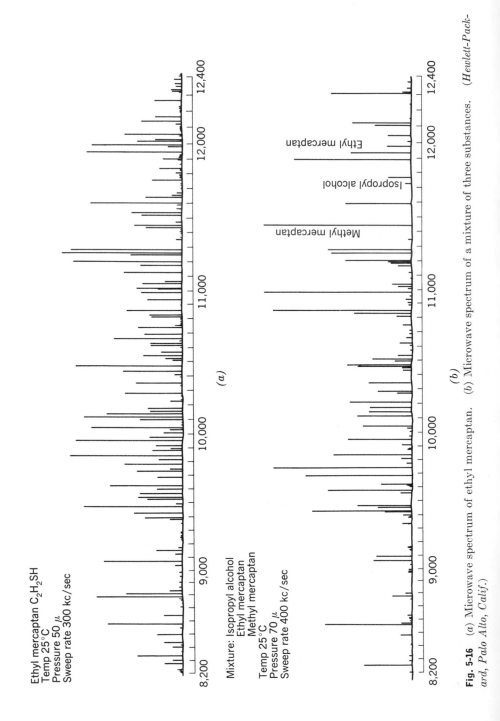

Ethyl mercaptan C₂H₂SH
Temp 25°C
Pressure 50 μ
Sweep rate 300 kc/sec

8,200   9,000   10,000   11,000   12,000   12,400

(a)

Mixture: Isopropyl alcohol
        Ethyl mercaptan
        Methyl mercaptan
Temp 25°C
Pressure 70 μ
Sweep rate 400 kc/sec

8,200   9,000   10,000   11,000   12,000   12,400

Ethyl mercaptan

Isopropyl alcohol

Methyl mercaptan

(b)

Fig. 5-16 *(a)* Microwave spectrum of ethyl mercaptan. *(b)* Microwave spectrum of a mixture of three substances. *(Hewlett–Packard, Palo Alto, Calif.)*

There are two microwave spectrometers on the American market.* Two typical spectra taken on one of them are reproduced as Fig. 5-16.

PROBLEMS

5-1 A sample of ethyl bromide suspected of containing trace amounts of water, ethanol, and benzene was examined in an infrared spectrophotometer. The following relative absorbances were obtained by a base-line method: at 2.65 μ, $A = 0.110$; at 2.75 μ, $A = 0.220$; at 14.7 μ, $A = 0.008$. The instrument employed, cells, etc., exactly duplicated those for which reference data have been reported.[18,24] Calculate the amounts of water and ethanol in parts per million and of benzene in percent.

5-2 A high-temperature gas stream in a plant manufacturing "water gas" contains as major constituents H_2, CO, CO_2, and H_2O, with a possible small quantity (0.5 percent or less by volume) of CH_4. It is desired to sensitize a nondispersive infrared analyzer to monitor the methane content. The gas is cooled prior to introduction into the analyzer so that excess water condenses out and is removed. Look up the spectra of these substances in an atlas, and describe in detail what should be placed in each compartment of the analyzer of Fig. 5-4.

5-3 Determine (in millimeters) the path length of the cell which produced the interference fringes shown in Fig. 5-7. At what wave number is this cell thickness equal to just 10 wavelengths?

5-4 The following observations have been reported[6] for the apparent absorbance of cyclohexanone in cyclohexane solution at 5.83 μ. (The cell thickness was 0.096 mm.)

Concentration, g/l	Absorbance
5	0.190
10	.244
15	.293
20	.345
25	.390
30	.444
35	.487
40	.532
45	.562
50	.585

Plot these data. (a) Over what range is Beer's law followed, after the necessary background correction is made? (b) Compute the value of ϵ, the molar absorptivity.

REFERENCES

1. American Petroleum Institute: Catalog of Infrared Spectrograms, *API Res. Proj.* 44, Texas A & M University, College Station, Tex.

* Made by Hewlett-Packard, Palo Alto, Calif., and the Tracerlab Division of the Laboratory for Electronics, Inc., Waltham, Mass.

2. Beebe, C. H., and M. D. Liston: *Instr. Soc. Am. Trans.*, **2**:331 (1963).
3. Bellamy, L. J.: "The Infrared Spectra of Complex Molecules," 2d ed., Methuen & Co., Ltd., London, and John Wiley & Sons, Inc., New York, 1958.
4. Bentley, F. F., and E. E. Wolfarth: *Spectrochim. Acta*, **15**:165 (1959).
5. Colthup, N. B.: *J. Opt. Soc. Am.*, **40**:397 (1950).
6. Conley, R. T.: "Infrared Spectroscopy," Allyn and Bacon, Inc., Boston, 1966.
7. Ewing, G. W.: *J. Chem. Educ.*, **43**:A683 (1966).
8. Gebbie, H. A.: *Proc. Second Intern. Conf. Quantum Electronics*, McGraw-Hill Book Company, New York, 1961.
9. Goddu, R. F., and D. A. Delker: *Anal. Chem.*, **32**:140 (1960).
10. Gordy, W.: Microwave and Radio Frequency Spectroscopy, in W. West (ed.), "Chemical Applications of Spectroscopy," p. 169, being vol. IX of A. Weissberger (ed.), "Technique of Organic Chemistry," Interscience Publishers (Division of John Wiley & Sons, Inc.), New York, 1956.
11. Gray, W. T.: *Instr. Soc. Am. J.*, **2**:189 (1955).
12. Hannah, R. W.: *Instr. News*, a publication of Perkin-Elmer Corporation, South Norwalk, Conn., **14**(3–4): 7 (1963).
13. Hansen, W. N., and J. A. Horton: *Anal. Chem.*, **36**:783 (1964).
14. Harrick, N. J.: *Anal. Chem.*, **36**:188 (1964).
15. Holter, M. R., S. Nudelman, G. H. Suits, W. L. Wolfe, and G. J. Zissis: "Fundamentals of Infrared Technology," The Macmillan Company, New York, 1962.
16. Hurley, W. J.: *J. Chem. Educ.*, **43**:236 (1966).
17. Jones, R. N., and C. Sandorfy: The Application of Infrared and Raman Spectrometry to the Elucidation of Molecular Structure, in W. West (ed.), "Chemical Applications of Spectroscopy," p. 308, being vol. IX of A. Weissberger (ed.), "Technique of Organic Chemistry," Interscience Publishers (Division of John Wiley & Sons, Inc.), New York, 1956.
18. McCrory, G. A., and R. T. Scheddel: *Anal. Chem.*, **30**:1162 (1958).
19. *Sadtler Standard Spectra*, a continually updated subscription service, Sadtler Research Laboratory, Inc., Philadelphia.
20. Silverstein, R. M., and G. C. Bassler: "Spectrometric Identification of Organic Compounds," 2d ed., John Wiley & Sons, Inc., New York, 1967.
21. Sloane, H. J., and R. Cavenah: *The Analyzer*, a publication of Beckman Instruments, Inc., Fullerton, Calif., **6** (1):15 (1965).
22. Sugden, T. M., and C. N. Kenney: "Microwave Spectroscopy of Gases," D. Van Nostrand Company, Inc., Princeton, N.J., 1965.
23. Szymanski, H. A.: "Infrared: Theory and Practice of Infrared Spectroscopy," Plenum Publishing Corp., New York, 1964.
24. Williams, V. Z.: *Anal. Chem.*, **29**:1551 (1957).

6

The Scattering of Radiation

The term "scattering," as applied to the interaction of radiant energy with matter, covers a variety of phenomena. The word always implies a more or less random change in the direction of propagation. The mechanism involved depends on the wavelength of the radiation, the size and shape of the particles responsible for the scattering, and sometimes their spacial arrangement. The scattered radiation may be the same frequency as the primary beam, or it may be changed in frequency; the latter (the Raman effect) will be taken up after the discussion of scattering without frequency change.

It was shown by Lord Rayleigh, in 1871, that radiation falling on a transparent particle small compared to the wavelength induces an electric dipole oscillating at a forced frequency equal to that of the incident radiation. This oscillating dipole then acts as a source, radiating at the same frequency in all directions, though not necessarily with equal power. This is to be distinguished from scattering caused by *reflection* at the surface of the particle, which requires a surface large compared to the wavelength.

If the particles are arranged in a regular manner in space, then the scattered radiation will show interference effects. The diffraction of x-rays from a crystal is an example of this, which will be discussed in a later chapter.

RAYLEIGH SCATTERING

From the standpoint of the analytical chemist, the systems of greatest interest consist of suspensions of solid or liquid particles in liquids (colloidal suspensions or emulsions). The particle size may or may not be small enough to allow the use of Rayleigh's theory. The electromagnetic theory of scattering has been worked out by Mie[1, 10, 16] to cover all cases, but is too complex to utilize directly. Scattering by relatively large particles is called *Tyndall* scattering.

Measurements are nearly always carried out with visible light. The sample is illuminated by an intense beam of power P_0 (Fig. 6-1). The transmitted power P_t can then be measured just as in spectrophotometry, or the power scattered at a specific angle (such as P_{90} at 90°) may be determined. The ratio P_t/P_0 decreases with increasing number of particles in suspension, while ratios such as P_{90}/P_0 will increase. For very dilute suspensions, measurement at an angle is much more sensitive than in-line measurement, as it involves observation of faint, scattered light against a black background, rather than the comparison of two large quantities of nearly equal values, a special case of an important principle of wide generality.

In-line measurement is called *turbidimetry*, and can be performed with any standard spectrophotometer or filter photometer. Measurement at an angle is usually restricted to 90°, for which standard cuvets can be used in the same configuration as that of a conventional fluorimeter, which is often used for the purpose. This technique is called *nephelometry*.

The rigorous mathematical treatment of these techniques is not easy. A quantity corresponding to absorbance can be obtained tur-

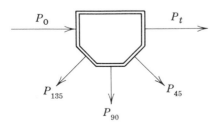

Fig. 6-1 Power relations in light scattering. P_0 is the power of the incident beam, P_t of the transmitted beam, P_{45}, P_{90}, and P_{135}, the powers scattered at the indicated angles.

bidimetrically, and we can write

$$S = \log \frac{P_0}{P_t} = kbc \qquad (6\text{-}1)$$

where S = "turbidance"

k = proportionality constant (which may be called *turbidity coefficient*)

b = path length

c = concentration in grams per liter

(The turbidity function τ, sometimes reported, is equal to 2.303k.) This expression is only valid for very dilute suspensions, because as c increases, more and more of the scattered light finds its way into the measuring photocell through multiple scattering. However, as indicated previously, very dilute suspensions lead to uncertainty in the photometric measurement because P_t is so nearly equal to P_0. Furthermore, this expression loses validity unless the particles are much smaller than the wavelength, since with larger particles an increasing fraction of the scattered light is propagated in the forward direction and may be seen by the detector.

It can be shown from the Rayleigh-Mie theory that scattering from *small* particles is proportional to the inverse fourth power of the wavelength; this neatly accounts for the blue of the sky and the redness of the setting sun, where scattering is predominantly due to particles of molecular dimensions. In chemical systems the exponent of the wavelength may vary from -4 to -2, principally because of larger particles, and partly because of nonuniformity of particle size.

MEASUREMENT DIFFICULTIES[6]

The 90° angle of observation in nephelometry is not necessarily the optimum. In some instances a large increase in sensitivity has been found at angles as small as it is physically possible to achieve, just a few degrees from "straight through." However, there are instrumental problems involved which make precision measurements difficult to achieve. If a cylindrical cuvet is employed, the curved surface will act like a lens to de-collimate the primary beam, and partially to collimate into the exit beam rays originating from a considerable region in the cuvet, so that the angle of scattering is uncertain and dependent on the refractive indices of both the solvent and the dispersed particles. On the other hand, prismatic cuvets, either square or multifaced (like that in Fig. 6-1) permit only specified angles to be observed. It is not practicable to derive a theoretical equation relating the scattered light at a given angle to the concentration and other variables. The best we can do is to write as a working relation

$$P_{90} = K_{90}cP_0 \qquad (6\text{-}2)$$

where K_{90} is an empirical constant for the system, and specify that measurements must be made under identical conditions. (Similar equations will apply for other angles.)

Maintaining identical conditions is not so easy as it might appear, especially constancy of particle size. The sols most often studied are formed by precipitation, and the resulting particle size tends to vary widely with variations in concentrations of reactants, temperature, agitation, presence of nonreactive materials which may affect the nucleation or crystal-growth patterns, etc. Hence it is important to control experimental conditions with utmost care.

If such precautions are taken, however, a considerable variety of analytical procedures of great sensitivity and usually adequate precision become available. Phosphorus, for example, can be detected at a concentration of 1 part in more than 300 million parts of water as a precipitate with a strychnine-molybdate reagent. One part of ammonia in 160 million parts of water can be detected by a mercuric chloride complex (Nessler's reagent).

A great many common precipitation reactions can be made to produce suspensions which will be stable at least long enough to permit measurements to be made. This can often be accomplished by adding a protective colloid such as gelatin or an ionic peptizing agent.

An example of analysis by turbidimetry is the determination of sulfur by conversion to barium sulfate under conditions which lead to a colloidal suspension. Such a suspension results when a dilute solution of sulfate containing sodium chloride and hydrochloric acid is shaken with excess solid barium chloride. The method is valid at concentrations as low as a few parts per million, but it is necessary to exercise close control over all variables in order to obtain consistent results. Thus the quantity and grain size of the crystalline barium chloride and the efficiency and duration of stirring must be uniform for samples and standards.

GASES

Nephelometry presents an important tool in the measurement of smokes, smog, and other aerosols. The theoretical equations generally give better results than for suspensions in liquid media, because the particles are usually smaller and farther apart.

MOLECULAR WEIGHTS[16]

Debye has shown that scattering can be used to advantage to determine the weight-average molecular weight of a polymer in solution. The Debye equation calls for knowledge of the turbidity, concentration, index of refraction, wavelength, derivative of the index with respect to concentration, and second virial coefficient, which is a measure of the nonideality of the solution. It is subject to the restriction that the molecules must be small compared to the wavelength; the details are beyond our present scope.

PARTICLE–SIZE DETERMINATION

The diameter of scattering particles can be determined with good precision, provided they are less than about one-twentieth of the wavelength, by means of the equations worked out by Gans from the basic Rayleigh theory. The distribution of scattered light with angle is required, as well as most of the factors previously mentioned.

INSTRUMENTATION: TURBIDIMETERS*

Nessler tubes can be used with good precision for turbidimetric analysis, as can a Duboscq colorimeter, either directly or modified for side illumination as a nephelometer.

Any filter photometer or spectrophotometer can be utilized for turbidimetric measurements, but with severely limited accuracy and sensitivity. If the solvent and dispersed particles are both free from color, then a wavelength in the blue or near ultraviolet should be selected for maximum sensitivity. If color is present, however, then the optimum wavelength had best be determined by trial and error.

A turbidimeter with unusual properties is the Du Pont Model 430. This is a double-beam instrument which depends for its operation on the relative degree of polarization of transmitted and scattered light. If a suspension is illuminated with plane-polarized light, the transmitted beam will be found to be depolarized in proportion to the concentration of particles in the suspension. In the 430 (Fig. 6-2) the beam, after any wavelength filtering that may be required, is passed through a polarizer followed by the sample. The transmitted light is split by a half-silvered

* The concentration of solids suspended in a liquid can be measured by the absorption of *sonic* energy, a method comparable to turbidimetry, though no electromagnetic radiation is involved. The Powertron Division of Giannini Controls Corporation, Plainview, N.Y., manufactures an instrument based on this principle, which can measure suspensions from a few tenths of a percent to a few percent concentration with precision fully adequate for many industrial purposes.

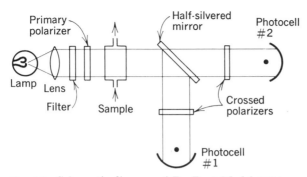

Fig. 6-2 Schematic diagram of Du Pont Model 430 turbidimeter. (*E. I. du Pont de Nemours and Co., Inc.,
Wilmington, Del.*)

mirror into two beams. One of these is passed through a polarizer with
its axis parallel to that of the primary polarizer, the other through a similar polarizer crossed with the primary. The response of photocell No. 1
is *decreased* by the scattering, while that of No. 2 is *increased*. The electronics of the instrument automatically determine the ratio between the
two signals, which is directly proportional to the concentration of suspended matter in the sample. This instrument is insensitive to color of
solvent or particles, or to lamp fluctuations, but it cannot be used with
solutions which contain optically active substances.

NEPHELOMETERS

A number of manufacturers have equipment in this field. Hach,* for
example, has several single-beam models for various ranges and applications. The most sensitive (0 to 0.2 JTU,† full scale) utilizes two
photocells on opposite sides of a sample-holder illuminated vertically.
The second model operates on a unique principle, wherein the light beam
impinges on a free surface of liquid gently overflowing from the top of a
vessel which may be part of a plant flow system; part of the light is
refracted into the body of the liquid and lost to sight, part is specularly
reflected into a light trap, and only that part scattered by particulate
matter in the liquid at a slightly backward solid angle will be seen by a
photocell. The most sensitive range is 0 to 5 JTU, full-scale, the least
sensitive extends to 4000 JTU. Hach also has a turbidimeter which
measures transmission through a freely falling liquid stream which may
be as thin as 1 or 2 mm; the scale reads up to a maximum of 10,000 JTU.

* Hach Chemical Company, Ames, Iowa.
† The JTU unit refers to an arbitrary scale of *Jackson turbidity units*, wherein zero
corresponds to complete clarity, and 100 originally signified a "just perceptible haze."

LIGHT-SCATTERING PHOTOMETERS

Several manufacturers have equipment in this field primarily intended for work with solutions of high-molecular-weight polymers. One of the best known is the Brice-Phoenix.* It provides a high degree of versatility, with a choice of cuvets (including the semioctagonal diagrammed in Fig. 6-1), several angles of observation, polarization-measuring devices which give information about the shapes of the particles, etc.

TURBIDIMETRIC TITRATIONS[4, 15]

Reactions in which the equivalence point is revealed by the appearance or the dissolution of a precipitate have long been part of the analyst's domain. A method for determining silver by a chloride titration to the point where no further precipitate forms was introduced by Gay-Lussac in 1832, and although time-consuming, is capable of giving excellent results. The procedures credited to Mohr, Volhard, and Fajans can be found in every quantitative analysis textbook.

These and similar procedures can be carried out in the cuvet of a photoelectric turbidimeter or nephelometer with lessened eye-strain, and no doubt in many instances increased sensitivity.

Titrations of the form $A + B \rightarrow C$, where C is insoluble, might be expected to give a curve of either turbidance or scattered intensity which would consist of two intersecting straight lines (curve 1 of Fig. 6-3), as the amount of precipitate must increase to a maximum, then stay constant. Meehan and Chiu,[9] however, have pointed out that this can only be true if the *number* of particles increases linearly to the equivalence point, while all remain the same size. This is not likely to be the case; more probably, added reagent will simultaneously form some new particles and add to those nuclei previously formed (in fact, the latter tendency is the more likely). In this situation, the prediction of titra-

* Phoenix Precision Instrument Co., Inc., Philadelphia.

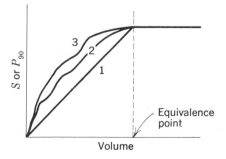

Volume

Equivalence point

Fig. 6-3 Turbidimetric titration curves; curve 1 is ideal, curves 2 and 3 might result from precipitates with mixed particle sizes, poor stirring, etc.

tion curves becomes highly complex.   If the particles become too large, a poorly defined equivalence point or none at all (curves 2 and 3 of Fig. 6-3) will result.   A complicating factor is that unless the suspension is very dilute the particles will continue to grow between added increments of titrant, so that excessively long times must be allowed between additions (6 or 7 minutes).

The method is usable in the 10^{-5} to 10^{-6} formal range, with an average relative error of ± 5 percent or more.

Bobtelsky and his coworkers[2,3] have been successful in carrying out many hundreds of turbidimetric titrations at somewhat higher concentrations, usually 10^{-3} to 10^{-4} formal.   They made no attempt to control the particle size, and ran the titrations at normal speed.   The curves cannot give meaningful data on turbidity as such, but in Bobtelsky's hands give reproducible and useful results.   As no linear relation between actual quantity of precipitate and apparent absorbance can be assumed, Bobtelsky is careful not to use the term turbidimetry, but calls his method *heterometry*, and his simplified photometer a *heterometer*.

PHASE TITRATIONS

Another technique which has been known in principle for some time, and only recently instrumented,[13] is the titration of a mixture of two liquids by a third which is miscible with one but not the other.   Addition of a

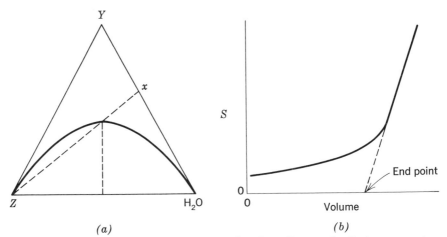

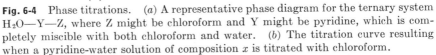

Fig. 6-4   Phase titrations.   (*a*) A representative phase diagram for the ternary system H_2O—Y—Z, where Z might be chloroform and Y might be pyridine, which is completely miscible with both chloroform and water.   (*b*) The titration curve resulting when a pyridine-water solution of composition x is titrated with chloroform.

sufficient quantity of the third liquid will cause a separation of phases visible as turbidity. In order to interpret the results, either the three-component phase diagram must be known, or the unknown must be titrated in comparison with known mixtures. Figure 6-4 illustrates the method for titration of water-pyridine mixtures by chloroform. The method could just as well be utilized for titrating chloroform-pyridine mixtures by water, or reversed, and a turbid mixture of chloroform and water titrated with pyridine to the point of clarity.

RAMAN SCATTERING[5,7,8,14]

Even a liquid which is clear of suspended matter scatters light, the major part of which follows the Rayleigh theory, where the scattering "particles" are the molecules. Since the particles are generally far smaller than the wavelength of the visible light employed, the theory is followed closely. The intensity is of course low compared with the more familiar Tyndall scattering of colloidal suspensions.

Simultaneously there may be scattering due to the Raman effect, in which a change of frequency occurs, as mentioned in Chap. 2. This effect appears in substances in which the molecular polarizability changes as the molecule vibrates. Thus Raman scattering occurs in connection with some vibrational modes which are not evident in infrared spectra. The intensity of both Rayleigh and Raman scattering varies inversely with the fourth power of the wavelength.

Whereas turbidimetry and nephelometry can be studied with relatively wide bands of wavelengths, or even with white light, Raman measurements require a source as nearly monochromatic as possible so that small shifts will be detectable. In the past the required mono-chromatic source was invariably a high-current dc mercury arc with a filter to isolate a single line, usually that at 4358 Å.

In the past few years, however, it has become apparent that the laser will eventually displace all other sources for Raman work. (See, for example, a preliminary report by Porto.[12]) At the time of writing the development of the laser is proceeding rapidly. So far, satisfactory operation in the blue region has not been achieved, but the helium-neon gas laser with radiation at 6328 Å in the red has been found satisfactory for Raman excitation. The advantages of an exceedingly narrow wave-length band at moderately high power outweigh the disadvantages of longer wavelength, even with its fourth-power dependency. One great advantage is that red light is far less likely to excite fluorescence in the sample (or impurities) than is the violet mercury line; even a trace of fluorescence will mask completely the Raman radiation.

The older Raman work was carried out with a glass-prism spectro-

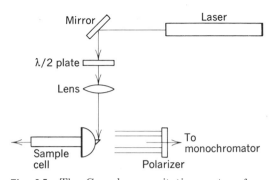

Fig. 6-5 The Cary laser excitation system for Raman spectrometry. The laser beam is reflected by a small prism (1 mm^2) into a hemispherical lens. The tubular sample cell is cemented to the flat side of the lens so that the laser beam passes through it and out the far end. The Raman-scattered light is largely kept within the tube by total reflection, and can escape principally through the hemispherical lens which directs it toward the entrance optics of a high-resolution monochromator. The half-wave plate and polarizer are useful in studies of depolarization phenomena. (*Cary Instruments Division of Varian Associates, Monrovia, Calif.*)

graph of large light-gathering power. Photographic exposures of many hours duration were frequently required. More recently, photoelectric Raman spectrometers have been constructed, wherein the faint Raman signals are amplified by sensitive photomultiplier tubes, instead of being integrated over long periods photographically. There are, at this time, several Raman spectrometers manufactured in the United States, all with provision for laser excitation. An example is shown schematically in Fig. 6-5. There can be no doubt that both research and routine analytical applications of the Raman effect are well on the way to a position of major importance.

Figure 6-6 shows for comparison both the Raman and the infrared spectra of 1,4-dioxane. One can easily pick out lines which are active in each method only, as well as some which can be seen in both.

APPLICATIONS OF RAMAN SPECTROSCOPY

The information required of a Raman spectrometer is the shift in frequency (or wave number) between the primary and scattered radiation. As Rayleigh scattering is always present, it makes a convenient reference point from which to measure Raman shifts.

The Raman shifts are associated with specific bond types, just as are infrared absorptions. The spectra are generally simpler, as fewer combination frequencies and overtones appear. Correlation tables and charts are available covering the range from zero to about 4000 cm^{-1}. Atlases of Raman spectra have also been compiled.

The samples are usually liquids, either pure or as solutions. Water is often an appropriate solvent, as its interferences in the infrared are inactive in the Raman mode. Solids are apt to give too much Rayleigh background. Gases can be handled, but require inconveniently long path cells.

Raman spectroscopy has been found particularly important in the

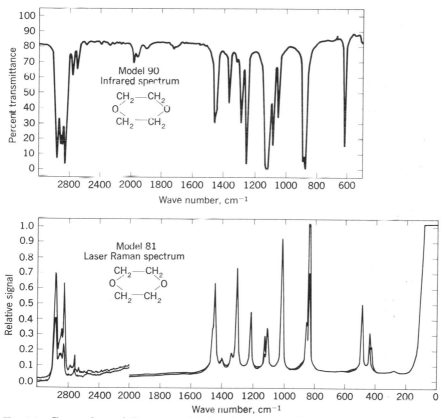

Fig. 6-6 Comparison of Raman and infrared spectra. The spectra of 1,4-dioxane, printed with the wave number scales in register, show many points of similarity. Of interest is the band at 1220 cm^{-1}, which is strong in the Raman and almost invisible in the infrared. Conversely, the band at 620 cm^{-1} is active in the infrared but not in the Raman. (*Cary Instruments Division of Varian Associates, Monrovia, Calif.*)

field of hydrocarbon analysis. An example is the work of Nicholson on the analysis of mixtures of benzene with isopropylbenzenes.[11] He used a Cary Raman spectrometer fitted with a mercury arc filtered to give only 4358 Å, and a double grating monochromator with a nominal slit width of 5 cm^{-1}. Nicholson followed other workers in reporting intensity data in the form of a *scattering coefficient*, the ratio of the observed signal to that given by the 459-cm^{-1} line of pure carbon tetrachloride measured in the same apparatus. He found scattering coefficients reproducible to within ± 0.5 percent for 10 pure liquid hydrocarbons. Analysis of synthetic mixtures gave results with an average deviation of about ± 1 percent absolute. The sensitivity of the spectrometer permitted the detection of about 10 ppm of benzene in carbon tetrachloride.

PROBLEMS

6-1　A Raman spectrum of a pure compound is observed, the exciting radiation being the 4358-Å line of mercury. Raman lines are observed at wavelengths 4420, 4435, 4498, and 4620 Å. (*a*) Compute the value of the Raman shift for each line in terms of wave numbers. (*b*) What would be the wavelength of the anti-Stokes line corresponding to the Stokes line at 4620 Å?

6-2　The sulfur content of organic sulfonates and sulfonamides can be determined turbidimetrically by the $BaSO_4$ precipitate following digestion to destroy the organic matter.[17] A 25.0-mg sample of a particular preparation of *p*-toluenesulfonic acid monohydrate, $C_7H_7SO_3H \cdot H_2O$ (formula weight 190.2), was subjected to the digestion, then exactly one-tenth of it made up to a volume of 10.00 ml with a prescribed conditioning solution, followed by 5.00 ml of 1.34 F $BaCl_2$ added with controlled shaking. After 25 min the turbidance at 355 nm in a Spectronic-20 was found to be 0.295. A standard ammonium sulfate containing 200 μg S, similarly treated, gave a turbidance of 0.322 and showed a linear relation upon dilution prior to treatment. Calculate the percent purity of the original preparation.

REFERENCES

1. Billmeyer, Jr., F. W.: Principles of Light Scattering, in I. M. Kolthoff and P. J. Elving (eds.), "Treatise on Analytical Chemistry," pt. I, vol. 5, chap. 56, Interscience Publishers (Division of John Wiley & Sons, Inc.), New York, 1964.
2. Bobtelsky, M.: *Anal. Chim. Acta*, **13**:172 (1955).
3. Bobtelsky, M.: "Heterometry," American Elsevier Publishing Company, Inc., New York, 1960.
4. Coetzee, J. F.: Equilibria in Precipitation Reactions and Precipitation Lines, in I. M. Kolthoff and P. J. Elving (eds.), "Treatise on Analytical Chemistry," pt. I, vol. 1, chap. 19, Interscience Publishers (Division of John Wiley & Sons, Inc.), New York, 1959.
5. Duncan, A. B. F.: Theory of Infrared and Raman Spectra, in W. West (ed.), "Chemical Applications of Spectroscopy," chap. 3, being vol. IX of A. Weissberger (ed.), "Technique of Organic Chemistry," Interscience Publishers (Division of John Wiley & Sons, Inc.), New York, 1956.

6. Hochgesang, F. P.: Nephelometry and Turbidimetry, in I. M. Kolthoff and P. J. Elving (eds.), "Treatise on Analytical Chemistry," pt. I, vol. 5, chap. 63, Interscience Publishers (Division of John Wiley & Sons, Inc.), New York, 1964.
7. Jones, R. N., and M. K. Jones: *Anal. Chem.*, **38**:393R (1966).
8. Jones, R. N., and C. Sandorfy: The Application of Infrared and Raman Spectrometry to the Elucidation of Molecular Structure, in W. West (ed.), "Chemical Applications of Spectroscopy," chap. 4, being vol. IX of A. Weissberger (ed.), "Technique of Organic Chemistry," Interscience Publishers (Division of John Wiley & Sons, Inc.), New York, 1956.
9. Meehan, E. J., and G. Chiu: *Anal. Chem.*, **36**:536 (1964).
10. Mie, G.: *Ann. Physik*, **25**:377 (1908).
11. Nicholson, D. E.: *Anal. Chem.*, **32**:1634 (1960).
12. Porto, S. P. S.: *Ann. N.Y. Acad. Sci.*, **122**:643 (1965).
13. Rogers, D. W., and A. Özsoğomonyan: *Talanta*, **11**:652 (1964).
14. Rosenbaum, E. J.: Raman Spectroscopy, in I. M. Kolthoff and P. J. Elving (eds.): "Treatise on Analytical Chemistry," pt. I, vol. 6, chap. 67, Interscience Publishers (Division of John Wiley & Sons, Inc.), New York, 1965.
15. Underwood, A. L.: Photometric Titrations, in C. N. Reilley (ed.), "Advances in Analytical Chemistry and Instrumentation," vol. 3, p. 31, Interscience Publishers (Division of John Wiley & Sons, Inc.), New York, 1964.
16. Van de Hulst, H. C.: "Light Scattering by Small Particles," John Wiley & Sons, Inc., New York, 1957.
17. Zdybek, G., D. S. McCann, and A. J. Boyle: *Anal. Chem.*, **32**:558 (1960).

7
Emission Spectroscopy

Ever since the work of Bunsen and Kirchhoff (1860), it has been known that many metallic elements under suitable excitation emit radiations of characteristic wavelengths. This fact is utilized in the familiar qualitative flame tests for the alkali and alkaline-earth metals. By employing more powerful electrical excitation in place of the flame, the method can be extended to all metallic and many nonmetallic elements. With some, such as sodium and potassium, the spectra are simple, consisting of only a few wavelengths, while with others, including iron and uranium, thousands of distinct, reproducible wavelengths are present. Elements giving complicated spectra cannot be identified by direct visual observation of the excited sample but can be recognized through the use of a spectroscope.

Quantitative analysis with the spectrograph is based on the relation between the power of the emitted radiation of some particular wavelength and the quantity of the corresponding element in the sample. This relation is empirical, as no adequate mathematical theory has ever been developed. The radiant power is influenced in a complicated way by many variables, including the temperature of the exciting arc and the size, shape, and material of the electrodes. For this reason, procedures must be rigidly standardized, and the spectra of unknowns must always be compared with those of standard samples prepared with the same apparatus under identical conditions.

The excitation of atomic spectra in a flame will be excluded from the present chapter and considered in the next. We will here be concerned primarily with spectra as excited by electrical means.

Spectrography utilizing photographic detection was developed to a high peak of analytical importance, largely in the period between the two World Wars. The present trend, however, is to replace the large and powerful emission spectrographs with other types of instruments which will achieve the same (or better) results more quickly, with a saving in laboratory space, and frequently without as exacting training of operating personnel. These newer techniques include notably x-ray fluorescent emission and atomic absorption, both of which will be treated in later chapters.

Emission spectroscopy retains a place in direct-reading photoelectric instrumentation and in flame photometry. Photographic instruments now manufactured are primarily small ones intended for the quick qualitative identification which must often precede quantitative analysis. For these reasons, then, only a brief account of classical spectrographic methodology will be included in this text. There are many excellent books available which give more complete treatments.[2,5,6]

EXCITATION OF SAMPLES

There are several ways in which emission spectra can be produced. Gaseous substances are easily excited by passing a high-voltage electric discharge through the sample contained in a glass tube.

More elaborate methods are necessary for solid samples. In the usual procedure, a powerful electric discharge is passed between two portions of the sample or between the sample and a *counterelectrode* which does not contain the elements being determined. If the sample is not obtainable in the form of a rod but rather as a powder or in solution, it may be placed in a recess drilled in the end of a graphite electrode, which is then connected as the lower electrode (usually positive). The upper electrode may be a pointed rod of graphite. Graphite has the advantage for this purpose of being highly refractory (that is, it does not melt or sublime at the temperature of the arc); it is a sufficiently good electric conductor; and it introduces no spectral lines of its own. A drawback, however, is that the hot carbon reacts very slightly with the nitrogen of the air to form cyanogen gas, which becomes excited to give

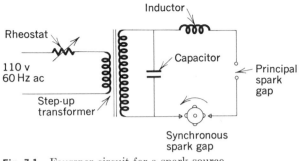

Fig. 7-1 Feussner circuit for a spark source.

bands of luminosity in the region 3600 to 4200 Å. This can be avoided, if necessary, by enclosing the arc in a mantle containing steam or an inert gas.

The *dc arc* is the most sensitive method of excitation and is widely used for qualitative analysis of metals. A current of 5 to 15 amp at 220 V is caused to flow through the arc in series with a variable resistor (10 to 40 Ω). The current may be obtained from a motor generator or an ac rectifier. The chief disadvantage of the dc arc is that it is less reproducible than might be desired. The discharge tends to become localized in "hot spots" on the surface of the electrode, which results in uneven sampling. This defect is overcome in the *ac arc*, wherein the discharge is automatically interrupted 120 times every second. A source of 2000 to 5000 V at 1 to 5 A is required. The flow of current is controlled by a variable inductor or resistor in the circuit. The ac arc is particularly suitable for the analysis of the residues from solutions which have been evaporated onto the surface of an electrode.

The *spark* source is energized from the ac power lines by means of the *Feussner* circuit (Fig. 7-1). High voltage (15,000 to 40,000) is obtained through a step-up transformer or Tesla coil. The secondary winding is connected across a capacitor and in series with the spark gap, an inductance coil, and an auxiliary spark gap operated by a motor whose rotation is synchronous with the alternations of the line current. The purpose of the synchronous gap is to make sure that the spark only passes at the moment of highest voltage, thereby insuring reproducibility. The spark excitation is better suited than the arc to precise timing of exposures, and in addition it does not destroy the sample, as only minute amounts are vaporized.

Many modifications of arc and spark circuits, including intermediate forms, have been described, some of which are quite ingenious and complex. An excellent summary is given by Jarrell in the "Encyclopedia of Spectroscopy."[3]

A recent development of significance is the application of the laser as a high-energy source. The laser beam can be focused on a minute area of the sample, causing localized vaporization of even the most refractory materials. The vapor may have received sufficient excitation thermally, or an electric discharge may be superimposed. The localization of the effect can be an advantage, permitting examination of areas as small as 50 μ in diameter, or it can be a disadvantage, rendering more difficult the representative analysis of a macro sample. An advantage of laser excitation is that the sample need not be electrically conducting. For literature references see page 304R of Ref. 4. (Excitation of small areas is also possible in x-ray emission with the electron-probe technique; see Chap. 9.)

PREPARATION OF ELECTRODES AND SAMPLES

The electrodes between which the arc or spark is to be struck can be of several kinds. If the sample is a metal and an adequate amount is available, it can be used directly as a pair of *self-electrodes*, with the arc passing between two pieces of the sample, usually in rod form. If only a single electrode can be fashioned from the sample, a *counterelectrode* of some noninterfering material may be employed. The counterelectrode may be pure carbon (graphite), or it may be one of the major constituents of the sample, in pure form. For example, in analyzing a steel for alloyed elements, the counterelectrode could be a simple steel, shown spectrographically to be free of the elements sought.

Samples other than metals are deposited in or on a noninterfering electrode. By one method, the sample is brought into solution, with acid or other reagent if necessary; a few drops are placed in a hole drilled in the end of a carbon electrode (Fig. 7-2e), and the solvent is evaporated by preheating the electrode in a small oven. This electrode is then positioned in the lower holder with a plain carbon counterelectrode (Fig. 7-2c) above it, for arcing. In another procedure, the dissolved sample is placed in a deep hole drilled nearly all the way through a porous carbon (Fig. 7-2g), which is made the upper electrode; the sample is allowed to seep out through the pores directly into the arc. In still another method, the sample in powder form is mixed with pure graphite powder and poured dry into the cup of an electrode (Fig. 7-2e). The added graphite, being an electric conductor, tends to stabilize the arc. Many modified procedures have been published which show advantages in particular kinds of work.

The spectroscopist must always keep in mind possible differences in volatility of components of the sample. In some cases one or more components may volatilize and be completely burned within a half-minute or

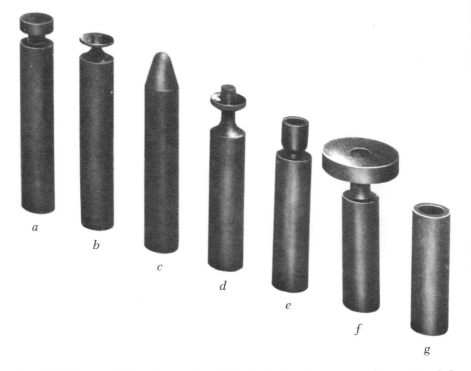

Fig. 7-2 Representative shapes of graphite electrodes. Forms *a* and *c* are intended as counterelectrodes rather than sample carriers. Forms *b*, *d*, *e*, and *f* are provided with cups into which the sample may be inserted; *b* and *d* have a central post against which the arc is struck. The narrow neck is to reduce thermal conductivity. Form *f* has a particularly large heat capacity. Form *g* is drilled nearly to the bottom and is porous. (*Ultra Carbon Corporation, Bay City, Mich.*)

so after the arc is started, before other substances present have been heated sufficiently to appear in the arc. This may be a disadvantage, especially in the case of trace amounts, where the sample may have lost some constituents before sufficient exposure has been given to the photographic plate. It is sometimes possible to take advantage of differential volatility, in that spectra of volatile constituents may be recorded without interference from the less volatile. An example is the determination of lithium, aluminum, and other oxides as impurities in uranium oxide;[7] uranium is especially rich in spectral lines, and hence determination of trace impurities is difficult. In this method, the uranium is first converted to the nonvolatile U_3O_8, and 2 percent of its weight of Ga_2O_3, a comparatively volatile oxide, is added. The gallium oxide acts as a carrier to sweep out even minute quantities of the impurities into the arc. This results in high sensitivity and accuracy even for impurities present only to the extent of a few parts per million.

IDENTIFICATION OF LINES

For qualitative spectrographic analysis it is necessary only to identify the element responsible for the emission of each wavelength present in the spectrum of the unknown. This is done by comparison with spectra obtained from authentic samples of pure elements. All the known wavelengths for all the elements are listed in reference tables, but to take advantage of the tables it must be possible to determine accurately the wavelengths of the lines produced by the unknown. Alternatively, spectra of various possible elements may be photographed with the same spectrograph and compared line by line with the unknown.

Wavelengths may be determined in an absolute sense by suitable geometric measurements and reference to a dispersion curve for a prism instrument or to the spacing distance for a grating. Such a measurement usually lacks sufficient precision unless made with great care, and this may be a prohibitively time-consuming task if many lines are to be identified. It is never undertaken in routine analytical work.

A more convenient method of determining wavelengths of unknown lines is the comparison of the unknown with the known lines of a standard photographed on the same plate. An example is given in Fig. 7-3. An arc struck between iron electrodes provides the most common standard. It is chosen because it produces thousands of lines fairly evenly spaced throughout the visible and ultraviolet. All these iron lines have been measured repeatedly with great care on both prism and grating spectrographs of high dispersion and thus can serve as a measuring scale for other spectra. In practice the spectrum of the unknown is often sandwiched between duplicate spectra of iron recorded immediately next to it on the same plate.

Qualitative analysis of steels or other alloys can be made visually with a simple spectroscope, usually in conjunction with a dc arc. One instrument designed for such service is the Steeloscope.* The manufacturer provides a series of comparison cards, each with a color-printed reproduction of the appearance of a portion of the iron spectrum—with exactly the magnification observed in the Steeloscope—and with the most suitable lines of common alloying elements clearly marked.

* Hilger and Watts, Ltd., London.

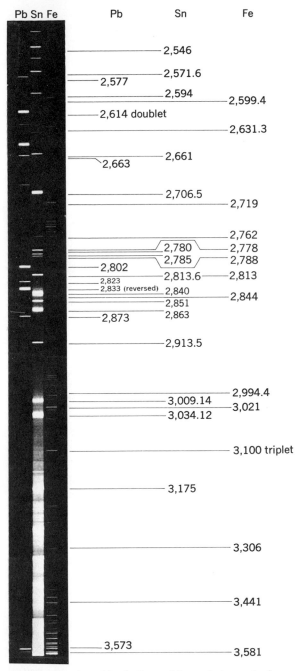

Fig. 7-3 Spectra of lead, tin, and iron photographed on a large Littrow prism spectrograph. Some of the brighter lead lines are visible in the tin spectrum (on the original plate) showing that the sample contained some lead. The 2833-Å line of lead is so intense that it shows photographic reversal (dark center). (*Bausch & Lomb, Inc., Rochester, N.Y.*)

QUANTITATIVE ANALYSIS

In a photographic instrument, a quantitative analysis can be carried out only by comparison of the optical density of the silver deposit caused by an emission line with the similar deposit derived from a standard. Because of the many variables introduced by the photographic process, this can be done to best advantage by the method of *internal standards*. This method depends upon the measurement of the ratio of the radiant power of a given line of the unknown to that of some line of another constituent of the sample which is present in known (or at least constant) amount. This standard may be an element already present in the specimen, such as the iron in a steel sample, or it may be an extraneous element added in known quantity to all samples. This procedure completely eliminates errors due to inequalities of plate characteristics and development. The line to be used as a standard should be as close as possible to the unknown in wavelength and also in power so that nonlinearity of the photographic emulsion with respect to these factors will not be a serious source of error. Two lines selected as particularly appropriate for this purpose are called an *homologous pair*.

A typical analytical procedure will be presented in some detail, as an illustration of this technique. The example chosen is the analysis of traces of magnesium in solution, with molybdenum as internal standard. It is taken from a series of similar procedures for many elements described by Nachtrieb,[5] to whose book the student is referred for further details.

A condensed spark discharge between copper electrodes has been found convenient for this work. The copper electrodes are rods 3-4 cm in length and about 5 mm in diameter. The tips must be carefully machined smooth and flat in a lathe, partly to ensure uniformity and partly to remove surface contamination. After machining, they must be handled only with forceps or between folds of filter paper and must be protected from dust. In preparation for use, the electrode is placed upright in a small electric heating coil, and a single drop (0.05 ml) of test or standard solution is placed on the tip by means of a special pipet. The solution is then carefully evaporated to dryness, and the electrode is ready for excitation.

A series of solutions which are to serve as standards are prepared from pure magnesium chloride at concentrations of the order of 0.0001 to 10.0 mg of magnesium per liter. To each solution (both standards and unknowns) is added ammonium molybdate to the extent of 20.0 mg of molybdenum per liter. The distilled water must be shown (spectroscopically) to be free of magnesium.

A pair of electrodes, both of which are coated with the sample, are clamped in position with an accurately measured spacing of 2 mm between

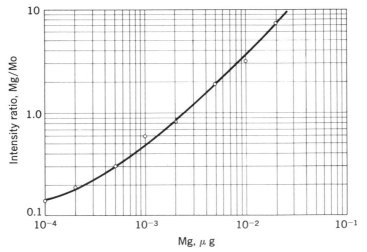

Fig. 7-4 Working curve for the analysis of magnesium by the copper spark method, with molybdenum as internal standard. (*McGraw-Hill Book Company, New York.*)

them. A 25,000-V spark is struck across the electrodes through a Feussner circuit.

The spectrum contains lines from both magnesium and molybdenum, as well as any other metals present.   Several homologous pairs are available and are about equally good.   The lines at 2798.1 Å (Mg) and 2816.2 (Mo) represent one such pair. The difference of the optical densities of these two lines is plotted against the concentration of magnesium (Fig. 7-4).

This method is fast and convenient and, at the same time, highly sensitive. It is possible to identify 1 ng of magnesium in a volume of 1 ml. The precision of the determination is of the order of 5 to 10 percent, which is quite satisfactory in view of the small quantities involved.

Another, more recent, example[1] of a quantitative spectrographic analysis is the determination of many transition metals in the parts-per-billion range in KCl intended for electrochemical purposes. The procedure includes a preconcentration by precipitation with 8-quinolinol in the presence of $InCl_3$ as carrier and $PdCl_2$ as internal standard.

PHOTOGRAPHIC INSTRUMENTS

A large part of the basic work in spectrography was accomplished with quartz prism instruments of either the Cornu or Littrow design.   Many large concave grating spectrographs have been employed also, particularly in the Eagle and Wadsworth mountings.   Each of the various types has been brought to a high degree of perfection by a number of manufacturers in the United States and abroad.

PHOTOELECTRIC INSTRUMENTS

Large spectrometers designed for rapid, routine determinations of several elements simultaneously are available.   One approach lies in the *direct-reading spectrometer* (Fig. 7-5).   This consists of a grating spectrograph in which the plate- or film-holder has been replaced with an opaque barrier following the focal curve and pierced by 12 or more slits located at wavelengths appropriate to the elements to be analyzed.   Behind each slit is mounted a photomultiplier tube.   The output of each is recorded automatically, so that the relative radiant powers of 12 elements can be read directly.   One of the elements observed is an internal standard with which the rest are compared.   Analysis for 11 elements can be completed in less than a minute.

Another photoelectric spectrometer is the Quantometer,* in which the results are automatically recorded on a strip chart.   A modified form of this instrument, the Quantovac, has been designed to extend the useful range to 1600 Å in the ultraviolet.   This permits measurements of sensitive wavelengths of carbon, phosphorus, sulfur, arsenic, and selenium, which are not accessible to conventional instruments.   In order to reach into the far ultraviolet, it is necessary to eliminate air from the optical path, as it absorbs too strongly in this region.   Hence the whole spectrometer must be enclosed in an evacuated housing.   The arc or

* Applied Research Laboratories, Glendale, Calif.

Fig. 7-5 A direct-reading photoelectric spectrometer.   (*Baird-Atomic, Inc., Cambridge, Mass.*)

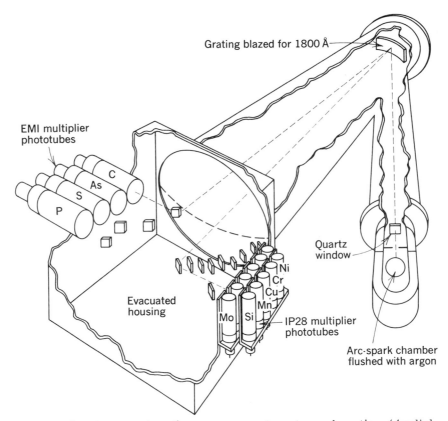

Fig. 7-6 Quantovac automatic vacuum spectrometer, schematic. (*Applied Research Laboratories, Inc., Glendale, Calif.*)

spark source cannot operate in a vacuum, so it is bathed in an atmosphere of argon, which does not absorb so strongly as air. Figure 7-6 shows schematically the arrangement of parts in the Quantovac. The grating is blazed for maximum energy at 1800 Å. Several small mirrors (as many as 23) are adjusted to intercept suitable spectral lines which are reflected onto photomultiplier tubes. This permits simultaneous determination of all the elements of interest in a steel analysis, both metals and nonmetals.

PROBLEMS

7-1 A shipment of "chemically pure" aluminum metal is to be analyzed for its magnesium content. A 1.000-g sample is dissolved in acid, and enough ammonium molybdate added to contain 2.000 mg of molybdenum. The solution is diluted to 100.0 ml, and a few drops are evaporated onto the tip of a copper electrode and sparked

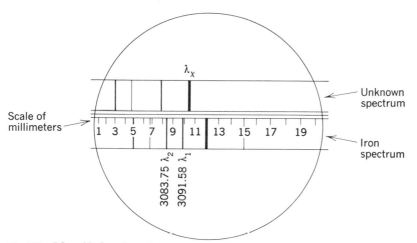

Fig. 7-7 Magnified region of a spectrum plate, with linear scale for interpolation. (*Bausch & Lomb, Inc., Rochester, N.Y.*)

before the spectrograph slit. The resulting plate, examined with a densitometer, shows a density of 1.83 for the 2795.5-Å line and 0.732 for the 2848.2-Å line. With the aid of Fig. 7-4, compute the percent of magnesium in the sample.

7-2 The sketch in Fig. 7-7 shows a portion of a spectrograph plate as seen through a magnifier with a built-in scale of millimeters. The field of view includes portions of the spectra produced by arcing, first, iron electrodes, then electrodes fashioned from an unknown thought to be an aluminum alloy. The iron lines with numbers 1 and 2 have been identified by comparison with a standard iron spectrum. Line x of the unknown spectrum is to be identified. The scale readings are as follows: $d_1 = 9.90$ mm, $d_2 = 8.37$ mm, $d_x = 10.25$ mm. (*a*) Calculate the dispersion of the spectrograph in this region in terms of angstroms per millimeter. (*b*) Determine the wavelength of the unknown line. (*c*) Refer to a table of wavelengths and make a tentative identification of this line, with due regard for the elements likely to be found in this type of alloy. In what regions of the spectrum would you suggest searching for strong lines to confirm your identification?

REFERENCES

1. Farquhar, M. C., J. A. Hill, and M. M. English: *Anal. Chem.*, **38**:208 (1966).
2. Harrison, G. R., R. C. Lord, and J. R. Loofbourow: "Practical Spectroscopy," Prentice-Hall, Inc., Englewood Cliffs, N.J., 1948.
3. Jarrell, R. F.: Excitation Units, p. 158, in G. L. Clark (ed.): "Encyclopedia of Spectroscopy," Reinhold Book Corporation, New York, 1960.
4. Margoshes, M., and B. F. Scribner: *Anal Chem.*, **38**:297R (1966).
5. Nachtrieb, N. H.: "Principles and Practice of Spectrochemical Analysis," McGraw-Hill Book Company, New York, 1950.
6. Sawyer, R. A.: "Experimental Spectroscopy," 3d ed., Dover Publications, Inc., New York, 1963.
7. Scribner, B. F., and H. R. Mullin: *J. Res. Natl. Bur. Std.*, **37**:379 (1946).

8

Flame Spectroscopy

In this chapter are considered several related methods of analytical importance concerned primarily with the chemical and physical phenomena taking place within a flame. A flame can serve effectively as a source of atomic emission lines and also as an absorbing medium for these same lines. The general relations can be visualized with the aid of Fig. 8-1. Fluorescence can also occur in the flame, as can enhancement of emission through chemiluminescence.

FLAME CHEMISTRY[5]

In the exothermic combination of two gases, such as hydrogen and oxygen, the initial process must be the endothermic breaking of two bonds:†

$$H_2 + Q \rightarrow 2H^{\cdot}$$
$$O_2 + Q \rightarrow 2O^{\cdot}$$

† Note the distinction between dot and asterisk: $H^{\cdot}$ represents an isolated hydrogen atom (not an ion) in its ground state; $HO^{\cdot}$ represents an hydroxyl free radical. $H^{\cdot *}$ represents an atom of hydrogen in an excited state. The symbol Q is used in a generic sense to indicate energy.

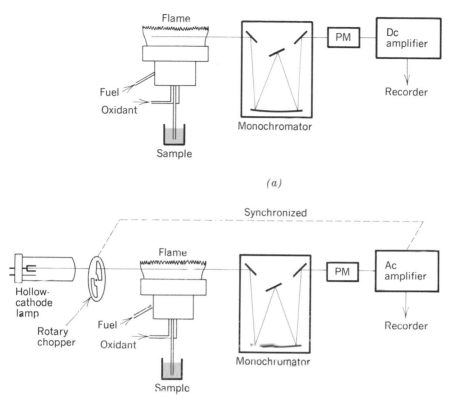

(a)

(b)

Fig. 8-1 Comparison of (*a*) flame emission and (*b*) atomic absorption photometers.

Following dissociation, various two-body reactions can occur, such as:

$$H\cdot + O_2 \rightarrow O\cdot + HO\cdot$$
$$O\cdot + H_2 \rightarrow H\cdot + HO\cdot$$
$$H\cdot + H_2O \rightarrow H_2 + HO\cdot$$

These reactions merely pass around the excess energy from one species to another, and cannot represent an approach to equilibrium. The energy can only be released by a collision with a third body B, which may be a molecule of a gas (such as N_2 or O_2), or of a salt (as $NaCl$), or a solid surface:

$$H\cdot + H\cdot + B \rightarrow H_2 + B + Q$$
$$H\cdot + HO\cdot + B \rightarrow H_2O + B + Q$$

(Part of the released energy must of course be returned to the system to dissociate incoming gases, thus maintaining the chain reaction.)

If B is a salt (possibly an analytical sample), we may have such reactions as

$$H \cdot + NaCl \rightarrow Na \cdot + HCl$$
$$H \cdot + HO \cdot + NaCl \rightarrow H_2O + Na \cdot + Cl \cdot$$

In the presence of sufficient available energy, the metal atom may be raised to an excited level:

$$Na \cdot + Q \rightarrow Na \cdot *$$

a process which will be followed immediately by the emission of the characteristic radiation:

$$Na \cdot * \rightarrow Na \cdot + h\nu$$

In practical flames which are hot enough to dissociate entrained salts, only a very small fraction of the metal atoms will become activated, the rest remaining in their ground states (see Table 8-1). Those

Table 8-1   **Fraction of atoms in excited states at various temperatures***

Element	Line, nm	Temperatures		
		2000°K	3000°K	4000°K
Cesium	852.1	4×10^{-4}	7×10^{-3}	3×10^{-2}
Sodium	589.0	1×10^{-5}	6×10^{-4}	4×10^{-3}
Calcium	422.7	1×10^{-7}	4×10^{-5}	6×10^{-4}
Zinc	213.9	7×10^{-15}	6×10^{-10}	2×10^{-7}

* Data from A. Walsh.[14]

which are activated will drop back to the ground state with the emission of a photon of radiation which can be identified and measured with a suitable *flame photometer*. On the other hand, the unexcited atoms are able to absorb radiation from an external source at the same characteristic wavelengths. The absorptivity can be measured, and serves as the basis of the technique known as *atomic absorption spectroscopy*.

The ratio of the number of atoms which become excited $N*$ to the number which remain in the ground state N^0 is given by a Boltzmann distribution:

$$\frac{N*}{N^0} = A \exp - \frac{E}{kT} \tag{8-1}$$

where A = constant for a particular system
    E = activation energy
    k = universal constant, 1.38054×10^{-16} erg/deg
    T = Kelvin temperature

This is a quantitative statement of the well-known fact that the higher the temperature, the greater the fraction of atoms which become activated.

For atomic absorption, then, it would appear that the temperature should be kept as low as possible, consistent with the ability to dissociate compounds to free atoms. However, as Table 8-1 indicates, the fraction of atoms activated is so small that its effect on N^0 is imperceptible; hence the temperature dependence as deduced from Eq. (8-1) is unimportant. It is desirable to keep the temperature *slightly* above the dissociation point for two reasons: (1) the absorption band broadens with temperature (Doppler broadening) and it is favorable to have it slightly broader than the line emitted by the source, and (2) if the flame is too hot, the dissociated atoms will tend to become ionized, and only neutral atoms in their ground states absorb at the required wavelength.

For atomic emission, on the other hand, a higher temperature is required, to increase the N^*/N^0 ratio, but the danger of ionization is still to be reckoned with.

All three temperature-dependent properties, dissociation, activation, and ionization, involve Boltzmann distributions comparable to Eq. (8-1), with appropriate values of A and E. Hence optimum temperatures for different experimental techniques are quite critical and often must be determined empirically. The tendency toward ionization can often be counteracted by the addition of a substance more easily ionized. Thus the ionization of calcium atoms will be negligible in the presence of potassium, but is likely to produce low results otherwise, in the determination of calcium by either absorption or emission.

The picture is even more complicated than this. There is often more resonance emission from a flame than can be accounted for on the basis of temperature alone, particularly when an organic solvent is present. It has been shown that this enhancement of emission is in part the result of fluorescence brought about by ultraviolet radiation produced in the flame itself. The enhancement may also be related to a transfer of electrons from stable molecular orbitals of a compound (an oxide for example) to an excited atomic orbital, when the compound is pyrolyzed; the excited electron then drops to its ground state with emission. This can be considered a case of *chemiluminescence*.[6]

BURNERS AND ASPIRATORS[8]

For any of the analytical methods based on the spectroscopy of flames, where the sample is a liquid solution, means must be found for introducing the liquid in finely divided form into the flame. The production of a suitable flame and the introduction of sample are two functions which are nearly always performed by a single instrumental component. For greatest versatility it would be desirable to utilize the same burner

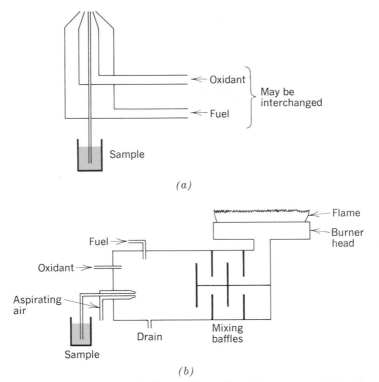

(a)

(b)

Fig. 8-2 Burners for emission or absorption photometers: (*a*) total consumption type; (*b*) premix type.

assembly for both emission and absorption, but this is seldom possible, particularly when one is pressing for high sensitivity, as the requirements are not identical for the two methods.

A great many burner designs have been suggested. They fall generally into two classes, according to whether or not the fuel and oxidant gases are mixed prior to entering the flame area. Figure 8-2 shows representative burners of the two classes. The type shown at (*a*) is called a *total consumption* burner; it has been in use for many years in emission flame photometry. The sample is aspirated by the Venturi effect into either fuel or oxidant, and enters the flame at its base as a fine mist.

A major difficulty inherent in the simple aspirator is that it produces droplets of widely differing sizes. Many of the larger droplets are frequently blown right through the flame, either without totally evaporating, or without complete pyrolysis of the solute. Thus the name total consumption is in a sense a misnomer; the sample all disappears, but not all is converted to a measurable state.

The total consumption burner concentrates the flame in a small area, an advantage in emission, where a maximum amount of light must be focused on the detector or the monochromator slit. It is a safe burner to operate, not subject to explosive strike-backs. It has the disadvantage for absorption work that the light beam has only a short path through the vaporized sample; this has been overcome in some instruments by using several (usually three) identical burners in a row, or by reflecting the beam back and forth several times through the flame. The turbulent flame from this kind of burner is often objectionably noisy.

The burner in Fig. 8-2(b) represents a more recent development, called a *laminar flow* or *premix* burner. The fuel and oxidant gases are mixed in the body of the burner assembly, along with the aspirated sample. The mixture then passes through an area containing one or more baffles to complete the mixing before entering the burner head. The larger drops from the aspirator impinge on the walls of the mixing chamber and baffles, and are drained off, so that only the fine droplets find their way to the flame. The burner orifice consists of a long narrow slit, which produces a nonturbulent, quiet, ribbon-shaped flame.

The premix burner must be designed with particular attention to the prevention of strike-back, which may produce a sharp explosion in the enclosed mixture. Its advantages are a long path length for the light beam in atomic absorption, and an inherent stability and lack of noise, both audible and electronic. A disadvantage, shared with the total consumption burner, is the loss of the larger droplets from the atomizer. As much as 90 percent of the sample may go down the drain. There is a possibility when mixed solvents are employed that rapid evaporation of the more volatile will tend to segregate the solute preferentially in the large droplets, introducing significant error.

A recent development with much to recommend it breaks with prior practice in that it divorces the conversion of the solution to a mist from the production of the flame. This is the *sonic nebulizer* reported by Kirsten and Bertilsson.[9] In this device the sample solution is pumped through a capillary tube onto the vibrating surface of an ultrasonic transducer. The intense high-frequency vibration is very effective at converting the liquid to a fine mist of uniform particle size. The mist is picked up by one of the gases and carried to the burner, which may be of either type. The only obvious drawback is the added expense of the power oscillator required to drive the transducer.

The last word on burner design has yet to be said. There is much room here for the exercise of ingenuity on the part of the instrument designer.

FUELS AND OXIDANTS

The only oxidants employed in practice are oxygen, air, and nitrous oxide. The latter can be considered intermediate between the other two; from about 500° to 900°C it decomposes to give a mixture of two parts of nitrogen to one part of oxygen, whereas air contains four of nitrogen to one of oxygen. The variety of fuel gases is greater, and the choice may depend on convenient availability. Table 8-2 lists the

Table 8-2   Maximum temperatures of various flames*

	Oxidants		
Fuels	*Air*, °C	*Oxygen* °C	*Nitrous oxide*, °C
Hydrogen	2100	2780	
Acetylene	2200	3050	2955
Propane	1925	2800	

* Values from Dean,[2] except for nitrous oxide, which is from Willis.[17]

maximum temperatures attainable from a number of combinations. Values for methane and butane are about the same as for propane, so this value can be taken as representative of "natural gas." Various kinds of manufactured gases can be used without difficulty.

For total consumption burners, oxygen and acetylene is the most widely used combination. It gives as high a temperature as is easily available, high enough to activate measurable fractions of atoms in a dozen or so elements. For flame photometers intended only for alkali metals (as in clinical work), compressed air with natural or manufactured gas is adequate.

For atomic absorption with premix burners oxygen-acetylene cannot be safely used because the flame propagates so rapidly that it is difficult to prevent its striking back and exploding the premixed gases in the body of the burner. Air and acetylene are most frequently chosen for this type of burner. Nitrous oxide[17] shows promise of great utility as oxidant with acetylene; it produces a temperature only slightly less than oxygen-acetylene, but without nearly so great an explosion hazard. Nitrous oxide is almost essential in the analysis of elements such as aluminum, titanium, and the lanthanides, which form refractory oxides in the air-acetylene flame.

The presence of certain anions in the sample can result in significant diminution in the concentration of dissociated metal atoms.[10] This appears to be due to the formation of chemical compounds of very low vapor pressure following evaporation of the solvent. The resulting microcrystalline particles are blown through the flame without decomposition. Thus the number of calcium atoms in the flame is markedly decreased by the presence of oxalate, sulfate, or phosphate, but is not affected by chloride or nitrate. Aluminum also removes calcium, presumably indicative of the formation of calcium aluminate.

Flame methods may be utilized with nonaqueous solutions. If the solvent is combustible, its presence may increase the flame temperature slightly, whereas water cools it by endothermic dissociation. The method is particularly convenient for metals which can be selectively extracted from aqueous solution by an organic solvent to produce a solution which can be aspirated directly into the flame. An example is the determination of iron in nonferrous alloys by flame emission following extraction from an acidified solution with acetylacetone. The emission line at 3720 A is six times more intense when the iron is dissolved in acetylacetone than in water.

ATOMIC ABSORPTION: SOURCES

In principle, a continuous source should be usable in atomic absorption, just as in the ultraviolet and infrared. Work has been reported along this line,[4,13] but so far it has not been commercialized. The difficulty is caused by the extreme narrowness of the absorption line, of the order of 0.01 Å half-width, which removes only a small fraction of the energy from the spectral region passed by the usual monochromator. Figure 8-3

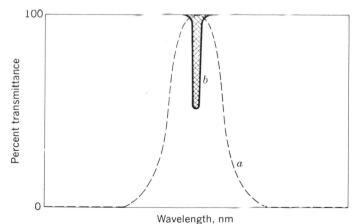

Fig. 8-3 Atomic absorption of radiation from a continuous source with monochromator; curve *a* represents the band of wavelengths passed by the monochromator, curve *b*, the absorption by an atomic species in a flame. (*Unicam Instruments Ltd.*)

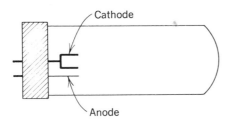

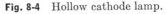

Fig. 8-4 Hollow cathode lamp.

shows this relation.   The area beneath curve *a* is only slightly diminished by the absorption of the narrow line at *b*.   In order to obtain results with adequate precision, a spectrophotometer with high dispersion is necessary, together with an expanded-scale recorder, so that integrated intensities can be read with precision.

More practical, at the present state of development, is the use of a series of sources which give sharp emission lines for specific elements. The most successful of such sources is the hollow-cathode glow-discharge lamp.[8] This consists of two electrodes, one of which is cup-shaped and made of the specified element (or an alloy of that element) (Fig. 8-4); the material of the anode is not critical.   The lamp is filled with a low pressure of a noble gas.   Application of 100 to 200 V will produce, after a short warm-up period, a glow discharge with most of the emission coming from within the hollow cathode.   The radiations consist of discrete lines of the metal, plus those of the fill gas.   The gas is selected by the manufacturer to give the least spectral interference with the metal concerned. The spectrum of a hollow-cathode lamp is shown in Fig. 8-5.   Note that

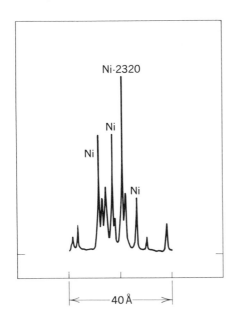

Fig. 8-5 Emission spectrum of a lamp with a hollow nickel cathode.   Only the line at 2320 Å is absorbed by nickel atoms in a flame.   (*Westinghouse Electric Corporation, Pittsburgh.*)

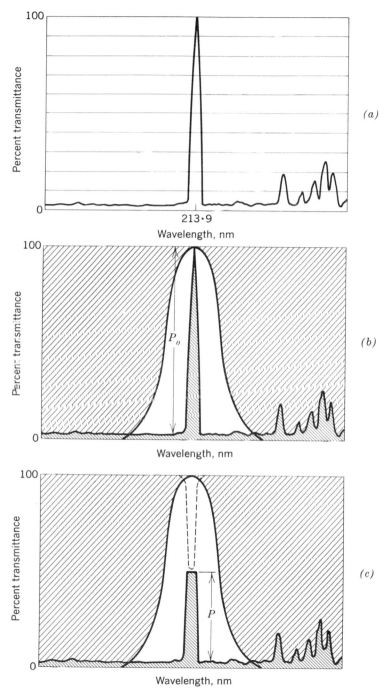

Fig. 8-6 Atomic absorption. (*a*) The spectrum of a zinc cathode lamp (the width of the 2139-Å line has been exaggerated for clarity). (*b*) Extraneous zinc lines are eliminated by a monochromator centered at 2139 Å. (*c*) The power of the 2139-Å line is sharply reduced by absorbing zinc atoms in a flame. (*Unicam Instruments Ltd.*)

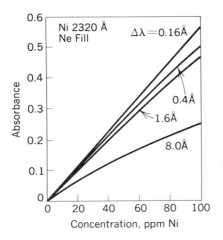

Fig. 8-7 Beer's law plot for flame absorption by nickel atoms of radiation from the lamp of Fig. 8-5; negative deviation increases with band width of the monochromator. (*Westinghouse Electric Corporation, Pittsburgh.*)

the sensitive 2320-Å nickel line is surrounded by numerous nonabsorbing nickel lines. The desired line can be isolated by a narrow band-pass monochromator.

Figure 8-6 shows schematically the relation between the emission spectrum of a hollow-cathode lamp (*a*), the same, as seen through a monochromator (*b*), and the effect of an absorption in the flame (*c*). Compare this result with Fig. 8-3. In the present case, only the peak heights need be measured, since the absorption and emission lines have nearly equal half-widths.

It will be apparent from the relations shown in Fig. 8-6(*c*) that measurement with an atomic absorption photometer is essentially identical procedurally to measurements with any single-beam absorption photometer. P_0 and P are determined successively, without any change in flame conditions, and Beer's law applies in its conventional form. The path length b is not easy to measure with precision, so precautions must be taken against its changing.

Figure 8-7 indicates the effect of monochromator band width on the working curve for nickel, where the radiation is taken from the lamp of Fig. 8-5. Commercial atomic absorption photometers are generally provided with monochromators capable of isolating 2 Å, but in which the slits can be opened wider for greater sensitivity when interference is not a problem.

Other light sources have been used, particularly for volatile metals such as mercury and the alkalis, in which conventional discharge tubes provide great brilliance. They generally have somewhat wider emission lines, however. The principal objection to any of these sources is the large number of lamps which must be kept on hand. There are some hollow-cathode lamps available with several metals contained in a single cathode structure, which alleviate this problem to some extent.   Examples are: Ca, Mg, and Al; Fe, Cu, and Mn; Cu, Zn, Pb, and Sn; and Cr, Co, Cu, Fe, Mn, and Ni.

ATOMIC ABSORPTION SPECTROPHOTOMETERS[3,12]

There are about a dozen distinct models of instruments in this category presently available.   About half of these are manufactured in the United States, with instruments also imported from Australia, where the method was originated, from England, and from Germany.   All models utilize sharp-line sources, roughly equivalent monochromators, and photomultiplier detectors.   Most use premix burners, though some offer a choice of burner type.   Beckman has a source and flame unit as an accessory to be attached to an existing spectrophotometer.   Many have provision for conversion to flame emission photometry.

The majority of atomic absorption spectrophotometers are single-beam instruments.   The reason for the relative scarcity of double-beam photometers is that, at least until recently, the major source of instability and noise has been the flame, the very segment of the instrument in which variations cannot be compensated for by the double-beam feature.

The Perkin-Elmer Model 303 is a double-beam spectrophotometer, which ensures compensation for lamp and detector fluctuations.   The reference beam bypasses the flame, rather than traversing a "blank" flame.   An important advantage in practice is that the warm-up time of the hollow-cathode lamp is practically eliminated as a source of delay, a great convenience when changing from one element to another.

Nearly all instruments provide modulation, either by supplying the lamp with pulsed current, or by chopping the light beam with a rotary shutter.   This is essential for precision results, because the flame always acts as a source as well as an absorber of resonance radiation.   In a modulated system, with an amplifier tuned to the chopper frequency, the emission from the flame will produce no response.   In a dc system, on the other hand, the detector would have no way of distinguishing between radiation originating in the flame and that coming through the flame from the hollow-cathode lamp.

NON-FLAME ATOMIC ABSORPTION

It is sometimes possible to produce a vapor of metal atoms by means other than a flame.   The oldest application of atomic absorption, for example, is the estimation of the concentration of mercury vapor in air by absorption of the 2537-Å resonance line.   There are several instruments on the market for this purpose which make use of a beam of radiation from a quartz germicidal lamp.

It is possible to dissociate metallic compounds by means of an electric discharge at radio frequency.[15] The discharge is produced in a gas by induction from a coil consisting of a few turns of heavy wire surrounding a silica tube which contains the gas. The coil is fed from the high-frequency generator of an induction heater. The sample is aspirated into the stream of carrier gas and is atomized in the discharge, where it can be examined by transmitted light from a hollow-cathode lamp. An advantage is that the carrier can be a nonreactive gas such as argon, so oxides and other refractory compounds will not be formed.

There have been a number of studies on methods of atomizing solid materials directly, bypassing the solution step. In one method, the sample is placed in a silica boat inside a silica combustion tube, and heated electrically. Monochromatic light is passed through the length of the combustion tube, and its absorption by vaporized metal is measured. In another procedure, the metallic sample is made the cathode in what amounts to a demountable hollow-cathode lamp.[7] The cathode is open at both ends (rather than cup-shaped), and the light beam from the source is passed through it. The sample is sputtered into the axial space by a high-voltage discharge. It appears possible, also, to vaporize a sample by a laser beam and measure its atomic absorption.

Except for the mercury-vapor detector, none of these non-flame methods has been developed commercially, and fundamental studies are still in progress.

APPLICATIONS OF ATOMIC ABSORPTION

A feeling for the wide applicability of analysis by atomic flame absorption can be gained by examination of a loose-leaf compendium entitled "Analytical Methods for Atomic Absorption Spectrophotometry," published by Perkin-Elmer Corporation. The May, 1966, edition includes 114 specific procedures for 29 elements, and statements of optimum operating conditions for another 40 elements. (The only nonmetal included is phosphorus.) As one might expect, settings are given for the two Perkin-Elmer instruments, but the procedures are easily adaptable to work with products of other manufacture.

As an example, consider lead. Besides general procedural remarks about the element (type of lamp, suitable gas mixture, optimum concentrations for standard solutions, etc.) details are given for the analysis of lead in copper-base alloys, gasoline, lubricating oils, blood, steel, urine, gold, and wine. Literature references are given in every case.

FLAME EMISSION PHOTOMETRY[1,2]

There are a large number of commercial instruments of this class, with a wide variation in sensitivity, sophistication, and price. Many are designed for clinical applications, where the major interest is in sodium, potassium, and calcium in body fluids, etc. Sufficient precision can be obtained with interference filters in place of a monochromator, as the lines are easily separated. Some instruments utilize the simplest of photometric circuits, a photovoltaic cell directly connected to a galvanometer. Others gain sensitivity by means of a photomultiplier, and stability by the application of added lithium as an internal standard. Still other instruments include a monochromator, or utilize an existing spectrophotometer by means of a flame adaptor.

The power of the radiation from the flame at a wavelength characteristic of a particular element is found to be very closely proportional to the concentration of the metal, if a background correction is first made. The background luminosity is caused largely by the presence of other metals, since in general each excitable cation will give some radiation over a wide spectral region, even at a considerable distance from its discrete lines. Any scattering in the monochromator or photometer will also contribute. The effect of background can best be eliminated by the application of a base-line technique of measurement analogous to that discussed previously in connection with absorption spectra. This is easy to do if the spectra are observed with a spectrophotometer, but is not so convenient with a filter flame photometer. In the latter case, corrections are usually made on an empirical basis.

In addition to the background effect, a specific interaction may sometimes be observed between metals, either enhancing or repressing the normal luminescence. This difficulty can be overcome, though with some loss in sensitivity, by intentionally adding a large excess of any likely contaminating cations before analysis. In the analysis of natural waters[16] containing sodium, potassium, calcium, and magnesium, interference with the analysis of each element by the other three can be avoided by the use of *radiation buffers*. For example, in the determination of sodium, to a 25-ml sample is added 1 ml of a solution which is saturated with respect to the chlorides of potassium, calcium, and magnesium. Any slight variation in the amounts of these elements in the sample will be negligible compared with the amount added. The mixed solution is then examined with the flame photometer, and the response at the wavelength of sodium radiation, corrected for background luminosity, is compared with a calibration curve prepared from standards. Concentration differences of 1 or 2 ppm can easily be detected for sodium or potassium, and 3 or 4 ppm for calcium. In the case of magnesium, the method is less sensitive.

Another procedure to eliminate the results of interferences is an application of the method of standard addition. After measurement of the emission from the sample, a known amount of the desired element is added to a second aliquot of the unknown and the measurement repeated.

The added standard will be subject to the same interferences as the sample constituent, hence analysis by direct comparison is possible.

COMPARISON OF FLAME ABSORPTION AND EMISSION[6,11]

The alkali metals and a few other easily excited elements can be detected and measured to lower concentrations by well-designed flame emission photometers than by atomic absorptiometers. Most other metals show either improved or equal sensitivity when measured by absorption (see Table 8-3). Where this is not the case, Kahn[8] has pointed out that the difference may well reflect the fact that absorption instrumentation is of more recent development, and some elements have not yet been studied exhaustively by this technique.

It should be noted that the response of an emission photometer is *linear* within limits, with respect to concentration, whereas the results from an atomic absorption photometer, following Beer's law, bear a logarithmic relation to concentration. Hence the absorption technique provides a much longer concentration range over which measurements can be made.

Another factor to consider is the wavelength effect in connection with the Boltzmann distribution. Recalling that $E = h\nu = h(c/\lambda)$, Eq. (2-2), we can rewrite Eq. (8-1) as

$$\frac{N^*}{N^0} = A \exp - \frac{hc}{\lambda k T} \tag{8-2}$$

From this it follows that the shorter the wavelength of the resonant radiation, the smaller will be the ratio N^*/N^0, and hence the less sensitive will be the emission method, while the absorption will be only slightly affected.

Atomic absorption is at present less convenient with respect to qualitative scanning of samples, because of the need to change lamps. On the other hand, quantitative analysis is usually more satisfactory, as atomic absorption does not suffer so greatly from chemical interferences.

FLUORESCENCE

Another aspect of flame spectroscopy is atomic fluorescence.[18] This presents another mechanism by which free atoms can be elevated to excited electronic levels. The atoms must first be liberated from their compounds by pyrolysis in the flame, as in previous methods. Excitation is then produced by intense illumination of the flame from an external

Table 8-3   Detection limits in flame emission and absorption*

Element	Emission, ppm	Absorption, ppm (conventional flames)	Absorption, ppm (N_2O/C_2H_2)
Aluminum	0.01	0.5†	0.1
Antimony	0.3	0.2	
Arsenic	1.0	0.04	
Barium	0.01	1.0	
Beryllium	0.1	0.03	0.02
Bismuth	1.0	0.15	
Boron	10.0; 0.1‡	250.	10.
Bromine	100.‡		
Cadmium	0.3	0.0004	
Calcium	0.001	0.01	
Carbon	10.‡		
Cerium	10.; 1.0‡		
Cesium	0.0003	0.05	
Chlorine	100.‡		
Chromium	0.003	0.001	
Cobalt	0.03	0.003	
Copper	0.01	0.0005	
Dysprosium	0.1	6.0	0.7
Erbium	0.3; 0.1‡	3.0	0.9
Europium	0.0026	2.5	
Fluorine	100.‡		
Gadolinium	0.2; 0.03‡		
Gallium	0.01	1.0	
Germanium	3.0	0.1	
Gold	0.1	0.1	
Hafnium	8.0		
Holmium	0.1	4.0	0.3
Indium	0.003	0.1	
Iodine	10.0‡		
Iridium	100.		
Iron	0.03	0.003	
Lanthanum	1.0; 0.01‡	10.	20.
Lead	0.03	0.01	
Lithium	0.000,01	0.005	
Lutetium	0.2; 0.1‡		
Magnesium	0.003	0.003	
Manganese	0.001	0.001	
Mercury	1.0	0.01	
Molybdenum	0.03	0.01	
Neodymium	1.0; 0.03‡		
Nickel	0.03	0.003	
Niobium	1.0	3.0	
Nitrogen	10.‡		
Osmium	100.		

Table 8-3   Detection limits in flame emission and absorption* (continued)

Element	Emission, ppm	Absorption, ppm (conventional flames)	Absorption, ppm (N_2O/C_2H_2)
Palladium	0.01	0.1	
Phosphorus	1.0‡	100. †	
Platinum	3.0	0.5	
Potassium	0.000,01	0.005	
Praseodymium	2.0; 0.3‡	. . . .	15.
Rhenium	0.3	25.	1.5
Rhodium	0.03	0.1	
Rubidium	0.0001	0.005	
Ruthenium	0.01	0.3	
Samarium	0.6; 0.1‡	50.	6.0
Scandium	0.06; 0.003‡	5.0	
Selenium	. . . .	0.015	
Silicon	3.0	0.1	1.0
Silver	0.01	0.02	
Sodium	0.000,01	0.005	
Strontium	0.001	0.02	
Sulfur	10. ‡		
Tantalum	10.	. . . .	6.0
Tellurium	2.0	0.3	
Terbium	1.0; 0.1‡	. . . .	3.0
Thallium	0.01	0.03	
Thorium	150.; 1.0§		
Thulium	0.2; 0.1‡	6.0	
Tin	0.3	0.02	
Titanium	0.5; 0.03‡	1.0	0.5
Tungsten	4.0	250.	9.0
Uranium	10.; 1.0§	10,000. (= 1%)	30.0
Vanadium	0.1	0.5	0.7
Ytterbium	0.01	2.0	0.05
Yttrium	0.3; 0.01‡	10.	0.5
Zinc	3.0	0.0002	
Zirconium	50.; 10. ‡	. . . .	5.0

* Data for first two columns taken from Gilbert,[6] who states that they represent the best detection limits reported in the literature, not necessarily attainable on commercial instruments; data for third column represent results of experiments with the Perkin-Elmer Model 303, as reported by Kahn.[8]
† Nonflame absorption method.
‡ Band emission (rather than line).
§ Continuum.

source. In principle, either a continuous source, such as a xenon lamp, or a discharge lamp corresponding to the element sought, can be utilized to excite the fluorescence.

Several types of fluorescence can be distinguished, of which probably the most useful is *resonance fluorescence*, in which the atoms are excited by radiation of the same wavelength they emit. A difficulty is apt to arise here due to scattering of the primary radiation by droplets of solution or other particulate matter in the flame. To avoid the effects of scattering, excitation at a shorter wavelength may be utilized; for example, the 5890-Å line of sodium may be excited by irradiation with the 3303-Å line. Then an ultraviolet-absorbing filter will prevent scattered radiation from reaching the detector. The results of scattering can be reduced also by the introduction of crossed polarizers in the primary and secondary beams.

The study of atomic fluorescence in flames is still too new to assess fully. It looks quite promising and may well turn out to exceed comparable methods in sensitivity and convenience, for applicable systems.

PROBLEMS

8-1 Sodium can be measured in the 0.5- to 2-percent range by atomic absorption using the 3302.59- and 3302.94-Å nonresonant radiation from a zinc hollow-cathode lamp. This is about one fiftieth as sensitive as the sodium secondary resonances at 3302.32 and 3302.99 Å from a sodium lamp, and the brightness at this wavelength is about half as great for the zinc lamp as for the sodium lamp. The presence of zinc in the absorption flame gives no interference in the determination of sodium. (a) Account for the seeming contradiction that although the lamp is half as bright, the sensitivity is only one-fiftieth, zinc lamp compared to sodium lamp. (b) Why does not zinc interfere?

8-2 If a series of dilutions of an element gave the flame emission readings listed below, what corresponding readings might be found on a properly adjusted atomic absorption photometer with the same solutions, in terms of percent absorption, relative to the most concentrated solution?

Concentration, ppm	Power emitted, arbitrary units
0.0	0
3.2	21
6.4	38
9.6	51
12.8	62
16.0	70

REFERENCES

1. Burriel-Martí, F., and J. Ramírez-Muñoz: "Flame Photometry: a Manual of Methods and Applications," American Elsevier Publishing Company, Inc., New York, 1957.
2. Dean, J. A.: "Flame Photometry," McGraw-Hill Book Company, New York, 1960.
3. Elwell, W. T., and J. A. F. Gidley: "Atomic-Absorption Spectrophotometry," 2d ed., Pergamon Press, New York, 1966.
4. Fassel, V. A., V. G. Mossotti, W. E. L. Grossman, and R. N. Kniseley: *Spectrochim. Acta*, **22**:347 (1966).
5. Fristrom, R. M., and A. A. Westenberg: "Flame Structure," McGraw-Hill Book Company, New York, 1965.
6. Gilbert, P. T., Jr.: Advances in Emission Flame Photometry, p. 193, in "Analysis Instrumentation—1964," Plenum Publishing Corp., New York, 1964.
7. Goleb, J. A., and J. K. Brody: *Anal. Chim. Acta*, **28**:457 (1963).
8. Kahn, H. L.: *J. Chem. Educ.*, **43**:A7, A103 (1966).
9. Kirsten, W. J., and G. O. B. Bertilsson: *Anal. Chem.*, **38**:648 (1966).
10. Koirtyohann, S. R., and E. E. Pickett: *Anal. Chem.*, **38**:585 (1966).
11. Parsons, M. L., W. J. McCarthy, and J. D. Winefordner: *J. Chem. Educ.*, **44**:214 (1967).
12. Robinson, J. W.: "Atomic Absorption Spectroscopy," Marcel Dekker, Inc., New York, 1966.
13. Veillon, C., J. M. Mansfield, M. L. Parsons, and J. D. Winefordner: *Anal. Chem.*, **38**:204 (1966).
14. Walsh, A.: *Spectrochim. Acta*, **7**:108 (1955).
15. Wendt, R. H., and V. A. Fassel: *Anal. Chem.*, **38**:337 (1966).
16. West, P. W., P. Folse, and D. Montgomery: *Anal. Chem.*, **22**:667 (1950).
17. Willis, J. B.: *Nature*, **207**:715 (1965).
18. Winefordner, J. D., and T. J. Vickers: *Anal. Chem.*, **36**:161 (correction p. 789) (1964); J. D. Winefordner and R. A. Staab, *Anal. Chem.*, **36**:165 (1964).

9
X-ray Methods

When a beam of electrons impinges on a target material, the electrons in general will be slowed down by multiple interactions with the electrons of the target. The energy lost will be converted into a continuum of x-radiation, with a sharp minimum wavelength λ_{min} (maximum frequency) corresponding to the maximum energy of the electrons, which cannot be exceeded. This cut-off wavelength is given (in angstroms) by

$$\lambda_{min} = \frac{hc}{Ve} = \frac{12,400}{V} \qquad (9\text{-}1)$$

where h = Planck's constant
c = velocity of electromagnetic radiation in vacuo
e = electronic charge
V = accelerating potential across x-ray tube, in volts

As the potential is increased, a point is reached where the energy is sufficient to knock a planetary electron completely out of the target atom. Then as another electron falls back into the vacancy, a photon

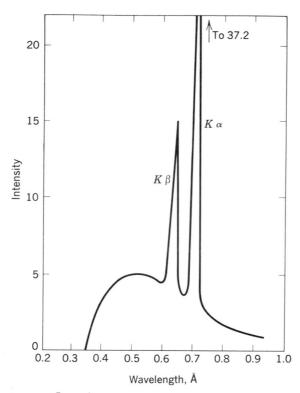

Fig. 9-1 Intensity curve for x-rays from a molybdenum target operated at 35 kV. (*John Wiley & Sons, Inc., New York.*)

of x-radiation is emitted with a wavelength dependent on the energy levels involved, and hence characteristic of the element. Since high energies are involved, the electrons closest to the nucleus are the principal ones affected. Thus a K electron may be ejected and its place taken by an electron from the L shell. Since these inner electrons are not concerned with the state of chemical combination of the atoms (except for the lighter elements), it follows that the x-ray properties of elements are independent of chemical combination or physical state. The wavelengths corresponding to such high energies are small, of the order of 10^{-2} to 10 Å.* The range of 0.7 to 2.0 Å includes the wavelengths most useful for analytical purposes.

The x-ray emission spectrum of a given target material will resemble that shown in Fig. 9-1, a continuum with superimposed discrete lines.

---

* X-ray wavelengths were formerly given in terms of *kX units*, where 1 kX unit = 1.00202 Å.

The sharp lines are designated $K\alpha$, $K\beta$, etc., for transitions to the K level.* At longer wavelengths other groups of lines corresponding to the L, M, and higher levels are found in the spectra of heavy elements. If the excitation is brought about by fluorescence, i.e., irradiation with x-rays of shorter wavelength, the continuum does not appear; only the characteristic lines are present. This is a favorable situation for x-ray emission analysis, greatly increasing the signal-to-noise ratio.

For analytical purposes x-radiation may be utilized in several distinct ways. (1) The absorption of x-rays will give information about the absorbing material, just as will absorption in other spectral regions. (2) The diffraction of x-rays permits analysis of crystalline substances with a high degree of specificity and accuracy. (3) Wavelength measurements will identify elements in the sample which is undergoing excitation. (4) Measurement of radiant power at a given wavelength can give a quantitative indication of the composition of the sample.

THE ABSORPTION OF X-RAYS

In common with other regions of the electromagnetic spectrum, x-rays can be absorbed by matter, and the degree of the absorption is controlled by the nature and amount of the absorbing material. The fundamental difference between the absorption of x-rays and that of radiation of longer wavelength is that the phenomenon is *atomic* rather than molecular. The absorption due to bromine, for example, depends only on the number of bromine atoms in the path of the rays, and this will be identical whether the bromine is in the form of a monatomic gas, a diatomic gas, a liquid, or a solid or is present in a compound, such as potassium bromide or bromobenzene.

The absorption follows Beer's law, which may be written in the form

$$P_x = P_0 e^{-\mu x} \tag{9-2}$$

or

$$\ln \frac{P_0}{P_x} = \mu x \tag{9-3}$$

for monochromatic x-rays, where P_0 is the initial power of the radiation and P_x is the power after passage through an absorbing sample, x cm in length. If the beam of x-rays has a cross section of 1 cm^2, then μ, which is called the *linear absorption coefficient*, represents the fraction of energy absorbed per centimeter. Frequently more convenient is a

* X-rays due to transitions from the L to the K shell are called $K\alpha$ x-rays, $K\alpha_1$ and $K\alpha_2$ corresponding to electrons originating in different sublevels of the L shell; x-rays due to transitions from the M to the K shell are called $K\beta$, etc.

mass absorption coefficient, defined as

$$\mu_m = \frac{\mu}{\rho} \tag{9-4}$$

where ρ is the density of the absorbing material. The coefficient μ_m is found empirically to be related to the wavelength and atomic properties of the absorbing substance by the formula

$$\mu_m = \frac{CNZ^4\lambda^n}{A} \tag{9-5}$$

where N is Avogadro's number, Z is the atomic number, A is the atomic weight of the absorbing element, λ is the wavelength, n is an exponent between 2.5 and 3.0, and C is a constant which is approximately the same for all elements, within limited regions, as will appear.

The variation of μ_m with wavelength follows an exponential law, so that if the logarithms are plotted, a straight line should result with slope equal to the exponent of λ. Figure 9-2 shows such a plot for the absorption coefficient of argon. The most striking feature of this graph is the discontinuity at $\lambda = 3.871$ Å. This is known as the K *critical absorption wavelength* of argon. Radiation of greater wavelength has insufficient energy to eject the K electrons of argon; hence, it is not absorbed so greatly as is radiation of slightly shorter wavelength. Larger atoms than argon show similar discontinuities at longer wavelengths corresponding to the photoelectric ejection of L and M electrons.

It is instructive to plot μ_m against the quantity Z^4/A for x-rays of a particular wavelength, as in Fig. 9-3, which shows the values for $CuK\alpha$ radiation for elements from sodium to osmium. The K edge and

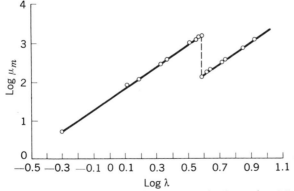

Fig. 9-2 The x-ray absorption of argon in the region 0.5 to 9.9 Å. *(From data of Compton and Allison.[6])*

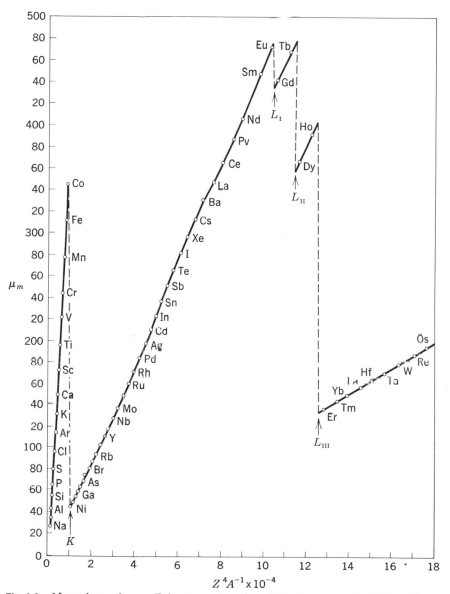

Fig. 9-3 Mass absorption coefficients as a function of atomic constants. The radiation is CuKα, 1.5418 Å. (*From data of Compton and Allison.*[6])

three L edges are evident.   The "constant" C of Eq. (9-5) changes value abruptly at the edge discontinuities, and is not perfectly constant between them, as shown by the slight curvature most easily visible between the K and L_I edges.   This curvature can alternatively be ascribed to variations in the exponent n.   Equation (9-5) will need refinement as our knowledge of the underlying phenomena improves.   Empirical absorption data for most elements are to be found in the literature.[6,13]

MONOCHROMATIC SOURCES

It is a simple matter to obtain narrow bands of wavelengths at various discrete points in the x-ray region, but not so simple to construct a monochromator which can be varied at will.   Narrow bands can be obtained by three methods: (1) by choosing characteristic emission

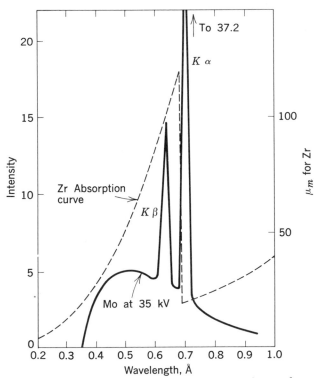

Fig. 9-4 The zirconium absorption curve superimposed on the 35-kV molybdenum emission spectrum results in the nearly monochromatic radiation of Mo$K\alpha$.   (*John Wiley & Sons, Inc., New York.*)

lines which are much stronger than the background, and isolating them with the aid of filters; (2) by means of a monochromator in which a crystal of known spacing acts as a diffraction grating, and (3) by using radioactive sources.

A monochromatic beam can often be obtained by means of a filter consisting of an element (or its compound) which has a critical absorption edge at just the right wavelength to isolate a characteristic line from a source target. For example, in Fig. 9-1, if a material could be found with a critical edge between the wavelengths of $MoK\alpha$ (0.709 Å) and $MoK\beta$ (0.632 Å), it would absorb the β- while passing the α-line. Such a filter can be made of zirconium, which has its K absorption edge at 0.689 Å (see Fig. 9-4). The ratio of the absorption coefficients of zirconium at 0.63 Å and 0.71 Å is only about 4.75, but the intensity of the $MoK\alpha$ is about 5.4 times that of $MoK\beta$, so that a fair degree of isolation can be achieved, the exact relation depending on the thickness of the filter. Table 9-1 lists filters suitable for $K\alpha$ isolation from several targets.

Table 9-1 Characteristic wavelengths and filters for x-ray tubes with commonly available targets[*]

Target	λ, $K\alpha_1$, Å	Element for filter	K-edge wavelength, Å	Thickness of foil, μ	Percent $K\beta_1$ absorbed
Cr	2.290	V	2.269	15.3	99.0
Fe	1.936	Mn	1.896	15.1	98.7
Co	1.789	Fe	1.743	14.7	98.9
Ni	1.658	Co	1.608	14.3	98.4
Cu	1.541	Ni	1.488	15.8	97.9
Mo	0.709	Zr	0.689	63.0	96.3
Ag	0.559	Pd	0.509	41.3	94.6
W	0.209	†			

[*] The data are from tabulations in Refs. 13 and 14.
† No suitable filter materials are commonly available for tungsten radiation.

CRYSTAL MONOCHROMATORS

If a greater degree of monochromaticity is required than can be obtained with filters, a grating monochromator must be employed. Design of these instruments will be detailed in connection with x-ray diffraction.

Such a monochromator could produce a narrow band of wavelengths at any point in the spectrum if a continuous source of x-rays were available. However, the continuous radiation produced at low tube voltages is not sufficiently powerful to be of much use, so for practical purposes we are still restricted to the various characteristic emission wavelengths. The advantage of the monochromator over the filter lies in its great reduction of background, which results in an improved signal-to-noise ratio in spite of a considerably attenuated signal.

RADIOACTIVE SOURCES

Radiations in the x-ray region are emitted from some radioactive elements by either of two mechanisms.[8] The first of these, correctly called *gamma radiation*, involves intranuclear energy levels, and so is not truly x-radiation. The other is known as *electron capture* (EC) or *K capture*. Since an electron in a K orbital has a finite probability of spending some of its time within or very close to the nucleus, there is in many atoms a finite probability of the capture of a K electron by the nucleus. This process lowers the atomic number by one unit and leaves a vacancy in the K shell. Hence true x-rays of the next lower element result, unaccompanied by any significant continuous radiation. Table 9-2 lists a few isotopes which have proven useful as x- or gamma-ray sources. Filters may still be needed to separate individual lines.

Table 9-2   Isotopes useful as sources of x-rays or gamma rays

Isotope	Half-life	Radiations*
^{55}Fe	2.60 yrs	EC: Mn$K\alpha$, 2.103 Å
^{57}Co	270 days	EC: Fe$K\alpha$, 1.937 Å; Fe$K\beta$, 1.757 Å
		Gamma: 0.861 Å, 0.1017 Å
^{60}Co	5.26 yrs	Gamma: 0.0106 Å, 0.00931 Å
^{241}Am	458 yrs	EC: Np$L\alpha$, 0.889 Å; Np$L\beta$, 0.698 Å; Np$L\gamma$, 0.597 Å
		Gamma: 0.208 Å, 0.470 Å

* The data are taken from References 4, 8, and 14, recalculated to angstroms where necessary. This table is not intended to be exhaustive.

Radioactive sources have the advantage of not requiring an elaborate high-voltage power supply and an evacuated tube with limited lifetime. On the other hand, they have the disadvantage that they cannot be turned off; radiation hazard is present at all times, whether an experiment is in progress or not.

ANALYSIS BY X-RAY ABSORPTION

As an analytical tool, x-ray absorption is of most value where the element to be determined is the sole heavy component in a material of low atomic weight. A number of important analyses fall in this category and make the method a significant one for industrial control purposes.

Lead, as the tetraethyl compound (TEL) in gasoline, can readily be determined by Mo$K\alpha$ x-ray absorption, for which the technique is admirably suited.[17] Lead absorbs x-rays more strongly than does carbon at this wavelength by a factor of about 250. The quantity of sulfur present will have an effect on the absorption, and generally the sulfur must be determined separately before the TEL analysis. A precision of ± 1 percent has been obtained on a routine basis on samples containing as little as 2 ppm of lead.

An example of absorption analysis with x-rays from [55]Fe is the determination of chlorine in organic compounds.[9] The mass absorption coefficients are 102 for chlorine, as compared with 0.2, 4.5, and 11.4 for hydrogen, carbon, and oxygen, respectively. Other heavy elements, such as phosphorus, sulfur, and bromine, must be absent or determined separately and corrected for. The results are reported to be accurate to within ± 4 percent (relative) for chlorine content below 0.5 percent. The minimum detectable concentration is of the order of 0.01 percent. The time required for an analysis, including preparation of the sample, the absorption measurement, and the calculation of results, but not including preparation of calibration curves, is commonly less than 5 min.

ABSORPTION EDGE ANALYSIS

Another method of applying x-ray absorption makes use of critical absorption edges as means of identification and quantitative analysis. Since the absorption of an element in a sample is markedly greater at a wavelength just below one of its absorption edges than just above it, and since the location of such edges on the wavelength scale is characteristic of the absorbing elements, a pair of absorption measurements bracketing

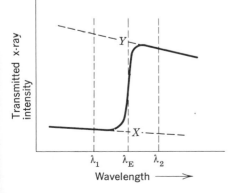

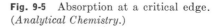

Fig. 9-5 Absorption at a critical edge. (*Analytical Chemistry.*)

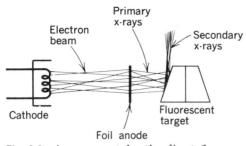

Fig. 9-6 Arrangement for the direct fluorescent excitation of x-rays. (*Review of Scientific Instruments.*)

the wavelength of the edge will serve to determine both the presence and the amount of the element sought.

If the power of an x-ray beam is plotted against wavelength in the vicinity of the K edge of an element, a curve like that of Fig. 9-5 will typically be obtained. The jump at λ_E, the wavelength of the absorption edge, is somewhat curved as shown, rather than vertical, because of the necessity of using slits of finite width to gain sensitivity. The quantity desired is the vertical distance between the intersections X and Y obtained by extrapolating respectively from lower and higher wavelengths. It is possible to estimate this height with good accuracy from measurements taken at two equally spaced wavelengths, λ_1 and λ_2. Mathematical details and justification for this short-cut procedure are given by Dunn,[7] who found that a relative error as low as 1 percent can be obtained at concentrations down to 0.1 percent for many elements.

ABSORPTION APPARATUS

There appears to be no general-purpose x-ray absorptiometer on the market, though General Electric formerly produced a double-beam, non-dispersive photometer with ionization-chamber detection. Laboratory analyses by x-ray absorption are carried out by the use of general-purpose equipment which will be described in a later section. For industrial-control purposes, apparatus is usually designed specifically for each installation.

A special fluorescent x-ray tube has been described which is particularly suited for the simultaneous absorptiometric analysis of several elements.[10] The usual electron beam strikes an anode of gold-plated copper foil (Fig. 9-6). The x-rays produced pass *through* the thin anode and strike a secondary target in the shape of the frustum of a cone. The cone is constructed of three segments of different elements with provision for turning it from one segment to another, thus providing a selection of wavelengths. The foil anode must be protected from overheating at the point of electron impact; this is accomplished by high-speed rotation of the entire ring-shaped anode. The elimination of the conventional window between the x-ray target and fluorescence generator increases the fluorescent power by as much as a hundredfold, particularly for light elements.

X-RAY DIFFRACTION[13, 16]

Since x-rays are electromagnetic waves of the same nature as light, they can be diffracted in a similar manner (Fig. 9-7). The equation given in Chap. 2 for diffraction by a grating

$$n\lambda = d \sin \theta \tag{9-6}$$

can also be applied to x-rays. In this case, however, the wavelength λ is smaller by a factor of 1000 or more, so that to get reasonable values of θ, the grating space d must also be made smaller by about the same factor. It is impracticable to rule a grating finely enough to meet this requirement, but it fortunately happens that the spacing between adjacent planes of atoms in crystals is of just the required order of magnitude. There is a wide variety of crystals suited for x-ray gratings; the most widely used include lithium fluoride, sodium chloride, calcite, gypsum, topaz, ethylenediamine d-tartrate (EDDT or EDT), and ammonium dihydrogen phosphate (ADP). In the simplest arrangement, the x-ray beam is reflected from a plane crystal, and the wavelength selected by varying the angle. Since the waves reflected at successive crystal planes must pass twice across the space between planes, Eq. (9-6) becomes the Bragg equation:

$$n\lambda = 2d \sin \theta \tag{9-7}$$

where d is now the distance between adjacent planes in the crystal. This equation can easily be derived by the student, with the aid of Fig. 9-7.

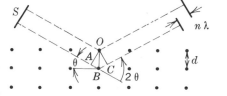

Fig. 9-7 Diffraction of x-rays by successive layers of atoms in a crystal.

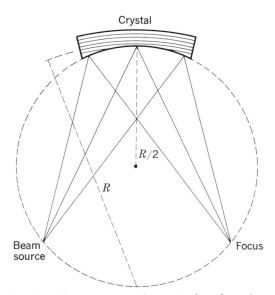

Fig. 9-8 An x-ray monochromator based on the Rowland circle. The crystal is bent and ground for accurate focus. (*John Wiley & Sons, Inc., New York.*)

It is also possible to employ concave gratings with Rowland-circle focusing.[13] For this purpose a crystal must be bent to conform to the geometric requirements. For improved focus, a crystal may first be bent so that its diffracting planes are curved with a radius equal to twice that of the Rowland circle, then ground so that the surface is given a curvature with radius equal to that of the circle (Fig. 9-8). The focusing monochromator is more expensive, largely because of the difficult hand-work required in shaping the crystal, but it can give monochromatic intensity perhaps 10 times that of a plane crystal.

The diffraction of x-rays is of greatest analytical interest as it is applied to the study of the crystalline material producing the diffraction. No two chemical substances would be expected to form crystals in which the spacing of planes is identical in all analogous directions; so a complete study, in which the sample assumes all possible angular positions in the path of the x-rays, should give a unique result for each substance.

DIFFRACTION APPARATUS

Equipment for x-ray diffraction is essentially comparable to an optical grating spectrometer. Lenses and mirrors cannot be used with x-rays, so the instrument is quite different in appearance from its optical analog.   A collimated beam (all rays parallel) can be obtained from an x-ray tube with an extended target by passage through a bundle of metal tubes or through a series of narrow slits, if collimation in one plane only is required. In some designs the emitting surface of the target is observed at a glancing angle (Fig. 9-9), which gives a close approximation to a line source with maximum intensity.   Such a tube can be mounted vertically, anode up, and beams can be taken through each of several ports in different horizontal directions, permitting two, three, or even four independent experiments to be carried on simultaneously.

The diffracted beam of x-rays may be detected photographically, electrically by means of the ionization which they produce in a gas, by scintillation counting, or by the photoelectric effect produced in the semiconductor elements germanium or silicon.   Detailed consideration of these electrical detectors will be postponed to Chap. 16.

The photographic method is typified by an apparatus known as a *Debye-Scherrer powder camera* (Figs. 9-10 and 9-11).   The sample is prepared in the form of a fine homogeneous powder, and a thin layer of it is inserted in the path of the x-rays.   The powder can be mounted on any noncrystalline supporting material, such as paper, with an organic mucilage or glue as adhesive.   The powdered sample contains so many particles that some are oriented in every possible direction relative to the beam of x-rays.   There will therefore be diffracted rays corresponding to all sets of planes in the crystals.   A strip of x-ray film is held in a circular position around the sample, as shown.   Upon development, this film will show a series of lines symmetrically arranged on both sides of the central spot produced by the undeviated beam.   The distance on the film from the central spot to any line will be a measure of the diffraction angle θ. Then, if the wavelength and order n are known, the spacing d can be calculated.

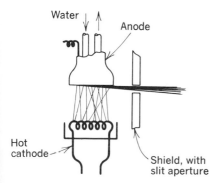

Fig. 9-9   The electrodes of a typical x-ray tube, showing the beam taken at a glancing angle.

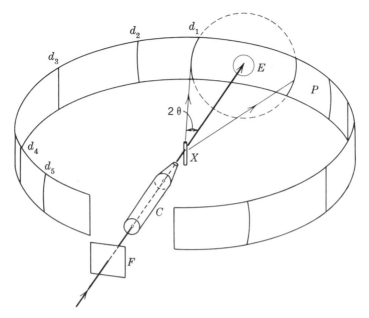

Fig. 9-10 Geometrical features of the Debye-Scherrer powder camera. (*John Wiley & Sons, Inc., New York.*)

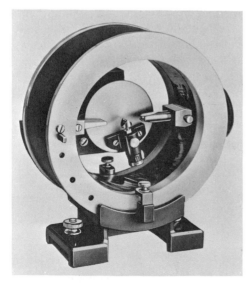

Fig. 9-11 Debye-Scherrer x-ray powder camera. (*General Electric Company, X-ray Department, Milwaukee, Wis.*)

Several examples of films from a powder camera are reproduced in Fig. 9-12. They are described specifically in the caption.

X-ray powder diffraction has been found a convenient means of identification of any compounds which can be obtained essentially pure in solid form.   It can provide a powerful tool in organic qualitative analysis, where both the compounds themselves and their chemical derivatives can be identified.

Ionization or scintillation detectors are most conveniently mounted in a housing which can be moved in a circular path around the sample. This apparatus is called a *goniometer* (Figs. 9-13 and 9-14).   In the model illustrated, the linear target of the x-ray tube provides a line source 0.06 by 10 mm in size with high intensity.   The angular aperture of the beam is indicated by the divergent lines in Fig. 9-14.   It is defined by a single divergence slit which also limits the primary beam to the specimen area.

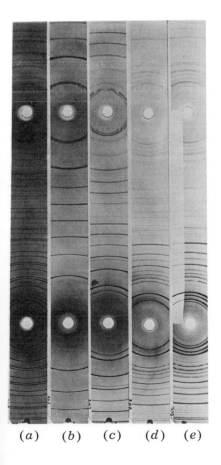

(a)   *(b)*   *(c)*   *(d)*   *(e)*

Fig. 9-12   X-ray diffraction patterns from films exposed in a powder camera.   The radiation was CuK; a nickel filter 0.0006 in. in thickness was employed for films *a* to *d* and for the left-hand portion of *e*. The specimen in each case was mounted on a 0.003-in. glass fiber, except in *e*, where it was packed in a 0.012-in.-diameter glass capillary.   (*a*) Pb(NO$_3$)$_2$; (*b*) W metal; (*c*) NaCl; (*d*) quartz; and (*e*) quartz.   (*Science.*)

An aperture of 1° is usually employed.   Flat specimens up to 10 by 20 mm can be accommodated, or a cylindrical sample can be rotated by a small motor if desired.   The receiving slit defines the width of the reflected beam detected by the counter.   Equally spaced thin metal foils (parallel-slit assembly) limit the divergence of the beam in any plane parallel to the

Fig. 9-13 Philips (Norelco) goniometer. The sample is mounted on the needle which projects to the left and is rotated by the motor which can be seen axially to the right.   The beam of x-rays enters from the slit beyond the sample (the x-ray tube is not shown).   The detector is located in the tubular housing at the top and can be turned in an arc around the sample by a motor not visible in the picture.   (*Philips Electronics, Inc., Mount Vernon, N.Y.*)

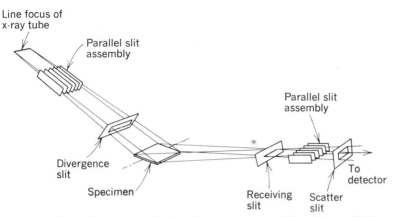

Line focus of
x-ray tube

Parallel slit
assembly

Parallel slit
assembly

Divergence
slit

Specimen

Receiving
slit

Scatter
slit

To
detector

Fig. 9-14 Optical system of the Norelco goniometer of Fig. 9-13. (*Philips Electronics, Inc., Mount Vernon, N.Y.*)

line source; two sets are used, as shown, which permit high resolution to be achieved. The scatter slit serves to reduce the background response caused by radiation other than that of the desired beam. The output from the detector is amplified and fed into a pen recorder, so that an automatic record is produced. As both the recording paper and the arm bearing the detector are turned by synchronous motors, the recorded graph may be interpreted as intensity of the diffracted beam plotted as a function of the angle of diffraction, usually denoted by 2θ. Several examples (to be described below) are given in Figs. 9-15 and 9-16.

X-ray diffraction leads primarily to the identification of *crystalline* compounds. Elements as such will be observed only if in the free crystalline state. This contrasts sharply with x-ray absorption and with emission spectroscopy, where the response is to the elements present, without much concern about their states of chemical combination.

For example, each of the oxides of iron gives its own particular pattern, and the appearance of a certain pattern proves the presence of that particular compound of oxygen and iron in the material being examined.

The power of a diffracted beam is dependent upon the quantity of the corresponding crystalline material in the sample. It is accordingly possible to obtain a quantitative determination of the relative amounts of the constituents of a mixture of solids.

An example of the identification of specific compounds in a mixture is shown in Figs. 9-15 and 9-16. In Fig. 9-15 are given the x-ray spectrograms of five minerals which may occur in soils, as determined on a Norelco Geiger-counter x-ray spectrometer. Figure 9-16 consists of

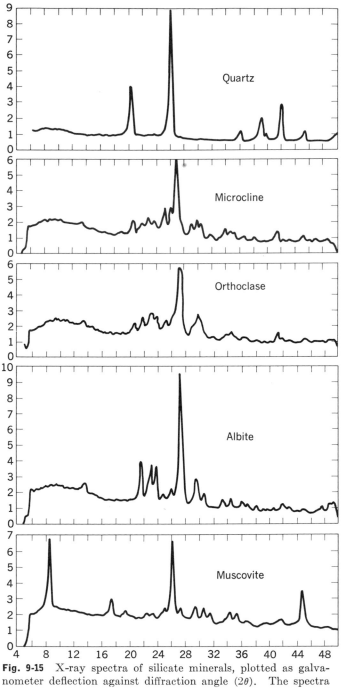

Fig. 9-15 X-ray spectra of silicate minerals, plotted as galvanometer deflection against diffraction angle (2θ). The spectra were taken with a Philips x-ray spectrometer. (*D. Van Nostrand Company, Inc., Princeton, N.J.*)

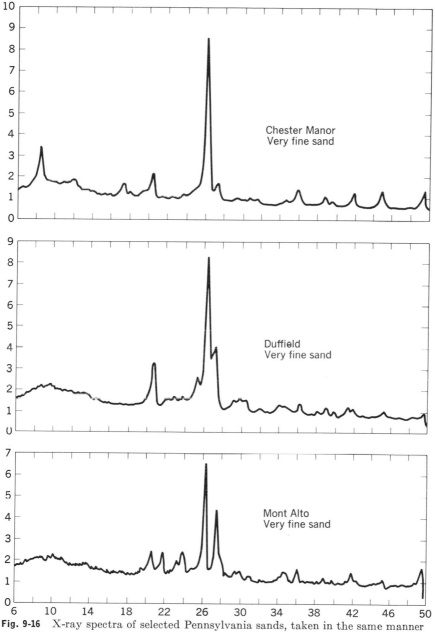

Fig. 9-16 X-ray spectra of selected Pennsylvania sands, taken in the same manner as the standards of Fig. 9-15. (*D. Van Nostrand Company, Inc., Princeton, N.J.*)

patterns of the fine sand components of representative soils, showing the considerable variations which may occur.   Quartz is a prominent mineral in all three, but the other constituent minerals are outstandingly different.

X-RAY EMISSION ANALYSIS

The fluorescent emission of x-rays provides one of the most potent tools available to the analyst for the study of metals and other massive samples. The chief limitation is in the analysis of light elements: those with atomic numbers less than 11 (sodium) so far are not detectable, and those below 20 (calcium), only with some difficulty.

Excitation of the sample is achieved by irradiation with a beam of primary x-rays of greater energy than the characteristic x-radiation which

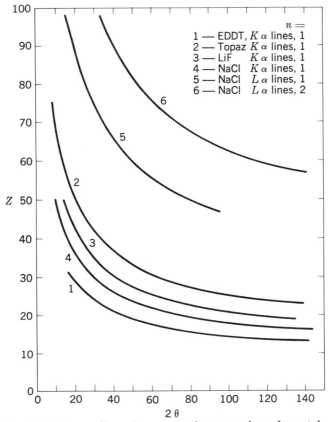

$$n =$$

1 — EDDT, $K\alpha$ lines, 1
2 — Topaz $K\alpha$ lines, 1
3 — LiF $K\alpha$ lines, 1
4 — NaCl $K\alpha$ lines, 1
5 — NaCl $L\alpha$ lines, 1
6 — NaCl $L\alpha$ lines, 2

Fig. 9-17 X-ray dispersion curves for a number of crystals. EDDT refers to ethylenediamine d-tartrate.

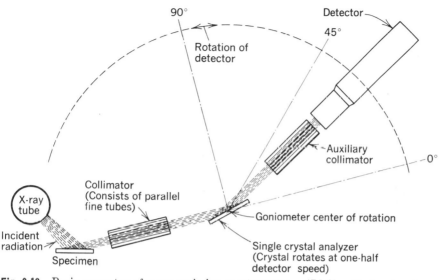

Fig. 9-18 Basic geometry of x-ray emission spectrometer. (*Philips Electronics, Inc., Mount Vernon, N.Y.*)

it is desired to excite in the sample. Because of the need for high-energy primary radiation, a tungsten-target tube is generally employed. The radiation need not be monochromatic.

Each heavy element in the sample is excited to emit the same frequencies which it would if it were made the target in a separate x-ray tube. Then by the methods of x-ray diffraction, the various wavelengths can be separated and measured. A diffraction apparatus operated with a crystal grating of known spacing is termed an *x-ray spectrometer*, and is strictly analogous to grating spectrometers for visible light. Several designs are possible which are similar in principle to the monochromators described earlier. Several analyzing crystals are frequently needed to cover different ranges, and must be provided with precision interchangeable mountings. The effectiveness of a crystal at separating radiations of the elements is given by a *dispersion curve*, a plot of the angle of deviation observed on the goniometer, as a function of the atomic number Z of the fluorescing element (Fig. 9-17). The data are usually presented in degrees of angle 2θ, which has the same significance as in Fig. 9-7. One design of spectrometer is shown in Fig. 9-18.

X-ray emission provides a convenient means of analysis for the high-alloy steels and for the heat-resistant alloys of the chromium-nickel-cobalt type. Analysis of low percentages is limited by the relationship between background radiation and the radiant power of the line

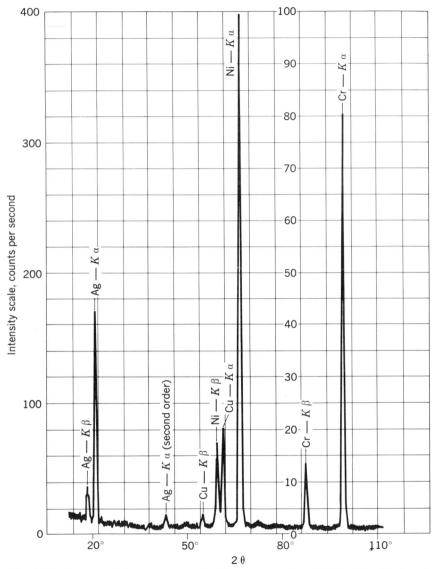

Fig. 9-19 X-ray emission spectrum of a silver-copper alloy plated with nickel and chromium. (*Philips Electronics, Inc., Mount Vernon, N.Y.*)

of the desired element. The effect of absorption of the emitted radiation of the element by the materials of the specimen is also a factor in determining the lowest percentages which can be determined. Samples in which the main constituent is an element of high atomic weight will absorb a higher percentage of the radiation than would be absorbed by a lighter element. Thus nickel in an aluminum alloy can be determined with greater sensitivity than is possible for nickel in steel or in a silver or lead alloy, where the absorption of the $NiK\alpha$ radiation is high. In favorable circumstances accuracy of the order of 0.5 percent of the element present can be achieved. The limit of detectability may be as low as a few parts per million.

Figure 9-19 is a reproduction of an automatic record showing an emission spectrum of a chromium-nickel plate on a silver-copper base. The method has been applied to the determination of hafnium in zirconium and tantalum in niobium.[1] It was possible, for example, to determine 1 percent of tantalum in niobium with a precision of ± 0.04 percent (i.e., ± 4 percent of the amount present).

Emission analysis has also been applied successfully to the determination of lead and bromine additives in aviation gasoline.[2] Figure 9-20 shows the results of this study. The liquid sample was held in a plastic container with windows of cellophane. The precision under the conditions employed was of the order of ± 10 percent for bromine and ± 1.5 percent for lead, both figures based on the amount present. This technique has the advantage over analysis by the absorption of x-rays in that it is much more specific; the presence of sulfur or chlorine has a negligible effect on the results.

The same authors[2] have reported a rapid method for analyzing uranium solutions down to 0.05 g per liter. In this case it was necessary to remove the water, which caused excessive scattering of the x-ray beam. They recommend evaporating the water by introducing a 1-ml

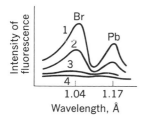

Fig. 9-20 Emission spectra of 2-methylheptane containing the following amounts of tetraethyl lead (TEL) and ethylene dibromide (EDB):

Curve	TEL, ml/gal	EDB, ml/gal
1	8.0	3.53
2	4.0	1.77
3	0.5	0.22
4	Blank	Blank

(*Analytical Chemistry.*)

sample dropwise into a heated shallow dish; each drop vaporizes almost immediately, leaving the solid salts in a suitable form for examination in the x-ray spectrometer.

NONDISPERSIVE X-RAY SPECTROMETERS[5]

Since the energy content of photons is reciprocally related to the wavelength:

$$E = h\nu = \frac{hc}{\lambda}$$

any method which will sort out photons as a function of energy will be capable of performing the equivalent of wavelength dispersion. Such sorting is possible with several types of electrical detectors known collectively as proportional counters (to be described in detail in Chap. 16).

This type of detector, based only on energy discrimination, receives all wavelengths simultaneously, rather than scanning them sequentially. The signals from the detector can be fed into an electronic integrating circuit, so that the corresponding voltages will steadily increase as the exposure to radiation is allowed to continue over a period of time. This means that the system can be made highly sensitive to x-rays of low power, several orders of magnitude better in sensitivity than the crystal spectrometer. Hence fluorescent emission excited by a less powerful source of primary radiation can be studied easily, and one of the less energetic radioactive sources, such as ^{241}Am, will be an appropriate choice. A small portion of active emitter is sufficient, and minimal safety shielding is required.

The resolution of a nondispersive instrument is only slightly less

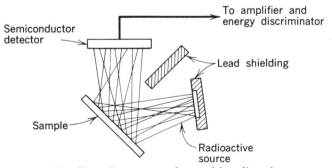

Fig. 9-21   Nondispersive x-ray analyzer with radioactive source. (*Science.*)

Fig. 9-22 X-ray photon spectroscopy system using the principle of Fig. 9-21. The large tank is a dewar which holds enough liquid nitrogen to last four days, to cool the lithium-drifted silicon detector. The sample is inserted in the housing below the dewar. The electronic unit is a pulse-height analyzer. (*Technical Measurement Corporation, North Haven, Conn.*)

than that of the crystal spectrometer, and it has considerable advantages in respect to lower cost and reduced power requirements, which make possible a battery-operated portable unit for field use;[12] it also permits observations of the K radiations of the heavy elements not usually accessible with an x-ray tube source, and it is not subject to overlapping spectral orders.

A nondispersive spectrometer has been introduced commerically by TII,* based on the original design published by Bowman et al.[4] An abbreviated schematic is shown in Fig. 9-21, a photograph of the TII unit in Fig. 9-22, and a representative spectrum in Fig. 9-23. This spectrometer uses a semiconductor detector (silicon or germanium, activated with lithium), which must be cooled for efficient operation. A

---

* Technical Instruments, Inc., North Haven, Conn., formerly Technical Measurement Corporation.

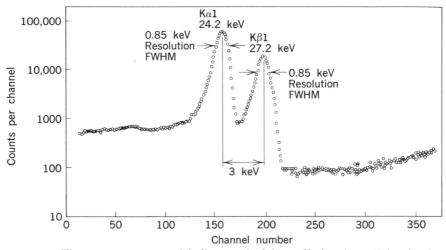

Fig. 9-23 The x-ray spectrum of indium excited by radiation from ^{241}Am in the apparatus of Fig. 9-22. The resolution is specified in terms of the full width of a peak at half the maximum height (FWHM). (*Technical Measurement Corporation, North Haven, Conn.*)

dewar flask, large enough to hold a four-day supply of liquid nitrogen, is incorporated.

This class of instrument is in its infancy at the time this is written, but it appears to have considerable potential for the future of x-ray analysis.

ELECTRON MICROPROBE ANALYSIS[18]

In the past few decades the methods of focusing electron beams (*electron optics*) have been developed to a high degree. It is possible to utilize a finely focused beam to excite x-rays in a solid sample as small as 1 μ^3 (10^{-12} cm^3). A schematic diagram of a typical apparatus is shown in Fig. 9-24. The electron beam originates at the electron gun at the top of the diagram, and is focused by specially shaped electromagnets onto the sample. (This section is very similar to an electron microscope used in reverse.) The x-rays produced are analyzed in a crystal spectrometer. An optical microscope must be incorporated, so that the operator can locate precisely the desired spot on the sample.

The detailed design is more complex than shown in the figure; it is not an easy problem to dovetail the electron system, the optical system, and the x-ray system so that they do not interfere with each other, and yet so that each can operate at its maximum efficiency. The electron beam must, of course, be in a highly evacuated chamber, which provides further complications. The associated electronics must include controls for the electron lenses, the electron gun, the x-ray detector, and a vacuum gage.

The complete instrument is therefore large, complex, and expensive. In spite of this, it provides such a wealth of information that a number of firms make such equipment. It is useful for phase studies in metallurgy and ceramics, for following the process of diffusion in the fabrication of transistors, for establishing the identity of impurities, inclusions, etc.

Until recently this was the only method available for qualitative and quantitative analysis of extremely small objects. With the advent of the laser an alternative method has become available; the laser beam can be focused on a tiny area, and carries enough power to vaporize most materials. The resulting puff of vapor can be analyzed by optical emission spectroscopy, by atomic absorption, or possibly by mass spectroscopy.

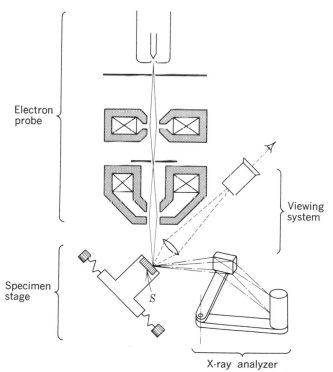

Fig. 9-24 Schematic view showing the basic elements of an electron probe microanalyzer. (*John Wiley & Sons, Inc., New York.*)

PROBLEMS

9-1 What are the dimensions of the constant C in Eq. (9-5)?

9-2 Determine from Figs. 9-15 and 9-16 what components are present in each of the three sands of Fig. 9-16 and estimate the approximate relative amounts, where possible.

9-3 A metallic sample is irradiated in an x-ray spectrograph with radiation from a tungsten-target tube. The spectrograph is equipped with a calcite crystal for which the grating space is 3.029 Å. A strong line is observed at an angle $2\theta = 34°23'$. Calculate the wavelength of the observed x-ray, and from a handbook listing, determine what element must be present. (The spectrum may be assumed to be in the first order.)

9-4 The μ_m values for cobalt at several wavelengths are given as

λ, Å	0.5	1.0	1.5	2.0	2.5	3.0
μ_m	15.5	110.0	345.0	87.0	177.0	270.0

The $NiK\beta$ radiation occurs at 1.50 Å. The density of cobalt is 8.9 g cm^{-3}. The intensity of $NiK\alpha$ is approximately 3 times that of $NiK\beta$ as emitted from the target.

By plotting the absorption coefficient for cobalt, determine its value at the wavelengths of $NiK\alpha$ and $NiK\beta$ radiations. With the data given above and that in Table 9-1, determine the ratio of $NiK\alpha$ to $NiK\beta$ which will penetrate a cobalt foil of 0.005-mm thickness.

REFERENCES

1. Birks, L. S., and E. J. Brooks: *Anal. Chem.*, **22**:1017 (1950).
2. Birks, L. S., and E. J. Brooks: *Anal. Chem.*, **23**:707 (1951).
3. Birks, L. S., E. J. Brooks, H. Friedman, and R. M. Roe: *Anal. Chem.*, **22**:1258 (1950).
4. Bowman, H. R., E. K. Hyde, S. G. Thompson, and R. C. Jared: *Science,* **151**:562 (1966); T. Hall, *Science,* **153**:320 (1966); H. R. Bowman and E. K. Hyde, *Science,* **153**:321 (1966).
5. Campbell, W. J., J. D. Brown, and J. W. Thatcher: *Anal. Chem.*, **38**:416R (1966).
6. Compton, A. H., and S. K. Allison: "X-rays in Theory and Experiment," D. Van Nostrand Company, Inc., Princeton, N.J., 1935.
7. Dunn, H. W.: *Anal. Chem.*, **34**:116 (1962).
8. Friedlander, G., J. W. Kennedy, and J. M. Miller: "Nuclear and Radiochemistry," 2d ed., John Wiley & Sons, Inc., New York, 1964.
9. Griffin, L. H.: *Anal. Chem.*, **34**:606 (1962).
10. Jacobson, B., and L. Nordberg: *Rev. Sci. Instr.*, **34**:383 (1963).
11. Jeffries, C. D.: The Minerals of the Soil, in D. E. H. Frear (ed.), "Agricultural Chemistry," vol. 1, chap. 22, D. Van Nostrand Company, Inc., Princeton, N.J., 1950.
12. Karttunen, J. O., H. B. Evans, D. J. Henderson, P. J. Markovich, and R. L. Niemann: *Anal. Chem.*, **36**:1277 (1964); J. O. Karttunen and D. J. Henderson, *Anal. Chem.*, **37**:307 (1965).

13. Klug, H. P., and L. E. Alexander: "X-ray Diffraction Procedures," John Wiley & Sons, Inc., New York, 1954.
14. Liebhafsky, H. A., H. G. Pfeiffer, E. H. Winslow, and P. D. Zemany: "X-ray Absorption and Emission in Analytical Chemistry," John Wiley & Sons, Inc., New York, 1960.
15. Parrish, W.: *Science*, **110**:368 (1949).
16. Rudman, R.: *J. Chem. Educ.*, **44**:A7, A99, A187, A289, A399, A499 (1967).
17. Vollmar, R. C., E. E. Petterson, and P. A. Petruzzelli: *Anal. Chem.*, **21**:1491 (1949).
18. Wittry, D. B.: X-ray Microanalysis by Means of Electron Probes, in I. M. Kolthoff and P. J. Elving (eds.), "Treatise on Analytical Chemistry," pt. I, vol. 5, chap. 61, Interscience Publishers (Division of John Wiley & Sons, Inc.), New York, 1964.

10

Polarimetry and Optical Rotatory Dispersion

In preceding chapters we have been concerned with radiant energy principally from the standpoint of its absorption and its wavelength distribution. Its wave nature has only interested us in connection with diffraction effects. We will now discuss phenomena concerned with polarized radiation.

Many transparent substances which are characterized by a lack of symmetry in their molecular or crystalline structure have the property of rotating the plane of polarized radiation. Such materials are said to be *optically active*. Probably the most familiar examples are quartz and the sugars, but many other organic and inorganic compounds also possess this property. The extent to which the plane is rotated varies widely from one active compound to another. The rotation is said to be *dextro-* (+) if it is clockwise to an observer looking toward the light source, and *levo-* (−) if counterclockwise. For any given compound, the extent of rotation depends upon the number of molecules in the path of the radiation or, in the case of solutions, upon the concentration and the length of the containing vessel. It is also dependent upon the wavelength of the radiation and the temperature. The *specific rotation*, represented by the symbol

$[\alpha]^t$, is defined by the formula

$$[\alpha]^t = \frac{\alpha}{dc} \tag{10-1}$$

where α is the angle (measured in degrees) through which the plane of the polarized light is rotated by a solution of concentration c grams of solute per milliliter of solution when contained in a cell of d decimeters in length. The wavelength is commonly specified as 5893 Å, the D line of a sodium-vapor lamp. Some representative values for specific rotation are given in Table 10-1.

Table 10-1 Specific rotations of solutions (at 20°C)

Active substance	Solvent	$[\alpha]_D^{20}$
Camphor	Alcohol	+ 43.8°
Calciferol (vitamin D_2)	Chloroform	+ 52.0
Calciferol (vitamin D_2)	Acetone	+ 82.6
Cholesterol	Chloroform	− 39.5
Quinine sulfate	0.5 F HCl	−220.
l-Tartaric acid	Water	+ 14.1
Sodium potassium tartrate (Rochelle salt)	Water	+ 29.8
Sucrose	Water	+ 66.5
β-d-Glucose	Water	+ 52.7
β-d-Fructose	Water	− 92.4
β-Lactose	Water	+ 55.4
β-Maltose	Water	+130.4

POLARIMETERS[1]

The most general instrument in this field is the *polarimeter*, Fig. 10-1. A typical manual instrument is shown schematically in Fig. 10-2. Monochromatic radiation from a sodium lamp is made parallel by a collimator and polarized by a calcite prism. Following the polarizer, there is placed a small auxiliary calcite arranged to intercept one-half of the beam. (The function of this feature will be explained below.) The radiation then passes through the sample, which is contained in a glass tube of known length, closed at both ends by clear glass plates, then through the analyzer to the eyepiece for visual observation.

In principle, the polarimeter could function without the small auxiliary prism. The polarizers would initially be crossed without any sample in the beam, and then again with the sample present. The angle through which the analyzer is turned would then be the quantity sought. However, this simple arrangement is unsatisfactory because it requires the observer to identify the position where the transmitted radiation is zero,

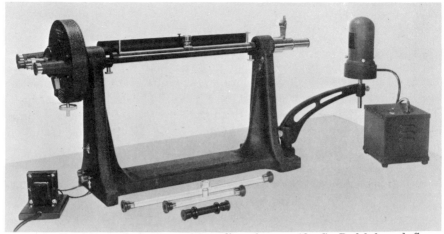

Fig. 10-1 Precision polarimeter, with sodium lamp. (*O. C. Rudolph and Sons, Fairfield, N.J.*)

which cannot be done with precision.   The added prism in half the beam makes it possible to avoid this difficulty.   It is permanently oriented with its polarizing axis at an angle of a few degrees from that of the polarizer. There is then a particular position of the analyzer at which the radiations passed in the two halves of the beam are just equal in power.   This provides a more satisfactory reference point than does the position of complete extinction, since the visual observation consists of matching exactly the powers of two half beams at some intermediate level, for which application the eye is well suited.

It is possible to design photoelectric polarimeters, with or without automatic recording features, and there are several such instruments on the market.   They are single-beam devices, differing primarily in the servomechanism or equivalent for rotating the plane of polarization to compensate for the rotation of the sample.

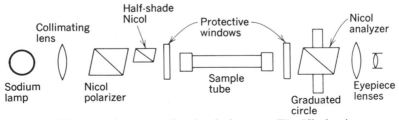

Fig. 10-2 Diagram of a conventional polarimeter.   The Nicol prisms are fabricated from calcite.

APPLICATIONS

The most extensive application of analysis by optical rotation is in the sugar industry.[2] In the absence of other optically active material, sucrose can be determined by a direct application of Eq. (10-1), which may be written in the form

$$c = \frac{\alpha}{d \cdot [\alpha]_D^{20}} = \frac{\alpha}{(2)(66.5)} = \frac{\alpha}{133.0} \tag{10-2}$$

for sucrose in the customary 2-dm tube, at a temperature of 20°C.

If, however, other active substances are present, a more elaborate treatment is required. Sucrose, alone among common sugars, can be made to undergo a hydrolysis reaction in the presence of acid, according to the equation

$$C_{12}H_{22}O_{11} + H_2O \xrightarrow{\text{acid}} C_6H_{12}O_6 + C_6H_{12}O_6$$

Sucrose	Glucose	Fructose
$[\alpha]_D^{20} = +66.5°$	$+52.7°$	$-92.4°$

The resulting mixture of glucose and fructose is called *invert sugar*, and the reaction *inversion*. During the inversion process, the specific rotation changes from $+66.5$ to $-19.8°$, corresponding to an equimolar mixture of the products. By measuring the rotation before and after inversion, it is possible to calculate the amount of sucrose present. The usual procedure is to start with a sample of 100-ml volume, measure its rotation, and then add 10 ml of concentrated hydrochloric acid. The acidified solution must stand at least 24 hr at 20°C, 10 hr at 25°C, or 10 min at 70°C, to ensure completion of the reaction. The rotation is then redetermined. It can be shown that under these conditions the change in observed angle of rotation $\Delta\alpha$ is given (at 20°C) by the relation

$$\Delta\alpha = \frac{360w_s[\alpha]_{D(I)}}{342(v + 10)} + \frac{w_x[\alpha]_{D(X)}}{(v + 10)} - \frac{w_s[\alpha]_{D(S)}}{v} - \frac{w_x[\alpha]_{D(X)}}{v} \tag{10-3}$$

where $[\alpha]_{D(I)}$ = specific rotation of invert sugar
$[\alpha]_{D(S)}$ = specific rotation of sucrose
$[\alpha]_{D(X)}$ = specific rotation of other active material which may be present
w_s = weight (grams) of pure sucrose in sample
w_x = weight of active impurity
v = original volume (100 ml)

From this relation, it follows that the weight of sucrose in the original sample is

$$w_s = -1.17\Delta\alpha - 0.00105[\alpha]_{D(X)}w_x \tag{10-4}$$

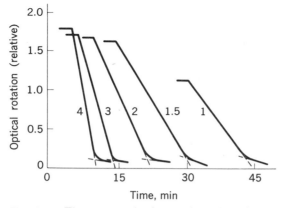

Fig. 10-3 The enzymatic destruction of penicillin with varying concentrations of penicillinase in a phosphate buffer at pH 7. Figures indicate relative concentrations of the enzyme. (*Analytical Chemistry.*)

If it is known that nearly all the active material present is sucrose, the second term becomes negligible. On the other hand, if one is determining the sucrose present as a minor constituent in a large portion of another active substance, the second term can be evaluated and the weight of sucrose calculated from the observed $\Delta\alpha$.

Another example of an analytical procedure based on optical rotation is the simultaneous determination of penicillin and the enzyme penicillinase.[8] Penicillin is destroyed quantitatively by the enzyme at a rate which is directly dependent upon the amount of enzyme present but independent of the penicillin concentration. A graph of rotation against time gives a straight line which terminates when the penicillin is all used up. The slope of the line is a measure of the concentration of the enzyme. Figure 10-3 shows such a graph for five values of enzyme concentration. It will be seen that the curves tail off, an effect due to secondary reactions. The true time of disappearance of penicillin is found by the intersection of the extrapolated straight portions. The penicillin concentration can be determined with a precision of about ± 1 percent, and the enzyme about ± 10 percent.

OPTICAL ROTATORY DISPERSION (ORD)[3, 4, 9]

The wavelength dependence of optical activity is a more fruitful source of structural information about asymmetric compounds than is the specific rotation at a single wavelength. It is closely related to the phenomenon called *circular dichroism* (*CD*).[1,10]

In Chap. 2 it was shown that an ordinary beam of light can be resolved into two plane-polarized beams. We will now carry that line of thought forward another step. Plane-polarized radiation can be further resolved into two beams said to be *circularly polarized* in opposite senses. The indices of refraction for a medium for the left- and right-hand circular components may not be the same, and will be designated n_L and n_R, respectively; the corresponding absorptivities are a_L and a_R. For an isotropic medium, such as glass or water, $n_L = n_R$ and $a_L = a_R$, and we say that the index and the absorptivity are independent of the state of polarization. If $n_L \neq n_R$, a phase difference appears between the two components, which is tantamount to saying that the plane of polarization has been rotated, i.e., that the angle $\alpha \neq 0$. If $a_L \neq a_R$, then one component is absorbed more strongly than the other, $a_L - a_R \neq 0$; the vector diagram representing the polarization then turns out to be elliptical, with the eccentricity θ a measure of $(a_L - a_R)$ (see Fig. 10-4).

It can be shown that the angle of rotation, as used in Eq. (10-1), is given by

$$\alpha = \frac{\pi}{\lambda} (n_L - n_R) \qquad\qquad (10\text{-}5)$$

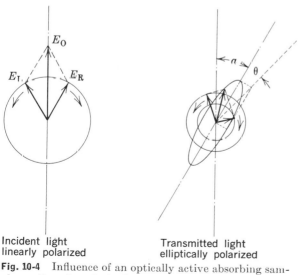

Incident light
linearly polarized

Transmitted light
elliptically polarized

Fig. 10-4 Influence of an optically active absorbing sample on linearly polarized light. (*Analytical Chemistry.*)

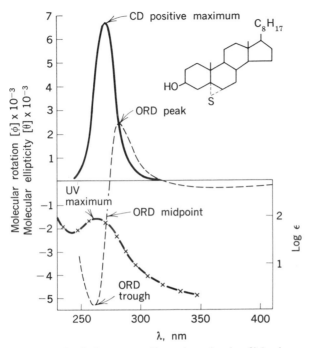

Fig. 10-5 Optical rotatory dispersion, circular dichroism, and ultraviolet absorption curves of 3β-hydroxycholestan-5α,6α-episulfide, illustrating current nomenclature practice. (*Journal of the Chemical Society, London.*)

and that the *ellipticity* θ is

$$\theta = \frac{1}{4}(a_L - a_R) \tag{10-6}$$

The quantity $(n_L - n_R)$ is called the *circular birefringence*, and $(a_L - a_R)$ the *circular dichroism* of the medium. The value of α can be measured directly with a polarimeter, but θ can only be determined indirectly through the circular dichroism.

It has been shown mathematically that the ORD curve and the circular dichroism are not independent; the relation between the two is the *Cotton effect*. Figure 10-5 shows the relation between these and also the ultraviolet absorption for the weak transition at 264 nm due to the episulfide group in an otherwise saturated steroid derivative. The relations are such that the midpoint between peak and trough of the ORD curve corresponds to the CD maximum, and that the ORD trough coincides in wavelength with the absorption maximum. The CD curves will often pinpoint "hidden" absorption maxima, and do so with less ambiguity than will the corresponding ORD curve.   On the other hand ORD curves are affected to a considerable degree by more distant chromophoric bands, and hence are more characteristic of specific compounds than CD. Both ORD and CD can give essential information about stereochemical features of optically active materials.   In practice, the relation between the several curves for the same substance may be more complex than the example in Fig. 10-5; as the theory is not within the scope of this book, the student is referred to the literature for further details.[3,6,10]

ORD PHOTOMETERS

There are several recording spectropolarimeters in present manufacture. They can be considered as modified single-beam spectrophotometers, as shown schematically in Fig. 10-6.   In this instrument (as in most others) the beam of light from a conventional monochromator passes sequentially through a polarizer, the sample, and an analyzer to a photomultiplier tube.   The beam is modulated with respect to its state of polarization at 12 Hz by a motor-driven device which causes the polarizer to rock back and forth through an angle of $\pm 1°$.   The servoamplifier responds only to the 12-Hz frequency, and causes the servomotor to adjust the analyzer continuously to the point where the 12-Hz signal will be symmetrically disposed about the null point.   The servomotor also positions the recording pen.

The Cary Model 60 spectropolarimeter is similar in principle, except in the method of modulating the beam, for which a *Faraday cell* is employed.   This is a vitreous silica rod surrounded by a coil carrying 60-Hz current, which causes a rotation of the plane of polarization through a few degrees at the 60-Hz frequency.   In this instrument, the analyzing

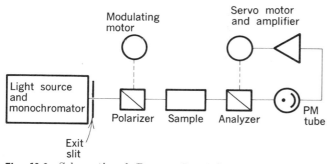

Fig. 10-6 Schematic of Durrum-Jasco Spectropolarimeter. (*Durrum Instrument Corporation, Palo Alto, Calif.*)

prism is immovable, and the servo system operates the polarizer to find the null point.

CD APPARATUS

Since conventional spectrophotometers are designed to determine the difference between the absorbances of a sample and a standard, they can easily be adapted to measure circular dichroism, which is likewise the difference between two absorbances.

Circular polarization is accomplished in two steps. Firstly the beam of radiation must be plane-polarized, and secondly the polarized beam must be passed through a device which will resolve it into right- and left-circularly polarized components, and retard one component relative to the other by exactly one-quarter wavelength. The most important circular resolvers are of two types: (1) those which depend on total internal reflections, such as a *Fresnel rhomb* (Fig. 10-7), and (2) the *Pockels electro-optical modulator*. In the latter, a high potential (in the kilovolt range) is applied across a plate of potassium dihydrogen phosphate or similar piezoelectric crystal cut perpendicularly to its optical axis. The first type has a somewhat limited wavelength range over which the retardation is near enough to one-quarter wavelength; in the Pockels type, the retardation can be altered by choice of potential, hence can be programmed to give continuously correct output as the wavelength is changed. The Pockels modulator with its power supply is much more elaborate and expensive than a Fresnel rhomb.

The Cary CD accessory for the Model 14 spectrophotometer utilizes two matched assemblies, each consisting of a polarizer, rhomb, and sample space, with the geometries so arranged that the radiation in the reference path is right-circularly polarized, while that in the sample path is left-

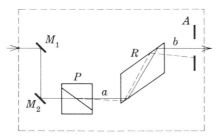

Fig. 10-7 Optical assembly for measurement of circular dichroism. Radiation enters at the left, is displaced downward by mirrors M_1 and M_2, plane-polarized by the compound prism P, and passed through the Fresnel rhomb R, where it undergoes two internal reflections which introduce a phase retardation of one-quarter wavelength, thus producing circular polarization. The mask A eliminates the extraordinary ray, while permitting the ordinary ray to pass. The entire unit fits into the sample chamber of a standard spectrophotometer; a second unit, oppositely oriented, is required for the reference path. The sample is placed at b for CD measurements, or at a to study plane-polarized transmission. (*Cary Instruments, Division of Varian Associates, Monrovia, Calif.*)

circularly polarized.   The CD adaptor for the Cary Model 60 spectropolarimeter makes use of a Pockels cell.

PROBLEMS

10-1   A 10.00-g sample of impure Rochelle salt is dissolved in enough water to make
100.0 ml of solution.   A 2-dm polarimeter tube is first filled with distilled water (to
establish the true zero of the scale), and then with a portion of the solution.   The
following observations are made at 20°C:

Scale setting for pure water	$+2.020°$
Scale setting for solution	$+5.750°$

What is the percent by weight of Rochelle salt in the sample?

10-2   Compute the percent error introduced into the determination of sucrose by
Eq. (10-4) with the second term omitted, in each of the following samples: (a) 30 g
sucrose + 2 g lactose; (b) 16 g sucrose + 16 g lactose; and (c) 2 g sucrose + 30 g
lactose.   The variation of the specific rotation of lactose between 15 and 20°C may
be neglected.

REFERENCES

1. Abu-Shumays, A., and J. J. Duffield: *Anal. Chem.*, **38**(7):29A (1966).
2. Bates, F. J., et al.: Polarimetry, Saccharimetry, and the Sugars, *Natl. Bur. Std.
   Circ.* C440, 1942.
3. Beychok, S.: *Science*, **154**:1288 (1966).
4. Djerassi, C.: "Optical Rotatory Dispersion: Applications to Organic Chemistry,"
   McGraw-Hill Book Company, New York, 1960.
5. Djerassi, C.: *Proc. Chem. Soc.* (London), **1964**:314.
6. Foss, J. G.: *J. Chem. Educ.*, **40**:592 (1963).
7. Gibb, Jr., T. R. P.: "Optical Methods of Chemical Analysis," McGraw-Hill
   Book Company, New York, 1942.
8. Levy, G. B.: *Anal. Chem.*, **23**:1089 (1951).
9. Struck, W. A., and E. C. Olson: Optical Rotation: Polarimetry, in I. M. Kolthoff
   and P. J. Elving (eds.), "Treatise on Analytical Chemistry," pt. I, vol. 6,
   chap. 71, Interscience Publishers (Division of John Wiley & Sons, Inc.), New
   York, 1965.
10. Velluz, L., M. Legrand, and M. Grosjean: "Optical Circular Dichroism," Aca-
    demic Press Inc., New York, 1965.

11

Introduction to Electrochemical Methods

A major series of analytical methods are based on the electrochemical properties of solutions. Consider a solution of an electrolyte contained in a glass vessel and in contact with two metallic conductors. It is possible to connect this cell to an outside source of electric power, and, unless the voltage is too low, this will usually cause a current to flow through the cell. On the other hand, the cell itself may act as a source of electrical energy and produce a current through the external connections. These effects for any specific cell may depend in both nature and magnitude on the composition of the solution, on the materials of which the electrodes are made, on mechanical features, such as electrode size and spacing, presence or absence of stirring, on the temperature, and on the properties of the external electrical circuit, direction of current flow, etc.

We shall consider first the properties of various types of electrodes and their interactions with solutions, then summarize the many possible ways of utilizing them.

ELECTRODES

For some applications an *inert electrode* is required solely to make electrical contact with the solution, without entering into a chemical reaction with any component. A noble metal, usually platinum, sometimes gold or silver, is most suitable, though in some instances an electrode of carbon will give good results.

In other situations, an *active electrode* is appropriate, that is to say, an electrode made of an element in its uncombined state, which will enter into a chemical equilibrium with ions of the same element in the solution. The magnitude of some of the electrical properties of the electrode (for example its potential) will be governed by the concentration of the corresponding ion. Among the commonly used electrodes of this class are silver, mercury, and hydrogen. A gas electrode (such as the hydrogen electrode) consists of a platinum wire or foil to conduct electricity with the gas bubbling over its surface; this combination acts as though it were made of the gaseous element alone. In principle, an electrode can be constructed for any element which is capable of existence in the form of simple ions, but in practice there are many restrictions. The more active elements are seldom employed as electrodes because of the obvious difficulty in preventing direct chemical attack. Hard metals such as chromium and iron tend to have inhomogeneous and nonreproducible surfaces, which impair their utility.

Other so-called "electrodes," such as the calomel and glass electrodes, are in reality combinations of inert or active elementary electrodes with appropriate compounds, which are sometimes fabricated into convenient units for insertion into a cell. They are more correctly termed "half-cells" and will be treated as such.

THE CELL REACTION

Whenever a direct current passes through an electrolytic cell, an oxidation-reduction reaction takes place. At one electrode, defined as the *anode*, oxidation occurs with transfer of electrons from the reduced species to the electrode; at the *cathode*, reduction takes place and electrons are transferred from the electrode to the oxidized species. The primary function of the external circuit is to convey the electrons from anode to cathode. The electric circuit is completed by ionic conduction through the solution.

A generalized redox reaction can be written as

$$rA_{red} + sB_{ox} + \cdots \rightleftharpoons pA_{ox} + qB_{red} + \cdots$$

where the subscripts "red" and "ox" refer respectively to the reduced and oxidized forms of substances A and B. For simplicity, we shall restrict the discussion to the case where A and B are the only substances oxidized or reduced (i.e., eliminate the "$+ \cdots$").

The equilibrium constant K is defined as

$$K = \frac{(A_{\text{ox}})_{\text{eq}}{}^p \, (B_{\text{red}})_{\text{eq}}{}^q}{(A_{\text{red}})_{\text{eq}}{}^r \, (B_{\text{ox}})_{\text{eq}}{}^s} \tag{11-1}$$

in which the parentheses denote molar activities, and the subscript "eq" emphasizes the fact that these are equilibrium quantities. We can also define a quantity Q, the activity quotient, as

$$Q = \frac{(A_{\text{ox}})_{\text{act}}^p \, (B_{\text{red}})_{\text{act}}^q}{(A_{\text{red}})_{\text{act}}^r \, (B_{\text{ox}})_{\text{act}}^s} \tag{11-2}$$

The notation "act" signifies that these quantities are the actual values in an experiment, not necessarily the equilibrium values. Now it can be shown from thermodynamic considerations that the change in free energy (maximum available work at constant temperature and pressure) is given by

$$\Delta G = RT \ln Q - RT \ln K \tag{11-3}$$

where R is the universal gas constant (8.316 J per mole-deg) and T is the Kelvin temperature. In electrochemical reactions, the free energy is electrical in nature and related to electrical quantities through the expression

$$\Delta G = -nFE_{\text{cell}} \tag{11-4}$$

where E_{cell} = potential of cell in volts

F = Faraday constant, approximately 96,500 coulombs per equivalent

n = number of electrons transferred for one formula unit of the reaction

Thus

$$E_{\text{cell}} = -\frac{\Delta G}{nF} = -\frac{RT}{nF} \ln Q + \frac{RT}{nF} \ln K \tag{11-5}$$

By substitution of the defined values of K and Q, Eqs. (11-1) and (11-2), into (11-5), followed by rearrangement of the logarithmic terms, we can show that

$$E_{\text{cell}} = \left[\frac{RT}{nF} \ln \frac{(A_{\text{ox}})_{\text{eq}}{}^p}{(A_{\text{red}})_{\text{eq}}{}^r} - \frac{RT}{nF} \ln \frac{(A_{\text{ox}})_{\text{act}}^p}{(A_{\text{red}})_{\text{act}}^r} \right]$$

$$- \left[\frac{RT}{nF} \ln \frac{(B_{\text{ox}})_{\text{eq}}{}^s}{(B_{\text{red}})_{\text{eq}}{}^q} - \frac{RT}{nF} \ln \frac{(B_{\text{ox}})_{\text{act}}^s}{(B_{\text{red}})_{\text{act}}^q} \right] \tag{11-6}$$

Now let us define $E_A^\circ = \dfrac{RT}{nF} \ln \dfrac{(A_{red})_{eq}{}^r}{(A_{ox})_{eq}{}^p}$, and similarly for E_B°, which gives

$$E_{cell} = \left[E_B^\circ - \frac{RT}{nF} \ln \frac{(B_{red})_{act}^q}{(B_{ox})_{act}^s} \right] - \left[E_A^\circ - \frac{RT}{nF} \ln \frac{(A_{red})_{act}^r}{(A_{ox})_{act}^p} \right] \qquad (11\text{-}7)$$

By this procedure, we have separated into two terms the effects of the two substances A and B on the potential of the cell. We can carry this one step further by defining what we may call *half-cell potentials*

$$E_A = E_A^\circ - \frac{RT}{nF} \ln \frac{(A_{red})_{act}^r}{(A_{ox})_{act}^p} \qquad (11\text{-}8)$$

$$E_B = E_B^\circ - \frac{RT}{nF} \ln \frac{(B_{red})_{act}^q}{(B_{ox})_{act}^s} \qquad (11\text{-}9)$$

Then, from Eq. (11-7)

$$E_{cell} = E_B - E_A \qquad (11\text{-}10)$$

Equations such as (11-8) and (11-9) were first introduced by Walther Nernst, and are frequently referred to as *Nernst equations*.

Just as the expression for the E of the cell has been broken down into two portions, so the chemical equation for the cell reaction can be separated into two portions, called *half-reactions*

$$rA_{red} \rightleftharpoons pA_{ox} + ne^-$$
$$sB_{ox} + ne^- \rightleftharpoons qB_{red}$$

where e^-, as usual, symbolizes an electron.

It is convenient, particularly for purposes of tabulation, to write all half-reactions with the electrons on the same side. We will follow recent practice and write them as *reductions:*

$$pA_{ox} + ne^- \rightleftharpoons rA_{red} \qquad (11\text{-}11)$$

$$qB_{ox} + ne^- \rightleftharpoons sB_{red} \qquad (11\text{-}12)$$

To obtain the complete cell reaction, the two half-reactions must be multiplied through by suitable numerical factors to make the n's equal. One half-reaction is then *subtracted* algebraically from the other.

The potentials E_A and E_B are, of course, associated with these half-reactions. The corresponding quantities E_A° and E_B° are called the *standard potentials* for the half-reactions, and in principle can be evaluated by setting up a half-cell in which the several activities are so chosen that their ratio as it occurs in the Nernst equation is unity and the logarithmic term becomes zero.

No valid method has been discovered for determining an *absolute* potential of a single electrode, since a measurement always requires a second electrode.   Therefore it is necessary to choose an electrode to be assigned arbitrarily to the zero position on the scale of potentials.   The *normal hydrogen electrode* (*NHE*) has been selected for this purpose.   This is an electrode in which hydrogen gas at a partial pressure of 1 atm is bubbled over a platinized platinum foil immersed in an aqueous solution in which the activity of the hydrogen ion is unity.   The NHE is defined to show zero potential at all temperatures.

A graphical display of the Nernst equation for a number of metals and hydrogen is presented in Fig. 11-1.   The potentials are referred for convenience to both the NHE and the saturated calomel electrode (SCE), which will be defined later.   The horizontal axis is marked off in units of pION, defined in a manner analogous to pH, the negative common logarithm of the activity of the ion.   The standard potentials occur at the intersections of the respective curves with the pION = 0 line.   The dashed lines indicate roughly the potential values which apply if pION is taken in terms of formal concentration rather than activity.   The deviations between straight and curved portions represent the effect of the activity corrections, and indeed potential measurements provide one of the most useful methods of determining activity coefficients.   It will be noted that activities and concentrations become indistinguishable at about 10^{-3} or 10^{-4} F.

The particular elements whose potentials are included in Fig. 11-1 are those which most closely follow the Nernst relation.   Others tend to show considerable deviation due to complexation, overvoltage, and other effects.   For transition metals, particularly, standard potentials are less useful because the determination of activities usually requires data which are not available.   For this reason *formal potentials* are often substituted.   Let us rewrite Eq. (11-8) showing activity coefficients γ:[*]

$$
\begin{aligned}
E_A &= E_A^\circ - \frac{RT}{nF} \ln \frac{[A_{\mathrm{red}}]^r \gamma_{A(\mathrm{red})}^r}{[A_{\mathrm{ox}}]^p \gamma_{A(\mathrm{ox})}^p} \\
&= E_A^\circ - \frac{RT}{nF} \ln \frac{\gamma_{A(\mathrm{red})}^r}{\gamma_{A(\mathrm{ox})}^p} - \frac{RT}{nF} \ln \frac{[A_{\mathrm{red}}]^r}{[A_{\mathrm{ox}}]^p} \qquad (11\text{-}13) \\
&= E_A^{\circ\prime} - \frac{RT}{nF} \ln \frac{[A_{\mathrm{red}}]^r}{[A_{\mathrm{ox}}]^p}
\end{aligned}
$$

The quantity $E^{\circ\prime}$ is defined as the formal potential.   The composition of the medium in which the measurements were made must be

---

[*] We will drop the subscript "act" from Eq. (11-8); square brackets [. . .] denote concentrations, while parentheses continue to signify activities.

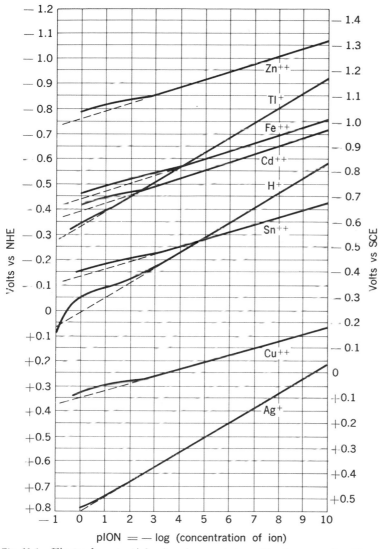

Fig. 11-1 Electrode potentials of various metals and hydrogen as functions of the ionic activities. For the significance of the dashed curves, see text.

specified carefully whenever formal potentials are employed. The values in Table 11-1, taken from the compilation by Meites,[5] show how greatly the formal potentials of a system can vary with the nature of the medium. The first entry is the accepted *standard* potential.

The standard and formal potentials for half-cells are such important

Table 11-1 Formal potentials of the system
Fe^{III}/Fe^{II} (at 25°C)

Medium	$E^{\circ\prime}$, V vs. NHE
Standard	$+0.771$
$1\ F$ $HClO_4$	$.735$
$0.5\ F$ HCl	$.71$
$1\ F$ H_2SO_4	$.68$
$5\ F$ HCl	$.64$
$10\ F$ HCl	$.53$
$2\ F$ H_3PO_4	$.46$
$0.5\ F$ Na_2 Tartrate (pH 5–6)	$.07$
$1\ F$ $K_2C_2O_4$ (pH 5)	$.01$
$10\ F$ NaOH	-0.68

and useful constants that a great many of them have been determined with precision and are available in tables.

The formal potentials of a number of redox systems are shown in Fig. 11-2, plotted as functions of the normal concentrations of HCl, H_2SO_4, or H_3PO_4.[1,6] Note that the sequence of potentials of various systems may change with acid concentration. For example, Fe(III) will oxidize As(III) in $2\ F$ HCl, but in $8\ F$ HCl, Fe(II) will be oxidized by As(V). It must be remembered that relative potentials are *thermodynamic* data, and have nothing to say about reaction rates. Although Ce(IV) has a much higher reduction potential than As(V), it will not oxidize As(III) unless a suitable catalyst is present.

In many situations it is convenient or even mandatory to separate physically the two electrodes of a cell, always maintaining electrolytic contact between them. This contact may be made in the pores of a porous cup or fritted glass barrier, or it may be through the medium of a salt bridge. Such a separation of the electrodes is necessary when the electrolytes of the two half-cells are incompatible, or when the redox reaction would occur directly without transfer of electrons via the external circuitry if the electrolyte of one half-cell were to come into contact with the other electrode.

Thus it happens that we frequently employ half-cells which not only have their own half-reactions and half-cell potentials but also have independent physical existence. It must always be remembered, however, that *one* half-cell is of no use; any application must involve at least two.

There are several types of half-cells of particular importance. In the general half-reaction, such as Eq. (11-11), one or both substances may be in solution, one may be insoluble (such as a free metal), one may be an

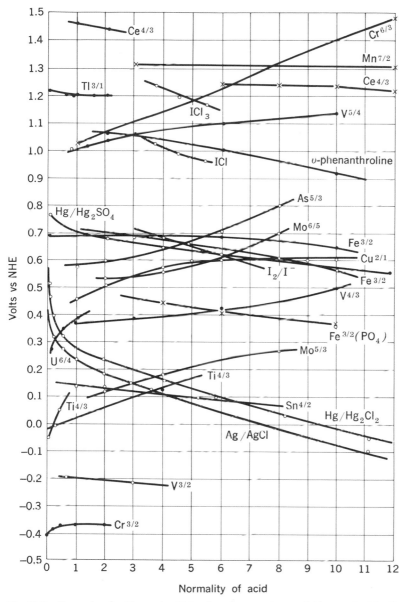

Fig. 11-2 Formal reduction potentials of various systems relative to the NHE. The notation $M^{4/3}$ denotes the potentials pertaining to the equilibrium $M(IV) + e^- \rightleftharpoons M(III)$. Open circles ($\circ$) denote chloride solutions, closed circles ($\bullet$) sulfates, and crosses ($\times$) phosphates. (*After Furman,*[1] *from John Wiley & Sons, Inc., New York, with addition of curves for* $Cr^{6/3}$, $Mn^{7/2}$, *and* $Ce^{4/3}$ *phosphates, after Rao and Rao,*[6] *from Talanta.*)

ion in equilibrium with a slightly soluble salt in the presence of excess solid phase, one or both may be participants in other equilibria, such as complex formation, with some substance which does not enter directly into the redox reaction. A few examples follow.

(1) *A Metal in Equilibrium with Its Ions (Class I Electrodes)*

$$Zn^{++} + 2e^- \rightleftharpoons Zn \qquad E° = -0.763 \text{ V}$$
$$Cu^{++} + 2e^- \rightleftharpoons Cu \qquad E° = +0.337 \text{ V}$$
$$Ag^+ + e^- \rightleftharpoons Ag \qquad E° = +0.799 \text{ V}$$

The oxidized form (A_{ox}) is the cation, the reduced form (A_{red}) the free metal. The $E°$ values quoted are in volts relative to the NHE at 25°C. The electrolyte for these half-cells is ordinarily a solution of a salt of the metal with the anion of a strong mineral acid, its selection dictated by solubilities and complexing tendencies: the sulfates, nitrates, and perchlorates are often appropriate. Equation (11-8) takes the form

$$E_A = E_A° + \frac{RT}{nF} \ln (A_{ox})$$

because the activity of a pure solid element is always taken as unity. The quantity (A_{ox}) refers to the activity of the simple cationic species; in the presence of a complex-forming substance this will be less, often much less, than the total amount of the metal in solution.

(2) *A Metal in Equilibrium with a Saturated Solution of a Slightly Soluble Salt (Class II Electrodes)*

$$AgCl_{(s)} + e^- \rightleftharpoons Ag + Cl^-_{(a=1)} \qquad E° = +0.2222 \text{ V}$$
$$Hg_2Cl_{2(s)} + 2e^- \rightleftharpoons 2Hg + 2Cl^-_{(a=1)} \qquad E° = +0.2676 \text{ V}$$

Half-cells of this type are widely used as *reference electrodes,* which in effect constitute secondary standards to replace the inconvenient NHE. For such service the activity (or concentration) of the anion is established at a selected value by the addition of a solution of a soluble salt with the same anion; in the examples cited, a solution of potassium chloride is usually chosen. The properties required of a practical reference electrode include ease of fabrication, reproducibility of potential, and low temperature coefficient. A few common reference electrodes are

Saturated calomel electrode (SCE):

$$Hg_2Cl_{2(s)} + 2e^- \rightleftharpoons 2Hg + 2Cl^-_{(sat'd\ KCl)} \qquad E = +0.246 \text{ V}$$

Normal calomel electrode (NCE):

$$Hg_2Cl_{2(s)} + 2e^- \rightleftharpoons 2Hg + 2Cl^-_{(1\ N\ KCl)} \qquad E = +0.280 \text{ V}$$

Normal silver–silver chloride electrode:

$$AgCl_{(s)} + e^- \rightleftharpoons Ag + Cl^-_{(1\ N\ KCl)} \qquad\qquad E = +0.237\ V$$

(3) *A Metal in Equilibrium with Two Slightly Soluble Salts with a Common Anion (Class III Electrodes)*

$$\begin{cases} Ag_2C_2O_{4(s)} \rightleftharpoons 2Ag^+ + C_2O_4^{--} \\ Ca^{++} + C_2O_4^{--} \rightleftharpoons CaC_2O_{4(s)} \end{cases}$$

A half-cell of this type can serve as a measure of the activity of Ca^{++}. It is a requirement that the second salt (CaC_2O_4) must be slightly more soluble than the first ($Ag_2C_2O_4$). The most widely applicable electrode which can be placed in this class is that involving the equilibria between EDTA, Hg^{++} ion, and the ion of a di-, tri-, or tetravalent metal; the slightly dissociated complexes play the same roles as the slightly soluble oxalate salts in the above example.

(4) *Two Soluble Species in Equilibrium at an Inert Electrode*

$$Fe^{3+} + e^- \rightleftharpoons Fe^{++} \qquad E^\circ = +0.771\ V$$
$$2Hg^{++} + 2e^- \rightleftharpoons Hg_2^{++} \qquad E^\circ = +0.920\ V$$
$$Ce^{4+} + e^- \rightleftharpoons Ce^{3+} \qquad E^\circ = +1.61\ V$$

The only function of the platinum electrode is to transport electrons to or from the ions in the solution. The E° values refer to conditions such that the activities of the two ionic species are equal, which does not necessarily mean that the total concentrations of oxidized and reduced forms of the element are equal. In the presence of complex formers, especially, the tendency of the metal in the two oxidation states to form complexes may not be the same, so that an E° value based on total concentrations rather than activities of free ions can be either larger or smaller than the quoted values.

SIGN CONVENTIONS

We have chosen to write half-reactions as reductions, which gives E° values which are negative with respect to the NHE for metals which are more powerful reducing agents than hydrogen, and positive E° values for those which are less powerful:

$$Zn^{++} + 2e^- \rightleftharpoons Zn \qquad E^\circ = -0.763\ V$$
$$2H^+ + 2e^- \rightleftharpoons H_2 \qquad E^\circ = 0\ V$$
$$Cu^{++} + 2e^- \rightleftharpoons Cu \qquad E^\circ = +0.337\ V$$

The E° values so stated are known as *standard electrode potentials*. This is consistent with experiment, for if a galvanic cell is constructed with electrodes of copper and zinc, each in contact with its own ions, the copper is the one which is found to be positive, the zinc negative, as observed with any ordinary voltmeter or potentiometer. If we write a half-reaction as an oxidation, then the sign of the associated potential must be reversed, but it should *not* be called the "electrode potential." There are two conflicting conventions for determining the sign of the potential of an electrode. The convention followed here is in accord with the recommendations of the International Union of Pure and Applied Chemistry (IUPAC) meeting in Stockholm in July, 1953. A complete account of the several sign conventions and of the IUPAC recommendations can be found in a paper by Licht and deBéthune.[4]

REVERSIBILITY

The terms *reversible* and *irreversible* are used with several different meanings, according to the context. In a purely chemical sense, a reaction is irreversible if its products either do not react with each other, or do so in such a way as to give products other than the original reactants. As an electrochemical example, consider a cell made up of zinc and silver–silver chloride electrodes in dilute hydrochloric acid. If this cell is short-circuited by an external connection, the half-reactions which occur are

$$Zn \rightarrow Zn^{++} + 2e^-$$
$$2AgCl + 2e^- \rightarrow 2Ag + 2Cl^-$$

and the entire cell reaction is

$$Zn + 2AgCl \rightarrow 2Ag + Zn^{++} + 2Cl^-$$

However, if this cell be connected to a source of electricity at a high enough voltage to force current through it in the reverse direction, the reactions will be

$$2H^+ + 2e^- \rightarrow H_2$$
$$2Ag + 2Cl^- \rightarrow 2AgCl + 2e^-$$

and the overall reaction is

$$2Ag + 2H^+ + 2Cl^- \rightarrow 2AgCl + H_2$$

Thus it is seen that the silver–silver chloride electrode is reversible while the zinc–hydrochloric acid half-cell is not.

In a *thermodynamic* sense, a reaction is reversible only if an infinitesimal change in driving force will cause a change in direction, which is equivalent to saying that the system is in thermodynamic equilibrium.

This implies that the reaction is fast enough to respond instantaneously to any small change in an independent variable. Thermodynamic reversibility is an ideal state which real systems may approximate more or less closely. If an electrochemical reaction is rapid enough that the departure from equilibrium is negligible, it can be considered reversible. A given reaction may be effectively reversible when observed by one technique (as potential measurement with no current flowing) and yet deviate noticeably from reversibility when studied under slowly changing conditions, as in polarography, and become "totally" irreversible when subjected to rapid changes, as in certain fast-scan or ac procedures.

POLARIZATION

An electrode (and hence a cell) is said to be *polarized* if its potential shows any departure from the value which would be predicted from the Nernst equation. This circumstance may occur, for example, when an arbitrary potential is impressed upon an electrode, or when a significant amount of current is drawn through it. Changes in potential due to actual changes in the concentrations of ions at the electrode surfaces are sometimes referred to as a form of polarization, *concentration polarization*. This expression is not recommended, however. The activities concerned should always be those of the ions at the electrode surface.

OVERVOLTAGE

According to thermodynamic definitions, any half-cell is operating irreversibly if appreciable current is flowing. The actual potential of a half-cell under such conditions cannot be calculated; but it is always greater than the corresponding reversible potential as computed from the Nernst equation (i.e., more negative for a cathode, more positive for an anode). The difference between the equilibrium potential and the actual potential is known as *overvoltage*. Overvoltage can then be thought of as the extra driving force necessary to cause a reaction to take place at an appreciable rate. Its magnitude varies with current density, temperature, and with the materials taking part in the reaction. Of especial importance is the overvoltage required to reduce H^+ ion (or water) to hydrogen gas, a process which, in the absence of overvoltage, would take place at zero volts (for activity of H^+ ion equal to unity, the NHE). Figure 11-3 gives representative values of hydrogen overvoltage on a number of cathodes, all in 1 F hydrochloric acid solution.[2]

Overvoltage can be of real advantage in some circumstances. Cations of metals such as iron and zinc, for example, can be reduced to the free metals at a mercury cathode, even though their standard poten-

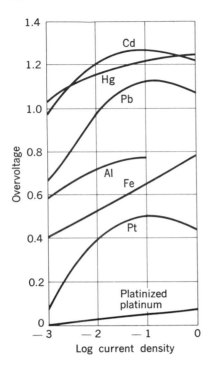

Fig. 11-3 Hydrogen overvoltage on various metals as a function of current density (A-cm^{-2}). The electrolyte is 1 F HCl (*John Wiley & Sons, Inc., New York.*)

tials are more negative than the NHE, because the high overvoltage of hydrogen on mercury prevents its liberation. At a platinum cathode these ions cannot be reduced from an aqueous solution, as the potential cannot exceed that required for hydrogen liberation.

ELECTROANALYTICAL METHODS

1. POTENTIOMETRY   This is a direct analytical application of the Nernst equation through measurement of the potentials of nonpolarized electrodes under conditions of zero current.
2. CHRONOPOTENTIOMETRY   According to this method, a known constant current is passed through the solution and the potential appearing across the electrodes is observed as a function of time. The interval of time starting with the closing of the switch until a steady state is reached relates to the composition of the solution. The related measurement of current changes following application of a constant potential is *chronoamperometry*.
3. VOLTAMMETRY AND POLAROGRAPHY   These are methods of studying the composition of dilute electrolytic solutions by plotting current-voltage curves. In the usual procedure the voltage applied to a

small polarizable electrode (relative to a reference electrode) is increased negatively over a span of one or two volts, and the resulting changes in current through the solution noted. *Voltammetry* is the general name for this method; the term *polarography* is usually restricted to applications of the dropping mercury electrode. *Amperometry* is similar to voltammetry, except that both electrodes may be polarizable.

4. CONDUCTIMETRY In this analytical method two identical inert electrodes are employed, and the conductance (reciprocal of resistance) between them is measured, usually with an ac-powered Wheatstone bridge. Specific effects of the electrodes are eliminated as far as possible.

5. OSCILLOMETRY This method permits one to observe changes in conductance, dielectric constant, or both, by the use of high-frequency alternating current (of the order of a few megahertz). At these frequencies it is not necessary to place the electrodes in direct contact with the solution, which is an advantage in some circumstances.

6. COULOMETRY This is a method of analysis which involves the application of Faraday's laws of electrolysis, the equivalence between quantity of electricity and quantity of chemical change.

7. CONTROLLED POTENTIAL SEPARATIONS It is frequently possible to achieve quantitative separations by means of electrolytic oxidation or reduction at an electrode the potential of which is carefully controlled. The quantity of separated substance may often be measured coulometrically or gravimetrically.

Interesting correlations of many of these electro-analytical methods have been presented in the literature.[3,7,8] These papers will be more readily appreciated *after* detailed study of the several methods individually.

REFERENCES

1. Furman, N. H.: Potentiometry, in I. M. Kolthoff and P. J. Elving (eds.), "Treatise on Analytical Chemistry," pt. I, vol. 4, p. 2294, Interscience Publishers (Division of John Wiley & Sons, Inc.), New York, 1963.
2. Daniels, F., and R. A. Alberty: "Physical Chemistry" (3d ed.), p. 266, John Wiley & Sons, Inc., New York, 1966.
3. Kolthoff, I. M.: *Anal. Chem.*, **26**:1685 (1954).
4. Licht, T. S., and A. J. deBéthune: *J. Chem. Educ.*, **34**:433 (1957).
5. Meites, L.: in L. Meites (ed.), "Handbook of Analytical Chemistry," table 5-1, pp. 5-6, McGraw-Hill Book Company, New York, 1963.
6. Rao, G. G., and P. K. Rao: *Talanta*, **10**:1251 (1963); **11**:825 (1964).
7. Reilley, C. N., W. D. Cooke, and N. H. Furman: *Anal. Chem.*, **23**:1226 (1951).
8. Reinmuth, W. H.: *Anal. Chem.*, **32**:1509 (1960).

12
Potentiometry

As discussed in the preceding chapter, the Nernst equation gives a simple relation between the relative potential of an electrode and the concentration of a corresponding ionic species in solution. Thus measurement of the potential of a reversible electrode permits calculation of the activity or concentration of a component of the solution.

As an example, suppose we construct a cell with a silver electrode dipping into a solution of silver nitrate, the latter connected by a salt bridge to a saturated calomel reference electrode. It is required to find the concentration of the silver salt solution. The cell assembly is diagramed in Fig. 12-1. The beaker on the left in the figure contains the unknown solution, the tube on the right contains the saturated calomel electrode. The central beaker is for the purpose of connecting the KCl solution of the SCE with the salt bridge from the silver solution. The salt bridge is filled with an agar gel which contains ammonium nitrate for electrolytic conduction. (This arrangement prevents contamination of the SCE by silver and prevents precipitation of silver chloride.)

If a potential measuring instrument be connected to the two elec-

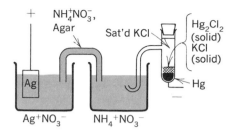

Fig. 12-1 Galvanic cell assembly with silver and saturated calomel electrodes, schematic.

trodes, the silver will be found to be positive; the SCE, negative. Suppose the meter indicates 0.400-V difference of potential between electrodes. We can write

$$E_{Ag} = E^o_{Ag} + \frac{RT}{nF} \ln (Ag^+)$$

$$E_{SCE} = +0.246 \text{ V}$$

and

$$E_{cell} = E_{Ag} - E_{SCE} = E^o_{Ag} - E_{SCE} + \frac{RT}{nF} \ln (Ag^+)$$

Solving for $\log (Ag^+)$, and, noting that $n = 1$, we have

$$\log (Ag^+) = \frac{E_{cell} - E^o_{Ag} + E_{SCE}}{2.303(RT/nF)}$$

$$= \frac{0.400 - 0.799 + 0.246}{0.0591} \quad \text{(at 25°C)}$$

$$= -\frac{0.153}{0.0591} = -2.59$$

$(Ag^+) = \text{antilog} (-2.59) = 2.57 \times 10^{-3}$ formal.

Note that in the first step, writing $E_{cell} = E_{Ag} - E_{SCE}$, we always subtract the more negative from the more positive. Note also that the quantity $2.303(RT/F)$ has the value 0.0591 at 25°C; this is a convenient figure to remember.

For another example, consider a cell consisting of a platinum electrode dipping into a solution nominally 0.1 F in ferrous sulfate. The KCl salt bridge of a calomel electrode is inserted directly into the same solution, since no harmful interaction can occur. The calomel electrode is the SCE; the temperature is 25°C. A potential difference of 0.395 V is observed. It is desired to find the percentage of Fe(II) which has been converted by air oxidation to Fe(III). The half-cell potentials are:

$$E_{Fe^{3+}/Fe^{++}} = E^{\circ}_{Fe^{3+}/Fe^{++}} + \frac{RT}{nF} \ln \frac{(Fe^{3+})}{(Fe^{++})}$$

$$E_{SCE} = +0.246 \text{ V}$$

$$E_{cell} = E_{Fe^{3+}/Fe^{++}} - E_{SCE}$$

$$= E^{\circ}_{Fe^{3+}/Fe^{++}} - E_{SCE} + 0.0591 \log \frac{(Fe^{3+})}{(Fe^{++})}$$

$$\log \frac{(Fe^{3+})}{(Fe^{++})} = \frac{E_{cell} - E^{\circ}_{Fe^{3+}/Fe^{++}} + E_{SCE}}{0.0591}$$

$$= \frac{0.395 - 0.771 + 0.246}{0.0591}$$

$$= -2.20$$

$$\frac{(Fe^{3+})}{(Fe^{++})} = \text{antilog } (-2.20) = 6.3 \times 10^{-3} = 0.63 \text{ percent}$$

which shows that 0.63 percent of the Fe(II) was converted to Fe(III).

THE CONCENTRATION CELL

If two identical electrodes are placed in beakers containing solutions similar in every way except concentration (connected by a salt bridge), the potential between the electrodes is related to the *ratio* of the two concentrations. For example, in a chloride analysis, one silver–silver chloride electrode may be dipped into a solution of unknown chloride concentration $[Cl^-]_x$, with the other being dipped into a standard solution, concentration $[Cl^-]_s$. Activity corrections can be neglected even at relatively high concentrations, as their effect will be similar in numerator and denominator of the logarithmic term.

Suppose the standard is $0.1000 \ F$. Its single electrode potential is

$$E_s = +0.222 - 0.0591 \log 0.1000 = +0.281 \text{ V}$$

If the unknown solution is more concentrated, say $0.1500 \ F$, then

$$E_x = +0.222 - 0.0591 \log 0.1500 = +0.271 \text{ V}$$

The measured potential difference is found, as usual, by subtracting the more negative from the more positive

$$E_{cell} = E_s - E_x = +0.010 \text{ V}$$

On the other hand, if the concentration of the unknown is less than that of the standard, say 0.0680 F, then

$$E_x = +0.222 - 0.0591 \log 0.0680 = +0.291 \text{ V}$$

and

$$E_{\text{cell}} = E_x - E_s = +0.010 \text{ V}$$

$E°$ is the same for both electrodes, its value cancels, and need not be included in the calculations. Since any measured potential (such as the +0.010 V above) may correspond to either of two unknown concentrations, the concentration cell may give rise to ambiguous interpretations. This may be avoided by noting carefully the relative signs of standard and unknown.

The concentration cell provides a highly sensitive analytical tool[6] because most sources of error arising within the cell cancel out in a method which is a comparison of two nearly identical solutions. This means that the overall precision of the method may very well be limited by the measuring instrument. As in any potentiometric method, the sensitivity (at 25°C) is $0.0591/n$ V for a change of a factor of 10 in concentrations. If the measurement were made relative to a universal reference electrode (i.e., *not* a concentration cell), a measuring instrument with a range of perhaps as much as 2 V would be required, whereas with a concentration cell a 20-mV instrument can be employed. The latter permits 100 times the precision of the former.

HYDROGEN-ION MEASUREMENT

One special analysis very frequently carried out by the potentiometric method is the determination of hydrogen ions. The theory is the same as that for metal cations; in this respect, hydrogen is classed with metals.

The calculation of hydrogen ion concentration is simplified if we adopt the *pH* as a measure of acidity or basicity. The pH of an aqueous solution is defined as the negative logarithm of the activity of hydrogen ions expressed in gram-ions per liter

$$\text{pH} = -\log (\text{H}^+) \tag{12-1}$$

Although the pH scale is defined by Eq. (12-1), it has been found expedient to set up a practical or working scale. This practical scale is based on the certified pH values of a series of buffer solutions prepared in a prescribed manner. In the United States the specified buffers are those listed in Table 12-1. Pure materials for their preparation are obtainable from the National Bureau of Standards. The British stand-

Table 12-1 pH Values of NBS standard buffers*

Solution (m = molality)	Temperature, °C			
	20	25	30	38
Potassium hydrogen tartrate (saturated at 25°C)	. . . . .	3.557	3.552	3.548
0.05 m Potassium hydrogen phthalate	4.002	4.008	4.015	4.030
0.025 m Potassium dihydrogen phosphate and 0.025 m disodium hydrogen phosphate	6.881	6.865	6.853	6.840
0.008695 m Potassium dihydrogen phosphate and 0.03043 m disodium hydrogen phosphate	7.429	7.413	7.400	7.384
0.01 m Borax	9.225	9.180	9.139	9.081

* Data from Bates,[1] by permission.

ard pH scale is defined on the basis of a single primary standard buffer, namely 0.05 F potassium hydrogen phthalate. The pH of this standard solution is defined by the relation

$$pH = 4.000 + \frac{1}{2}\left(\frac{t - 15}{100}\right)^2 \tag{12-2}$$

for temperatures in the range of t = 0 to t = 55°C. This scale and the American standards give almost exactly concordant values. A full discussion of the theoretical as well as practical aspects of the pH concept will be found in the monograph by Bates.[1]

For the measurement of pH, an electrode is required which will respond reversibly to the concentration of hydrogen ions. The most obvious electrode to meet this requirement is a hydrogen electrode dipping into the unknown solution.

Two hydrogen electrodes employed for this purpose constitute a concentration cell. The potential is given by the relation

$$E_{cell} = 0.0591 \log \frac{[H^+]_{std}}{[H^+]_{unkn}} \qquad (at\ 25°C)$$

If the standard is the NHE, this becomes

$$E_{cell} = -0.0591 \log [H^+]_{unkn} = 0.0591\ pH$$

or

$$pH = \frac{E_{cell}}{0.0591} \tag{12-3}$$

In case a hydrogen electrode is used as indicator against a saturated calomel reference electrode, correction must be made for the potential of the SCE, and we can write

$$pH = \frac{E_{cell} - 0.246}{0.0591} \tag{12-4}$$

The gaseous hydrogen electrode is inconvenient to use, and is therefore usually replaced by some other electrode which is responsive to hydrogen ions. Several such electrodes will be considered in the following pages. It should not be forgotten, however, that the hydrogen electrode remains the defined standard of reference.

QUINHYDRONE ELECTRODE

An equimolar mixture of quinone and hydroquinone in solution establishes an equilibrium which involves both electrons and hydrogen ions, according to the equation:

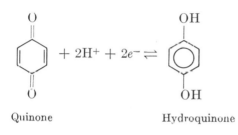

Quinone   Hydroquinone

This represents a reversible redox system with a characteristic potential which can be sensed by an inert electrode. The Nernst equation for the half-cell is

$$E_{QH} = E^{\circ}_{QH} + \frac{0.0591}{2} \log \frac{[Q][H^+]^2}{[H_2Q]}$$

where Q denotes quinone, H_2Q, hydroquinone, and QH, the 1:1 addition compound, quinhydrone. In practice, sufficient solid quinhydrone is added to the test solution to saturate it with respect to both Q and H_2Q, which have nearly equal solubilities and activity coefficients. Hence the ratio of activities of Q and H_2Q is not significantly different from unity. The value of E°_{QH} has been determined to be +0.700, so that the Nernst equation above can be rewritten in the form

$$E_{QH} = 0.700 + 0.0591 \log [H^+] = 0.700 - 0.0591 \, pH$$

valid at 25°C. If this electrode is to be measured with reference to the

SCE, we can deduce the relation

$$pH = \frac{0.454 - E_{cell}}{0.0591} \tag{12-5}$$

The system cannot be applied to solutions more basic than about pH 9, because hydroquinone, a weak acid, is neutralized by base. Strongly oxidizing or reducing solutions must also be avoided. It is somewhat less convenient than the glass electrode, but also much less expensive, so that it still finds some application.

ANTIMONY ELECTRODE

An electrode of antimony metal inserted into an aqueous solution becomes coated with its oxide and responds to the hydrogen-ion concentration, presumably according to the equilibrium

$$2Sb + 3H_2O \rightleftharpoons Sb_2O_3 + 6H^+ + 6e^-$$

The relation between the potential of the antimony-SCE cell and the pH depends to some extent on the method of preparation of the electrode and the nature of the solution, and therefore must be determined experimentally for any given installation. The electrode is not so fragile as the glass electrode, and is useful in some industrial applications for that reason. It is seldom employed in the laboratory. A thorough study of the antimony electrode has been reported by Bishop and Short.[2]

A number of transition metals when employed as electrodes establish equilibria with the corresponding oxides just as antimony does, and thus in principle can be used to measure pH. Among such metals are manganese, tungsten, molybdenum, mercury, and germanium. None has practical importance.

THE GLASS ELECTRODE

Without question the most important pH-sensitive electrode is the glass electrode. This device is based upon the fact that thin membranes of certain varieties of glass are responsive to hydrogen ions. If two solutions are separated by this membrane, a difference of potential will arise between its two surfaces which is, ideally, proportional to the logarithm of the ratio of hydrogen-ion activities of the two solutions:

$$E_g = \frac{RT}{F} \ln \frac{(H^+)_1}{(H^+)_2} = (0.0591)(pH_2 - pH_1) \qquad \text{(at } 25°C\text{)}$$

A typical form of glass electrode is shown in Fig. 12-2. It is in the

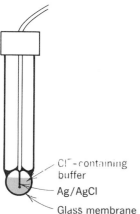

Cl⁻-containing buffer

Ag/AgCl

Glass membrane   **Fig. 12-2**   Glass electrode for H⁺-ion measurement.

form of a test tube terminating in a thin-walled bulb of pH-sensitive glass. Inside is a chloride-containing buffer solution and a reference electrode, usually either a silver–silver chloride or a calomel. The tube is permanently sealed at the top. In use this assembled electrode is dipped into the unknown, as is also the salt bridge from a calomel or other reference electrode.

The glass electrode against any suitable reference electrode gives a potential related to the pH by an expression of the form

$$\text{pH} = \frac{E_{\text{cell}} - E_g}{0.0591} \qquad \text{(at 25°C)} \tag{12-6}$$

The term E_g includes the potential of the reference electrodes, both internal and external, and in addition small, spurious potentials which are often present, called *asymmetry potentials*, possibly resulting from unequal strains in the glass. E_g is a constant for a particular electrode assembly, but it cannot be evaluated theoretically. For this reason it is customary to standardize the electrode assembly by measuring the potential produced when the electrodes are dipped into a standard buffer.

Consider now the situation which arises when a glass electrode which has an internal SCE in a pH 7 buffered solution together with an external SCE are dipped into a pH 7 buffer. An essentially symmetrical system results:

Inner SCE	pH 7 buffer	Glass membrane	pH 7 buffer	External SCE

"Glass electrode"   Test   Reference
                        solution   electrode

From this symmetry it is evident that, ideally, no voltage will be developed under these conditions.   Then it follows from Eq. (12-6) rewritten as

$$E_g = E_{cell} - 0.0591 \text{ pH}$$

that

$$E_g = 0 - (0.0591)(7) = -0.4137 \text{ V}$$

and the pH of the test solution is given by

$$\text{pH} = \frac{E_{cell} - [-(0.0591)(7)]}{0.0591} = \frac{E_{cell}}{0.0591} + 7$$

This treatment could be repeated for a glass electrode with an internal pH of any other value, say 4, in which case the resulting expression would be

$$\text{pH} = \frac{E_{cell}}{0.0591} + 4$$

Thus any combination of glass and reference electrodes is characterized by a definite "pH of zero potential."

It should be noted that the temperature coefficient of a glass–reference electrode pair includes not only the Nernst coefficient, but also the change with temperature of the solubility of the slightly soluble salts involved.   The overall coefficient will therefore follow the Nernst slope only if the inner and outer reference electrodes are identical (e.g., both SCE's or both silver–silver chloride saturated KCl electrodes).

The glass electrode has become exceedingly important in modern analytical and industrial practice, and has almost entirely replaced other pH-responsive systems.   However, it has some definite limitations.   It may give readings which are too high by as much as 1 pH unit when the solution is at a pH of 10 or above in the presence of a high concentration of sodium ions.   Special electrodes are available which have unusually low sodium errors.   The glass surface can selectively adsorb some specific ions, which may cause erroneous measurements.   Thorough rinsing will usually prevent difficulties.   The glass electrode is seriously in error in fluoride solutions more acid than about pH 6.

Combination glass and reference electrodes are fabricated by many manufacturers.   In most designs the reference electrode makes contact with its salt-bridge solution (KCl or other) which is held in an annular container surrounding the glass electrode.   The salt solution acts as an electrostatic shield for the high-resistance glass electrode.   The combination electrode has become quite popular, mostly because of its convenience.   The cost is less than for separate glass and reference electrodes, but more than for the glass electrode alone.

GLASS ELECTRODES FOR METAL IONS

Some glasses, particularly those containing a relatively large proportion of alumina in their compositions, show a usable response to the activity of ions of alkali metals, as well as to hydrogen ions. Eisenman and coworkers[3,4] have shown that the potential of the electrode in the presence of sodium and hydrogen ions is given by

$$E_g = E_g^\circ + \frac{RT}{F} \ln\left[(H^+) + k(Na^+)\right] \tag{12-7}$$

in which the constant k, called the *selectivity ratio*, depends upon the formulation of the glass. In case the hydrogen-ion activity is large compared with that of sodium ion, this gives rise to the usual glass-electrode relation, Eq. (12-6). If the relative activities are reversed, we obtain the analogous expression

$$E_g = E_g^\circ + k' + \frac{RT}{F} \ln(Na^+) \tag{12-8}$$

where $k' = (RT/F) \ln k$.

Equation (12-7) suggests the nature of the sodium error in pH electrodes; k is small for electrodes with low sodium error. This equation does not fully describe the response found experimentally when the two ions have comparable activities. The point is discussed by Bates.[1] The electrodes show varying degrees of response to other alkali metals also.

Glass electrodes are available commercially for sodium-ion measurements. Their use has been reported in connection with the titration of alkali metals by tetraphenyl borate, and in the study of sodium complexes.

OTHER MEMBRANE ELECTRODES

The glass electrodes discussed in the previous sections can be considered to belong to a more extensive group, called *membrane electrodes*.[3,4,13] They are actually half-cells separated from a test solution by a membrane which acts reversibly to a specific ion or group of ions. There are three commercial sources of electrodes in this area (in addition to those usually designated "glass electrodes"). Corning* produces a series based on a liquid ion-exchange material in an aqueous chloride solution. Contact is made with the test solution in the interstitial spaces of a porous glass membrane. The electric circuit is completed through a silver–silver chloride inner electrode and any desired external reference. The ion-exchange material maintains a constant activity of the specified cation (for example, Ca^{++}) in the inner space. The response of the electrode

* Corning Glass Works, Corning, N.Y.

is Nernstian with respect to the activity of the corresponding ion in the external solution over a span of at least three decades (pCa 1 to 4), and is at least 100 times less sensitive to the presence of other ions.

Another series of electrodes is manufactured by Orion.* In these the membrane is fabricated from a sheet of an insoluble silver salt, for example, silver iodide. The inner aqueous solution has a constant activity of silver ion and is in contact with a silver electrode. The membrane acts as an ionic conductor with a specific response to silver

* Orion Research, Inc., Cambridge, Mass.

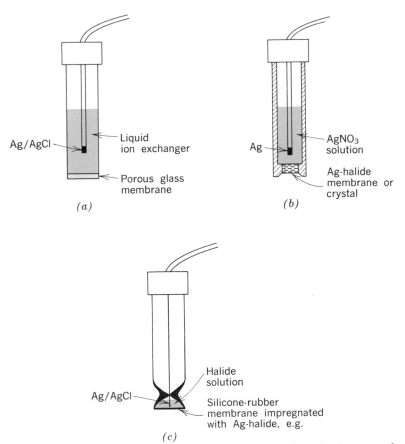

Fig. 12-3 Representative types of membrane electrodes selective toward cations other than H^+. (*a*) Liquid ion-exchanger type. (*b*) Electrode in which the membrane consists of single or multiple crystals of a nearly insoluble salt. (*c*) The type using an impregnated rubber membrane. Those depending on a glass membrane resemble the H^+ electrode of Fig. 12-2.

Table 12-2 Membrane electrodes*

$Mfr.$†	Type of membrane	Designation	Ion to be determined	Range $pION$	Permissible pH	Interferences‡
Many	Glass	pH	H^+	0–14	0–14	(Na^+)
B	Glass	sodium	Na^+	0–6	7–10	Ag^+, (K^+)
		sodium	Ag^+	0–7	4–8	Na^+, $[(K^+)]$
B	Glass†	cation		0–6	4–10	
	Responds to Ag^+, K^+, NH_4^+, Na^+, Li^+, in that order					
C	Porous Glass§	calcium	Ca^{++}	0–5	7–11	Ba^{++}, Sr^{++}, Ni^{++}, Mg^{++}, (Na^+)
O	AgCl	chloride	Cl^-	0–4	0–14	$[S^=, I^-, CN^-$ abs], Br^-, (NH_3), $[(OH^-)]$
O	AgBr	bromide	Br^-	0–5	0–14	$[S^=$ abs], I^-, CN^-, (Cl^-), $[(OH^-)]$
O	AgI	iodide	I^-	0–7	0–14	$[S^=$ abs], CN^-, (Br^-), $[(Cl^-)]$
O	Ag_2S	sulfide	$S^=$	0–20	0–14	None
O	Crystal	fluoride	F^-	0–6	0–8	None
O	Porous§	calcium	Ca^{++}	1–5	5–12	Mg^{++}, (Na^+), (K^+)
O	Porous§	divalent cation		1–5	5–11	
	Responds to Pb^{++}, Ni^{++}, Zn^{++}, Fe^{++}, Ca^{++}, Mg^{++}, Ba^{++}, Sr^{++}, in that order					
O	Porous§	cupric	Cu^{++}	1–5	5–8	Fe^{++}, Ni^{++}, Zn^{++}, (Ba^{++}), (Sr^{++}), (Ca^{++}), (Mg^{++}), $[(K^+)]$, $[(Na^+)]$
O	Porous§	perchlorate	ClO_4^-	1–4	4–12	(I^-), (Br^-), (NO_3^-), $(SO_4^=)$, (Cl^-), $[(OAc^-)]$, $[(F^-)]$, $[(HCO_3^-)]$
NIL	Silicone	iodide	I^-	1–7	....	$[S^=$ abs], (Cl^-), $[(SO_4^=)]$
NIL	Silicone	sulfate	$SO_4^=$	1–5	....	Cl^-, PO_4^{---}
NIL	Silicone	phosphate	PO_4^{---}	1–5	....	Cl^-, $SO_4^=$
NIL	Silicone	nickel	Ni^{++}	1–5	....	(Co^{++})
NIL	Silicone	chloride	Cl^-	1–5	....	$[S^=$ abs], Br^-, I^-
NIL	Silicone	bromide	Br^-	1–6	....	$[S^=$ abs], I^-, (Cl^-)
	NIL has comparable electrodes sensitive to Bi^{+++}, $S^=$, F^-, K^+, Al^{+++}, Ag^+, Ba^{++}, Cu^{++}, Sb^{+++}					

* The information in this table has been taken from the manufacturers' literature and altered where necessary to give a unified format; it is to be considered illustrative only.

† B = Beckman; C = Corning; O = Orion; NIL = National Instrument Laboratories.

‡ [. . . abs] designates an interferent which must be absent; (. . .) designates slight interference, [(. . .)] very slight.

§ Electrode contains a liquid ion-exchange material, proprietary in nature.

NOTE: The inserted lines following the third and eleventh entries refer to the items immediately preceding.

ion in accordance with the Nernst equation, so that the potential developed is logarithmically related to the ratio of outer to inner silver-ion activities. It can act as a class II electrode, and indicate the activity of halide ions in solution. Orion also manufactures electrodes with a liquid ion-exchanger and porous membrane, similar in principle to those of Corning.

The third source of electrodes in this category is N.I.L.* which imports those invented by E. Pungor and manufactured in Hungary. The Pungor membranes consist of silicone rubber with an insoluble salt such as silver iodide imbedded in it like a "filler." The loaded membrane must be soaked for several hours in a solution of the ion to which it is sensitized.

Electrodes of these several kinds are shown diagrammatically in Fig. 12-3, and those available at the time of writing are listed in Table 12-2.

POTENTIOMETRIC TITRATIONS

A great many titrations can be followed by potentiometric measurements. The only requirement is that the reaction involve the addition or removal of some ion for which an electrode is available. The potential is either measured after successive additions of small volumes of titrant, or continuously, with automatic recording. It is assumed that the student is already familiar with titration curves and their calculation from ionic equilibria and other pertinent data.

The precision of a potentiometric end point can be improved, though with loss of convenience perhaps, by the use of a concentration cell. The two vessels of the cell will contain identical electrodes reversible to the ion being titrated. In the reference side is placed a solution identical to that expected at the end point. Then, when the titration in the indicator vessel reaches equivalence, the potential will be zero. The increased precision results from the possibility of using the measuring instrument on its most sensitive range.

NEUTRALIZATION REACTIONS

Acid-base titrations can be performed with a glass indicator electrode measured against a suitable reference. In Fig. 12-4 are shown the curves obtained in the titration of a series of acids of varying K_a by sodium hydroxide. For classical titrations to a visual indicator end point, the statement is sometimes made that to be "feasible," the product of the concentration of the titrant (NaOH) and the ionization constant K_a

---

* National Instrument Laboratories, Inc., Rockville, Md.

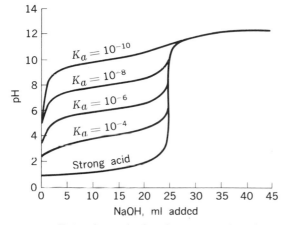

Fig. 12-4 Potentiometric titration curves of equivalent amounts of strong and weak acids with the indicated K_a values, plotted as pH against volume of base.

of the acid must not be less than about 10^{-8}. This empirical relation is based on the fact that under usual conditions it requires a change of about 2 pH units to convert an indicator from one color to the other, and therefore the nearly vertical portion of the curve must cover a 2-unit span; obviously this depends on numerous variables, some of which are subjective. In potentiometric titration with a pH meter sensitive to tenths of pH units, this practical limit can be reduced typically by a factor of 20. Hence the concentration-dissociation constant product may be as low as 5×10^{-10}.

PRECIPITATION AND COMPLEXATION REACTIONS

These two classes are similar as far as their electrochemistry is concerned, as the formation either of a precipitate or a complex will remove from solution the simple hydrated ion. Electrodes of Classes I, II, or III (cf. p. 242) can be used, as appropriate. Probably the most widely applicable in precipitation reactions are electrodes of silver and of mercury, since they are highly reversible, and because so many anions can be precipitated quantitatively by solutions of silver or mercury salts. As long as any precipitate is present, the distinction between electrodes of Classes I and II is lost.

A large number of metals can be titrated with EDTA.*[14,15] The electrode is a mercury pool; a single drop of $0.01F$ solution of the mercury-EDTA complex added before starting the titration is sufficient to give

* EDTA = ethylenediaminetetraacetic acid or ethylenedinitrilotetraacetic acid.

well poised potentials.   Halide ions must be absent.   The chief difficulty is with respect to interferences, since so many metals form EDTA
chelates.   Careful buffering will reduce the problem but not eliminate it.

REDOX REACTIONS

Oxidation-reduction titrations, where both oxidized and reduced forms
are soluble, can be followed by a simple indicating electrode of platinum
or other inert element.   Difficulties may be encountered in the presence
of such powerful oxidizing agents as permanganate, dichromate, cobaltic,
and ceric ions, due to the formation of an oxide film on the metal.   This
effect is most noticeable when submicrogram quantities of oxidant or
reductant are titrated.   The oxide is easily removed by cathodic reduction in a dilute acid solution.

TITRATIONS IN NONAQUEOUS SOLVENTS

The Brønsted theory of acids and bases predicts that weak acids should
appear stronger in a solvent such as liquid ammonia or pyridine which has
a greater tendency to accept a proton than does water.   Similarly a weak
base appears stronger when dissolved in such a solvent as glacial acetic
acid.   For this reason and also because of the limited aqueous solubility
of many weak acids and bases, it is desirable to be able to perform titrations in various nonaqueous media as well as in water.   This can be done
in many instances with organic indicators, but often a potentiometric
titration is desired.

Most of the electrodes discussed above, including the glass electrode,
can be employed in nonaqueous or mixed solvents.   The thermodynamic
interpretation of the results is not always clear; particularly, it has not
been found possible to make a quantitative theoretical correlation between
electrode potentials in one solvent and those in another.   However, this is
no great deterrent to the application of such potentials in titrations.

A good example of nonaqueous titrations can be found in the work of
Sensabaugh et al.[16]   They observed that 2,4-dinitrophenylhydrazine and
the corresponding hydrazones of aldehydes and ketones:

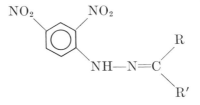

act as acids when dissolved in pyridine.   The titrant was tetrabutylammonium hydroxide in a benzene-methanol mixture. A glass electrode was measured against a modified calomel electrode in which the usual aqueous solution was replaced by a saturated solution of potassium chloride in methanol. Well-defined titration curves were obtained.

Apparently no comparable systematic studies have been made of potentiometric redox titrations in nonaqueous systems.[9] Perhaps this merely reflects the lack of special solvent effects. Examples of work in this field have been reported by Piccardi et al.,[12] who used Pb(IV) and ICl_3 as oxidants in glacial acetic acid.

APPARATUS

Although potentials are measured in volts, the potentials of electrodes cannot be determined simply by connecting a voltmeter across the terminals of the cell and observing the deflection of the needle. This is not satisfactory because the voltmeter would draw an appreciable amount of current from the cell being measured, thus introducing an error due both to the electrical resistance of the cell and to the changes in concentration of ions resulting from electrolysis. To avoid such difficulties, a more complicated apparatus, such as a potentiometer or electronic voltmeter, is required.

THE POTENTIOMETER

The principle of the potentiometer is shown in Fig. 12-5, and the circuit of a typical instrument in Fig. 12-6. A storage battery (Fig. 12-5) sends a current through the wire AB and the variable resistor R. The setting of R is adjusted until the current flowing is of the correct value to cause a selected potential, say 2 V, to appear across the terminals of the wire AB. Then each point along that wire must have a potential intermediate between A and B, and if the wire is perfectly uniform in cross sec-

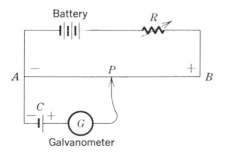

Fig. 12-5 Elementary potentiometer.

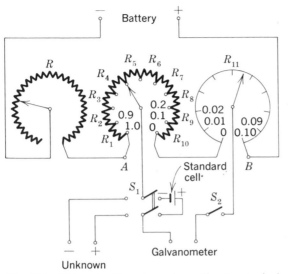

Fig. 12-6 The potentiometer: schematic of a typical instrument.

tion, the potential of any point P will be proportional to its distance from the end of the wire.   To measure the potential of an unknown cell C, it is connected in series with a sensitive galvanometer G between A and a sliding contact on the wire.   The contact is moved along the wire until a point is found at which the galvanometer gives no deflection (the *null point*).   Then the difference of potential between A and P must be the same as the potential of the cell and can be determined by measurement of the distance AP.

In practice, the slide wire is usually broken up into several components, as in Fig. 12-6, where 11 separate resistors are shown, all having equal resistance values.   The first 10 are wired to a multiple-tap switch with 11 positions, while the eleventh, R_{11}, is a uniform wire wound on a drum and provided with a continuously movable contact.   The battery and resistor R have the same functions as in Fig. 12-5.   The two knobs or dials controlling resistors R_1 to R_{11} are provided with calibrations reading directly in volts.   To adjust the potentiometer, resistor R must be varied until the current flowing is exactly that for which the instrument was designed, commonly 1 or 2 mA.   This standardization is best performed by connecting a cell of known potential in the position of the unknown cell.   Then the dials are set at the known potential of the cell, and switches S_1 and S_2 are closed.   Resistor R is then adjusted until the galvanometer shows no deflection.   Ordinarily, it need not be varied again for the duration of the experiment. The unknown cell to be measured is connected, switch S_1 is thrown to the left, S_2 is closed, and the two main dials are adjusted till the galvanometer does not deflect. The setting of these dials added together is the potential of the cell.

The advantage of the potentiometer is that, at the moment when the potential of the cell is read, the current drawn from it is very nearly zero, within limits set by the sensitivity of the galvanometer. Hence the value determined is very close to the true (reversible) potential; residual errors due to the resistance of the cell or polarization effects are negligible.

An excellent example of a precision potentiometer is the Leeds & Northrup Type K-3.* This instrument has three voltage ranges: 0 to 0.0161100, 0 to 0.161100, and 0 to 1.61100 V, with limits of error specified as ± 0.015 percent (± 0.010 percent for the highest range).

The *standard cell* used to validate the calibrations of the potentiometer must be carefully selected and constructed. It should be such that it is easily reproduced by various manufacturers or technicians, it should have a very low temperature coefficient, and it should be perfectly reversible. The most common is the *saturated Weston cell*, the composition of which is shown in Fig. 12-7. Its potential is 1.01864 V at 20°C, and decreases by 4×10^{-5} V for each 1°C rise in temperature. Since the amalgam at the negative terminal consists of two phases, a small change in ratio of cadmium to mercury will change only the relative quantities of

* Leeds & Northrup Company, North Wales, Pa.

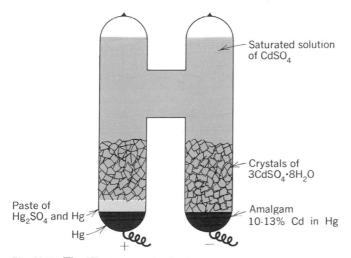

Fig. 12-7 The Weston standard cell.

the phases, not their composition. Hence the small currents passed through the cell during the process of balancing the potentiometer will have no observable effect on the potential. The cell should be protected against short circuit, even momentary, because the reattainment of equilibrium may be slow. Bates[1] gives a short critique of other cells usable as standards.

THE ELECTRONIC VOLTMETER

Another device for measuring the potentials of electrodes without the flow of appreciable current depends upon the power-amplifying properties of electronic vacuum tubes or transistors. The design of amplifiers suitable for this service is considered in Chap. 26. It will suffice for the present to state that the output from an amplifier can be controlled in magnitude by a *potential* applied to its input. The circuit can be designed so that the effective resistance offered to the potential being measured is as much as 10^{11} Ω or more. This means that if the potential is, say, 1 V, then the current drawn from the cell will be only 10^{-11} A, which is completely negligible. This feature is especially important in glass-electrode measurements, where the electrode may have a resistance of as much as 10^{8} Ω, so that a current of 10^{-11} A will cause an internal drop of 10^{-3} V, which means an error of the order of 0.1 percent.

An electronic amplifier for use with a glass electrode is called a *pH meter*. There are many manufacturers of such instruments in the United States who offer a large variety of models. Some of these are battery-operated, some plug into the ac power lines. They fall roughly into three categories, with high, medium, and relatively low precision and accuracy (and price). The first group is primarily designed for research purposes, the second for general laboratory use, the third for field use where small size and rugged construction are more important than a high degree of precision.

A line-operated laboratory pH meter of the deflection type must have three controls on its panel, and may have a fourth. These are (1) a switch with "standby" and "operate" positions; (2) a calibration or standardization adjustment, which amounts to a zero offset; with it one adjusts the instrument to read the correct value when the electrodes are immersed in a standard buffer; (3) a temperature compensator which permits alteration of the sensitivity to account for the temperature dependance of the Nernst potential. Some pH meters also have a scale selector which allows the instrument to cover the whole pH range (usually 0 to 14) or to fill the scale with a selected portion of that range, perhaps 2 or 3 pH units; this type is called an *expanded-scale* pH meter. Some meters have a folded scale, with 0 to 8 pH on one switch setting, 6 to 14 pH on another.

TITRATION APPARATUS

A potentiometer such as that of Fig. 12-6 can be used in titrations, but some simplification is possible since only relative values are required. Potential readings can be made on any arbitrary scale, not necessarily referred to the hydrogen or other standard electrode. Hence the standard (Weston) cell can be omitted.

A satisfactory potentiometer for titrations can be constructed from simple components (Fig. 12-8). The working current is supplied by the battery B, which consists of two dry cells in series. The variable resistor R may be an inexpensive radio component. V is a voltmeter, range 0 to 2 V. G is a galvanometer of intermediate sensitivity, critically damped by the resistor R_D. S_2 is a reversing switch which permits connecting the reference and indicator electrodes in either polarity. S_2 is essential for convenience, as the sign of the potential may change during a titration. To operate, insert the electrodes into the sample solution, close S_1, adjust R until the galvanometer does not deflect, and then read the voltmeter. If no balance can be obtained, reverse S_2. Repeat after each addition of reagent.

As previously mentioned, the glass electrode cannot be used with a simple potentiometer on account of its high resistance. However, an electronic pH meter is admirably suited for titrations. In fact, it is so convenient that it is commonly employed even where its high input resistance is not required, as in the case of electrodes other than glass. For such applications, the scale is read in millivolts rather than in pH units.

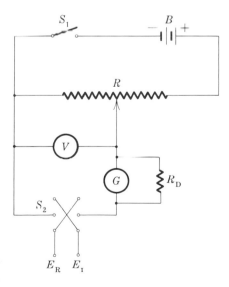

Fig. 12-8   Titration potentiometer.

AUTOMATIC TITRATORS

For some types of research problems and for industrial analytical laboratories, automatic operation is widely applied to potentiometric titrations. There are two classes of potentiometric titrators, those in which the instrument plots a complete titration curve on chart paper, and those which act to close an electrically operated buret valve exactly at the equivalence point. Those of the first type consist in essence of a pH meter connected to a strip-chart recorder. The chief difficulty arises from the need for delivery of titrant at a constant rate, which is not possible from a conventional gravity-flow buret. This can be overcome by substituting a constant-head reservoir (e.g., a mariotte bottle) and measuring time of flow rather than volume or by using a constant-flow pump or a motor-driven hypodermic syringe.

Typical of the second type is the Beckman Automatic Titrator, which employs a vacuum-tube amplifier to control an electrically operated valve that takes the place of the usual buret stopcock. It is necessary to

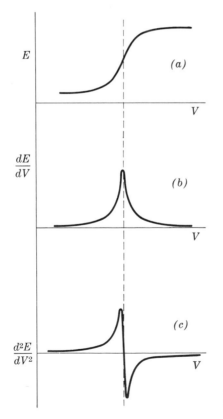

Fig. 12-9 (*a*) A potentiometric titration curve; (*b*) its first derivative curve; and (*c*) its second derivative curve.

know in advance the potential difference which will appear between the electrodes at the equivalence point.   The potential dial on the instrument is set manually at the equivalence potential.   The amplifier then permits reagent to flow from the buret until the preset potential is attained.   The flow of reagent is rapid at first and then is automatically slowed down as the equivalence potential is approached, to avoid overshooting.   A warning light appears on the panel when the titration is completed.   Any desired electrodes may be used with this outfit, and it may also be operated for conventional point-by-point potentiometric titrations.

Another important example is the Sargent-Malmstadt titrator.* This employs a special capacitive electronic circuit which automatically doubly differentiates the variations in potential which are fed into it from the electrodes.   If the potential as a function of volume $E = f(V)$ is given by the familiar curve a of Fig. 12-9, then the derivative dE/dV is represented by curve b, and the second derivative d^2E/dV^2, by curve c.   In the titrator, advantage is taken of the fact that the positive peak in curve c comes very slightly *before* the equivalence point.   The first *decrease* in the second derivative current energizes a relay which closes the buret stopcock.   The mechanical inertia in the moving parts causes a delay which just offsets the advance warning given by the second derivative curve, so that the flow of titrant is effectively stopped at just the right moment.

SPECIAL TECHNIQUES

Several methods of improving the convenience and speed of potentiometric titrations have been described.   Many of these are valid only for a limited number of reactions, and in none is the precision increased materially over standard methods.   Two such techniques will be described briefly.

DERIVATIVE TITRATIONS

An apparatus can be constructed which will give a derivative curve directly.   One of the simplest is shown in Fig. 12-10.   It consists of two simple platinum electrodes, one of which is a wire sealed inside an ordinary medicine dropper.   As long as the solution inside the dropper is the same as that outside, no potential difference will be observed. The addition of a few drops of reagent to the beaker will change the solution only slightly at points remote from the equivalence point, but much more markedly near that point.   In operation, the potential is read after each addition of reagent, the bulb is squeezed several times to mix the whole solution, then another portion is added, etc.   The observed potentials plotted against the volume of reagent will then pro-

* E. H. Sargent & Co., Chicago.

duce a derivative curve from which the equivalence point or points can be read directly.

BIMETALLIC ELECTRODE PAIRS

It is found that supposedly inert metals respond at different rates to the ratio of oxidized to reduced forms of a substance in solution. Platinum responds quickly, but some other metals, particularly tungsten, reach the equilibrium potential at a slower rate. Hence, since only relative values are required in a titration, tungsten can be used in place of the usual reference electrode, if the titration is performed at a constant rate. This permits the elimination of the calomel electrode with its salt bridge, so that a more sturdy assembly, which needs less attention and replacement, results. The curve obtained is similar to that for a standard potentiometric titration, but frequently the "break" is somewhat sharper.

An example is the titration of ferrous sulfate by potassium dichromate in the presence of acid (Fig. 12-11). Curves 1 and 2 give the potentials of platinum and tungsten electrodes individually against the SCE. Curve 3 represents the point-by-point difference between the first two curves. It is also the titration curve which actually will be observed if only the platinum and tungsten electrodes are used. The figure includes only the region in the vicinity of the equivalence point. At the start of the titration, there is usually a noticeable potential difference between the electrodes, but addition of reagent reduces it

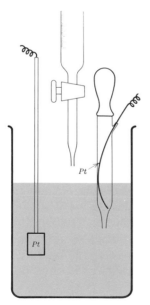

Fig. 12-10   Apparatus for derivative titration.

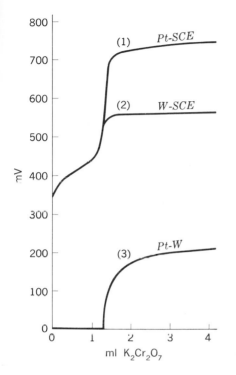

Fig. 12-11 Potentiometric titration of Fe(II) by dichromate: bimetallic electrode system.

rapidly to zero, where it remains until equivalence is nearly attained. The curves for the individual electrodes, after the equivalence point, may or may not be parallel, so that the bimetallic titration curve may rise, descend, or remain horizontal after equivalence is passed.

CONSTANT-POTENTIAL TITRATIONS[7,8,10]

Another approach to potentiometric titration is to measure the amount of titrant required to maintain the indicator electrode at a constant potential. The titration curve then becomes a plot of volume of standard solution added as a function of time. This procedure has been employed rather extensively in the field of enzymology.

For example, the enzyme cholinesterase acts to decompose acetylcholine, producing acetic acid in the process. The enzyme is highly sensitive to pH, and the medium must be held very close to pH 7.4 for the reaction to proceed optimally. A bicarbonate buffer has been used in the past, and the carbon dioxide liberated by the acetic acid measured manometrically. This is awkward, and in addition has been attacked on the grounds that physiological conditions are not reproduced.

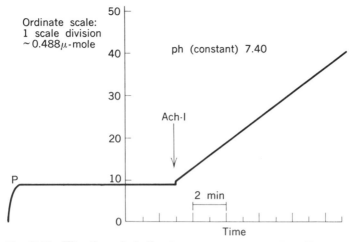

Fig. 12-12 Titration of cholinesterase at constant pH. (For details, see text.)

The present method makes use of a pH meter connected through a servo system to control a motor-driven syringe buret supplying sodium hydroxide solution at a constant rate, as needed to maintain constant pH. A pH meter and associated equipment used in this manner are called a *pH stat*. Figure 12-12 shows an actual example of a curve by which the cholinesterase activity of a sample of animal tissue was determined.[7] A 0.3-g portion of tissue was homogenized in physiological saline solution (0.9 percent NaCl), the pH adjusted to 7.4 (at point P on the curve) and held for some 10 min to establish a base line showing that no acid was being liberated spontaneously. Then an excess of acetylcholine iodide (Ach·I) was added, whereupon the inflow of sodium hydroxide commenced. For this experiment the cholinesterase activity was determined from the slope to be 4.96×10^{-6} mole-g^{-1}min^{-1}.

This technique does not seem to have been applied outside of biochemical work, but is certainly of general applicability. The apathy may be due partly to lack of suitable instruments, but recently several have appeared on the American market.

PROBLEMS

12-1 A voltaic cell is formed with a lead electrode in 0.015 F lead acetate against a cadmium electrode in 0.021 F cadmium sulfate. The two solutions are joined by a salt bridge containing ammonium nitrate. What is the potential of the cell at 25°C if the activity coefficients in the two solutions are considered equal?

12-2 What is the pH of a solution at 25°C if a potential of 0.703 V is observed between a hydrogen electrode and an NCE placed therein?

12-3 What potential would be shown by a quinhydrone electrode relative to the SCE in a solution at pH 5.5?

12-4 Exactly 25.00 ml of 0.1000 F silver nitrate was pipetted into each of two beakers. Identical silver electrodes were inserted, and the two solutions joined by a potassium nitrate salt bridge. The potential difference between the electrodes was, as expected, zero. A 10.00-ml aliquot portion of a lead nitrate solution was added to one beaker and 10.00 ml distilled water to the other, whereupon a potential of 0.070 mV appeared between the silver electrodes. If the lead nitrate solution had been prepared by dissolving a 10.00-g sample of metallic lead in nitric acid and diluting to a volume of 100 ml, calculate the percentage of silver impurity in the lead. Estimate the precision of the analysis if the potentiometer has a maximum uncertainty of $\pm 1\mu V$.

12-5 A cell is set up as follows: silver electrode, unknown solution, salt bridge, saturated KCl solution, $Hg_2Cl_{2(s)}$, mercury electrode. (a) Which electrode is the reference electrode, and which the indicator? (b) What is the purpose of the salt bridge, and what electrolyte should it contain? (c) If the potential of the cell is found to be 0.300 V, with the silver electrode more positive than the mercury, what is the concentration of silver ion in the unknown?

12-6 A cell is assembled with two platinum wires as electrodes dipping into separate beakers which are connected by a salt bridge, each beaker containing 25.00 ml of a mixture of Fe^{++} and Fe^{3+} ions, each ion 1 F. Now 1.00 ml of a solution of a reducing agent is added to one side, whereupon the potential difference changes from zero to 0.0260 V. Compute the normality of the solution of reducing agent.

12-7 The fact that fluoride ion complexes strongly with Ce(IV) but not appreciably with Ce(III) can be used in the potentiometric determination of fluoride.[11] (a) What effect, qualitatively, should the addition of F^- ion have upon the potential of a platinum-wire electrode in the presence of equal concentrations of Ce(III) and Ce(IV)? (b) Show how a concentration cell could be set up to facilitate the determination of fluoride. (c) Compare this method with the direct determination of F^- ion with an Orion fluoride electrode, with respect to sensitivity, specificity, and convenience.

12-8 What is the maximum concentration of Cu^{++} ion which can exist in solution in contact with metallic zinc?

12-9 According to the modified Liebig method for the titration of cyanide by silver nitrate in the presence of iodide and ammonia, the following reactions take place successively:

$$Ag^+ + 2CN^- \rightarrow Ag(CN)_2^-$$
$$Ag^+ + I^- \rightarrow AgI_{(s)}$$

AgCN cannot precipitate so long as any ammonia is present. (See Vogel[17] for details of the procedure.) Plot qualitatively the potential of a silver electrode against an isolated Ag–AgCl reference during the titration of equimolar amounts of cyanide and iodide.

REFERENCES

1. Bates, R. G.: "Determination of pH: Theory and Practice," John Wiley & Sons, Inc., New York, 1964.

2. Bishop, E., and G. D. Short: *Talanta*, **11**:393 (1964).
3. Eisenman, G. (ed.): "Glass Electrodes for Hydrogen and Other Cations," Marcel Dekker, Inc., New York, 1967.
4. Eisenman, G., D. O. Rudin, and J. U. Casby: *Science*, **126**:831 (1957).
5. Furman, N. H.: Potentiometry, in I. M. Kolthoff and P. J. Elving (eds.), "Treatise on Analytical Chemistry," pt. I, vol. 4, chap. 45, Interscience Publishers (Division of John Wiley & Sons, Inc.), New York, 1963.
6. Furman, N. H.: Potentiometry, in J. H. Yoe and H. J. Koch, Jr., "Trace Analysis," chap. 9, John Wiley & Sons, Inc., New York, 1957.
7. Jensen-Holm, J., H. H. Lausen, K. Milthers, and K. O. Møller: *Acta Pharmacol. Toxicol.*, **15**:384 (1959).
8. Jørgensen, K.: *Scand. J. Clin. Lab. Invest.*, **11**:282 (1959).
9. Kratochvil, B.: *Record Chem. Progr.*, **27**:253 (1966).
10. Malmstadt, H. V., and E. H. Piepmeier: *Anal. Chem.*, **37**:34 (1965).
11. O'Donnell, T. A., and D. F. Stewart: *Anal. Chem.*, **33**:337 (1961).
12. Piccardi, G.: *Talanta*, **11**:1087 (1964); G. Piccardi and P. Legittimo, *Anal. Chim. Acta*, **31**:45 (1964).
13. Rechnitz, G. A.: *Anal. Chem.*, **37**:29A (1965).
14. Reilley, C. N., and R. W. Schmid: *Anal. Chem.*, **30**:947 (1958).
15. Reilley, C. N., R. W. Schmid, and D. W. Lamson: *Anal. Chem.*, **30**:953 (1958).
16. Sensabaugh, A. J., R. H. Cundiff, and P. C. Markunas: *Anal. Chem.*, **30**:1445 (1958).
17. Vogel, A. I.: "Quantitative Inorganic Analysis," 3d ed., p. 74, John Wiley & Sons, Inc., New York, 1961.

13

Voltammetry, Polarography, and Related Techniques

In the previous chapter we dealt with potentials of unpolarized electrodes. We shall now investigate phenomena which occur in an electrolysis cell wherein one electrode is polarizable, and one is not. Such a system is conveniently studied by means of current-voltage curves. This general method of studying the composition of a solution is called *voltammetry*. The term *polarography* is used chiefly where the polarizable electrode consists of dropping mercury, but the distinction is not always followed.

ELECTRODES

The nonpolarized electrode is the reference electrode and is usually saturated calomel. A simple mercury pool is sometimes substituted for the SCE; it can be counted on to be nonpolarized if the solution contains an appreciable concentration of chloride or other ion which forms a nearly insoluble salt with mercury(I), but the mercury pool is not otherwise reliable as a reference electrode. The reference electrode must have large enough dimensions that its electrical resistance is low, as it may be required to pass currents of the order of 100 μA.

The polarizable electrode is made much the smaller of the pair, and is sometimes referred to as a *microelectrode*. It is generally a noble metal such as mercury or platinum.

The electrodes are connected in series with a microammeter to a source of variable known potential. Figure 13-1a shows one possible arrangement, wherein the potential impressed on the microelectrode can be varied from 0 to -3 V. The voltmeter indicates the exact potential at any setting of the control. The current flowing is read from the microammeter (μa). A more versatile arrangement is shown in Fig. 13-1b, in which the potential may be varied continuously from $+3$ through 0 to -3 V. These circuits are intended as a basis for discussion. To be useful in practice, a number of elaborations are required, as will be considered later.

The choice of microelectrode depends largely upon the range of potentials it is desired to investigate. For potentials positive to the SCE the best choice is platinum. Mercury cannot be made more positive than about $+0.25$ V versus SCE because of the ease of its anodic dissolution. Platinum is limited in the positive direction only by the oxidation of water $(2H_2O \rightarrow O_2 + 4H^+ + 4e^-)$, which occurs at about $+0.65$ V. On the other hand, for negative potentials platinum can only be used to about -0.45 V, at which potential hydrogen is liberated $(2H^+ + 2e^- \rightarrow H_2$ or $2H_2O + 2e^- \rightarrow H_2 + 2OH^-)$, while mercury, due to its high overvoltage for hydrogen (see Chap. 11), can be utilized as

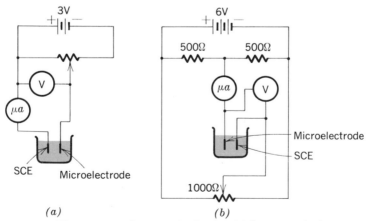

Fig. 13-1 Elementary voltammetric circuits: (a) for convenient operation on one side of zero; (b) continuously adjustable from one sign to the other.

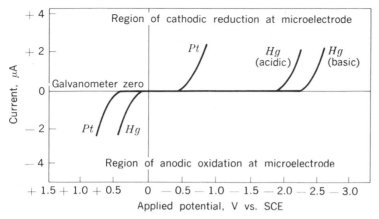

Fig. 13-2 Conventions of plotting voltammetric curves. The curves labeled Pt or Hg indicate the approximate potential limits attainable at these electrodes.

far as -1.8 V in acid or about -2.3 V in basic media. Figure 13-2 shows graphically these limiting potentials and also indicates the conventions regarding the sign of the current and method of plotting voltammetric curves.

DIFFUSION–LIMITED CURRENT

Consider first a plane platinum microelectrode in a deaerated, unstirred solution containing 0.1 F KCl and 10^{-3} F $PbCl_2$ in contact with a low-resistance salt bridge leading to an SCE reference. Suppose that a switch is closed in an external circuit, suddenly applying 0.50 V across the electrodes, with the microelectrode negative. This potential is great enough to reduce Pb^{++} ion to the metal, but not to reduce H_2O (at pH 7). K^+ ion would of course require a much higher potential, and there is nothing else present which is reducible. (We need not concern ourselves about the simultaneous anodic oxidation, which is $Hg + Cl^- \rightarrow \frac{1}{2}Hg_2Cl_2 + e^-$ at the SCE.)

Several things will happen: (1) K^+ and Pb^{++} ions will start to move in the electric field toward the microelectrode, and Cl^- ions in the opposite direction. K^+ ions, not being reducible, will almost instantaneously form a sheath (about one ion thick) around the electrode, which will result in almost completely neutralizing the field so far as the bulk of the solution is concerned. The Pb^{++} ions will not contribute to this sheath, because they *are* reducible; every Pb^{++} ion that approaches the electrode will immediately be discharged and deposited on the surface. But since the field has been neutralized by K^+ ions, the only way the lead ions can get to the vicinity of the electrode is by *diffusion*. (Because of the requirement of overall electrical neutrality, the reverse movement of chloride ions will have dropped to the low level required to match the reduction of lead ions.)

The rate of movement of any species because of diffusion is proportional to the concentration gradient, the difference between the concentrations at two points divided by the distance between them. In calculus terms,

$$\frac{dC}{dt} = D\frac{dC}{dx} \tag{13-1}$$

where C refers to the concentration of the diffusing species, and D is a constant of proportionality called the *diffusion coefficient*. (This is called *Fick's first law*.[1])

Application of Fick's law to the electrolysis problem we are considering leads to the relation

$$i_d = nFA\left(\frac{D}{\pi\tau}\right)^{\frac{1}{2}} C \tag{13-2}$$

in which i_d is the electrolysis diffusion current (in microamperes) flowing at time τ (seconds) from the start of an experiment, n is the number of electrons involved in the electrode reaction, F is the Faraday constant (approximately 96,500 Cb per faraday), A is the area (in square centimeters) of the electrode, D is Fick's diffusion coefficient (square centimeters per second), and C is the bulk concentration of electroactive species (in millimoles per liter). (The concentration is assumed to be zero at the electrode surface.) Both i_d and n are taken as positive for cathodic reductions and negative for anodic oxidations. It is important to note the proportionality between the current and concentration; however, since the current falls off with the square root of time, rather than assuming a fixed value, this equation is not a convenient basis for analytical work.*

THE DROPPING–MERCURY ELECTRODE (DME)

The most widely used microelectrode is mercury in the form of a succession of droplets emerging from a very fine-bore glass capillary (Fig. 13-3). This has several major advantages to offset the inconvenience of handling mercury. One is the high hydrogen overvoltage of a mercury cathode.

---

* Analysis by means of current-time curves at constant potential is called *chronoamperometry*. It has some value in studies of the electron-transfer kinetics in irreversible systems.[4,5]

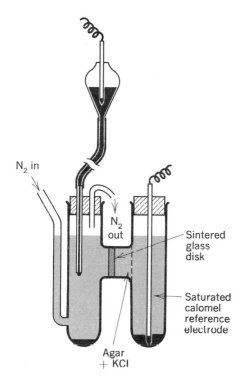

N₂ in

N_2 in

N₂ out

N_2 out

Sintered glass disk

Saturated calomel reference electrode

Agar + KCl

Fig. 13-3 Polarographic H cell.

Another advantage is the fact that the electrode surface is continually being renewed, and hence cannot become fouled or poisoned. In comparison with the solid electrode of the preceding paragraphs, the DME has the advantage that the increasing area of the electrode during the lifetime of a drop more than offsets the decreasing current observed with an electrode of fixed size; this, as we will see shortly, makes quantitative analysis much more practicable.

A variety of electrolysis cells for polarography have been proposed. The simplest is merely a small beaker into which the DME is inserted and a connection made to a mercury pool in the bottom. The beaker should be covered and a tube inserted to permit nitrogen to be bubbled through the solution to remove oxygen. The more complicated cell shown in Fig. 13-3 is employed by many polarographers. The built-in SCE is permanently attached through the cross member of the H, which is plugged with a sintered-glass disk, backed up by an agar gel containing potassium chloride.

The capillary for the dropping electrode is made from a section several centimeters long, cut from glass tubing of 0.03 to 0.05 mm internal diameter. It must be handled with great care to prevent any water or aqueous solution from entering the capillary, since it is practically impossible to clean it out if it once becomes fouled. After use the capillary should be removed from the cell and rinsed with distilled water while the mercury is still flowing from the tip. It can then be clamped in the air (protected from dust), and the mercury reservoir lowered sufficiently to stop the flow. It should not be left immersed in water.

To derive an equation for the DME corresponding to Eq. (13-2), we make the assumption that the flow of mercury is constant, and the approximations that the drop is spherical in shape right up to the moment of separation, and that the law of linear diffusion applies to the spherical surface. This leads to the expression

$$i_d = 708.2nD^{1/2}m^{2/3}\tau^{1/6}C \tag{13-3}$$

where m is the *rate* of flow of mercury (in milligrams per second). The numerical coefficient includes goemetrical factors, the Faraday constant, and the density of mercury. If the current is plotted as a function of time, a fluctuating curve is obtained, as in Fig. 13-4,[15] with a period of a few seconds.

This fluctuating current is inconvenient to read. One way to obtain a valid reading is to arrange an electronic timing circuit triggered by the falling of a drop, and blanking out all but the final second or so of the lifetime of the drop; this means that the actual measurement will be made during the time that the current is changing least rapidly. An

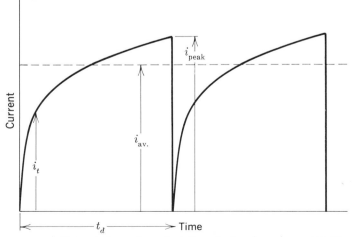

Fig. 13-4 Current-time characteristics at the dropping-mercury electrode. (*John Wiley & Sons, Inc., New York.*)

instrument for accomplishing this is the Tast (German: touch) or *strobe* polarograph.[*15]

The more common method of minimizing the drop fluctuation is by measuring the *average* current $\bar{\imath}_d$ as indicated in Fig. 13-4. This is obtained mathematically by integrating over one drop time and dividing by the time interval:

$$\bar{\imath}_d = \frac{1}{t_d} \int_0^{t_d} i_d \, dt = \frac{6}{7} i_d \tag{13-4}$$

$$\bar{\imath}_d = 607.0 n D^{1/2} m^{2/3} t_d^{1/6} C \tag{13-5}$$

In these equations, t_d is the time between the falling of successive drops (in seconds). Note that $\bar{\imath}_d$ is still proportional to C. This has the advantage that the average current can be measured easily, merely by damping out the oscillations in the meter or recorder. Equation (13-5) is well-known as the *Ilkovič* equation. The value of m for any one capillary will remain constant if the hydrostatic pressure on the mercury is kept the same. The drop time t_d varies not only with the pressure of mercury but also as a function of the applied potential; it is a maximum at a potential of about -0.5 V versus SCE.[†]

Although the temperature does not enter explicitly into the Ilkovič equation, it is nevertheless important, as each factor of the equation (except n) is to some extent temperature-dependent. The chief effect is through the temperature coefficient of the diffusion constant D. The value of the diffusion current $\bar{\imath}_d$ increases at the rate of 1 or 2 percent per degree in the vicinity of room temperature. Therefore the temperature of the electrolysis cell should in practice be controlled to within a few tenths of a degree.

Under certain conditions the Ilkovič equation fails to describe adequately the factors which determine the diffusion current. The current becomes much larger than predicted and less reproducible if the drop time becomes less than 3 or 4 sec. The effect is due to the stirring action of the fast-falling drops, which disturb the diffusion layer.

The diffusion coefficient D varies with the viscosity of the medium, and hence such a variation is found in the diffusion current. If other factors are held constant, the current should be inversely proportional to the square root of the relative viscosity. This relation is found to be valid in the absence of colloidal material, but the proportionality fails when the viscosity is increased through addition of gelatin or other hydrophilic colloid.

* Manufactured by Atlas Mess- und Analysentechnik GMBH, Bremen, Germany.
† This is the potential at which a mercury-water interface shows a maximum interfacial tension, known as the *electrocapillary maximum*.

The diffusion current also varies from its normal value if the concentration of supporting electrolyte is less than 25 or 30 times that of the reducible substance. This effect results from the fact that under these conditions the reducible ions carry an appreciable fraction of the current; this fraction is known as the *migration current*. The migration current is due to the electrostatic attraction or repulsion between the DME and the ions. Hence the observed diffusion current is increased slightly for the reduction of cations and diminished for the reduction of anions if the concentration of supporting electrolyte is lowered, while it is unchanged for reduction of nonionic species.

VOLTAGE–SCANNING POLAROGRAPHY

Up to this point we have considered current-time curves produced at a constant impressed potential. Now we will take up the effect of changing the potential. This can be done stepwise or continuously. The stepwise procedure is generally manual. The operator sets the potential with a circuit like one of those of Fig. 13-1, and waits perhaps a minute for a steady state to be attained, records the average current, then moves to the next desired potential.

To facilitate determinations of entire current-voltage curves, automatically recording instruments, generally known as *polarographs*,* have been developed. In the simplest of these, the potential applied to the cell is continuously increased, usually in the negative direction (i.e., the DME negative to the SCE), and the current is recorded. As the potential is linear with time, and the recording paper moves with constant speed, the resulting curve, called a *polarogram*, can be labeled in terms of current and potential units.

Consider a cell, such as that of Fig. 13-3, provided with DME indicator and SCE reference electrodes and filled with an oxygen-free solution which is 0.1 F in KCl and 0.001 F in $CdCl_2$. The polarogram obtained by the usual procedure resembles that shown in Fig. 13-5, which is idealized for clarity of discussion.

The curve divides itself into three regions. In region A the potential is too low to permit reduction of any of the substances known to

---

* The name Polarograph is a registered United States trademark of E. H. Sargent & Company, Chicago, and properly designates only their manufacture of voltammetric instruments. However, the word *polarograph*, without capitalization, is in common use in reference to instruments of all makes. It is interesting to note that the polarograph as invented by Heyrovský and Shikata in 1925[8] represents one of the earliest examples of an automated analytical instrument.

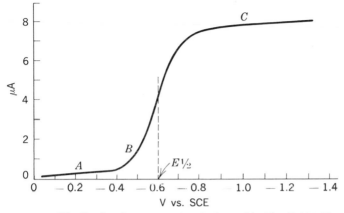

Fig. 13-5 Idealized polarogram of cadmium chloride (0.001 F) in 0.1 F potassium chloride.

be present. The small current which does flow is called the *residual current* and can be explained as the sum total of currents due to the reduction of traces of impurities (possibly iron, copper, or oxygen), and a so-called *charging current* which results from the fact that the mercury-solution interface, with its sheath of unreducible ions, acts like a capacitor of continually increasing area. The residual current is small and is reproducible if care is taken to eliminate reducible impurities.

In the vicinity of -0.5 V (B in the above example), the current starts to increase above the value of the residual current alone. This added current removes Cd^{++} ions from the layers of electrolyte in contact with the surface of the electrode, by reduction to the metal; it is replaced by diffusion from the body of the solution.

At C the current shows a saturation effect which is caused by the total depletion of Cd^{++} ions in the near vicinity of the DME. More cadmium ions continually reach the electrode by diffusion, and are reduced immediately upon arrival. The rate of diffusion is determined solely by the difference between the bulk concentration of the solution and zero, the concentration at the electrode surface. The value of the limiting current at C, called the *diffusion current*, is directly dependent upon the concentration of the reducible species; it must, of course, be corrected for the residual current.

If the negative potential is increased beyond C in Fig. 13-5, the current will rise slowly and uniformly (parallel to the residual current curve) up to about -2 V, at which point it again increases rapidly, corresponding to the reduction of hydrogen ions or water.

THE SHAPE OF THE POLAROGRAPHIC WAVE

The equation for the current as a function of potential can be derived from the Nernst equation, provided that the reaction at the microelectrode proceeds reversibly.   We will carry out this derivation for the important special case of the reduction of a simple cation to a metal soluble in mercury.   Since solutions studied polarographically are almost always very dilute, we can assume that the activity coefficient of the cation is not appreciably different from unity.   The same assumption applies to the activity coefficient of the metal in the amalgam.

Let us write the half-reaction in the form

$$M^{n+} + ne^- \rightarrow M_{(\text{Hg})}$$

where $M_{(\text{Hg})}$ denotes the metallic M dissolved in mercury.   The potential of the electrode must then be that given by the Nernst equation:

$$E_{\text{DME}} = E^\circ - \frac{RT}{nF} \ln \left(\frac{[M]_{\text{Hg}}}{[M^{n+}]_{aq}} \right) \tag{13-6}$$

Since the current $\bar{\imath}$ is limited by diffusion, it follows that

$$\bar{\imath} = k([M^{n+}]_{aq} - [M^{n+}]_{aq}{}^0)$$

where the superscript signifies conditions at the point of contact with the mercury surface.   For the limiting current $\bar{\imath}_d$, $[M^{n+}]_{aq}{}^0$ becomes very small, and we have

$$\bar{\imath}_d = k[M^{n+}]_{aq}$$

From the Ilkovič equation, it follows that the constant k is

$$\frac{\bar{\imath}_d}{C} = k = 607 n D^{\frac{1}{2}} m^{\frac{2}{3}} t_d^{\frac{1}{6}}$$

The concentration of M in the amalgam is proportional to the current, or

$$\bar{\imath} = k'[M]_{\text{Hg}}$$

The constant k' is identical with k, except that the diffusion constant D is replaced by D', a similar diffusion constant for M within the amalgam, so that the ratio $k/k' = \sqrt{D/D'}$.

These several equations can be combined to give

$$E_{\text{DME}} = E^\circ - \frac{RT}{2nF} \ln \left(\frac{D}{D'} \right) - \frac{RT}{nF} \ln \left(\frac{\bar{\imath}}{\bar{\imath}_d - \bar{\imath}} \right) \tag{13-7}$$

At the point where $\bar{\imath} = \frac{1}{2}\bar{\imath}_d$, the last term drops out, and the potential is designated as $E_{\frac{1}{2}}$, the *half-wave potential:*

$$E_{\frac{1}{2}} = E^\circ - \frac{RT}{2nF} \ln \left(\frac{D}{D'} \right) \tag{13-8}$$

Hence,

$$E_{\text{DME}} = E_{\frac{1}{2}} - \frac{RT}{nF} \ln\left(\frac{\bar{\imath}}{\bar{\imath}_d - \bar{\imath}}\right) \tag{13-9}$$

Equation (13-8) shows that the half-wave potential, which is an easily measured quantity, is simply related to the standard potential $E°$. The diffusion constants D and D' are usually not greatly different, so that $E_{\frac{1}{2}}$ is always nearly equal to $E°$ (in the absence of complexing agents; see below).

Equation (13-9) gives the form of the polarographic wave in terms of the parameters $\bar{\imath}_d$ and $E_{\frac{1}{2}}$, and provides a convenient method for establishing the value of n. The most direct measure of n is the slope of the tangent to the curve at the half-wave point. Another, more precise, method is to plot values of $\log \bar{\imath}/(\bar{\imath}_d - \bar{\imath})$ against the potential $-E_{\text{DME}}$. The equation predicts a straight line with slope given by $2.303RT/nF$, or (at 25°C) $0.0591/n$. The point on this curve corresponding to $\bar{\imath} = \frac{1}{2}\bar{\imath}_d$ will give a precise measure of $E_{\frac{1}{2}}$.

The preceding derivation is concerned with the reduction of a simple (aquo) ion. In the presence of a complex-former, the half-wave potential (Eq. 13-8) is shifted to more negative values in accordance with the relation

$$E_{\frac{1}{2}} = E° + \frac{RT}{nF} \ln K_c - \frac{pRT}{nF} \ln [X] \tag{13-10}$$

where $K_c =$ instability constant of complex
$[X] =$ concentration of complexing agent
$p =$ number of moles of X which combine with 1 g-atom of metal M

(This equation is based on the assumption that the diffusion coefficients involved are nearly equal.) Equation (13-9) is essentially unaltered by the presence of a complex, though the value of $\bar{\imath}_d$ for a given concentration may change slightly.

For the case of reduction to a species which is insoluble in both water and mercury, e.g., iron and chromium metals, the equations must be modified somewhat. This calculation has little practical importance, because in all known examples such a reduction is irreversible, which rules out application of the Nernst equation. Analyses based on such reductions may be just as useful and correct as though the process were reversible, but they must be treated empirically.

Another case which is important is the reduction of one water-soluble species to another, for example, the reduction of ferric to ferrous ions. If we return to the notation employed in Chap. 11 [Eq. (11-11), etc.] we may write as a general equation

$$A_{ox} + ne^- \rightleftharpoons A_{red}$$

If both forms of A are present in solution, the potential of the DME is given by

$$E_{DME} = E_{1/2} - \frac{RT}{nF} \ln \left(\frac{i - i_{d(a)}}{i_{d(c)} - i} \right) \tag{13-11}$$

and the half-wave potential is

$$E_{1/2} = E^\circ - \frac{RT}{2nF} \ln \left(\frac{D_{ox}}{D_{red}} \right) \tag{13-12}$$

where $i_{d(a)}$ represents an *anodic* diffusion current due to oxidation of A_{red} at the DME, $i_{d(c)}$ is the *cathodic* diffusion current corresponding to reduction of A_{ox}. D_{ox} and D_{red} are the diffusion coefficients of the respective forms, which may safely be assumed very nearly equal to each other. Thus E°, the standard potential for the redox couple, is very closely equal to $E_{1/2}$. The DME acts as an inert electrode in this case. If either A_{ox} or A_{red} is absent from the solution, the corresponding i_d becomes zero in Eq. (13-11), but the value of $E_{1/2}$ remains unchanged.

The above discussion was given for the most part in terms of cathodic reduction taking place at the DME, as this is the most widely applicable condition. However, the same discussion holds true where the electrode process is anodic oxidation.

Figures 13-6 to 13-8 show a few polarograms to illustrate the above

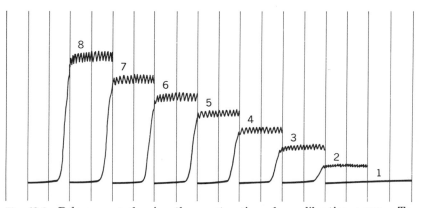

Fig. 13-6 Polarograms showing the construction of a calibration curve. To 10 ml of 1 F NH$_3$, 1 F NH$_4$Cl (curve 1) have been added successive 0.05-ml increments of 0.05 F Cd^{++} (curves 2 to 8); each curve starts at -0.2 V; each scale division corresponds to 200 mV. (*Academic Press Inc., New York.*)

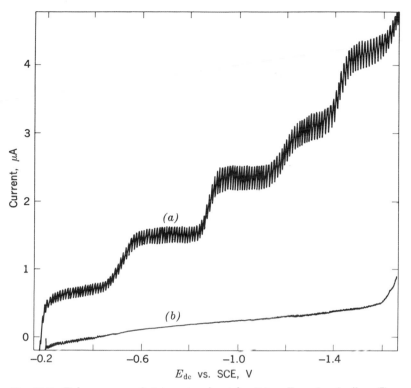

Fig. 13-7 Polarograms of (*a*) approximately 0.1 m*F* each of silver(I), thallium(I), cadmium(II), nickel(II), and zinc(II), listed in the order in which their waves appear, in 1 *F* NH₃, 1 *F* NH₄Cl containing 0.002 percent Triton X-100; (*b*) the supporting electrolyte alone. (*John Wiley & Sons, Inc., New York.*)

points. Figure 13-6 shows waves of one reducible species at several concentrations, together with the residual curve.[7]

Figure 13-7 is a polarogram[13] showing the sequential reduction of five cations with various properties: (1) Ag^+ is reduced so easily that its wave cannot be formed completely, though the corresponding diffusion plateau is well defined; (2) Tl^+ shows a reversible reduction with $n = 1$; (3) Cd^{++} is also reversible, but the slope shows that $n = 2$; (4) Ni^{++} shows a more drawn-out curve, though $n = 2$, which indicates irreversibility; and (5) Zn^{++} resembles cadmium in showing a reversible wave with $n = 2$.

Figure 13-8 shows the anodic oxidation of ferrous ion (curve *c*) compared with the reduction of ferric (curve *a*).[13] Curve *b* is obtained when both forms are present in equivalent concentrations. The vertical lines indicate the observed positions of the half-wave potentials, which should be identical.

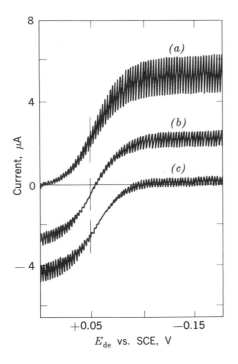

Fig. 13-8 Polarograms of (*a*) 1.4 m*F* iron(III), (*b*) 0.7 m*F* each of iron(II) and iron(III), (*c*) 1.4 m*F* iron(II); the supporting electrolyte is saturated oxalic acid containing 0.0002 percent methyl red. (*John Wiley & Sons, Inc., New York.*)

MAXIMA

Frequently, as the applied potential is raised, the current, after increasing as an ion is discharged, fails to level off, but decreases again, leaving a maximum in the curve. Such a maximum may be only a slight hump, or it may be a very sharp peak, exceeding the true wave height by a factor of 2 or more. This phenomenon is not fully understood, though it appears to be related to the tangential streaming motion of solution past the surface of the drop.

Maxima can usually (but not always) be eliminated by the addition of an organic surfactant. Gelatin has often been used for this purpose, but has the disadvantage that its solutions are prone to deterioration and must be made up fresh every day or two. Certain dyes, such as methyl red, are sometimes effective. The nonionic detergent Triton X-100* has been found to be particularly useful as a maximum suppressor, and is widely employed. A stock solution of 0.2 percent Triton X-100 is convenient; 0.1 ml of this for each 10 ml of solution in the polarograph cell is usually satisfactory.

Care must be taken not to use too much suppressor, or the desired

* Rohm & Haas Co., Philadelphia.

wave may be distorted or suppressed along with the maximum. The half-wave potential is apt to be shifted a few millivolts from its normal value.

Figure 13-9 shows polarograms obtained on a solution containing $3 \times 10^{-3} F$ Pb(II) and $2.5 \times 10^{-4} F$ Zn(II) in $2 F$ NaOH, with and without the addition of Triton.[13] The lead ($E_{1/2} = -0.8$ V) shows a sharp peak, while the zinc ($E_{1/2} = -1.5$ V) shows a more gradual maximum. The diffusion current for lead could be measured without great error by ignoring the maximum, but no valid wave height could be obtained for zinc, partly because of its smaller concentration. The suppressor effectively eliminates both maxima.

OXYGEN INTERFERENCE

Dissolved oxygen is reducible at the micro-electrode in many media. A typical example of the effect of oxygen on a polarogram is shown in Fig. 13-10.[13] Two waves are observed. The first ($E_{1/2} = -0.05$ V versus SCE) is caused by the reduction of O_2 to H_2O_2, the second ($E_{1/2} = -0.9$ V)

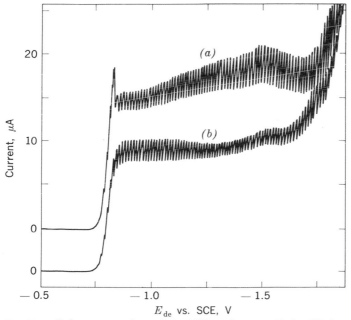

Fig. 13-9 Polarograms of 3 mF lead(II) and 0.25 mF zinc(II) in 2 F sodium hydroxide: (*a*) in the absence of a maximum suppressor, and (*b*) after the addition of 0.002 percent Triton X-100. (*John Wiley & Sons, Inc., New York.*)

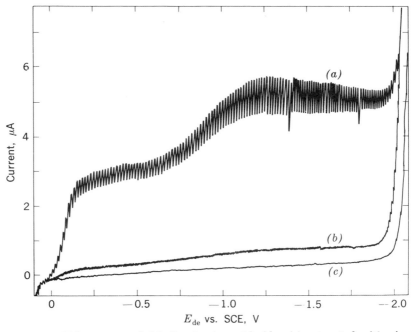

Fig. 13-10 Polarograms of 0.1 F potassium chloride: (a) saturated with air, showing the double wave of oxygen, (b) partially deaerated, and (c) after complete deaeration. (*John Wiley & Sons, Inc., New York.*)

corresponds to the reduction of O_2 to H_2O. The first of these waves shows an intense, sharp maximum in the absence of a suppressor and particularly in a dilute supporting solution. This wave can be used in the analytical determination of dissolved oxygen, but more often appears as an interference of high nuisance value. So in most polarographic work, provision must be made for removal of oxygen. In alkaline solutions this can often be easily accomplished by adding a small amount of potassium sulfite which reduces the oxygen quantitatively. In any case the oxygen can be removed by flushing the solution with a nonreducible gas, such as nitrogen. Using a simple constricted glass tube as bubbler, this may take 20 to 30 min. The time required may be cut to 2 to 3 min by the substitution of a fritted-glass gas disperser. The stream of gas must be discontinued while data are being taken because of the undesirable stirring action. Mercury should be excluded from the solution until after the deaeration, since it may be oxidized by the dissolved oxygen in certain media. This means that the capillary should not be inserted into the cell prior to the removal of oxygen.

INSTRUMENTS

A satisfactory manual polarograph can easily be constructed from electrical components commonly available in the laboratory, with a circuit such as those of Fig. 13-1. The indicating instrument, a galvanometer or microammeter, should have a full-scale range of not more than 50 μA, preferably less, and it should be equipped with a series of calibrated shunts to permit measurement of larger currents. For many applications true current values are not needed, and the readings may be taken simply in units of galvanometer deflection. There are a number of manual instruments on the market, all of which are essentially similar

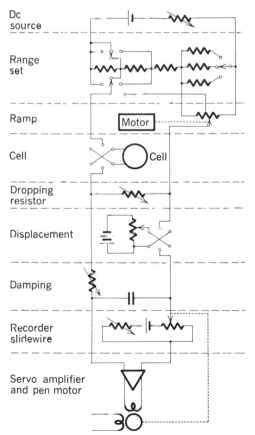

Fig. 13-11 Basic schematic of the Sargent Model XV Polarograph. Standardization circuits and portions of switching details have been omitted. (*E. H. Sargent & Co., Chicago.*)

to each other, varying somewhat in precision, auxiliary features, and convenience of use.

A modernized version of the original recording polarograph of Heyrovský and Shikata is now manufactured in the United States by Sargent as their Model XII. The record is made photographically on a sheet of sensitized paper clipped to a drum which is rotated by a small motor. The motor also drives the slide-wire contact in the voltage source. A beam of light is reflected by the mirror of a galvanometer to fall upon the moving paper. Upon development the paper shows a clear trace of the polarographic curve.

Current practice in instrument design is to include a strip-chart recorder, either as an integral part of the instrument, or as an essential accessory. A widely distributed example is the Sargent Model XV, diagrammed in Fig. 13-11. Omitted from this diagram are auxilliary circuits for standardizing both current and voltage scales by comparison with a self-contained standard cell; details of switching for damping and current ranging; and the internal design of the vacuum-tube servoamplifier.

In this instrument there is no choice of paper speed (1 in./min)*, and the voltage scan covers its range in 10 min., so that every completed polarogram is 10 in. long. The range selection switching permits choice of a span of 1, 2, or 3 V, to start at any of several specified voltages from +1.5 to −4.0, and scan in either direction within these limits.

The cell current passes through a precision dropping resistor, one of a bank of 19, so selected that full-scale deflection will always correspond to 2.5 mV. To the potential drop across the resistor may be added an adjustable displacement voltage which permits the zero point to be positioned anywhere on the recorder scale or suppressed in either direction by as much as six scale lengths (in some polarographs the corresponding control is called the "bias" adjustment). Across the circuit at this point is placed an optional RC damping network which permits increase of the time constant to reduce fluctuations due to the falling drops of mercury. The potential is then compared with the voltage drop across part of a precision slide-wire, the slider of which is positioned by a servoamplifier and motor to give a potentiometric balance. The servomotor also drives the recording pen.

CORRECTION FOR RESISTANCE

The voltage V applied to the polarographic cell circuit may be equated to the potential of the anode E_A, of the cathode E_C, and the summation

---

* It is curious and mildly unfortunate that Sargent has chosen to mix mensural systems, so that chart paper is marked off in inches in one direction and millimeters in the other.

of all the ohmic-potential drops in the system:

$$V = E_A - E_C + I \Sigma R \tag{13-13}$$

The resistance includes that of the solution itself, of each electrode, of the column of mercury, of that portion of the polarizing slidewire included at a given moment, and of the current-measuring resistor. All of these except the resistance of the solution are under the control of the instrument designer and can be minimized satisfactorily.

However, circumstances may require a high-resistance solution, especially in nonaqueous or mixed solvents, so that the IR term cannot be neglected. A number of methods of correcting for this potential drop have been suggested. The resistance can be measured independently, and a correction term calculated for each measured point, but this is awkward and time-consuming. It is possible to install an auxiliary potentiometer on the pen motor of the recorder, so that its position will be determined by the magnitude of the current; it can be connected into the circuit in such a way as to counteract the IR drop. The difficulty with this arrangement is that it must be differently adjusted for each new solution if the resistance is not identical.

A better way of eliminating the effect of the resistive drop is through the use of a *three-electrode cell*, which contains the DME, a reference electrode, and a counter electrode.[2] The reference electrode (SCE) does not carry the cell current, hence does not need to be of a low-resistance construction. The tip of the salt bridge from the SCE must be placed close to the DME, and not in the direct path between it and the counter electrode, so as not to sense part of the IR drop. An isolating amplifier (usually an "operational" amplifier) is required to take cognizance of the potential of the reference electrode without drawing current through it. Such a circuit has been reported as a simple and successful accessory for the Sargent XV (presumably equally usable

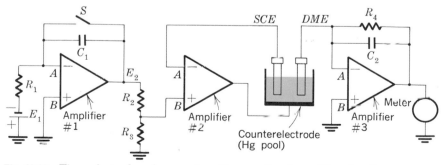

Fig. 13-12 Three-electrode polarograph, with operational amplifier circuitry.

with other commercial instruments).   Figure 13-12 shows such a three-
electrode cell with a completely redesigned circuit with which to operate it.

This circuit includes three *operational amplifiers* (triangular symbols).   Operational amplifiers will be discussed in some detail in a later
chapter.   For present purposes, their salient features are the following:
(1) no current can flow into or out of the amplifier inputs (marked A
and B); (2) the amplifier operates in such a way as to maintain *equal*
potentials at its two inputs.

An amplifier connected with only a capacitor from its output back
to its input acts as an integrator with respect to time.   Amplifier No. 1
in Fig. 13-12 is such an integrator, and if E_1, R_1, and C_1 are constant,
its output E_2 will be

$$E_2 = -\frac{E_1}{R_1C_1}\int_0^t dt = -\frac{E_1}{R_1C_1}t \tag{13-14}$$

This indicates that the output of this amplifier will increase linearly
with time, with a slope determined only by circuit constants.   Because
of the negative sign, if E_1 is negative, then E_2 will be positive-going.
This increasing potential is called a *ramp* function.

An appropriate fraction of the ramp (determined by the magnitudes
of R_2 and R_3) is applied to the B input of amplifier No. 2, which is
thus kept at the ramp potential above ground.   The only response that
is available to amplifier No. 2 is to emit current from its output to the
counter electrode.   This current cannot go to the SCE, because the
SCE connects only to the input of an amplifier, and hence cannot carry
current; therefore, the current must flow through the DME, resistor
R_4, and the meter to ground.   But the DME is connected to one input
of amplifier No. 3, the other of which is grounded, so that because of the
equality of inputs, the DME is essentially at ground potential at all
times.   The SCE, for the same reason, is at ramp potential above
ground.   The result is that the No. 2 amplifier allows precisely the
required amount of current to flow to maintain the SCE-DME potential difference at the desired linearly increasing potential, with the DME
negative-going.   Hence the meter (or recorder) will give the true polarogram.   The function of capacitor C_2 is to lengthen the time constant,
to give a desired amount of damping; R_4 relates to the scale of the
meter.   The switch S across the integrating capacitor is used to start
the polarogram; the moment of opening S corresponds to time zero.
Range switching and zero offset (bias) are easily supplied, but omitted
from the figure for clarity.

This type of polarograph is convenient to use, has no moving
parts, and is quite versatile.   It can be used with aqueous solutions of

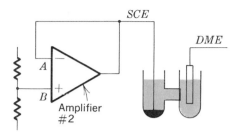

Fig. 13-13 Circuit of the preceding figure connected for two-electrode polarography.

low resistance as well as with nonaqueous, high-resistance media. The same electronic system can be operated with a conventional two-electrode cell, as in Fig. 13-13. Here the current flows through the SCE, which must therefore be a low-resistance type, as in the H cell. Other features of Figs. 13-12 and 13-13 are identical.

QUALITATIVE ANALYSIS

Since the half-wave potential is characteristic of the substance undergoing reduction or oxidation at the microelectrode, this parameter can be utilized for its identification. The value of $E_{1/2}$ for a given substance depends on the nature of the supporting electrolyte, largely because of variation in the tendency to form complex ions. A few representative values are listed in Table 13-1. The importance of wise selection of electrolyte can be seen by comparing the data for lead and cadmium. These cations have identical half-wave potentials in NaOH, but are separated fairly well in KCl or H_3PO_4, and even further in KCN.

Many half-wave potential data are to be found in handbooks and monographs on polarography (see bibliography). However it is frequently expeditious to plot polarograms of known substances for direct comparison with similar curves for unknowns. Published values for which information regarding the exact composition of the supporting electrolyte and nature of the reference electrode is lacking are of little value for purposes of identification.

The half-wave potential can be determined graphically as shown in Fig. 13-14. Portions AB and DF of the curve are extended as shown, and a tangent is drawn to the curve at its inflection point C. The line GH is bisected, and a line JK is drawn parallel to AB and DF. The abscissa of the point of intersection of JK with the curve gives the value of $E_{1/2}$. This rather complicated procedure is necessary in the frequently encountered, nonideal case where DF is not so nearly parallel with AB as might be desired. According to the described procedure, slight errors of judgment in the location of the tangent line GH will have negligible effect on the ultimate value of $E_{1/2}$.

Table 13-1   Half-wave potentials of some common cations in various supporting electrolytes, V versus SCE*

Cation		*Supporting electrolyte*			
	KCl (0.1 F)	NH$_3$ (1 F) NH$_4$Cl (1 F)	NaOH (1 F)	H$_3$PO$_4$ (7.3 F)	KCN (1 F)
Cd^{++}	−0.60	−0.81	−0.78	−0.77	−1.18†
Co^{++}	−1.20†	−1.29†	−1.46†	−1.20†	−1.13† [to Co(I)]
Cr^{3+}	....‡	−1.43† [to Cr(II)] −1.71† [to Cr(0)]	....	−1.02† [to Cr(II)]	−1.38 [to Cr(II)]
Cu^{++}	+0.04 [to Cu(I)] −0.22 [to Cu(0)]	−0.24 [to Cu(I)] −0.51 [to Cu(0)]	−0.41†	−0.09	NR§
Fe^{++}	−1.3†	−1.49†	....	....	....
Fe^{3+}	....	....	−1.12¶ [to Fe(II)] −1.74¶ [to Fe(0)]	+0.06 [to Fe(II)]	....
Ni^{++}	−1.1†	−1.10†	....	−1.18	−1.36
Pb^{++}	−1.40	....	−0.76	−0.53	−0.72
Zn^{++}	−1.00	−1.35†	−1.53	−1.13†	NR§

* From data published by Meites.[13]
† Irreversible reduction.
‡ .... indicates insufficient solubility or lacking information.
§ NR indicates that the ion is not reducible in this medium.
¶ 3 F KOH solution plus 3 percent mannitol.

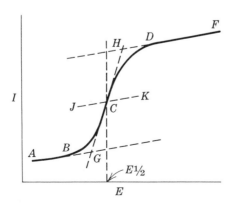

Fig. 13-14 Graphical method of locating the half-wave potential.

QUANTITATIVE ANALYSIS

The magnitude of the diffusion current is related to the concentration of reducible species by the Ilkovič equation. Hence if the several factors of that equation are known or can be measured, the concentration of the species can theoretically be computed from the observed current. The only factor which is difficult to evaluate independently is the diffusion coefficient. It can in some instances be determined from measurements of the electric conductivity of the solution, or by chronopotentiometry. It can sometimes be estimated by comparison with known values for other ions of comparable size. It must be remembered that the diffusion coefficient is sensitive to the viscosity and temperature of the solution. Coefficients for many species can be found in the literature, but must be applied with great caution.

Many factors which are difficult to handle in a polarographic analysis may be avoided by making only comparative measurements. To do this it is necessary to prepare standard solutions of the desired substance, and determine their current-voltage curves under the same conditions as the unknown. Then the Ilkovič equation can be applied in the simplified form

$$i_d = KC \qquad (13\text{-}15)$$

The proportionality constant can be evaluated graphically and need not be resolved into its theoretical equivalent.

An example of a typical analysis will make this clear. A sample of zinc is suspected of containing some cadmium, and a quantitative analysis is required. A 0.1-g sample is dissolved in hydrochloric acid, enough Triton added to eliminate maxima, and the solution diluted to a known volume with 1.0 F potassium chloride. A small portion is placed in the polarographic cell, and oxygen is swept out with a current of nitrogen. A preliminary current-voltage curve is plotted, with the galvanometer operating at about one-tenth its full sensitivity. The potential range from about -0.4 to -0.8 V is adequate for this analysis; it is not always necessary to run a complete curve. The trial curve will show qualitatively whether or not cadmium is present, according to whether a wave is found at about -0.64 V. The reduction potential of zinc is so great that it will not interfere with the wave for cadmium. If cadmium proves to be present, the preliminary curve will give an idea of the relative dilution needed to give the best analysis. The optimum range of concentration for the reducible ion is approximately 10^{-3} to 10^{-5} F. If the solution is already too dilute, the sensitivity of the galvanometer can be increased; if too concentrated, an aliquot can be diluted further with potassium chloride solution. Another current-

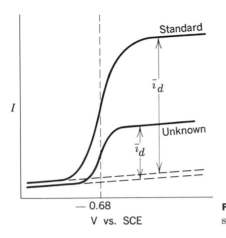

Fig. 13-15 Polarograms of an unknown and standard.

voltage curve is now plotted for the final analysis. A standard solution of cadmium chloride in 1.0 F potassium chloride must also be prepared at about the same final dilution, and a polarogram must be plotted for it. The resulting curves will resemble Fig. 13-15. The values of $\bar{i}_d$ for both standard and unknown are measured from the graph. The concentration C_x of cadmium in the final dilution of the unknown can be calculated by a simple proportion. Alternatively, a calibration graph (Fig. 13-16) can be constructed, from which concentrations of this and any future cadmium unknowns can be read. It is, of course, wise to

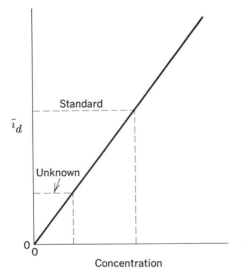

Fig. 13-16 Calibration curve for a polarographic analysis. The indicated heights for standard and unknown correspond to the curves of Fig. 13-15.

check such a calibration curve by analyzing two or more standard solutions of varying concentrations.

In a modification of this procedure, a known quantity of standard solution can be added to a measured portion of the unknown. Polarograms taken before and after the addition give all the information needed to calculate the concentration of the unknown. This procedure is known as the method of *standard addition*. It has the advantage that unknown and standard are certain to be measured under identical conditions (assuming constant temperature). The calculation is slightly more involved, however.

All quantitative polarographic measurements must take into account the appropriate residual current. If a considerable voltage is scanned before the start of the wave, the residual current can be approximated by extrapolation, as in Fig. 13-15. Otherwise it may be necessary to run a blank; this would be the only way, for example, to measure the height of the first oxygen wave in Fig. 13-10. Error from this source will become larger as one goes to smaller concentrations, and some automatic method of compensation is called for. Such compensation can be provided in a recording polarograph by a device which feeds a small current proportional to the scanning voltage in opposition to the cell current. This cannot give complete compensation, because the residual current is not strictly linear, but it can permit polarographic analysis down to 10^{-5} or even 10^{-6} formal.

Another limitation to quantitative precision in polarography appears when the waves corresponding to two reducible species are relatively close together. If they are so close that no horizontal portion is evident between them, then there is no way in which the individual diffusion currents can be measured, though the sum of the two species can be determined. The only way to improve the situation without resorting to chemical separation is to seek another supporting electrolyte; for example, according to Table 13-1, Cd^{++} and Pb^{++} cannot be distinguished in NaOH, but are easily resolved in KCN medium.

PILOT-ION PROCEDURE

Since the diffusion coefficients vary from ion to ion, there is no uniformity in the heights of the waves obtained with equivalent concentrations of different reducible ions. Figure 13-17 shows the waves of a number of ions at equal concentrations.[12] The ordinates are given in terms of the quantity $\bar{\imath}_d/(Cm^{2/3}t^{1/6})$, which is sometimes defined as the *diffusion-current constant*. The Ilkovič equation predicts constancy for this ratio for any given temperature. Its value gives the relation between diffusion current and concentration, so that a knowledge of this quantity obviates the need

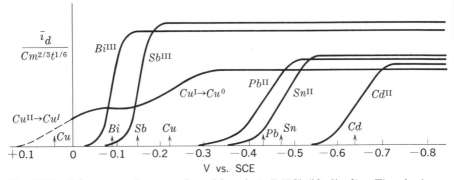

Fig. 13-17 Polarograms for a number of ions in 1 F HCl (idealized). The abscissas are volts, the ordinates are relative current values calculated on the basis of equal concentrations of ions, measured with a constant value of $m^{2/3}t^{1/6}$. Gelatin is present in each case. (*Analytical Chemistry.*)

for repeated calibrations. This possibility cannot be fully realized since the values of m and t must be redetermined for each new capillary. However, the values for the various ions all change in the same proportion. Hence, once a series of these constants are determined for one capillary, it is necessary to repeat only the determination of the constant for *one* ion to establish those of the entire series for a new capillary. This makes it necessary to maintain only a single standard stock solution for each supporting electrolyte likely to be needed. This is called the *pilot-ion* method.

ORGANIC POLAROGRAPHY

Polarography presents an important tool for analysis and structure determination in organic chemistry. The principles are no different from those discussed above. The product of electrochemical action is, of course, insoluble in mercury, but is almost always soluble in whatever solvent or solvent mixture is suitable for the original material. Any solvent which will dissolve an electrolyte is potentially useful for polarography. Various alcohols and ketones, pure or mixed with water, have been used, as have molten urea, ammonium formate, dimethylformamide, ethylenediamine and others. Certain substituted ammonium salts, such as tetrabutylammonium iodide, are readily soluble in nonaqueous solvents and serve admirably as supporting electrolytes.

Many classes of organic compounds can be reduced at the DME: conjugated unsaturated compounds; certain carbonyl compounds; organic halogen compounds; quinones; hydroxylamines; nitro, nitroso, azo, and azoxy compounds; amine oxides; diazonium salts; certain sulfur compounds; certain heterocyclic compounds; peroxides; and reducing sugars. For details the student is referred to the literature.[19, 20]   A typical example is the analysis of a mixture of aromatic carbonyl compounds.   The following half-wave potentials (vs. SCE) were observed in a supporting electrolyte of lithium hydroxide in aqueous ethanol: benzaldehyde, -1.51 V; *n*-propyl phenyl ketone, -1.75; isopropyl phenyl ketone, -1.82; *tert*-butyl phenyl ketone, -1.92.   The waves were clearly defined, and the diffusion currents were linear with concentration over the range 0.2 to 2.5 mM.   It is found that the pH and the ionic strength of the solution are more important in organic polarography than is usual in the inorganic field.

RAPID-SCAN POLAROGRAPHY

One possible method of avoiding the inconvenient serrations caused by dropping mercury is to scan the voltage range at such a high rate that a complete polarogram is obtained within the lifetime of a single drop. The rate of change of voltage required is too great for practicable motor drives, but can be obtained without difficulty by an electronic ramp generator like that in Fig. 13-12.   The recorder must be able to handle a rapidly changing signal, and this is well within the reach of many modern recorders.

The curve obtained by this method does not resemble a conventional polarogram, but shows a peak of a characteristic shape, as in Fig. 13-18, which is a rapid-scan polarogram of a solution with two reducible species.[16]

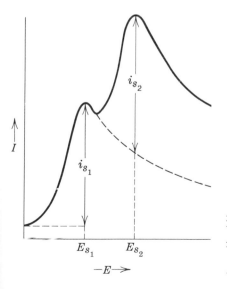

Fig. **13-18**   Rapid-scan polarogram of a mixture of two reducible species.   The background current for the second peak must be extrapolated from the first peak (dashed line).   (*Academic Press Inc., New York.*)

The reason for the peak shape (which is not related to the maxima appearing in conventional polarograms) is that the slow process of diffusion is unable to supply reducible material to the electrode fast enough to keep up with the rapidly increasing potential, so that a steady state is never attained. It can be shown mathematically that E_s, the *summit potential*, is related to the half-wave potential:

$$E_s = E_{1/2} - 1.1 \frac{RT}{nF} \tag{13-16}$$

which at 25°C becomes

$$E_s = E_{1/2} - \frac{0.028}{n} \quad \text{volts} \tag{13-17}$$

The value of the current at the summit for a reversible system is given by an equation derived independently by Randles and by Ševčik, which is analogous to the Ilkovič equation, but includes the derivative dE/dt:

$$i_s = k n^{3/2} m^{2/3} t^{2/3} D^{1/2} \left(\frac{dE}{dt}\right)^{1/2} C \tag{13-18}$$

The linear relation between i_s and the concentration C is to be noted. The constancy of the ratio of i_s to $(dE/dt)^{1/2}$ is a good test for the degree of reversibility. The Randles-Ševčik equation is valid only when both oxidized and reduced species are soluble (in water or mercury).

CYCLIC VOLTAMMETRY

This is a modification of the rapid-scan technique, wherein the direction of scanning is reversed following the reduction of interest. To accomplish this, a so-called *triangular wave* voltage is applied to the electrolytic cell (Fig. 13-19a) rather than the simple ramp function. A typical curve obtained by this method is shown in Fig. 13-19b.[16] The entire process takes place in a second or two near the end of the lifetime of a mercury drop. When the voltage scan is first applied, the current will start near the origin (A) and only residual current will flow until the potential is negative enough to effect the reduction of Zn(II), whereupon a maximum appears, exactly analogous to one of those in Fig. 13-18. At point D the *direction* of scan is reversed, so that the voltage proceeds back toward zero with the same rate at which it had previously increased. The sudden drop (D to E) is caused by the reversal of the capacitive current. In the region E to F, the current, still cathodic, continues

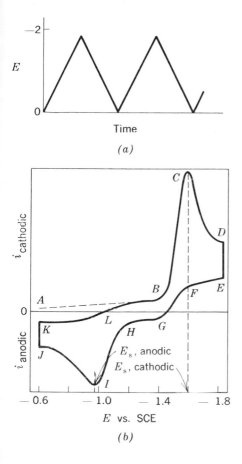

Fig. **13-19** Cyclic voltammetry. (*a*) Triangular-wave excitation function. (*b*) Voltammogram obtained with 10^{-3} F Zn(II) in 2 F NH$_3$, NH$_4$Cl buffer. (*Academic Press Inc., New York.*)

the process of depletion of Zn(II) from the vicinity of the electrode which had been taking place in the C to D region. As the voltage gets to the point F, a diffusion steady state again becomes the controlling factor, and the current drops to the G-H area, where it is essentially the residual capacitive current. In the H-I-J region, metallic zinc, the product of prior reduction, is reoxidized, with a large anodic current, diffusion-controlled. At J the voltage scan is again reversed. The sudden rise in current, J to K, followed by a slow rise, K to L, represents the inverse of the D-to-G decrease. At L the curve merges with the original A to B section, and from there the entire sequence is repeated. The fact that the current at I is not equal in magnitude to that at C may be due to incomplete removal of zinc metal from the mercury drop, and to a difference between diffusion coefficients for the aqueous species containing

Zn(II), presumably $[Zn(NH_3)_4]^{++}$ ion, and metallic Zn in the amalgam. The concentration gradient causing diffusion is, of course, very unequal for the two.

The electrode reaction used as an example here is highly irreversible, as shown by the large difference (about 0.6 V) between $E_{s,\,cathodic}$ and $E_{s,\,anodic}$. For a reversible process with $n = 2$, this difference (at 25°C) would be 0.28 V, predictable from Eq. (13-17). This makes another convenient way to determine the degree of reversibility.

OTHER MODIFICATIONS

Several additional methods have been developed which are modifications of polarography, but which are less frequently employed and require more involved apparatus, so that we will not discuss them in detail. The first of these is *ac polarography*, in which the ramp voltage is modulated with a small alternating potential (a few millivolts); the ac component of the electrode current is then measured. The rapid, small variations in potential, via the Nernst equation, produce a corresponding rapid variation in the ratio of oxidized to reduced form of the electroactive species, which means that an alternating current must flow. The magnitude of the current is greatest where the two concentrations are equal, at the half-wave potential (for a reversible system).

In *square-wave polarography*, the ramp is modulated with a square rather than a sine wave. The chief advantage is that the residual (capacitive) current can be made negligible, because after the initiation of each voltage pulse, the capacitive current dies out logarithmically in a very short interval, leaving the faradaic current at a finite value which can be read with great precision. The timing circuits and related instrumentation become quite complex. Closely related is *pulse polar-*

Table 13-2   Comparison of different types of polarography*

Polarography	*Limiting sensitivity, F*	*Resolvability, mV*	*Separability*
Conventional direct current	1×10^{-5}	100	10 : 1
Rapid-scan	3×10^{-7}	40	400 : 1
Alternating current	5×10^{-7}	40	1000 : 1
Square-wave	5×10^{-8}	40	20,000 : 1
Pulse	1×10^{-8}	40	10,000 : 1

* The values, taken from Schmidt and von Stackelberg,[16] are approximate only, and correspond to the most favorable situations in each case.

ography, in which the magnitude, duration, and spacing of voltage pulses are all variable independently, with corresponding increase in complexity as well as flexibility.

Table 13-2 gives some comparative figures for these variations of polarography. *Resolvability* is defined as the minimum voltage interval between two half-wave potentials which will permit measurements of both wave heights; *separability* refers to the maximum ratio of concentration of substance A to that of B, where the $E_{1/2}$ for B is more negative than that of A by the amount in the resolvability column. Each figure represents the best observed example, which may not be attainable in routine practice.

AMPEROMETRIC TITRATION

It is possible to carry out a titration in a voltammetric electrolysis cell and to follow its progress by observing the diffusion current after successive additions of reagent. This is analogous to potentiometric and conductometric titrations, and is known as *amperometric titration*. Since the diffusion current is generally proportional to concentration, the titration curve is found to consist of two straight-line segments, the intersection of which corresponds to the equivalence point. Three types of curves may be distinguished (Fig. 13-20). Curve *a* results from the titration of a reducible ion by a reagent which does not itself yield a polarographic wave. An example is the titration of lead ion by oxalate with the dropping-mercury cathode at a potential of -1.0 V versus SCE, with potassium nitrate as the supporting electrolyte. The initial diffusion current is relatively high, and it decreases regularly as the Pb^{++} ion is removed by reaction with added oxalate. After the equivalence point, further increments of reagent have no effect on the current.

The reverse titration, oxalate by lead ion, under similar conditions will give a curve such as *b* in Fig. 13-20. The Pb^{++} ion cannot accumulate in the solution and give a diffusion current until all the oxalate is precipitated. Another example is the precipitation of lead, as the

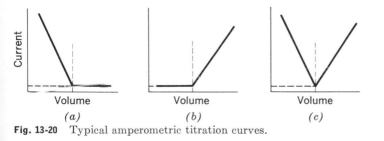

Fig. 13-20 Typical amperometric titration curves.

chromate, by potassium dichromate in an acid solution, with zero applied volts, that is, with the dropping-mercury and calomel electrodes both directly connected to the galvanometer. Under these conditions the lead ion is not reduced, but the dichromate ion gives a well-defined diffusion current.

Curve c results from the titration of lead by dichromate at an applied potential of -1.0 V, where both lead and dichromate ions are capable of being reduced at the dropping cathode.

Since the diffusion current is lowered by the addition of solvent, any amperometric titration data must be corrected for the *dilution effect* resulting from the addition of reagent, if straight lines are to be obtained. This correction may be accomplished by multiplying each diffusion current reading by the factor $(V + v)/V$, where V denotes the original volume of the solution titrated and v the volume of reagent added. The need for applying this correction may be avoided by employing a reagent solution of 10 times or more greater concentration than the solution titrated, so that the volume added will be negligible compared with the total volume. If these procedures are not followed, the graph will consist of lines which are curved rather than straight, though the location of the end point is unchanged.

Appreciable solubility of the precipitate causes rounding of the graph in the region of the equivalence point. If the solubility is too great, of course, the method will fail. The limit of permissible solubility lies in the range of 10^{-4} to 10^{-5} F, but it varies with the original concentration of the solution being titrated and with other factors. It may be advantageous to carry out the titration in 50 percent alcohol, to reduce the solubility of the product.

Amperometric titration is applicable not only to precipitation reactions but to many redox and complexometric titrations as well. It is inherently capable of greater accuracy than nontitrative polarographic methods, since each analysis involves a number of separate determinations so related that individual errors tend to cancel.

TITRATION APPARATUS

Any standard polarographic instrument can be used to follow titrations. However, just as in the potentiometric method, simplification is possible. In the first place, it is necessary to adjust the applied potential only to the nearest tenth of a volt, which makes a highly precise voltage divider and meter unnecessary. Secondly, as only relative current values are needed, calibration of the galvanometer can be omitted. The specific characteristics of the capillary are also without effect. The temperature must remain essentially constant throughout a titration, but need not be recorded. As a titration requires perhaps only 10 min, temperature control is not a serious problem.

There are a number of reducible substances which produce a diffusion current at zero applied potential relative to the SCE. Silver and chromate ions are two of these. A considerable simplification in apparatus can be effected in such cases, as no polarizing unit is needed. Since, for titrations, potentials need not be selected closer than about 0.1 V, it is theoretically possible to choose a reference electrode for any given titration so that the diffusion potential will apear without any external voltage source. A number of suitable half-cells for this purpose are listed in Table 13-3, in the order of their potentials. They can be

Table 13-3 Selected reference half-cells for amperometric titration

Half-cell	Volts	
	vs. NHE	vs. SCE
KI (1 F); AgI; Ag	−0.137	−0.383
KI (0.1 F); AgI; Ag	−0.084	−0.330
KBr (1 F); AgBr; Ag	0.086	−0.160
NaOH (0.5 F); HgO; Hg	0.150	−0.096
SCE	0.246	0.000
KCl (0.1 F); Hg_2Cl_2; Hg	0.335	0.089
H^+ (pH = 5); quinhydrone; Pt	0.405	0.159
K_2CrO_4 (0.2 F); Ag_2CrO_4; Ag	0.500	0.254
H^+ (pH = 2); quinhydrone; Pt	0.582	0.336
H_2SO_4 (1 F); Hg_2SO_4; Hg	0.682	0.436

prepared in advance and kept on hand ready for use as occasion demands. A longer table of suitable half-cells has been given by Harris.[6]

An example of the application of this principle is the titration of mercaptans with ammoniacal silver nitrate.[9] A reference electrode, −0.23 V versus SCE, was prepared by dissolving 4.2 g of potassium iodide and 1.3 g of mercuric iodide in 100 ml of saturated potassium chloride solution. The resulting solution was placed in contact with metallic mercury and connected with the titration vessel by a potassium chloride salt bridge.

ROTATING PLATINUM ELECTRODE

The substitution of a rotating platinum electrode for the DME gives increased sensitivity, because of the partial disruption of the diffusion layer by stirring. The electrode is usually fabricated with a 2 to 3 mm length of platinum wire extending horizontally from its vertical glass

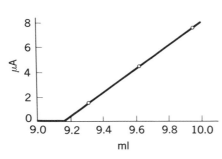

Fig. 13-21 Amperometric titration of arsenite by bromate. (*Journal of Physical Chemistry.*)

supporting tube. The tube is rotated at a few hundred rpm, which rate must be held quite constant to obtain consistent results. Another advantage of the platinum electrode is the greatly reduced residual current.

An excellent example is the titration of arsenite by potassium bromate in the presence of bromide.[10] The arsenite solution is made 1 F in hydrochloric acid and 0.05 F in potassium bromide. The applied potential is +0.2 to 0.3 V versus SCE. The reaction is

$$3AsO_2^- + BrO_3^- \rightarrow 3AsO_3^- + Br^-$$

The bromide serves as an indicator, since only after the arsenite is completely reacted can the bromide be oxidized

$$BrO_3^- + 5Br^- + 6H^+ \rightarrow 3Br_2 + 3H_2O$$

Of all the substances present, free Br_2 is the only one which gives a polarographic wave. Figure 13-21 shows the curve resulting from the titration of 100 ml of a 9.18×10^{-4} N solution of arsenious acid with 0.0100 N potassium bromate. At this applied potential, oxygen is not reduced at the cathode; hence it is not necessary to remove oxygen before analysis.

In this titration with a rotating electrode, it is not necessary to make any readings prior to the equivalence point. The operator adds reagent continuously at a moderate rate until he sees some deflection. Then three or four readings in the presence of excess reagent will establish the sloping portion of the curve, which is extrapolated back to the zero axis to determine the equivalence point.

The number of titrations to which the amperometric method can be applied is much greater than that for potentiometric titration because the electrodes are nonspecific. There are so many ions and molecules which can yield polarographic waves at either a mercury or a platinum microelectrode that there is a good possibility of finding a suitable reagent for direct or indirect titration of nearly any substance. The method is best suited for precise determination of low concentrations of the unknown.

BIAMPEROMETRIC TITRATIONS

A simplified procedure is made possible by impressing a small potential across two identical inert electrodes. The apparatus required consists merely of a source of about 50 to 100 mV, and a galvanometer in series with the two platinum electrodes. No current can flow between the electrodes unless there are present both a substance which can be oxidized at the anode and a substance which can be reduced at the cathode. Any redox couple which is easily reversible will permit electrolysis to take place. In the ferric-ferrous system, for example, Fe^{3+} ions can be reduced at the cathode, and Fe^{++} oxidized simultaneously at the anode. Some systems other than reversible couples will also permit current to flow. Hydrogen peroxide, for example, is oxidized at the anode to oxygen and reduced at the cathode to hydroxide ions. The permanganate-manganous couple is not reversible but, nevertheless, permits electrolysis, as Mn^{++} is anodically oxidized to MnO_2 and MnO_4^- undergoes cathodic reduction to MnO_4^{--} or to MnO_2. Some other electrolyzable systems are: I_2-I^-, Br_2-Br^-, Ce^{4+}-Ce^{3+}, $Fe(CN)_6^{3-}$-$Fe(CN)_6^{4-}$, Ti^{4+}-Ti^{3+}, VO_3^--VO^{++}.

A few titrations performed in this manner are shown graphically in Fig. 13-22.[17] The titration of iodine by thiosulfate (curve *a*) is one of the earliest reactions to be followed by this method; the abrupt cessation of current at the end point gave rise to the name *dead-stop*

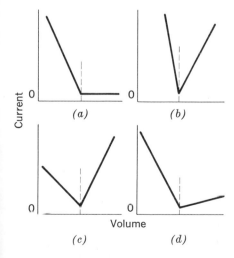

Fig. 13-22 Typical biamperometric titration curves: (*a*) iodine by thiosulfate, (*b*) iron(II) by cerium(IV), (*c*) vanadium(V) by iron(II), and (*d*) hexacyanoferrate(II) by cerium(IV). (*Analytical Chemistry.*)

titration. Prior to the equivalence point, both free iodine and iodide ions are present in solution, and therefore current can flow, even with as little as 15 mV applied potential. As the titration proceeds, iodine is reduced to iodide; at the equivalence point no free iodine remains, and the solution cannot conduct. Beyond the equivalence point, as more thiosulfate is added, no current can flow, as thiosulfate-tetrathionate does not constitute a reversible couple. This is widely used in the Karl Fischer moisture titration.

In the titration of ferrous iron by cerate (ceric sulfate) (curve *b*), on the other hand, current can flow on both sides of equivalence, as both ferrous-ferric and cerous-ceric pairs form reversible couples. There is a point of (nearly) zero current corresponding to the complete removal of ferrous ions before excess cerate ions have been added.

The other curves can be interpreted along similar lines. The end points in all these titrations are sharply defined, as the current drops essentially to zero.

CHRONOPOTENTIOMETRY[1, 4, 5]

This is an electrochemical method which is closely related to polarography in that the phenomena are diffusion-controlled, but it does not involve the study of current-voltage curves. It may be classed as an example of potentiometry at constant current.

Consider a three-electrode cell (Fig. 13-23) with a small constant current passing through the working and counter electrodes, while the potential of the working electrode is monitored against a reference. The working electrode can be a mercury pool a centimeter or so in diameter, the counter electrode a platinum foil, and the reference electrode an SCE with the capillary tip of its salt bridge very close to the working electrode. The solution must be quiet (unstirred). Let the solution

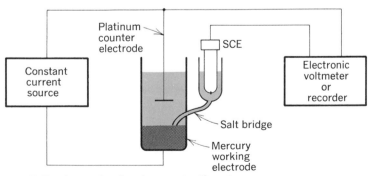

Fig. 13-23 Apparatus for chronopotentiometry.

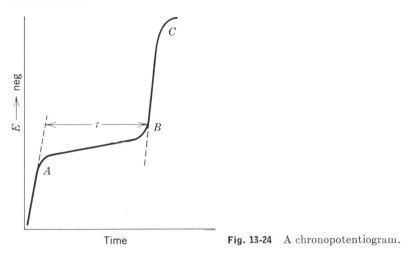

Fig. 13-24   A chronopotentiogram.

consist of an inert supporting electrolyte with a small concentration of cadmium ion. At the moment of closing the switch in the current source, the potential of the (cathodic) working electrode will jump from zero to a value negative enough to allow reduction of Cd^{++} to take place (point A, Fig. 13-24). It will stay at nearly this value, but rise slightly, until the Cd^{++} ion is depleted in the vicinity of the electrode (point B), when it will again rise sharply, until it is negative enough to reduce the next most easily reducible material present, in this case water, at point C.

The period of time required to deplete the surface layer (A to B) is called the *transition time* τ and is related to the bulk concentration of the reducible species by the *Sand equation*, analogous to the Ilkovič equation of polarography:

$$\tau^{1/2} = \frac{\pi^{1/2} n F A D^{1/2} C}{2I} \tag{13-19}$$

in which all symbols have previously been defined. This permits calculation of the concentration in terms of known or measurable parameters, which validates the method for quantitative analytical applications.

The potential corresponding to the A-B portion of the curve can be related to the reduction potential of the electroactive species. An equation derived by *Karaoglanoff*, and known by his name, shows that for a system where both oxidized and reduced species are soluble (in either solvent or electrode) the potential of the working electrode E_{we} is

$$E_{we} = E^{\circ} - \frac{RT}{2nF} \ln\left(\frac{D}{D'}\right) - \frac{RT}{nF} \ln\left(\frac{t^{1/2}}{\tau^{1/2} - t^{1/2}}\right) \tag{13-20}$$

The last term vanishes when

$$t^{1/2} = \tau^{1/2} - t^{1/2}$$

or

$$t = \frac{\tau}{4} \tag{13-21}$$

Note that $E° - (RT/2nF) \ln (D/D')$ has already been defined as $E_{1/2}$ by Eq. (13-8), so if we write $E_{\tau/4}$ for the value of E_{we} when $t = \tau/4$, it follows that

$$E_{we} = E_{\tau/4} - \frac{RT}{nF} \ln \left(\frac{t^{1/2}}{\tau^{1/2} - t^{1/2}} \right) \tag{13-22}$$

which is analogous to Eq. (13-9) for polarography, and shows that $E_{\tau/4}$ is identical to $E_{1/2}$.

Chronopotentiometry has several points of advantage and of disadvantage. The cell cannot be improvised easily, as can a cell for polarography, because the geometry of the working electrode and the relation to it of the capillary from the reference half-cell are quite critical. This is principally because the equations are derived on the assumption of linear diffusion, and this must be enforced by careful cell design. Any change in the electrode surface, as adsorption or deposition of an insoluble material, will tend to interfere with the diffusion process, and hence must be guarded against.

The technique is especially useful in kinetic studies. It is easy to establish by chronopotentiometry whether or not an electrode process is diffusion-controlled; if so, the product $I\tau^{1/2}$ should be constant even though the current is altered. The current density (I/A) can be varied in practice by several orders of magnitude, and its effect on the electrode reaction studied.

For solely analytical purposes chronopotentiometry does not usually offer enough advantage to outweigh its inconvenience.

PROBLEMS

13-1 A 1.000-g sample of zinc metal is dissolved in 50 ml of 6 F hydrochloric acid, diluted to the mark in a 250-ml volumetric flask, and a drop of maximum suppressor added. A 25.00-ml portion is transferred to a polarographic cell, and oxygen is flushed out. A polarogram in the range 0 to 1 V (versus mercury-pool electrode) shows a wave at $E_{1/2} = -0.65$ V, $i_d = 32.0$ units of galvanometer deflection. A 5.00-ml portion of $5 \times 10^{-4} F$ cadmium chloride is added directly to the polarograph cell which already contains the zinc solution, oxygen is again flushed out, and a second polarogram taken. The wave shows the same $E_{1/2}$, but the i_d is 77.5 units. Calculate the percent by weight of cadmium impurity in the zinc metal. Do not overlook the dilution effect. (Note that the supporting electrolyte is zinc chloride–hydrochloric acid.)

13-2 A 5×10^{-3} F solution of $CdCl_2$ in 0.1 F KCl shows a diffusion current at -0.8 V versus SCE of 50.0 μA. The mercury is dropping at a rate of 18.0 drops per min. Ten drops are collected and found to weigh 3.82×10^{-2} g. (*a*) Calculate the diffusion coefficient D. (*b*) If the capillary were replaced by another, for which the drop-time is 3.0 sec, and 10 drops weigh 4.20×10^{-2} g, what will be the new value of the diffusion current?

13-3 Alpha-benzoinoxime (cupron) is a precipitating agent for copper.[11] In a medium consisting of 0.1 F NH_4Cl and 0.05 F NH_3 (pH $= 9$), cupron is reduced at the DME, giving a wave with $E_{1/2} = -1.63$ V versus SCE (curve *a*, Fig. 13-25). The double wave of copper in the same medium is shown as curve *b*. Sketch the curves which would be found in the amperometric titration of copper by cupron at applied potentials of -1.0 V and -1.8 V versus SCE. Which potential would be preferred for the titration of copper in the absence of interfering substances? Which would be more liable to interference by reducible impurities?

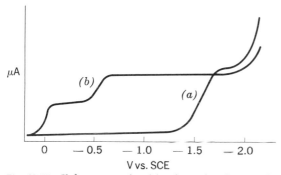

Fig. 13-25 Polarograms for (*a*) α-benzoinoxime, and (*b*) copper, in ammonia–ammonium chloride buffer. (*After Langer.*[11])

13-4 Small concentrations (0.1 to 10 ppm) of nitrate ion can be determined polarographically in 0.1 F zirconyl chloride ($ZrOCl_2$) as supporting electrolyte, by the difference in diffusion currents before and after reduction of nitrate by ferrous ammonium sulfate, with the DME 1.2 V negative to the SCE.[14] The following data are recorded for two standards and an unknown:

Solution, NO_3^-	Diffusion current, μA	
	Before reduction	After reduction
10.0 ppm	87.0	22.0
5.0 ppm	48.5	15.2
Unknown	59.0	17.0

Calculate the concentration of nitrate in the unknown.

13-5   Calculate values of E_{DME} corresponding to currents of 1, 2, 3 μA, etc., up to 9 μA for three hypothetical reductions in each of which $E_{\frac{1}{2}} = -1.000$ V versus SCE, and $\bar{i}_d = 10$ μA, but for which $n = 1$, 2, and 3, respectively (temperature, 25°C).   Plot these data on a single graph so that the three curves intersect at the half-wave point. For the same data, plot on another sheet the quantity $\log [\bar{i}/(\bar{i}_d - \bar{i})]$, as abscissas against E_{DME} as ordinates.

13-6   Tin(IV) in acidic pyrogallol solution gives a polarogram with two steps of equal height having $E_{\frac{1}{2}}$ values of -0.20 and -0.40 V versus SCE.[3]   Note that $E°$ for $Sn^{4+} + 2e^- \rightarrow Sn^{++}$ is -0.10 V, and for $Sn^{++} + 2e^- \rightarrow Sn^0$ is -0.38 V versus SCE.   (*a*) Explain why two equal steps are observed.   (*b*) Which form of tin, Sn(IV) or Sn(II), do you conclude is more strongly complexed by pyrogallol, and why?

13-7 Ag^+ ion in $NaClO_4$ as supporting electrolyte is reducible at the DME at the potential of the SCE. Cl^- ion in the same medium gives an anodic wave with $E_{\frac{1}{2}} = +0.25$ V versus SCE.   It is possible to determine whether the complex $AgCl_2^-$ is reducible under these conditions by titrating Ag^+ by Cl^-, amperometrically, at the DME.   Sketch and explain the titration curves which might result.

13-8   The following data were obtained in the chronopotentiometry of anthracene in acetonitrile containing 0.1 F $NaClO_4$:[18]

Concentration (C)	Current (I)	$\tau^{\frac{1}{2}}$
1.04×10^{-3} F	3.23 mA	0.591 sec$^{\frac{1}{2}}$
1.10×10^{-3} F	3.24 mA	0.610 sec$^{\frac{1}{2}}$
2.28×10^{-4} F	890 μA	0.474 sec$^{\frac{1}{2}}$

Determine the value of K in the abbreviated form of Eq. (13-19): $\tau^{\frac{1}{2}} = KCI^{-1}$. An unknown solution, at 1.00 mA, gave a value of $\tau^{\frac{1}{2}} = 3.29$ sec$^{\frac{1}{2}}$.   What was its concentration?

REFERENCES

1. Anderson, L. B., and C. N. Reilley: *J. Chem. Educ.*, **44**:9 (1967).
2. Annino, R., and K. J. Hagler: *Anal. Chem.*, **35**:1555 (1963).
3. Bard, A. J.: *Anal. Chem.*, **34**:266 (1962).
4. Delahay, P.: "New Instrumental Methods in Electrochemistry," Interscience Publishers (Division of John Wiley & Sons, Inc.), New York, 1954.
5. Delahay, P.: Chronoamperometry and Chronopotentiometry, in I. M. Kolthoff and P. J. Elving (eds.), "Treatise on Analytical Chemistry," pt. I, vol. 4, chap. 44, Interscience Publishers (Division of John Wiley & Sons, Inc.), New York, 1963.
6. Harris, W. E.: *J. Chem. Educ.*, **35**:408 (1958).
7. Heyrovský, J., and J. Kůta: "Principles of Polarography," Academic Press Inc., New York, 1966.
8. Heyrovský, J., and M. Shikata: *Rec. Trav. Chim.*, **44**:496 (1925).
9. Kolthoff, I. M., and W. E. Harris: *Ind. Eng. Chem., Anal. Edition*, **18**:161 (1946).

10. Laitinen, H. A., and I. M. Kolthoff: *J. Phys. Chem.*, **45**:1079 (1941).
11. Langer, A.: *Ind. Eng. Chem., Anal. Edition,* **14**:283 (1942).
12. Lingane, J. J.: *Ind. Eng. Chem., Anal. Edition,* **15**:583 (1943).
13. Meites, L.: "Polarographic Techniques," 2d ed., Interscience Publishers (Division of John Wiley & Sons, Inc.), New York, 1965.
14. Rand, M. C., and H. Heukelekian: *Anal. Chem.*, **25**:878 (1953).
15. Reilley, C. N., and R. W. Murray: Introduction to Electrochemical Techniques, in I. M. Kolthoff and P. J. Elving (eds.), "Treatise on Analytical Chemistry," pt. I, vol. 4, chap. 43, Interscience Publishers (Division of John Wiley & Sons, Inc.), New York, 1963.
16. Schmidt, H., and M. von Stackelberg: "Modern Polarographic Methods," Academic Press Inc., New York, 1963.
17. Stone, K. G., and H. G. Scholten: *Anal. Chem.*, **24**:671 (1952).
18. Voorhies, J. D., and N. H. Furman: *Anal. Chem.*, **31**:381 (1959).
19. Wawzonek, S.: *Anal. Chem.*, **30**:661 (1958).
20. Zuman, P.: "Organic Polarographic Analysis," The Macmillan Company, New York, 1964.

14

Electrodeposition and Coulometry

In Chap. 12, we considered the significance of the potentials of two completely *nonpolarized* electrodes in an electrolytic cell. In Chap. 13, we investigated the diffusion-current phenomena which arise when *one* of the electrodes is polarized. In the present chapter, we shall be concerned with the analytical applications of electrolysis involving the passage of such considerable currents that *both* electrodes are polarized.

Let us consider an electrolytic cell set up as in Fig. 14-1, with a pair of platinum electrodes immersed in a solution of copper sulfate. As the applied potential is increased from zero, essentially no current flows until the decomposition potential is reached. Beyond this point, larger amounts of current will flow as copper is deposited on the cathode and oxygen liberated at the anode.

The total potential V across the cell during the passage of current may be broken down into the following terms:

$$V = (E_a + \omega_a) - (E_c + \omega_c) + IR \tag{14-1}$$

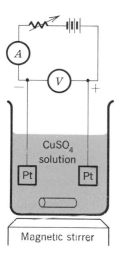

Fig. 14-1 Simple electrolysis cell.

in which E_a and E_c are the reversible half-cell emf's of anode and cathode, respectively, ω_a and ω_c represent the added potentials of the two electrodes due in part to overvoltage with respect to the deposition of copper and oxygen and in part to the concentration polarization resulting from the passage of current, and IR is the potential drop through the solution itself.   Since we are primarily interested in the cathodic deposition of metals, it is convenient to maintain conditions such that the reaction occurring at the anode is always the same.   Then the values of E_a and ω_a will not change appreciably during an experiment.   The deposition at the cathode is determined by the quantity $(E_c + \omega_c)$. E_c is calculable from the Nernst equation and the tabulated value of E° or $E^{\circ\prime}$.*

  In much electrodeposition work (in contrast with polarography) comparatively concentrated solutions are encountered, and therefore for precise work, activity corrections cannot be neglected.   These are shown as approximations for several cations in Fig. 11-1.   For other ions, if specific information about activities is lacking, the activity correction can be taken as the same for ions of the same charge, without likelihood of great error (H^+ ion is an exception).   It should be emphasized that the activity corrections of Fig. 11-1 cannot be considered exact, since

* This is valid if one can assume that the platinum cathode has become covered with copper and hence acts as a copper cathode.

the activity coefficients are greatly affected by the presence of other materials in solution.

This chart can be used to select appropriate potentials for the electrodeposition of one metal from a mixture. Thus a cathode at 0.6 V negative to the SCE would be negative enough to deposit tin without interference from thallium or cadmium. The concentration of tin can be diminished by electrolysis to less than 10^{-6} F (pSn = 6), whereas thallous ion cannot be discharged unless its concentration is higher than 0.8 F (pTl = 0.1), and cadmium not at all. Hydrogen will not be evolved at this potential in the presence of tin, because its overvoltage on a tinplated electrode is too high.

The polarization term ω_c cannot be calculated with exactness. It can, however, be minimized in an experiment by providing efficient stirring and by preventing the current density from becoming excessive.

In the classical procedure for the electrolytic separation of metals a sufficiently great potential is applied so that a considerable current will flow (a few amperes), and the electrolysis is allowed to continue with only occasional adjustment of current until deposition upon a weighed cathode is complete. Only the simplest of apparatus is required. A 6-V storage battery or rectifier connected in series with a rheostat, an ammeter, and the electrolysis cell (with a voltmeter across the cell) is entirely adequate. Mechanical stirring should be provided. Since this method has very wide application, there are a number of self-contained units on the market which make it possible to carry out two, four, or even more electrolyses simultaneously. This type of procedure is commonly considered in texts on elementary quantitative analysis, and so will not be discussed further.

In any technique of electrodeposition, precautions must be taken to ensure a firmly adherent deposit which will not flake off. Efficient stirring is the most important factor, as it prevents local depletion of the solution around the cathode. For similar reasons, the current density should not be too large. Sometimes the introduction of a colloid such as gelatin produces smaller crystals and a firmer deposit.

CHOICE OF ANODE

Although the primary concern in electrodeposition of metals is usually the cathode, the anode must also be given attention. Platinum is most commonly used, but one must be certain that no product of anodic oxidation is capable of interfering with the desired cathodic process or is in any other way objectionable. If the principal anions present are nitrate, sulfate, phosphate, or other ions which are not oxidizable at the potentials attained in the experiment, then the process at the anode will result in the liberation of oxygen, which is not objectionable.   If a large concentration of chloride is present, chlorine may be liberated, which is liable to attack the platinum anode, besides being disagreeable to the operator.   Generation of chlorine can be prevented by adding hydrazine to the solution; in preference to chloride, hydrazine is oxidized to produce free nitrogen.   Another way to prevent the generation of chlorine is to select a silver anode.   The anodic process will then be $Ag + Cl^- \rightarrow AgCl + e^-$.   The silver chloride is formed as an adherent layer, which is easily reduced back to silver by standing overnight in a zinc chloride solution in electric contact with a piece of zinc.

CONTROLLED POTENTIAL ELECTROLYSIS

To achieve complete separation of metals which have standard potentials relatively close together, such as copper and tin, it is necessary to introduce some device for measuring the potential of the cathode indepen-

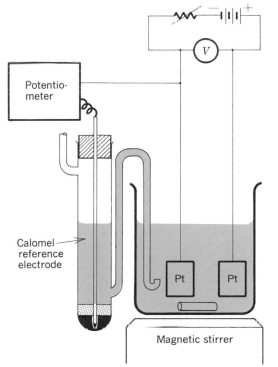

Fig. 14-2   Apparatus for electrodeposition with manually controlled cathode potential.

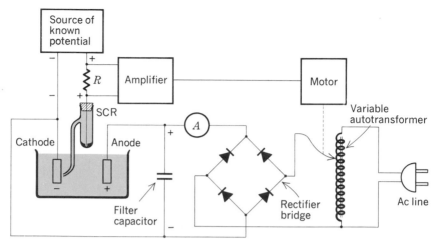

Fig. 14-3 Apparatus for electrodeposition with automatically controlled cathode potential (potentiostat).

dently of the other terms of Eq. (14-1). This can be done by inserting a calomel reference electrode into the electrolysis cell (Fig. 14-2). The potential between it and the cathode can be measured by a potentiometer or electronic voltmeter. It is then possible to maintain the cathode at just the correct potential to carry out the desired deposition without interference. This is the method of *controlled cathode-potential electrolysis*. The potential measured is $E_{\mathrm{SCE}} - (E_c + \omega_c)$.

This type of apparatus lends itself readily to automatic control. The principle of operation can be followed with the aid of Fig. 14-3. A source of known reference potential (which may be a calibrated potentiometer) is connected in opposition to the potential of the cell formed by the electrolysis cathode and the reference electrode (SCE). A resistor R is inserted in the circuit as shown. Any difference between the potentials of the reference source and the cathode–SCE will appear as a potential drop across R. This unbalance potential is amplified electronically and controls the operation of a reversible motor driving a variable autotransformer, which in turn controls the ac input to a rectifier and hence the dc output, which is the electrolysis current.

Suppose that the cathode potential is to be held at -0.35 V versus SCE but that momentarily it becomes slightly more negative than that. The reference potential has been adjusted to -0.35, so the difference between it and the cathode–SCE potential will be impressed across R, with the negative sign at the top. This *error potential* is amplified and applied to the motor so as to drive the autotransformer in the direction to reduce the voltage applied to the cell. This has the effect of restoring the potential of the cathode to the desired value, at which time the potential across R becomes zero and the motor stops. If a positive unbalance potential appears across R, the action will be just the opposite. Only when the two potentials are exactly balanced will the motor remain at rest. This automatic apparatus is known as a *potentiostat*.

Many ingenious designs for potentiostats have been published. These range in principle from an electromechanical unit with no electronics to modern all-electronic instruments with no moving parts. As one might judge from Fig. 11-1, it is not ordinarily necessary to regulate the cathode potential closer than about $\pm$ 10 mV, which is not a very exacting requirement for a well-designed control system.

It must be emphasized that the reduction potentials required for deposition of metals may vary to a greater or lesser extent, depending on the other materials present in the solution. The presence of complex-forming substances is particularly important, reducing the effective concentration of the free metal ion.

This method is equally suited to control of the anode potential for selective electrolytic oxidations, but this has not found so much analytical application.

In controlled-cathode electrolysis, at a potential where only a single species is reducible, the *current* is limited by diffusion, hence is proportional to the concentration of the reducible species. It is therefore evident that both concentration and current will fall off exponentially with time. We can write

$$\frac{C_t}{C_0} = \frac{i_t}{i_0} = 10^{-kt} \tag{14-2}$$

where C_t represents the concentration at time t, C_0 that at time $t = 0$, i_t and i_0 the corresponding currents, and k is a constant. It can be shown that k is proportional to DAS/V, where D is the diffusion coefficient, A the area of the cathode, S the rate of stirring, and V the volume of solution. Thus a plot of $\log i$ against t will give a straight line with a negative slope equal to k. From such a plot one can determine the time required to deposit any given fraction of the desired species. In an experiment quoted by Lingane[8] for the deposition of copper, k was found to be 0.15 min^{-1}, from which it follows arithmetically that deposition was 99 percent complete in 13 min, and 99.9 percent complete in 20 min.

Controlled potential electrolysis has been found extremely useful in some fields of chemistry other than analytical. In both organic and inorganic preparations involving electrooxidation or electroreduction, improved operation has been found to result from potential control. Selenium and tellurium have been prepared in this way in the -2 oxidation state, also W(III) and W(V); various pinacols, hydroxylamines, etc., have been reported in 100 percent yields.   It provides a valuable method of separation of radioactive nuclides in submicrogram amounts.

MERCURY–CATHODE ELECTROLYSIS

A particularly convenient and important technique for the separation of metals is electrodeposition upon a mercury cathode.[10] Since the hydrogen overvoltage on mercury is very high (greater than 1 V), any metal with a deposition potential less than this can be deposited on a mercury surface, but those requiring a more negative potential will remain in solution.   The elements *not* deposited include aluminum, the metals of the scandium, titanium, and vanadium subgroups, tungsten, and uranium.   The alkali and alkaline earth metals are deposited only

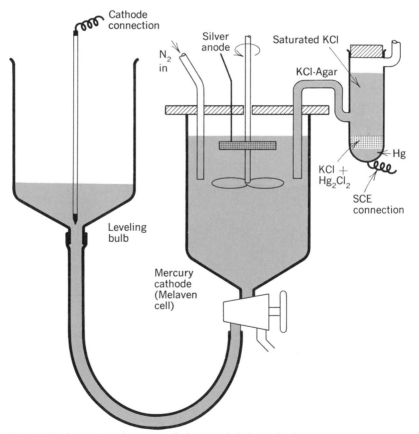

Fig. 14-4   Apparatus for controlled-potential electrolysis at a mercury cathode.

if the solution is basic. The method is applied with great success to the removal of iron and similar metals from solutions of aluminum alloys prior to the determination of that element by gravimetric or other means. It has also been extensively applied to the purification of uranium solutions.[1]

As has been demonstrated by Lingane,[8] the mercury cathode is particularly valuable for controlled potential electrolysis because of its close relationship to polarography. One can determine the optimum potential for electrolysis by inspecting a polarogram run in the same supporting electrolyte. Ordinarily the potential to select is that at which the diffusion-current plateau is just attained, which is often about 0.15 V more negative than the half-wave potential. The apparatus may resemble that of Fig. 14-4. The obvious difficulty lies in the need to weigh accurately a large amount of mercury in order to determine the quantity of a metal reduced at its surface. As we will see in the next section, this problem can be circumvented by a coulometric rather than gravimetric determination.

COULOMETRY

According to Faraday's law of electrolysis, a given amount of electricity passed through an electrolyte will result in a proportionate quantity of chemical change at the electrodes. This relation is stated mathematically as

$$W = \frac{QM}{nF} \tag{14-3}$$

in which W is the weight of substance of formula weight M oxidized or reduced by the passage of Q coulombs, and n is the number of equivalents per formula weight. F is a universal constant, the *faraday*, with a value of $96,487.0 \pm 1.6$ coulombs per equivalent (frequently rounded off to $96,500$). Application of this law permits the quantitative determination of any substance which can be made to undergo an electrochemical reaction with 100 percent current efficiency (i.e., no side reactions). This is the method known as *coulometric analysis*.

Two procedures are possible: operation at constant current, so that the amount of material deposited is proportional to the elapsed time, and operation at constant potential, in which case the current decreases exponentially from a relatively large value to practically zero.

In either technique, the quantity to be measured is the integral $\int i \, dt$, where i is the current in amperes flowing at any instant and t is the time in seconds. The integration may be performed graphically by measuring the area beneath a current-time curve, or mechanically with some type of integrator.

The classical method of measuring the quantity of electricity is by the use of a *chemical coulometer*. The electrolytic cell containing the unknown is connected in series with another cell so designed that the amount of electrochemical action is easily and precisely determinable. One of the most precise is the silver coulometer, which consists of a pure silver anode suspended inside a platinum crucible which acts as cathode. The crucible is filled with silver nitrate solution. The silver anode is surrounded with a porous cup to catch any particles of metallic silver which may drop off. The crucible is carefully washed, dried, and weighed, both before and after the experiment. The weight of the metal deposited in the analysis cell can be calculated from the weight of silver deposited in the coulometer. It is obvious that this procedure is of little advantage if the deposit in the analysis cell can be weighed directly. It is useful, however, in cases where the primary deposit cannot be weighed conveniently, such as mercury-cathode electrolysis.

Another type of chemical coulometer depends upon the volume of gas (at known temperature and pressure) produced by passing the current through a cell where electrolysis of water takes place. This *gas coulometer* (or *water coulometer*)[7] is a highly sensitive device (1 ml of combined H_2 and O_2 at STP corresponds to 0.05950 meq, or 16.80 ml per meq). The coulometer consists of a graduated tube (Fig. 14-5)

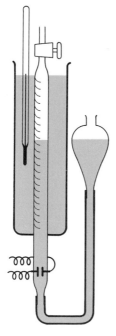

Fig. 14-5 Water coulometer. (*Journal of the American Chemical Society.*)

with two platinum electrodes sealed into its lower end.* The upper part of the tube is provided with a water jacket to define the temperature of the gas. The tube and its leveling bulb are filled with 0.5 F sodium sulfate solution. To be precisely reproducible, the electrolyte must be freshly saturated with H_2-O_2 mixture before each use. This is done by a 5-min electrolysis with the stopcock open. The electrolyte must be of the highest purity to ensure 100 percent current efficiency. In particular, even a slight impurity of iron or other metal with two soluble valence states will cause a considerable error, as the foreign ions undergo oxidation at the anode followed by reduction at the cathode, the process repeating indefinitely.

A similar gas coulometer containing hydrazinium sulfate, which evolves nitrogen and hydrogen upon electrolysis, has been described.[11] The sensitivity is identical, but a slight negative error found at low currents with the water coulometer is absent with the hydrazine unit.

Several electronic or electromechanical integrators have been described which can replace the chemical coulometer. They are, in general, capable of equal precision and are more convenient to operate. One of the best types makes use of an operational amplifier with a feedback capacitor, which acts as an integrator, as mentioned in the preceding chapter and discussed more fully in Chap. 26.

COULOMETRIC PROCEDURES

Analytical coulometric procedures may be based on almost any type of electrode reaction, including dissolution, electrodeposition, and oxidation or reduction of one soluble species to another. Some of these may be called "direct" processes, when the electrode reaction involves the substance to be determined, as in the electrodeposition of metals. Others are "indirect," in that the electrode reaction consumes or liberates a substance which is not itself to be determined, but which can be related quantitatively to such a substance.

Direct coulometric analyses often merely substitute electrical measurements for weighing. The only additional precaution necessary results from the requirement of 100 percent current efficiency. In electrodepositions this may often be established by judicious use of cathode-potential control. In any case it is wise to check the efficiency by analyzing known samples or by comparison of coulometric and gravimetric data for the same experiment.

---

* A Warburg manometer can serve as a highly sensitive and convenient water coulometer. A simple thermostated cell with platinum electrodes is constructed to fit the ground-glass connection to the manometer. A change of 1 cm in the liquid level in the manometer may correspond to about 0.05 Cb, or 6×10^{-4} meq of chemical action.

COULOMETRIC TITRATION

An indirect coulometric analysis normally consists in the electrolytic generation of a soluble species which is capable of reacting quantitatively with the substance sought. This falls within the broad definition of titration, as the reagent is added to the solution gradually (by electrolytic generation rather than from a buret), and some independent property must be observed to establish the equivalence point in the reaction. A process of this nature is commonly a hybrid between a direct and an indirect determination.

As an example, let us consider the coulometric determination of the concentration of ferric iron, in a solution containing hydrobromic acid, by electrolytic reduction to the ferrous state. This is accomplished with the aid of a platinum cathode and a silver anode. The anodic half-reaction is $Ag + Br^- \rightarrow AgBr + e^-$; we are therefore only concerned with processes taking place at the cathode. Suppose that we first attempt a *direct* reduction. The situation will be clarified by reference to the current-voltage curves of Fig. 14-6. Curve 1 includes the reduction waves of the $FeBr_2^+$ ion and of the H^+ ion in the HBr solution. If we force a constant current of magnitude a through this cell, the cathode potential will assume the value $+0.40$ V (approximately) vs. the SCE, where curves a and 1 intersect. As the electrolysis proceeds and iron is reduced, the plateau corresponding to the diffusion current of $FeBr_2^+$ is progressively lowered until the cathode potential suddenly jumps to approximately -0.3 V (curve 2), corresponding to reduction of H^+ ions. From this point on, ferric iron is reduced and hydrogen liberated simultaneously, so that the current efficiency with regard to iron reduction is less than 100 percent, and the analysis is no longer valid.

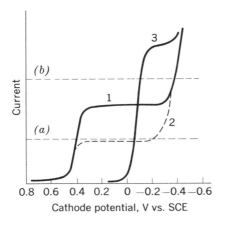

Fig. 14-6 Current-voltage curves as observed with a platinum cathode versus the SCE.

Let us now repeat the experiment with the addition of a considerable excess of Ti(IV), which is reducible in the presence of acid to Ti(III) at a platinum cathode with a potential slightly more negative than that of the SCE, curve 3. If again the current is established at level a, the iron will be reduced as before until its diffusion current is lowered to a, but at that point, the potential will jump, not to -0.3 V, but to -0.05 V versus SCE. From then on both iron and titanium will be reduced, but this does not result in any loss, because the Ti(III) reacts in the stirred solution to reduce ferric iron, so that the net result is the reduction of one iron atom per electron, no matter whether directly or indirectly. Thus the overall current efficiency is maintained at 100 percent, as required. (If the current had been set at level b in Fig. 14-6, then iron and titanium would have been reduced simultaneously from the very beginning, but the net result would have been the same.)

The equivalence point in this reaction is the point in time when the quantity of electricity is just equivalent to the total amount of ferric iron originally present in the sample. It can be identified (1) potentiometrically, (2) by an amperometric observation at about 0.25 V positive to the SCE, (3) by a biamperometric method, (4) by a photometric method involving addition of KSCN or other chromogenic reagent for Fe(III), (5) by a photometric method with a redox indicator which will be reduced by Ti(III) only after all the iron is reduced, or possibly by other methods.

In the experiment just described, no interference resulted from the electrochemical reaction at the anode. In many titrations, however, a soluble product will be formed at the counter electrode, and if no precaution is taken it will react unfavorably either at the generator electrode or with the intermediate in solution. To avoid this kind of difficulty, the counter electrode is often shielded by a glass tube with a fritted tip to discourage convection. Sometimes this is fully adequate, but in other cases it falls short of eliminating the error; it may be necessary to resort to an agar-gel salt bridge or other device.

A shield closed by an ion-exchange membrane instead of a glass frit will serve excellently in many situations.[3,4,6] Such membranes are available in two forms: cation exchangers which will not permit cations to pass through the membrane, and anion exchangers which exclude anions. For example, if we wish to titrate base coulometrically by electrogeneration of H^+ ions, the generator electrode will be the anode, and its half-reaction the familiar $H_2O \rightarrow \frac{1}{2}O_2 + 2H^+ + 2e^-$. At the cathode the complementary reaction, $2e^- + 2H_2O \rightarrow H_2 + 2OH^-$, will take place. Obviously the hydroxide ion produced at the cathode must not be allowed to mix with the solution being titrated (the anolyte); an anion-exchange membrane separating the two compartments will prevent such mixing. This principle appears not to have received the study it deserves.

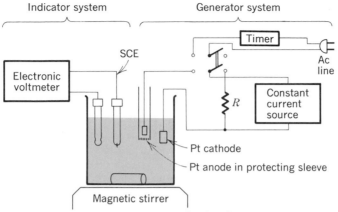

Fig. 14-7   Apparatus for coulometric titration at constant current, with potentiometric end-point detection.

In Fig. 14-7 is shown schematically an apparatus suitable for the titration of base.   The generator electrodes are connected to a constant-current source with an associated timer, while the indicator electrodes are the conventional glass-reference system of a pH meter.   Constant-current supplies will be discussed in Chap. 26.

A large number of reagents have been prepared by electrolytic generation, including H^+, OH^-, Ag^+, and other metal ions, oxidants such as $Ce(IV)$, $Mn(III)$, $Ag(II)$, Br_2, Cl_2, I_2, and $Fe(CN)_6^{3-}$, reductants such as $Fe(II)$, $Fe(CN)_6^{4-}$, $Ti(III)$, $CuBr_2^-$, and $Sn(II)$, complexogens such as EDTA and CN^- ions.

By one or another of these reagents it has become possible to substitute for practically all of the procedures of classical volumetric titrimetry their coulometric counterparts.   This presents the great practical advantage that it is not necessary to prepare and store standardized solutions.   The "primary standard" for coulometric titration is the combination of a constant-current source and an electric timer, which are applied to *all* titrations, no matter what their chemical nature.   Coulometric titration also has the advantage that it is applicable to samples one or two orders of magnitude smaller than conventional procedures (samples of 0.1 down to 0.001 meq are usual, compared with 1 to 10 meq for volumetric titrations).   Furthermore, as can be seen from the list above, some reagents can be employed which are unstable, or for other reasons not suitable for volumetric use, such as $Mn(III)$, $Ag(II)$, $CuBr_2^-$, and Cl_2.

The precision possible in coulometric titration can easily equal, and with precautions can exceed, that attainable by volumetric means.   Eckfeld and Shaffer[2] have reported a careful study of precision in coulometric neutralization. They could measure coulombs (as micro-equivalents) to about ± 0.004 percent. One simple but effective pre-caution was to provide a slow flow of indifferent electrolyte in a salt bridge, so as to eliminate all possibility of contamination of solution or loss of sample through the fritted glass. This paper should be studied by anyone attempting precise work in coulometric titrations.

The coulometric method is less useful for larger concentrations, and this is its chief limitation. The reason is that it would be necessary to operate with much larger currents if unduly long times are to be avoided, and this tends to reduce the current efficiency from the required 100 percent, except in the unusual case where only one electrode reaction is possible. A current of 5 to 10 mA is about as large as can safely be used.

Pretitration is often advisable. After the apparatus is assembled a small portion of the material to be analyzed is inserted and the electrolysis allowed to proceed until the desired end point is observed. Then the measured sample is added and titrated until the end point is again reached. This ensures that any impurities which can react with the generated titrant are removed in advance. It also obviates any uncertainty about the surface condition of the electrodes (formation of an oxide film, for example). The titration curve will have the appearance shown in Fig. 14-8. The material pretitrated reacts with generated reagent from time t_0 to t_1, following which reagent accumulates (at A). The sample is then added; it reacts immediately with the reagent which has just been generated, returning the curve to zero (at B), then continues to react with generated titrant until it is all consumed, when the curve will rise again at C. The two sloping portions of the curve are extrapo-

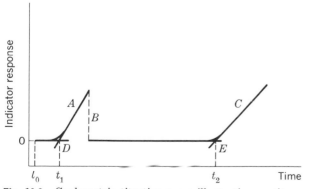

Fig. 14-8 Coulometric titration curve illustrating pretitration. The vertical axis refers to the indication of any detector (photometric, amperometric, etc.) the output of which is directly proportional to concentration.

lated back to the zero level at D and E, and the time between these two points, $t_2 - t_1$, is taken as the electrolysis time. The slopes at A and C will differ if appreciable dilution has occurred with addition of the sample.

Coulometric titration is easily automated. As mentioned in Chap. 12 in connection with automatic potentiometric titrators, a difficulty in the mechanization of volumetric titrations lies in the nonuniform flow of reagent from a buret as induced by gravity alone. Such a difficulty does not exist with electrolytic generation, as passage of a constant current provides a linear time base for the titration. Several automatic or semiautomatic coulometric titrators are commercially available. There are also a number of continuous coulometric analyzers, in which a recorder indicates the magnitude of current required to maintain constant the concentration of some component by causing it to react with an electrogenerated reagent.

Coulometric generation can readily be adapted to an all-electronic version of the pH stat discussed in Chap. 12.

STRIPPING ANALYSIS[12]

We turn now to a two-step analytical method applicable principally to trace quantities of transition metals. The metals are first deposited at a solid or mercury cathode, then successively stripped from the electrode anodically. The final analysis may be by coulometric or polarographic techniques. The major advantage lies in the possibility of concentrating the trace metals from a large volume of solution.

The most prevalent arrangement makes use of a hanging mercury drop as working electrode in a cell resembling that of Fig. 14-9. A Teflon scoop is positioned between a polarographic capillary and a mercury-plated platinum contact wire, so that the operator can catch one or more drops of mercury and transfer it to the platinum, where it will adhere. The platinum is best very slightly recessed above the level of the glass holder, as shown in the inset, so that there will be no direct contact between the platinum and the solution.

There are several variations of the techniques both for electrodeposition and for stripping, but we will describe only one. The electrodes are connected to a potentiostat, and the solution electrolyzed at a potential sufficiently cathodic to reduce all the metals present which are of interest. Electrolysis is continued for an exactly specified period of time, of the order of 5 to 30 min, with reproducible stirring conditions. By analogy with Eq. (14-2) we can see that an equal fraction of the existing quantity of each species will be reduced at the hanging mercury drop to form a drop of mixed amalgam.

After the electrolysis step, the stirring is stopped, and as soon as

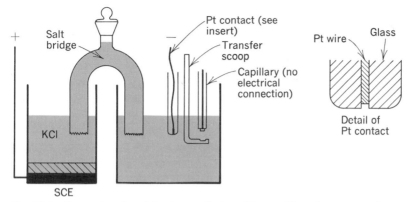

Fig. 14-9 Apparatus for stripping analysis. The capillary is connected to a mercury reservoir, as though it were a DME.

the solution is quiet (30 sec), the electrode is subjected to a potential scan, as in voltammetry, starting at the voltage at which the electrolysis was conducted and proceeding at a constant rate toward less negative values. Figure 14-10 shows the resulting anodic current-voltage curve for stripping cadmium from a mercury drop following a 15-min electrolysis in 10^{-8} F CdCl$_2$ in 0.1 F KCl. If this solution had been analyzed

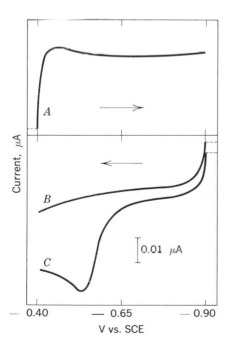

Fig. 14-10 Current-voltage curves for anodic stripping of 10^{-8} F cadmium solution in 0.1 F potassium chloride, using voltammetry with linearly varying potential. The rate of voltage scan is 21 mV-sec^{-1}. A is the cathodic residual current, B is the anodic residual current after the 15-min preelectrolysis, and C is the anodic stripping current for 10^{-8} F cadmium after the 15-min preelectrolysis. (*Analytical Chemistry.*)

by conventional rapid-scan voltammetry, it would have produced a "peak" of about 1.5×10^{-8} μA, which would be completely undetectable.

Stripping analysis presents a very powerful tool in the important field of trace analysis.

PROBLEMS

14-1 A constant current of 1.500 A is allowed to flow for a period of 1 hr (to the nearest second) through a number of electrolytic cells connected in series. All electrodes are platinum. The cells contain an excess of the electrolytes listed. For each, tell what substance is deposited at the cathode, and calculate the quantities in grams (if solid or liquid) or in milliliters at STP (if gaseous).

(a) $Cu(NO_3)_2$		(d) HgI_2	
(b) $NaOH$		(e) $Pb(NO_3)_2$	
(c) $K_4Fe(CN)_6$		(f) $Ag(NH_3)_2Cl$	

14-2 A hydrogen-oxygen coulometer (Fig. 14-5) has a cross-sectional area of 2.3 mm². It can be read with optimum precision at a linear vertical displacement of 20.0 cm. What weight of impure As_2O_3 should be taken for analysis, by coulometric titration with electrolytically generated bromine, so that this displacement will be obtained if the sample contains 90 percent As_2O_3? Room temperature is 25°C; barometric pressure may be assumed to be normal.

14-3 A cadmium amalgam is to be prepared for use in a Weston standard cell by electrolyzing cadmium chloride with a mercury cathode and silver anode. The desired concentration is 12 percent cadmium by weight. Starting with 20 g of mercury, how long should electrolysis continue at 5 A to attain this concentration?

14-4 Ceric ion is to be determined by coulometric titration with electrolytically reduced ferrous ion. The end point is observed potentiometrically with platinum and saturated calomel electrodes connected to a pH meter. Preliminary studies show that the potential of the platinum wire at the end point should be +0.800

Time, sec	Coulometer reading	Potential, Pt-SCE
Preliminary titration		
. . . . .	20	+0.830
436.0	120	0.801
472.8	170	0.790
Analysis titration		
472.8	170	0.893
. . . . .	470	0.861
. . . . .	720	0.820
. . . . .	750	0.814
. . . . .	770	0.810
925.0	790	0.803
939.5	810	0.794

V versus SCE. The acidified solution contained 0.0005 mole of ferric iron in a volume of 250 ml. Interfering substances were removed by a preliminary titration of the same type. For this purpose about 0.2 μeq of ceric ion was added, and current was passed through the generating electrodes ($Fe^{3+} + e^- \rightarrow Fe^{++}$) until the indicator showed 0.800 V. Then a 1.00-ml portion of the unknown ceric solution was added, and generation continued until the end point was reached again. Actually, in both titrations, the end point was overrun intentionally, and the time and coulometer readings at the equivalence points were determined by interpolation. The tabulated data were taken. The coulometer readings are in arbitrary units such that a change of one unit corresponds to 4.79×10^{-4} microfaraday. Note that the slight excess of ferrous ion from the preliminary titration was used to react with a portion of the unknown sample, so that it was not lost. The change in coulometer readings between the two end points then corresponds to the generation of ferrous ion equivalent to the ceric in the unknown.

Calculate the concentration of the unknown in micrograms of cerium per milliliter.

14-5 The copper from a 0.400-g sample of a brass was deposited electrolytically with a suitably controlled cathode potential. The cell was placed in series with a recording ammeter. The record obtained is reproduced in Fig. 14-11. The quantity of electricity involved can be determined by integration, approximately, by counting the squares beneath the curve, or by the method of trapezoids. Perform this integration and compute the percentage of copper in the alloy. Why does this curve not follow the exponential law?

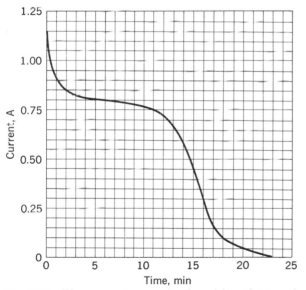

Fig. 14-11 Time-current curve for deposition of a sample of copper.

14-6 A solid sample consists of a mixture of chlorides, bromides, and iodides of alkali metals. A 0.5000-g portion is weighed out for analysis and dissolved in water containing a little dilute nitric acid. It is then transferred to an electrolysis cell provided with silver electrodes and electrolyzed with the anode maintained at -0.06 V relative to an SCE, a suitable potential for the deposition of iodide, but not for the other halides, to form $AgI_{(s)}$. A gas coulometer (like that of Fig. 14-5) in series shows a volume of 19.84 ml of mixed gas (at 25°C and 760-torr barometric pressure) collected during the deposition of all the iodide. The silver anode was then replaced with a new one weighing 2.3772 g. At the same time the potential was raised to +0.25 V, sufficient to cause simultaneous deposition of chloride and bromide. A volume of 79.32 ml of gas (same conditions) was collected during the deposition of these two halides. After the electrolysis was complete, the anode was removed, rinsed, dried, and found to weigh 2.5472 g. Compute the percentage of each halogen in the sample.[9]

14-7 What value of current (in milliamperes) would be required for coulometry at constant current so that a seconds timer would read directly in microequivalents of electrode reaction?

14-8 Anthranilic acid (o-aminobenzoic acid) can be brominated by electrolytically generated bromine at pH 4 to produce tribromoaniline. Hargis and Boltz[5] have shown that small amounts of copper can be determined by precipitating Cu(II) anthranilate, $Cu(C_6H_4NH_2CO_2)_2$, dissolving the precipitate, and coulometrically titrating the liberated anthranilic acid.

The copper in a 1.000-g sample of a biological material is converted to ionic form, and precipitated with excess anthranilic acid. The precipitate is filtered, washed, redissolved, and the acid brominated with a constant current of 6.43 mA. The time required is 22.40 min. Calculate the amount of copper in the original sample, in parts per million.

14-9 It is desired to analyze a factory effluent water for Cu, Pb, and Cd by means of stripping analysis at an electrode consisting of a thin film of mercury plated onto a supporting metal. In one experiment, the potential of the electrode was held at −1.0 V versus SCE for 90 min in a 200-ml sample, by which time the current had dropped to a negligible value. It was then scanned anodically, and three peaks were obtained corresponding to the dissolution of the three metals. Integration indicated 12.5 μCb at the potential relating to Cd, 41.0 μCb for Pb, and 38.2 μCb for Cu. Calculate the concentration of each metal in the effluent, in terms of parts per billion (parts per 10^9).

REFERENCES

1. Casto, C. C.: in C. J. Rodden (ed.), "Analytical Chemistry of the Manhattan Project," p. 511, McGraw-Hill Book Company, New York, 1950.
2. Eckfeld, E. L., and E. W. Shaffer, Jr.: *Anal. Chem.*, **37**:1534 (1965).
3. Eisner, U., J. M. Rottschafer, F. J. Berlandi, and H. B. Mark, Jr.: *Anal. Chem.*, **39**:1466 (1967).
4. Feldberg, S. W., and C. E. Bricker: *Anal. Chem.*, **31**:1852 (1959).
5. Hargis, L. G., and D. F. Boltz: *Talanta*, **11**:57 (1964).
6. Ho, P. P. L., and M. M. Marsh: *Anal. Chem.*, **35**:618 (1963).
7. Lingane, J. J.: *J. Am. Chem. Soc.*, **67**:1916 (1945).
8. Lingane, J. J.: "Electroanalytical Chemistry," 2d ed., Interscience Publishers (Division of John Wiley & Sons, Inc.), New York, 1958.
9. MacNevin, W. M., B. B. Baker, and R. D. McIver: *Anal. Chem.*, **25**:274 (1953).
10. Maxwell, J. A., and R. P. Graham: *Chem. Rev.*, **46**:471 (1950).
11. Page, J. A., and J. J. Lingane: *Anal. Chim. Acta*, **16**:175 (1957).
12. Shain, I.: Stripping Analysis, in I. M. Kolthoff and P. J. Elving (eds.), "Treatise on Analytical Chemistry," pt. I, vol. 4, chap. 50, Interscience Publishers (Division of John Wiley & Sons, Inc.), New York, 1963.

15

Conductimetry

In this chapter we shall investigate the analytical significance of those electrical properties of solutions which do *not* depend on the occurrence of electrode reactions.

If two platinum electrodes are inserted into a solution of an electrolyte and connected to a source of electricity, the current which flows is determined both by the applied voltage E and by the electrical resistance R of that portion of the solution between the electrodes. This relation is expressed mathematically by Ohm's law, $I = E/R$, where I is the current in amperes if E is in volts and R in ohms. (Ohm's law is obeyed only if specific electrode reactions and diffusion limitation are eliminated. How this can be accomplished will be considered subsequently.)

The *conductance* L is defined as the reciprocal of the resistance, so that $I = EL$. The unit of conductance is the reciprocal ohm (ohm^{-1} or mho).

The observed conductance L of a solution depends inversely on the distance d between the electrodes and directly upon their area a;

it also depends upon the concentration c_i of ions per unit volume of the solution and upon the *equivalent ionic conductance* λ_i of these ions. We can write

$$L = \frac{a}{d} \sum_i c_i \lambda_i \tag{15-1}$$

The summation symbol Σ points to the fact that the contributions to the conductance of the various ions present are additive. The units of concentration must be equivalents per cubic centimeter (rather than per liter)* because a and d are in terms of centimeters. It is usual to express the concentration as normality, equivalents per liter, and hence we must introduce a factor of 1000, and let C denote the normality

$$C = 1000c \tag{15-2}$$

In usual conductivity vessels, the geometry of the electrodes is not convenient to measure, and it is customary to replace the ratio d/a by a single symbol θ which has a constant value for any given pair of electrodes, and is termed the *cell constant*. We can then combine this constant with Eqs. (15-1) and (15-2), to obtain

$$L = \frac{\Sigma C_i \lambda_i}{1000\theta} \tag{15-3}$$

The summation $\Sigma C_i \lambda_i$ can be replaced by $C\Lambda$, where Λ is the *equivalent conductivity*, for the case of a single ionized compound in solution. Thus the equivalent conductivity is equal to the sum of the equivalent ionic conductivities, $\Lambda = \Sigma \lambda_i$.

In order to compare the values of conductance obtained with various electrode assemblies, the conductance may be replaced by the *conductivity* or *specific conductance*, denoted by κ, and defined as

$$\kappa = L \frac{d}{a} = L\theta \tag{15-4}$$

The cell constant is experimentally determined by means of Eq. (15-4), written as $\theta = \kappa/L$. Conductance measurements are carried out on a solution of known κ. Solutions of potassium chloride are most commonly employed, as their specific conductances have been precisely determined (see Table 15-1).

Electrolytic conductance is temperature-dependent, increasing its value by about 2 percent per degree rise in temperature, so that for precise work conductance cells must be immersed in a constant-temperature bath. It is usual to select 25°C for all measurements. For direct

* For the precision required in analytical applications of conductance, the liter can be considered equal to 1000 cm³; actually 1 liter = 1000.027 cm³.

Table 15-1 Specific conductances of potassium chloride solutions (25°C)[5]

Concentration, g/kg of solution	κ, mho cm^{-1}
71.1352	0.11134
7.41913	0.012856
0.74526	0.0014088

comparison with a standard, any temperature is satisfactory, provided that it is held constant throughout the experiment.

The equivalent ionic conductivity λ is an important property of ions which gives quantitative information concerning their relative contributions to conductance measurements. The value of λ is to some extent dependent on the total ionic concentration of the solution, increasing with increasing dilution. It is convenient to tabulate numerical values of λ^0, i.e., the limiting magnitude of λ as the concentration approaches zero (infinite dilution). This is essentially the same as the "mobility" of ions which is sometimes quoted. A compilation of such data is given in Table 15-2.

Table 15-2 Equivalent ionic conductivity at infinite dilution (25°C)*

Cations	λ^0, mhos	Anions	λ^0, mhos
H^+	349.8	OH^-	198.6
K^+	73.5	$\frac{1}{4}Fe(CN)_6^{4-}$	110.5
NH_4^+	73.5	$\frac{1}{3}Fe(CN)_6^{3-}$	101.0
$\frac{1}{2}Pb^{2+}$	69.5	$\frac{1}{2}SO_4^{2-}$	80.0
$\frac{1}{3}La^{3+}$	69.5	Br^-	78.1
$\frac{1}{3}Fe^{3+}$	68.	I^-	76.8
$\frac{1}{2}Ba^{2+}$	63.6	Cl^-	76.4
Ag^+	61.9	NO_3^-	71.4
$\frac{1}{2}Ca^{2+}$	59.5	$\frac{1}{2}CO_3^{2-}$	69.3
$\frac{1}{2}Sr^{2+}$	59.5	$\frac{1}{2}C_2O_4^{2-}$	74.2
$\frac{1}{2}Cu^{2+}$	53.6	ClO_4^-	67.3
$\frac{1}{2}Fe^{2+}$	54.	HCO_3^-	44.5
$\frac{1}{2}Mg^{2+}$	53.1	$CH_3CO_2^-$	40.9
$\frac{1}{2}Zn^{2+}$	52.8	$HC_2O_4^-$	40.2
Na^+	50.1	$C_6H_5CO_2^-$	32.4
Li^+	38.7		
$(n\text{-}Bu)_4N^+$	19.5		

* Data mainly from Frankenthal.[1]

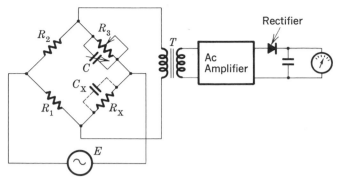

Fig. 15-1   Ac Wheatstone bridge for conductance measurements.

APPARATUS

To avoid electrolytic complications, conductance measurements are almost invariably taken with alternating current. The frequency is not critical; 1000 Hz is often selected, but line frequency is satisfactory for titrations. Efficient stirring is essential.

The usual circuit for the determination of electrolytic conductance is the Wheatstone bridge (Fig. 15-1), modified to operate on alternating current. The source E energizes the bridge, either from the line or from an oscillator. R_1 and R_2 are the "ratio arms," which may be equal or may be equipped with a selector switch (not shown) so that the ratio R_1/R_2 may be set at any of several values, such as 0.1, 1, and 10. R_x represents the resistance of the conductance cell, and R_3 that of the balance or standard arm, a precision variable with a calibrated dial. The capacitance C_x shown in parallel with R_x is the capacitance of the conductance cell and its connecting wires. The presence of this capacitance causes a phase shift in the alternating potential across R_x, and this must be balanced out by adjustment of a small variable air capacitor C across the balance arm. This capacitance effect is more important at 1000 Hz than at 60, more important in the measurement of solutions of low conductance than of high (because a cell must be selected with a small cell constant, i.e., with relatively large electrodes closely spaced, which results in higher capacitance). The transformer T is included to permit connection of oscillator and amplifier to a common "ground" or reference point.

The alternating potential appearing across the diagonal of the bridge may be observed directly with earphones or with an ac galvanometer, or for greater sensitivity it may be amplified (as shown in Fig. 15-1), and the output of the amplifier rectified and metered. A condition of balance in the bridge exists if the output of the amplifier (or the sound in the earphones) is zero, at which point

$$R_x = \frac{R_1}{R_2} R_3 \qquad\qquad (15\text{-}5)$$

A bridge such as that described above can readily be assembled from standard electric components. Several complete instruments are commercially available. In one of the most widely distributed (Beckman Instruments, Inc.) the condition of balance is indicated by a cathode-ray ("magic-eye") tube. As the variable resistor R_3 is turned from one side of the balance point to the other, the shadow segment of the eye tube is seen first to open then close again. The point of maximum shadow angle corresponds to the balance point.

A more elaborate model, the Serfass bridge (Fig. 15-2), contains a switching arrangement which permits the operator to read directly either the resistance of the sample in ohms or the conductance in mhos, so that he need not calculate reciprocals. The change-over from ohms to mhos is effected by means of a *dpdt* switch which connects the calibrated variable resistor R_1 as either the adjacent or opposite arm of the bridge, with respect to the unknown. A separate control makes it possible to correct for the cell constant so that specific conductances can be obtained directly.

Beckman also manufactures several abridged models intended for specific applications. They may be provided, for example, with scales calibrated "0–30 ppm NaCl," "2–30 percent H_2SO_4," or "0–10 gr/gal

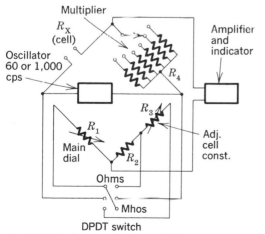

Fig. 15-2 Serfass conductance bridge, schematic. (*A. H. Thomas Co., Philadelphia.*)

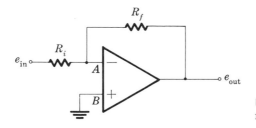

Fig. 15-3 Operational amplifier for measuring conductance.

sea water." Connection of such units to pen recorders, electrically operated valves, etc., is standard practice.

Another approach to the measurement of conductance is based on a straightforward application of an operational amplifier. As mentioned in Chap. 13, an operational amplifier acts to maintain its two inputs (A and B in Fig. 15-3) at the same potential, which in this circuit must be at ground. Hence the current through the input resistor R_i must be $i = e_{in}/R_i$, and this must be identical to that passing through the feedback resistor R_f. It follows that the output voltage e_{out} is given by

$$e_{out} = -\frac{R_f}{R_i} e_{in} \qquad (15\text{-}6)$$

For conductance measurements the cell is conveniently substituted for R_i while R_f is held constant or varied stepwise for range control. A source of constant-amplitude alternating voltage supplies e_{in}, so that the output e_{out} is directly proportional to the conductance of the solution $1/R_i$. The minus sign may be neglected.

Another method of measuring conductance at audio frequencies, described by Griffiths,[2] is unique in that no physical contact is required between electrodes and solution. The circuit is shown in slightly simplified form in Fig. 15-4. The device consists of two coils so arranged that their magnetic fields, though shielded from each other, both link with a closed loop of glass or plastic tubing containing the electrolytic solution. The signal observed by the electronic voltmeter is propor-

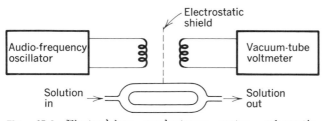

Fig. 15-4 Electrodeless conductance meter, schematic. (*Beckman Instruments, Inc., Cedar Grove, N.J.*)

tional to the conductance of the solution, provided the output of the oscillator is constant. The unit is recommended for use with abrasive slurries, highly corrosive liquids, etc., where excessive damage might be incurred by conventional electrode systems.

The same principle has been applied successfully to the measurement of the salinity of sea water. For this purpose no container at all is needed; the two coils are molded into a block of plastic with a smooth hole through the axis, so that the liquid within the hole together with the external body of solution constitutes the connecting loop.

APPLICATIONS

Nearly all the applications of conductance measurements are concerned with aqueous solutions, though one can easily imagine its usefulness with other solvents or with molten salts. Water itself is a very poor conductor. The theoretical specific conductance of water, due to its dissociation into hydronium and hydroxide ions, is approximately 5×10^{-8} mho per cm (at 25°C). Typical distilled water may have a value a hundredfold larger than this, though only repeated distillation from specially designed equipment results in a water approaching the theoretical.

Solutions of strong electrolytes in general show an almost linear increase in conductance with increasing concentration, up to the vicinity of 10 to 20 percent by weight. At higher concentrations the conductance passes through a maximum, as the interionic attraction hinders free motion through the solution. A few examples are shown in Fig. 15-5.[12]

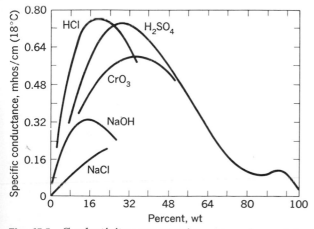

Fig. 15-5 Conductivity-concentration curves for certain electrolytes. (*McGraw-Hill Book Company, N.Y.*)

An excellent example of conductimetric analysis based on pre-determined calibration curves or tables is the analysis of fuming nitric acid to determine the ratio of NO_2 to H_2O.[8] It is necessary to make two conductance measurements, one on the untreated sample, another on a duplicate sample which has been saturated with KNO_3. By this method NO_2 can be determined in the range 0 to 20 percent NO_2 (by weight) and water in the range 0 to 6 percent, each with an accuracy of ± 0.3 percent.

A rapid and convenient procedure has been devised[13] for checking conventional analyses of natural waters and brines. A series of factors has been computed for those ions likely to be present, such that each factor, multiplied by the concentration of the corresponding ion, gives the contribution of that ion to the specific conductance of the solution. The factors are listed in the second column of Table 15-3. In the third

Table 15-3   Conductimetric check on the ionic content of waters and brines

| | | Hypothetical water sample | |
| | | | |
Ion	Factor, $\mu mhos/meq/liter$	Concentration, $meq/liter$	Partial specific conductance, $\mu mhos$
Chloride	75.9	0.36	27.3
Sulfate	73.9	0.05	3.7
Carbonate	84.6	0.00	0.0
Bicarbonate	43.6	0.47	20.7
Nitrate	71.0	0.02	1.4
Calcium	52.0	0.44	22.9
Magnesium	46.6	0.26	12.3
Sodium	49.6	0.20	9.9
Total			98.2

column are a series of ionic concentrations as determined by other analytical means for a hypothetical sample of water. The figures in the fourth column are obtained by multiplying the concentrations by the relative factors. The total is the calculated specific conductance for the sample. The observed specific conductance should agree within 2 percent. If it does not, an error in one or more of the analyses is indicated.

The conductance method has been found suitable for the determination of small amounts of free ammonia in biological materials.[3] The ammonia is removed from the sample by distillation, or swept out with a current of air, and absorbed in a solution of boric acid. The specific conductance of the solution is then determined and compared with previously measured standards. Boric acid is chosen as absorbent because its ionization is so slight that the specific conductance of the solution is a linear function of the concentration of the ammonium salt. This method for the determination of ammonia is applicable to the important Kjeldahl procedure for amine nitrogen.

It is possible to use conductance measurements to determine a specified ion in the presence of moderate concentrations of others if a reagent is available which will selectively remove the desired ion as a precipitate or un-ionized complex.[9] The specific conductance of the solution is measured both before and after the addition of a known quantity of the reagent. The addition of the reagent itself (if ionized) causes an increase in conductance, while the removal of the desired ion combined with the equivalent amount of reagent causes a decrease. The reagent must always be added in sufficient excess so that the final conductance is greater than the original. The net change in conductance is plotted against the concentration of the desired ion for a calibration graph. A correction must be applied, based on the value of the initial conductance. This is because the contribution of the reagent to the conductance is smaller, the greater the total ionic concentration. The correction factors are readily determined by experiment and, once evaluated, can be relied on for all future determinations of the ion in question.

CONDUCTOMETRIC TITRATIONS*

The conductance method can be employed to follow the course of a titration, provided that there is a significant difference in specific conductance between the original solution and the reagent or the products of reaction. The only additional apparatus required is a buret. It is not necessary to know the cell constant, since relative values are sufficient to permit locating the equivalence point. It is essential, however, that the spacing of the electrodes does not vary during a titration.

The conductance produced by any ion is proportional to its concentration (at constant temperature), but the conductance of a particular solution will generally not vary linearly with added volume of reagent because of the dilution effect of the water which is being added along with the reagent. Hydrolysis of reactants or products, or partial solubility of a precipitated product, will also cause departures from linearity.

* Note that the term *conductometric* refers to titration procedures, while *conductimetric* refers to nontitrative measurements. The whole subject is called *conductimetry*. The distinction in spelling is to be preferred, but is not universally followed.

We shall now discuss in detail the relation between the chemical reaction occurring during titration and the shape of the resulting graph.

NEUTRALIZATION REACTIONS

Consider what happens when 0.01 F hydrochloric acid is titrated with 0.1 F sodium hydroxide solution. Initially, the conductance is quite high because of the high equivalent conductance (mobility) of hydrogen ions. Reference to Table 15-2 shows that 82 percent of the conductance is contributed by the hydrogen ions and only 18 percent by chloride. The contribution of the chloride ions remains constant throughout the entire titration, while that of the hydrogen ions diminishes to zero at the equivalence point. The hydrogen ions are replaced by the same number of sodium ions, but these have a very low mobility, so that the overall conductance decreases steeply to the equivalence point. After this point is passed, the conductance increases again, as both the sodium and hydroxide ions accumulate in the solution.

The simple case described above will produce a graph like that of Fig. 15-6. Since conductances are additive, we can divide up the area beneath the curves into segments representing the amounts of the various ions present at successive stages in the titration. As the amount ˈof chloride ion does not change, it is represented by an area with constant height. The amount of sodium ion is zero to start with and increases uniformly throughout the titration, as indicated by a continuously expanding area. (In these figures, the dotted vertical lines indicate the equivalence points.) The contribution of hydrogen ions, initially very high, decreases to zero at the equivalence point, while that due to hydroxide ions starts at zero at the equivalence point and rises from there on.

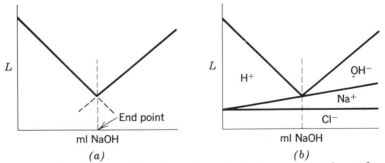

Fig. 15-6 Conductometric titration of HCl by NaOH: (a) experimental, the end point located by the intersection of two nearly straight lines; (b) interpretation, the conductance after the addition of each increment of reagent being the sum of the contributions of all ions present.

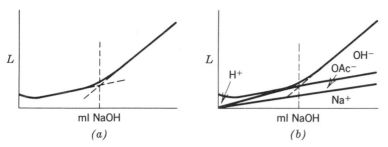

Fig. 15-7 Conductometric titration of acetic acid by sodium hydroxide: (a) experimental, (b) interpretation.

The observed increase in conductance after the equivalence point is less steep than the previous decline because of the lesser mobility of the hydroxide ion as compared with the hydrogen ion.

The curve obtained in the titration of a moderately weak acid, such as acetic, by sodium hydroxide is shown in Fig. 15-7. (The symbol OAc⁻ represents the acetate ion.) In this case no ion remains at constant concentration throughout. Initially both acetate and hydrogen ions are present in only small amounts, but the conductance contributed by the hydrogen ion is the greater because of its high mobility. As the reaction proceeds, the amount of acetate ion increases regularly as it is released from the acetic acid, until the equivalence point is reached; from there on, it remains constant. At the same time, the sodium ion continues to increase in quantity. The concentration of hydrogen ion changes in a somewhat complicated fashion, in that it disappears rapidly at first and then more slowly to zero at the equivalence point. This effect is caused by the suppression of the already low ionization of the acetic acid by the acetate ion formed in the reaction. Added to the increasing conductances of the sodium and acetate ions, this produces an initial dip followed by a nearly linear rise to the equivalence point.

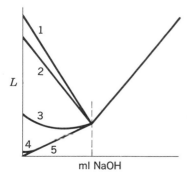

Fig. 15-8 Conductometric titrations of various acids by sodium hydroxide. Curve 1 represents a strong acid, and curve 5 an extremely weak acid, while the others are intermediate. The acids and their ionization constants (at 25°C) are (1) hydrochloric acid; (2) dichloroacetic acid, $K = 5 \times 10^{-2}$; (3) monochloroacetic acid, $K = 1.4 \times 10^{-3}$; (4) acetic acid, $K = 1.8 \times 10^{-5}$; (5) boric acid, $K = 6.4 \times 10^{-10}$.

After this point, the conductance increases more sharply owing to the accumulation of sodium and hydroxide ions.

In this titration, the graph is rounded in the vicinity of the equivalence point, with higher conductance values than predicted by the simple theory. This is a result of the hydrolysis of the salt formed in the reaction, such that a significant excess of hydroxide ions is present at equivalence.

In Fig. 15-8 are represented the titration curves resulting from the titration with a strong base of a series of five acids, ranging in degree of dissociation from a strong acid (curve 1) through progressively weaker acids (curves 2 to 4) to an extremely weak acid (curve 5). In the case of curve 5, the ionization of the acid is so slight that it contributes nothing to the conductance, which is therefore due solely to the salt formed in the reaction. This entire treatment applies equally well to the analogous titrations of strong and weak bases by a strong acid.

It is also possible to titrate conductometrically a weak acid by a weak base, or vice versa, an operation extremely difficult with other techniques. In the titration of acetic acid, for example, by aqueous ammonia, the curve prior to the equivalence point is similar to the analogous portion of Fig. 15-7. The addition of excess reagent, however, scarcely increases the conductance at all. The effect of hydrolysis is negligible, since it does not result in an excess of either hydrogen or hydroxide ions.

DISPLACEMENT REACTIONS

A reaction such as that between ammonium chloride and sodium hydroxide can be carried out as a conductometric titration, since the hydroxide ion cannot accumulate in the solution until all the ammonium ion has reacted to form ammonia and water. Similarly sodium acetate can be titrated by strong acid, pyridinium acetate by strong base, etc.

PRECIPITATION REACTIONS

The conductance method is well suited to the determination of equivalence points in precipitation titrations. The method of analysis of the curve, in terms of ions, is similar to the reactions previously discussed. A degree of solubility of the precipitate can be tolerated which would render it unfit for a gravimetric analysis. The effect of appreciable solubility of the precipitate is to round off the intersection of the two arms of the graph. If the effect is not too great, adequate precision can still be obtained by extending the straight portions to their intersection. Sometimes the solubility can be reduced by the addition of alcohol, and the precision of the results is increased accordingly.

A particularly favorable titration is that of silver sulfate by barium chloride, where two substances are precipitated simultaneously. The conductance drops to a very low value at the equivalence point. Magnesium and other sulfates can be similarly titrated by barium hydroxide.

GENERAL CONSIDERATIONS

Conductometric titrations are potentially useful in any reaction where the ionic content is markedly less at the equivalence point than either before or after it. The method is not conveniently applicable where the total ionic content is large, and changes only slightly at equivalence. For example, in the titration of iron by permanganate following reduction of iron by stannous chloride, the solution typically contains manganous, stannic, mercurous, mercuric, potassium, hydrogen, chloride, phosphate, and sulfate ions in addition to the ferrous and permanganate ions of primary interest. The change in overall conductance during the titration would be insignificant compared with the total conductance.

In favorable instances, titrations should give a precision of ± 2 to 3 parts per thousand.

HIGH-FREQUENCY TITRATIONS

The conductance procedures discussed in the preceding pages depend upon the motion of ions in an electric field. The use of alternating current to avoid electrochemical deposition still permits the ions to move a short distance between alternations. However, if the frequency is increased, a point will eventually be reached beyond which the ions will not have time to acquire their full speed. At such high frequencies, the phenomenon of *molecular polarization* becomes important.

When any molecule is subjected to an external electric field, the electrons within it are drawn toward the positive electrode, while the nuclei are attracted in the opposite direction. There results an actual motion of the two kinds of particles relative to each other, giving rise to a distortion of the molecule. This effect is temporary and disappears upon removal of the field.

Some molecules possess permanent electric dipoles, which is to say that the centers of negative and of positive electric charge within the molecules do not coincide. This is true of most unsymmetrical molecules. For example, water, acetone, nitrobenzene, and chloroform possess dipole moments, while methane, carbon tetrachloride, benzene, and *p*-dinitrobenzene do not. In an applied electric field, dipolar molecules tend to orient themselves, positive toward negative and negative toward positive. Such molecules are said to show orientation polarization in addition to the temporary distortion polarization common to all molecules.

Both types of polarization result in the flow of electric current (defined as electric charges in motion) for an extremely short time following application of the field.   The duration of the polarizing current is considerably less than a millionth of a second and thus is completely insignificant as compared with ordinary ionic conduction at low frequencies (60 or 1000 Hz).   At radio frequencies, measured in millions of hertz (megahertz), the polarization and conduction currents become of the same order of magnitude, and neither can be neglected.

Instruments for measuring the properties of solutions at high frequencies require relatively complex electronic circuitry and show very little resemblance to the simple circuits for low-frequency work.   In general the sample is placed between the plates of a capacitor—less often within an inductance coil—in such manner that the resonant frequency of a tunable circuit is altered by the absorption of energy in the sample.   This tunable "tank" circuit may be the frequency-determining element of an oscillator and the final observation a frequency shift caused by the sample.   Alternatively, it may be incorporated into a tuned amplifier, the output of which will decrease as its resonant frequency departs from the original value.   Usually the tunable circuit includes a calibrated variable capacitor which can be used to compensate for changes in reactance of the cell, returning it to its original value.

This apparatus is equivalent to that employed in the measurement of dielectric constants of nonconducting samples.   For liquids with appreciable conductivity, the absorption of energy is largely due to the motions

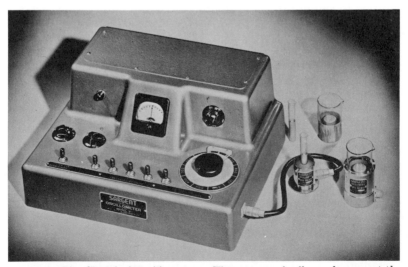

Fig. 15-9   The Chemical Oscillometer.   Three types of cells can be seen at the right.   (*E. H. Sargent & Co., Chicago.*)

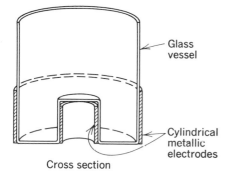

Cross section

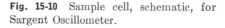

Fig. 15-10 Sample cell, schematic, for Sargent Oscillometer.

of ions, and the method resembles, in its interpretation if not in technique, conventional low-frequency conductimetry. The theory of the high-frequency method, as applied both to conducting and dielectric liquids, is too complex to summarize in the space available here. The student is referred to the excellent treatments by Reilley[11] and by Pungor,[10] which include references to many papers describing specific circuit designs and applications.

An instrument which is commercially available is the Sargent Chemical Oscillometer, (Fig. 15-9). Standard capacitors are provided for returning the frequency to a constant value. The sample cells are of the capacitive type, in which the solution occupies an annular space between two concentric cylindrical electrodes cemented to the outside of the glass surfaces (Fig. 15-10).

Direct high-frequency analyses of binary mixtures can be carried out by means of a calibration curve prepared from known solutions. Thus mixtures of o- and p-xylenes, of hexane and benzene, of water and acetone, and many others have been analyzed successfully. More often the technique is used to follow the course of a titration. Various types of reactions can be studied, including neutralization and precipitation. The shapes of the titration curves in general cannot be predicted by simple summation methods such as apply in low-frequency conductometric titration, but they show sharp breaks corresponding to equivalence points. The shape is dependent in a complicated manner on the frequency. An example is given in Fig. 15-11.[15]

One significant advantage of the method is that no electrodes are placed in contact with the solution. It would seem that the most promising field for the exploitation of high-frequency analysis is in connection with systems depending upon the presence or absence of molecules possessing permanent dipoles, since other techniques are not available in this area.

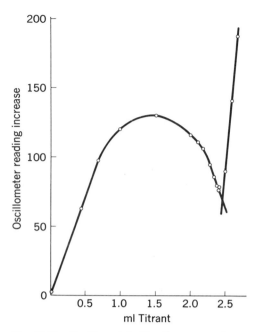

Fig. 15-11   Oscillometric titration curve of *sec*-butyllithium in hexane solution, with acetone as titrant.   This is a difficult analysis by other methods, but easily done by oscillometry, with an average deviation of ± 0.7 percent or better.   (*Analytical Chemistry.*)

PROBLEMS

15-1   A conductance cell containing a 0.0100 molal solution of potassium chloride is connected in series with a 1000-Ω standard resistor and a battery which has an output of 6.00 V.   The current which flows immediately after closing the circuit is shown by an ammeter to be 0.00571 A.   By means of Ohm's law, calculate the conductance of the sample.   From this and the known specific conductance, compute the cell constant.   (Assume a temperature of 25°C.)

15-2   A particular Wheatstone bridge is balanced with 100 Ω in each arm.   The sensitivity of the galvanometer is 1.00×10^{-7} A per mm, and its internal resistance is 50.0 Ω.   The bridge is energized by a 4.00-V battery.   What is the smallest detectable change in the resistance of the unknown arm if the least observable galvanometer deflection is 0.2 mm?

15-3   It has been established[14] that the specific conductance κ of sea water at 25°C depends upon the salinity S according to the relation

$$\kappa = 1.82 \times 10^{-3}S - 1.28 \times 10^{-5}S^2 + 1.18 \times 10^{-7}S^3$$

where S is expressed in grams of salts per kilogram of sea water.   The average value of S for undiluted sea water is 35.

A sample of water taken near the mouth of a river showed a specific conductance of 1.47×10^{-2} mho per cm. What is its salinity? How much fresh water has been mixed with each kilogram of sea water at this location? (Assume that the conductance of the river water is negligible and that differences in density have no effect.)

The cubic equation can be solved by the method of successive approximations. Calculate a value S' by neglecting the terms in S^2 and S^3. Calculate a second value S'' by substituting S' in the higher terms, and a third value S''' by using S'' in the higher terms. S''' should represent a sufficiently good approximation to the true value for S.

15-4 A 25.00-ml portion of 0.1000 F sodium hydroxide was placed in a conductance cell. It was observed that the electrodes were covered, so no water was added. The sample was titrated with 0.100 F hydrochloric acid. The following data were recorded:

Buret reading, ml	Resistance, Ω	Buret reading, ml	Resistance, Ω
0.00	45.3	30.00	149.9
5.00	60.3	35.00	128.4
10.00	79.6	40.00	114.4
15.00	103.8	45.00	104.9
20.00	137.5	50.00	97.7
25.00	186.9		

For each observation, calculate the conductance, and apply the appropriate correction for dilution. Plot both the corrected and uncorrected values against added volume of acid. Discuss.

15-5 Predict the curve which would be obtained in the conductometric titration of $AgNO_3$ by HCl. Explain. The pertinent reactions are:

$$Ag^+ + Cl^- \rightarrow AgCl \ (ppt)$$
$$AgCl + Cl^- \rightarrow AgCl_2^- \ (soluble)$$

The approximate equivalent ionic conductance of $AgCl_2^-$ is estimated to be 80.

15-6 By means of graphs comparable to Figs. 15-6b and 15-7b, predict qualitatively the curves obtainable from the following conductometric titrations: (a) $NH_3(aq)$ titrated by HCl; (b) $NH_3(aq)$ titrated by HOAc (acetic acid); (c) $AgNO_3$ titrated by HCl; (d) HCl titrated by $AgNO_3$; and (e) Na_2CO_3 titrated by HCl in a hot, stirred solution.

15-7 The concentration of a very weak acid, such as boric, can be determined with increased precision by a double titration. One portion of the unknown is titrated with standard sodium hydroxide, and another aliquot with an aqueous ammonia solution which is of the same concentration as the sodium hydroxide. The two curves are superimposed on a sheet of graph paper. Sketch the resulting graph for such an analysis, and show the location of the end point. Explain fully.

15-8 Glycine, NH_2—CH_2—COOH, is amphoteric in that it reacts with a base such as sodium hydroxide to form a salt, $Na^+(NH_2$—CH_2—$COO)^-$, and with an acid such as hydrochloric to form a substituted ammonium salt, $(NH_3$—CH_2—$COOH)^+Cl^-$.

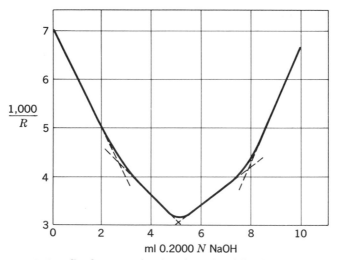

Fig. 15-12 Conductometric titration of acidified glycine by sodium hydroxide. (*After Loiseleur.*[7])

To a 10.00-ml portion of a glycine solution is added 1.000 ml of 1.000 F hydrochloric acid, and the mixture is rinsed into a conductance cell with a small quantity of water. It is then titrated conductometrically with 0.2000 F sodium hydroxide. The resulting curve is shown in Fig. 15-12. (*a*) Explain the observed shape of the curve. (*b*) Calculate the weight of glycine in the 10-ml portion.

15-9 Sketch on one graph the curves which you would predict for the conductometric titration of 0.1 N AgNO$_3$ solution by 1 N solutions of each of the following: HCl, KCl, NH$_4$Cl, CaCl$_2$, NaCl, and LiCl. Which would result in the greatest precision? Explain.

15-10 Potassium can be precipitated quantitatively as the dipicrylamine salt, KC$_{12}$H$_4$N$_7$O$_{12}$, from a slightly alkaline medium. The precipitate after washing can be dissolved in acetone-water mixture and titrated conductometrically with hydrochloric acid according to the reaction KR + HCl → HR + KCl, where R represents the C$_{12}$H$_4$N$_7$O$_{12}$ radical.[6] (*a*) Sketch the curve which would result from such a titration. (*b*) Sketch the curve which would result if the precipitate were first treated with a slight excess of hydrochloric acid and then titrated with sodium hydroxide. Neglect dilution factors.

15-11 Henry, Hazel, and McNabb[4] have reported the conductometric titration of organic bases by boron tribromide or vice versa in such aprotic solvents as nitrobenzene. For example, an inverted V-shaped conductance curve is obtained when BBr$_3$ in nitrobenzene is titrated by the addition of quinoline. Outline the reactions involved and sketch a titration curve showing graphically the contributions of the various species present to the total conductance. (It will be necessary to refer to the original paper for details.)

REFERENCES

1. Frankenthal, R. P.: Conductometry, in L. Meites (ed.), "Handbook of Analytical Chemistry," pp. 5-29ff, McGraw-Hill Book Company, New York, 1963.
2. Griffiths, V. S.: Newer Methods of Determining Electrolytic Conductivity, in G. Charlot (ed.), "Modern Electroanalytical Methods," p. 174, American Elsevier Publishing Company, New York, 1958.
3. Hendricks, R. H., M. D. Thomas, M. Stout, and B. Tolman: *Ind. Eng. Chem., Anal. Edition,* **14**:23 (1942).
4. Henry, M. C., J. F. Hazel, and W. M. McNabb: *Anal. Chim. Acta,* **15**:187 (1956).
5. Jones, G., and B. C. Bradshaw: *J. Am. Chem. Soc.,* **55**:1780 (1933).
6. Kolthoff, I. M., and G. H. Bendix: *Ind. Eng. Chem., Anal. Edition,* **11**:94 (1939).
7. Loiseleur, J.: *Compt. Rend.,* **221**:136 (1945).
8. Mason, D. M., L. L. Taylor, and S. P. Vango: *Anal. Chem.,* **27**:1135 (1955).
9. Polsky, J. W.: *Anal. Chem.,* **19**:657 (1947).
10. Pungor, E.: "Oscillometry and Conductometry," Pergamon Press, New York, 1965.
11. Reilley, C. N.: High-frequency Methods, in P. Delahay (ed.), "New Instrumental Methods in Electrochemistry," chap. 15, Interscience Publishers (Division of John Wiley & Sons, Inc.), New York, 1954.
12. Rosenthal, R.: Electrical-conductivity Measurements in D. M. Considine (ed.), "Process Instruments and Controls Handbook," pp. 6-159ff, McGraw-Hill Book Company, New York, 1957.
13. Rossum, J. R.: *Anal. Chem.,* **21**:631 (1949).
14. Ruppin, E.: *Z. Anorg. Chem.,* **49**:190 (1908).
15. Watson, S. C., and J. F. Eastham: *Anal. Chem.,* **39**:171 (1967).

16

Radioactivity as an Analytical Tool

INTRODUCTION TO NUCLEAR METHODS

Up to this point we have considered analytical methods which are in one way or another concerned with the *electrons* in the substances examined. No consideration has been given to atomic nuclei per se.

Now we shall examine several techniques which can provide analytical and structural information, based on *nuclear* properties. In elementary work, the only such properties usually considered are mass and charge, because only these have a direct bearing on the number and spatial distribution of orbital electrons, hence on the chemistry of the elements. However, it has long been known to physicists that a complex series of energy levels exist within nuclei. This was first evidenced in the late nineteenth century by observations of radiations of definite energy content emanating from atomic nuclei.

The properties which can be ascribed to nuclei include the following of interest in the present context: mass, net charge, spin, magnetic moment, energy levels with respect to neutrons and to protons, stability with respect to radioactive decay processes, and the ability to combine with neutrons or other particles.

Each of these properties, or combinations of them, can be studied profitably by analytical chemists. In this and succeeding chapters we shall take up in turn radioactivity, both natural and induced, in which the emission of particles or radiation by nuclei provide a number of powerful analytical tools; Mössbauer spectroscopy, based on resonance absorption of gamma rays; mass spectroscopy, whereby elements or molecules in ionized form can be sorted out, identified, and measured according to their mass-to-charge ratio; and nuclear magnetic resonance, which depends on the interaction of quantized nuclear spin with an applied magnetic field. (Electron-spin resonance is included in this section because of its close relation to nuclear magnetic resonance.)

RADIOACTIVITY

Although it may be assumed that the reader has a knowledge of the rudiments of radioactive phenomena, we shall summarize those principles briefly for reference purposes.

Nearly all the known elements exist in several isotopic forms. Many of these isotopes do not occur in nature but can be formed artificially by various procedures from suitable isotopes of the same or other elements. Most of the artificially prepared isotopes and many of the naturally occurring ones are unstable in that their nuclei tend to disintegrate spontaneously with the ejection of energetic particles or, in some cases, with the emission of radiant energy. The other product of the disintegration is a residual nucleus slightly lighter in mass than before. This is the phenomenon we call *radioactivity*.

Several different types of particles are ejected by radioactive substances. Those which are of importance for our purposes (see Table 16-1) are the electron (negative), the positron (or positive electron), the alpha particle, and the neutron. The emission of these particles is frequently, but not always, accompanied by the radiation of energy as gamma rays. Another mode of radioactive decomposition sometimes encountered is the spontaneous capture by the nucleus of an electron from the K level (or, less frequently, from the L or higher levels). This process, known as *electron capture*, is most commonly evidenced by the emission of the characteristic x-rays produced by electrons of higher quantum level falling in to fill the vacancy created by the capture.

The particles and radiations from different radioactive nuclei vary widely in their energy content and in the frequency with which they are produced. Both of these properties are characteristic of the particular isotope which is disintegrating, and hence their measurement under suit-

Table 16-1   Particles produced in radioactive decay

Particle	Symbol	Mass*	Charge†	Penetrating power	Ionizing power
Negatron‡	β^-	5.439×10^{-4}	-1	Medium	Medium
Positron‡	β^+	5.439×10^{-4}	$+1$	Medium	Medium
Alpha particle	α	3.9948	$+2$	Low	High
Neutron	n	1.0000	0	Very high	Very low
Photon (gamma ray)	γ	0	0	Very high	Very low

* In units of 1.675×10^{-24} g.

† In units of 4.80298×10^{-10} electrostatic cgs units of charge.

‡ Both negatrons and positrons are considered beta particles, and are often called electrons.

ably standardized conditions will serve to prove the presence of that isotope.

The frequency of occurrence of atomic disintegrations is related to a constant characteristic of each active isotope, namely, its *half-life*, the time required for any given sample of the isotope to be reduced to one-half its initial quantity.   This varies among the known active materials from millionths of a second to millions of years.   The half-lives at either extreme, of course, cannot be measured directly but are inferred from other evidence.   Isotopes of very short life cannot be useful for analytical purposes simply because any experiment takes a considerable amount of time, and these isotopes disappear too quickly.   On the other hand, isotopes of very long life are difficult to apply because the disintegrations are too infrequent.   Isotopes useful for analytical applications are those with half-lives roughly between a few hours and a few thousand years.   If the experiments can be carried out quickly in the same laboratory where the radioactive material is prepared, the lower limit can be reduced to perhaps 10 min.   The length of any experiment generally cannot exceed about ten times the half-life of the isotope employed.

A selection of radioisotopes which have been found useful in analytical applications is given in Table 16-2.   Radioisotopes may be useful as sources of radiations or as tracers to follow some reaction or process and assist in determining its extent.   Before turning to these applications, we shall consider the methods of detection and measurement.

DETECTION OF RADIATIONS

The principal methods of detection of importance in analytical work are (1) the observation of visible light produced by the radiation (scintilla-

Table 16-2 Radioisotopes used in analysis*

| Isotope | Type of decay† | Half-life | Energy of radiation, MeV | |
			Particles	Gamma transitions
^{3}H	β^-	12.26 yrs	0.0186	None
^{14}C	β^-	5720 yrs	0.155	None
^{22}Na	β^+ (90%) EC (10%)	2.58 yrs	0.545	1.27
^{32}P	β^-	14.3 days	1.71	None
^{35}S	β^-	87 days	0.167	None
^{36}Cl	β^-	3.0×10^5 yrs	0.714	None
^{40}K	β^- (89%) EC (11%)	1.27×10^9 yrs	1.32	1.46
^{42}K	β^-	12.36 hrs	3.55 (75%) 1.98 (25%)	1.52 (25%)
^{45}Ca	β^-	165 days	0.255	0.32
^{51}Cr	EC	27.8 days	. . . .	0.32 (8%)
^{55}Fe	EC	2.60 yrs	. . . .	None
^{59}Fe	β^-	45 days	0.460 (50%) 0.27 (50%)	1.29, 1.10
^{57}Co	EC	270 days	. . . .	0.122, 0.0144, 0.136
^{60}Co	β^-	5.26 yrs	0.32	1.333, 1.173
^{65}Zn	EC (97.5%) β^+ (2.5%)	245 days	0.33	1.11
^{85}Kr	β^-	10.6 yrs	0.67	(γ)
^{90}Sr	β^-	29 yrs	0.54	None
^{90}Y	β^-	64 hrs	2.27	(γ)
^{95}Zr	β^-	65 days	0.36, 0.40	0.72, 0.76
^{95}Nb	β^-	35.1 days	0.16	0.77
^{110}Ag	β^-	253 days	0.085 (58%)	0.44, 2.46
^{119m}Sn	IT	250 days	. . . .	0.065, 0.024
^{131}I	β^-	8.06 days	0.60 (87.2%) (others)	0.364 (80.9%) (others)
^{137}Cs	β^-	30 yrs	0.51 (92%) 1.17 (8%)	0.662 (from ^{137m} Ba daughter)
^{133}Ba	EC	7.2 yrs	. . . .	0.360, 0.292, 0.081, 0.070
^{140}La	β^-	40.2 hrs	1.34 (70%) (others)	0.49, 0.82, 1.60 (others)
^{147}Pm	β^-	2.65 yrs	0.225	(γ)
^{170}Tm	β^-	127 days	0.97 (76%) 0.88 (24%)	0.084
^{203}Hg	β^-	47 days	0.21	0.28
^{198}Au	β^-	2.70 days	0.96 (others)	0.412 (others)
^{204}Tl	β^- (98%) EC (2%)	3.80 yrs	0.76	None
^{210}Pb	β^-	22 yrs	0.015, 0.061	0.046

* Data selected from extensive tabulation by Friedlander, Kennedy, and Miller.[8]
† EC = Electron capture; IT = Internal transition.

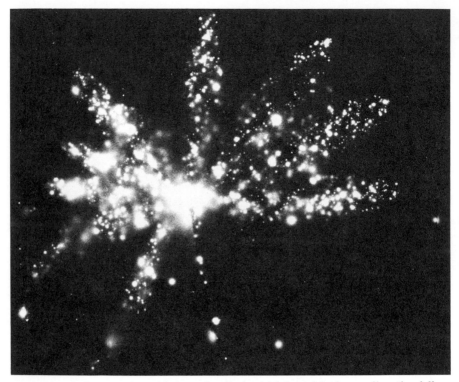

Fig. 16-1 Radioautograph of leaves of a shrub which has picked up radioactive fall-out; collected near the A-bomb test site in Nevada. (*Courtesy of Dr. Lora M. Shields.*)

tions), (2) the electrical measurement of ionization in a gas, (3) the displacement of electrons in semiconductor crystals, and (4) photography. The direct action of nuclear radiations on photographic materials is especially useful for mapping the distribution of a radioactive substance in the surface layers of a solid material, such as a mineral or biological specimen. This procedure is known as *radioautography* or *autoradiography* (Fig. 16-1). The only other common application of the direct photographic method is in film badges worn by nuclear workers to detect excessive exposure to radiation.

SCINTILLATION COUNTERS

The quantitative determination of radioactivity by means of scintillations has long been known, but only in the last decade has it been modernized to provide a method with a high degree of convenience and accuracy. When a ray or particle strikes the surface of a suitable crystal, a tiny flash of light is emitted. A means of counting such flashes therefore provides a means of counting the particles. In the early days the flashes were counted directly by an observer with a microscope. In modern practice, the light from the crystalline target is detected by a photomultiplier tube which converts the radiant energy to an electric signal which is fed into an amplifier for measurement. Figure 16-2 shows a typical detector and housing assembly.

Among the materials which have been found useful as scintillation crystals are anthracene, stilbene, terphenyl, and sodium iodide. The last named must be activated, i.e., rendered more sensitive, by the admixture of a trace of thallous iodide. These detectors will respond to alpha, beta, or gamma radiations. The iodide crystal is particularly desirable for gamma rays because its density ensures absorption of a relatively high fraction of the incident radiation.

The geometry of the photomultiplier tube and of the optical elements which couple the scintillator to it is important to the efficiency of the assembly. The dimensions of the scintillator ideally should be large enough to contain the whole range of the most energetic beta particles, and of the electrons produced photoelectrically by gamma rays. The scintillator should be surrounded by a reflecting surface to minimize

Fig. 16-2 Iron shield and sample holder for planchet counting. Any type of end-window detector can be installed. The overall height is about 40 cm, the weight 90 kg. (*Radiation Counter Laboratories, Inc., Skokie, Ill.*)

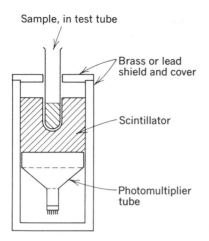

Fig. 16-3 Well scintillation counter, schematic.

light loss.   A well-designed counter exhibits a response which is linear with respect to the energy of the radiation.

A considerably higher efficiency can be obtained with a *well counter* (as compared with a window counter like that of Fig. 16-2).   This variety consists of a block of crystalline or plastic scintillator with a hole drilled into it to accept a test tube (Figs. 16-3 and 16-4).   For reproducible results, a fixed volume of solution must always be taken for well-counting.

Also available are high-efficiency units in which a pair of scintillators encloses the relatively small sample as in a sandwich, so that essentially all the emitted radiation enters the detector.   These are called 4-π detectors.

For quantitative measurements of isotopes which emit low-energy beta particles, such as ^{14}C, ^{35}S, and especially ^{3}H (tritium, sometimes symbolized as T), a *liquid scintillator* is to be preferred, into which the active compound can be directly incorporated.   This ensures maximum efficiency in the production of scintillations from betas.   There are a number of organic compounds which will act as scintillators when dissolved in suitable solvents.   These include anthracene, *p*-terphenyl, 2,5-diphenyloxazole (PPO), α-naphthylphenyloxazole (NPO), and phenylbiphenyloxadiazole (PBD).   Of these, PPO is the most effective, but its emitted radiation is ultraviolet.   It is customary to mix with PPO a *secondary scintillator*, which translates through a fluorescent mechanism the ultraviolet scintillations into the visible range.   The most used secondary scintillator is 1,4-bis-2-(5-phenyloxazolyl)-benzene (POPOP) or its dimethyl derivative (dimethyl-POPOP).   A recommended solution contains 5 g per liter PPO and 0.3 g per liter dimethyl-POPOP in toluene.

When working with such low-level activity, special precautions must be taken to distinguish those pulses which are produced by scintillations from spurious pulses arising in the photomultiplier, caused either by stray or cosmic radiation, or by shot-effect noise. This can be accomplished by employing two identical photomultipliers looking at the same scintillator, and connected in a *coincidence circuit* (Fig. 16-5). This consists of an electronic unit called an *AND-gate*, which will transmit a signal to the counting equipment only when it receives *simultaneous* pulses from both photomultipliers. The shot noise, being random, will occasionally produce simultaneous signals, and if necessary this source of undesired counts can be reduced still further either by a triple coincidence circuit with three photomultipliers, or by reducing the thermal motion of electrons by refrigeration.

Scintillation counting is inherently proportional, in that the energy in each flash of light is determined by the energy of the particle which originated it. This proportionality can be maintained through the detector and amplifier to the ultimate record.

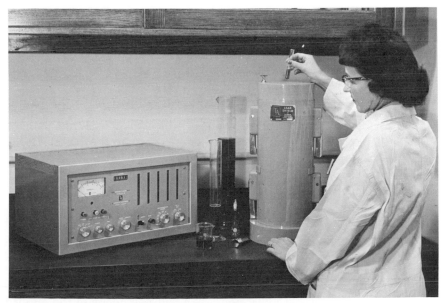

Fig. 16-4 Laboratory counting assembly. The electronic unit is a decade scaler. To the right is shown a lead shield which encloses a well-type scintillation counter; samples are inserted from the top. (*Technical Associates, Burbank, Calif.*)

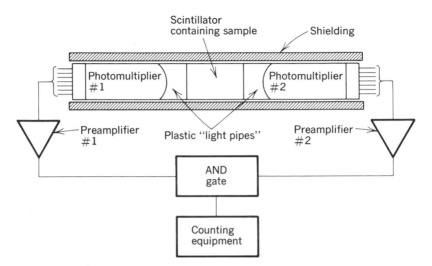

Fig. 16-5 Liquid scintillation counting assembly, schematic. The sample, photomultipliers, and preamplifiers may be refrigerated to decrease electrical noise.

GAS–IONIZATION DETECTORS

As indicated in Table 16-1, all types of nuclear radiation (except neutrons) produce significant ionization in materials into or through which they penetrate. (It is this ionization which is the direct cause of radiation damage to living material.) Ionization due to radiation is most easily measured in gases.

Let us consider the phenomena which occur in a gas-filled glass vessel provided with two electrodes, one of which is a metal tube perhaps 2 cm in diameter by 10 cm in length, the other a wire passing along the axis of the cylinder (Fig. 16-6). Let the central wire be connected through a high resistance R to the positive terminal of a variable-voltage dc supply, while the outer electrode is held at ground potential, as is the negative of the power supply. The positive electrode is connected also, normally through a capacitor C, to the input of an amplifier (indicated in Fig. 16-6 as the grid of a vacuum tube). The output of the amplifier, not shown, operates a deflecting meter or an electromechanical counting register. Let us subject the tube to a constant small source of energetic beta particles, which we will assume to be few enough per second to cause individual pulses of ionization. We will now increase the applied potential gradually from zero to several thousand volts. At small voltages (region A of Fig. 16-7),[12] the ions produced are accelerated only slowly by the electric field, and many of them recombine to form neutral molecules before they can reach the electrodes. As the potential is increased, the number of ions per pulse reaching the electrodes also increases, until a condition of saturation is attained, where essentially *all* the ions formed are discharged at the electrodes, and the observed size of the pulses is constant over a region of 100 V or more (region B in Fig. 16-7). This is called the *ionization chamber* region.

Upon increasing the voltage beyond this region, at C, the pulse size again increases, because the ions are accelerated to such a degree that they cause secondary ionization by collision with other molecules of the gas. In this portion of the curve, the height of a pulse is still dependent on the energy of the original ionizing event. The pulse height is now a *multiple* of that caused by the primary ionization; hence this part of the curve is called the proportional region, and a device operating under these conditions is a *proportional counter*. The curve of Fig. 16-7 is steeply sloping in the proportional region, which means that for truly proportional response the voltage must be precisely controlled.

Increasing the voltage into region D produces greatly increased secondary ionization, so that proportionality is lost. At E, a plateau is attained where over a band of 100 or 200 V all pulses have equal magnitude, regardless of the energy of the ionizing particle. This is the Geiger region, and the counter used in this manner is a *Geiger counter* (sometimes called Geiger Müller, or G-M counter). Beyond this plateau (at F) the tube breaks into a continuous glow discharge.

Thus detection and counting of pulses are possible in three modes, corresponding to areas B, C, and E. The relative advantages and disadvantages of each can be summarized as follows.

The ionization chamber has the advantage of requiring only a low voltage (100 to 200 V), but the current which it passes is very small (perhaps 10^{-8} A) and therefore requires either a high-gain amplifier or a highly sensitive electrometer. In practice, it is most useful for pulse

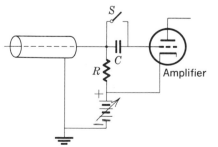

Fig. 16-6 Elementary circuit for an ionization chamber.

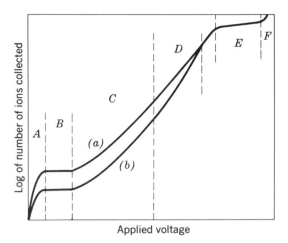

Fig. 16-7 Number of ions collected as a function
of applied voltage; curve (*a*), alpha particles; curve
(*b*), beta particles. (*D. Van Nostrand Company,
Inc., Princeton, N.J.*)

work with the highly ionizing alpha particle. With beta and gamma
particles it is ordinarily restricted to the measurement of relatively intense
beams of radiation, where the current is continuous rather than pulsed.
For this application, the capacitor C of Fig. 16-2 is shorted out by switch
S, and the amplifier must be of the direct-coupled type or its equivalent,
which can respond to slow changes in a minute direct current.

The proportional counter operates at a much higher potential, perhaps 1000 to 2000 V. As with the ionization chamber, the size of pulse
is proportional to the number of ions produced by the primary particle,
but the ion-multiplication process results in internal amplification which
may be as great as 10^4. The external amplifier therefore need not have
unusually high gain, but, if observed pulse heights are to reflect the
energy of the original particle, it must be linear (i.e., distortion-free,
output proportional to input). A counter operated in this region has
an extremely short recovery time, and can count more than 10^5 pulses
per min. A well-regulated voltage supply must be provided.

The Geiger counter takes a somewhat higher voltage (though still
in the 1000 to 2000 V range), but the pulses are so large that little or no
amplification is required. This favorable situation is often more than
offset by the fact that the pulses are all uniform in size and therefore
give no information about the energy of the ionizing particles. Another
disadvantage is the slow speed which limits counting rates to about 10^4
per min. The Geiger counter, which formerly almost monopolized the

field, is now rapidly yielding to the proportional counter and to the scintillation counter. The voltage required for proportional or Geiger-Müller counting depends on the geometry of the tube and on the nature and pressure of the filling gas.

The operating regions for Geiger and proportional counters are determined by plotting the counting rate, in counts per minute, for a fixed source of radiation, against applied volts. In both types of counters precautions must be taken to prevent the positive ions which are accelerated toward the outer (negative) electrode from causing the emission of secondary electrons. Such electrons would move toward the central wire, produce secondary ionization of the gas, and a continuous discharge would result. The accepted method of avoiding this difficulty is to add to the filling gas a few percent of an organic compound (alcohol or methane) or a halogen. The positive ions transfer their excess energy to molecules of the additive rather than to the electrode.

A modification of the proportional counter which is particularly convenient, especially for weakly penetrating radiations, is the *flow counter*, in which the gas (frequently argon–10 percent methane) is allowed to flow slowly through the tube during use. This obviates the deterioration of the organic additive and also permits a sample to be placed directly inside the counter, often an important advantage.

SEMICONDUCTOR DETECTORS

Excellent proportional detection can be achieved with detectors of silicon or germanium.[1,14] Crystals of these elements are prepared by one of two techniques so as to have a sensitive volume within which absorbed radiation will displace electrons from their normal locations.

One such technique requires the preparation of a p-n (or n-p) junction* on or just barely beneath the open surface of the crystal. The upper and lower surfaces are then made conducting with thin, deposited metal films. The resulting diode is connected electrically in the reverse-biased mode, so that electrons are pulled away from the junction on the n side, while positive holes are pulled away on the p side. This procedure causes the formation of a *depletion region*, the thickness of which is variable from zero to about 1 mm by variation of the applied voltage.

According to the second technique, ions of lithium are diffused or "drifted" into the crystal under the influence of an electric field; the lithium has the effect of "cleaning up" or compensating the natural charge carriers, so that when a field is applied across the treated crystal, the

---

* The nature and properties of semiconductor junctions will be discussed in more detail in Chap. 26.

depletion layer will be many times deeper, perhaps as thick as 1 cm.

It is necessary that lithium-drifted germanium detectors be operated at liquid-nitrogen temperature (77°K), because the lithium ions are
sufficiently mobile at room temperature to destroy the geometry of the
depletion layer.   This cooling is not required for silicon detectors nor
for germanium ones which do not contain lithium.

These detectors are used in a manner quite analogous to gas-filled
ionization chambers.   Radiation which is absorbed in the depletion
region produces pairs of electrons and holes which are accelerated toward
the respective electrodes, forming a current proportional to the energy
of the ionizing particle or photon.   A sensitive and noise-free electrometer
amplifier is required because the currents are of only a few microamperes.

The thin depletion layers of the less expensive surface-junction
detectors are sufficient for the detection of beta and heavier particles,
which cannot penetrate deeply, but the thicker layer of a lithium-drifted
detector, together with the larger absorption coefficient of germanium
as compared to silicon, makes the germanium-lithium detector by far the
best for gamma spectroscopy.   (Mention was made in Chap. 9 of the
use of this detector with x-rays.)

BACKGROUND

There is a significant amount of radiation always present in the atmosphere, so that even in the absence of a sample any detector of radioactivity will show a finite response.   This radiation is due in part to the
natural radioactivity of the surroundings, and in part to cosmic rays.
By shielding the counter with 2 or 3 in. of lead, it can be reduced markedly
to perhaps 15 to 20 cpm (counts per minute).   A much higher background
count, or a sudden increase, may indicate accidental contamination of the
immediate surroundings with radioactive matter, or it may indicate incip-
ient failure of the counter itself.   All measurements of activities must be
corrected for background count before any use or interpretation can be
undertaken.

AUXILIARY INSTRUMENTATION

Since scintillation, proportional, and Geiger counters all deliver their
information as pulses, a facility must be provided for counting these
pulses.   An electromechanical counter is satisfactory for slow counting,
less than 10 to 100 cps (counts per second) but has too much mechanical
inertia to permit faster operation.   For faster counting, one may employ
a *glow-transfer tube*, a gas-filled glow-discharge tube in which 10
successive pulses transfer the glow successively to 10 positions in the

tube. The fastest counting, however, must utilize an electronic scaling circuit, which is an amplifier so connected that it will transmit to its output terminals only every *second* pulse which comes to its input connections. A number of these stages can be used in succession to deliver one output count for every 2, 4, 8, 16, etc., pulses from the counter tube. Units with scaling factors of 64 or 128 are common. A sixteenfold scaler can be modified by suitable electronic circuitry to reduce its factor to 10, which permits the convenience of decimal scaling.

COUNTING ERRORS

Radioactive decay is statistical in nature, which is to say that the exact number of atoms which will distintegrate and eject particles in any particular second is governed by the laws of probability. Observed counts are significant only when a large enough number have been accumulated to permit valid statistical analysis. The most convenient criterion is the *standard deviation* σ, sometimes called the *root-mean-square deviation*. It can be shown that for radioactivity measurements when the half-life of the decaying isotope is long compared to the duration of the experiment, the standard deviation is simply the square root of n, the total number of counts. One is thus justified in expressing the results of the experiment as $n \pm \sqrt{n}$.

The background count is also statistical in nature. If we let σ_s represent the standard deviation of the sample count, σ_b that of the background, and σ_t that of the sample with background (total), then it can be shown that

$$\sigma_s = (\sigma_t^2 + \sigma_b^2)^{1/2} \tag{16-1}$$

There are two ways of handling this: The background count can be run long enough that $\sigma_b \ll \sigma_t$, and hence can be neglected, or the ratio of counting times t_t for the sample and t_b for the background which will give the best precision in the least time can be estimated by the relation

$$\frac{t_t}{t_b} = \left(\frac{R_t}{R_b}\right)^{1/2} \tag{16-2}$$

where the R's refer to the respective count rates, which need only be known roughly from a preliminary measurement.[5]

As an example, suppose in an experiment the approximate activities for sample and background are $R_t = 1000$ cpm and $R_b = 40$ cpm. The time ratio should be $t_t/t_b = (1000/40)^{1/2} = 5$. If a precision of 1 percent standard deviation is desired, the total count must be 10,000 counts, $R_t = 10,000 \pm \sqrt{10,000} = 10,000 \pm 100$, which means counting for 10 min. Therefore the background count must be taken for at least $1/5$ of

10 min, or 2 min. The value of σ_s is given by $(10^4 + 80)^{1/2} = 100.4$ counts per 10 min, so the final result is

$$R_s = (1000 - 40) \pm 10 \text{ cpm} = 960 \pm 10 \text{ cpm}$$

The duration of a counting experiment must thus depend on the level of activity and the desired precision. Some scalers have provision for automatic stopping when the count reaches a predetermined value. If, for example, 2 percent standard deviation is desired, which corresponds to 2500 counts, the scaler will be preset to turn itself off when this value is reached. A reading of the timer will then give the information necessary to compute the counts per minute. An interesting discussion of counting statistics can be found in a paper by Kuyper.[11]

Another source of error is due to the occurrence of *coincidences*, here defined as two pulses coming so close together that the counter does not have sufficient time to recover from one pulse in time to respond to the next. The recovery time varies considerably from one type of detector to another. If the recovery time be designated by τ, the observed counting rate by R, and the true pulse rate by R', the governing relation is

$$R' = R + R^2\tau \tag{16-3}$$

It must be realized that in general only a fraction of the radiation from any sample can enter the counter (unless, of course, the sample is *inside* the counter). With a single end-window counting tube, the geometric efficiency must be less than 50 percent, and may be much less. With thin-walled and dipping counters, similarly low efficiencies must be expected. The figure can be raised to much better than 50 percent by making use of the 4π-geometry previously mentioned. However, an efficiency of less than 50 percent is often tolerated because the layout of the equipment may be much simpler. It is always important to maintain constant geometry for determinations which are to be compared with one another. This is facilitated by suitable design of sample-holding equipment such as that shown in Fig. 16-2.

Precautions must be taken against the effects of partial absorption of radiations by the sample itself or by other materials which may be present, such as solvent or filter paper. Samples of solids are best prepared as thin films deposited on a firm, smooth support (*planchet*). Liquids are most conveniently counted in a well scintillator.

PULSE-HEIGHT ANALYSIS

Ionizing particles and photons vary widely in energy content, as illustrated in Table 16-2. Individual active species can often be identified

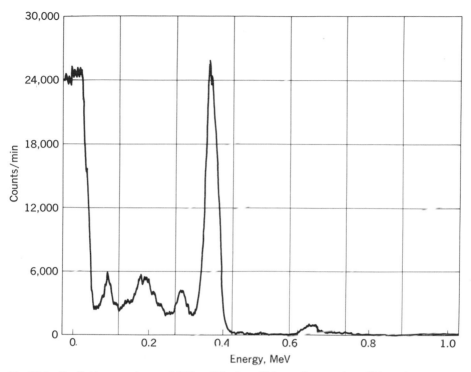

Fig. 16-8 Radiation spectrum of ^{131}I. (*Nuclear-Chicago Corporation, Chicago.*)

by observation of these energy values. One method is by the use of *standard absorbers*—known thicknesses of aluminum or copper for the less energetic particles, or lead for gamma rays. The *half-thickness*, i.e., the thickness of metal which reduces the activity to half its value, can be translated into energy by means of predetermined calibration curves.

A more elegant method is by the observation of pulse heights from a semiconductor, proportional gas, or scintillation counter.[14] The semiconductor detector is especially valuable in this service because of its high intrinsic resolution and short recovery time. An apparatus for this purpose is a *spectrometer;* it sorts out radiation according to its energy content, just as an optical spectrometer does. The spectrometer consists of a voltage-discriminating circuit, together with the requisite amplifiers and counting registers. The discriminator is a kind of electronic gate which permits passage of pulses of only a specified range of heights. The instrument first counts those pulses with heights between zero and 1 V (for example), then those in the range 1 to 2 V, then 2 to 3, etc., and plots these data on a recorder. The result is a spectral curve such as that shown in Fig. 16-8.

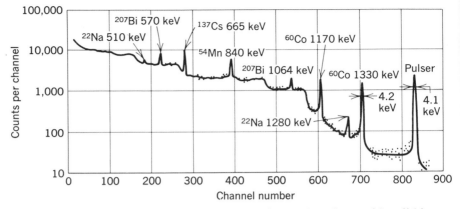

Fig. 16-9 Gamma-ray spectrum taken on a multichannel analyzer with a lithium-drifted germanium detector. (*Isotopes, Inc., Westwood, N.J.*)

A number of manufacturers produce elaborate instruments which carry out such counting and plotting automatically. They are called *multichannel analyzers*, and vary in resolution from 100 to 500 or even more channels. Detection operates simultaneously at all levels, so that the count in each channel increases as long as the experiment is continued. In some analyzers provision is made for printing out on paper tape the total count in consecutive channels at the conclusion of the experiment. Figure 16-9 shows the spectrum of a mixture of gamma emitters taken with a multichannel analyzer equipped with a lithium-drifted germanium detector.

NEUTRON COUNTING

Since neutrons are uncharged, they do not produce ionization in a gas by any direct process. They may however be detected in a counter filled with gaseous BF_3, because neutrons react very readily with the ^{10}B nucleus to produce ^7Li and alpha particles. The resultant alphas trigger the counter in the usual way. Neutrons can be detected in an analogous manner by a scintillation counter: a boron compound is added to an alpha-sensitive scintillator.

Several different types of neutron spectrometers have been designed. Low-energy neutrons can be diffracted in a crystal spectrometer very much as can x-rays, since their deBroglie wavelength ($\lambda = h/mv$) is in the x-ray range. Other methods involve selection according to velocities, whereby all neutrons which pass over a measured course in a vacuum tube in a given length of time are observed. Then scanning of the time for a fixed distance will yield a spectrum. The details of these instruments cannot be described in the space available here.

The selective absorption of neutrons of varying energy content is an extremely powerful analytical tool, but, unfortunately, for general application it requires a source of neutrons of such high flux that only a nuclear reactor can provide them. A less powerful source can be applied to the absorptiometric determination of a few elements which have unusually high absorptive power ("cross section"), particularly B, Cd, Li, Hg, Ir, In, Au, Ag, and several of the lanthanides. An extensive introduction to the subject has been published by Taylor, Anderson, and Havens.[17] More recent data on nuclear cross sections are available.[8]

ANALYTICAL APPLICATIONS OF RADIOACTIVE SOURCES

Gamma rays are physically indistinguishable from x-rays of similar wavelength, and hence in principle gamma-ray sources can be substituted for the more elaborate x-ray tube in many of its applications as discussed in Chap. 9.

The absorption of alpha and beta rays is also potentially useful for analytical purposes. Because of their low penetrating power, alpha rays are best suited for the analysis of gases. This process has been studied by Deisler, McHenry, and Wilhelm,[7] who mounted an alpha-ray source (polonium, ^{210}Po, in an aged preparation of radium-D, ^{210}Pb) inside an ionization chamber which also served as the sample container. Under conditions of constant applied potential and constant gas pressure, the current through the chamber is a function only of gas composition. In favorable cases, binary gas mixtures can be analyzed with a precision of ± 0.2 to 0.3 mole percent, by reference to a calibration curve prepared from measurements on known mixtures. An instrument based on this principle, the Billion-Aire, is manufactured by Mine Safety Appliances Co., Pittsburgh, Pa., for the detection of such toxic gases as $Ni(CO)_4$ and TEL in air in the parts-per-billion range.

The absorption of beta rays has also been applied to analysis,[10,16] this time primarily with liquids, though the principle is applicable also to solids and gases. It can be shown on theoretical grounds that the absorption of beta rays by matter is almost entirely due to electron-electron collisions. Elementary considerations show that hydrogen has a greater number of electrons per unit of weight than any other element, by a factor of at least 2. This renders the method particularly sensitive to the presence of hydrogen.

RADIOACTIVE TRACERS

The ease with which the presence of active isotopes can be detected and the precision with which they can be measured, even in very small quantities, lead to a variety of analytical procedures of great versatility. The most important general techniques are activation analysis, isotope dilution, and radiometric analysis.

ACTIVATION ANALYSIS

A great many elements become radioactive when bombarded by energetic particles such as protons, deuterons, alpha particles, or neutrons. The resulting activity can provide data for quantitative analyses.[12] As an example, we shall consider activation by exposure to neutrons of *thermal* velocities.

The neutron may be considered to be captured by the atomic nucleus to give a larger nucleus with the same positive charge, which is therefore an isotope of the same element. This new nucleus is in many cases unstable and spontaneously decomposes by the emission of a particle or a gamma ray; in other words, it is radioactive. The active isotopes formed in this way from the various elements vary widely in half-life and in many instances can be identified by the determination of this constant, along with other pertinent information, such as the gamma-ray spectrum described previously.

This phenomenon can be used for analysis by subjecting a sample to neutron bombardment either in a nuclear reactor (pile) in which uranium is undergoing fission or by other means. Radioactivity will be induced in each of the elements present which are capable of being activated by neutrons. The intensity of radioactivity in the sample is then plotted against the time to give a so-called *decay curve*. This curve is generally complex in nature, being the summation of the activities of all the active elements present. The half-life of the longest-lived component can be determined from the later portions of the curve after the more transient substances have virtually vanished. The activity due to this element can then be subtracted from the readings for shorter times. Then the next longest-lived substance can similarly be identified and its effects subtracted, then the next, and so on.

For example, an alloy of manganese and aluminum was analyzed by this method.[2] A small, carefully cleaned and weighed sample of the sheet alloy (about 25 mg) was subjected to the action of neutrons in an atomic reactor for a period of 5 min. The subsequent radioactivity of the sample was followed with a Geiger counter for a period of some 60 hr. The results are plotted in Fig. 16-10. The activity due to the aluminum (the principal constituent of the alloy) was very intense, but of short duration (2.3 min half-life). Since it was not necessary to analyze for aluminum, it was convenient to wait for half an hour after irradiation before commencing measurements with the counter, so that practically all the active

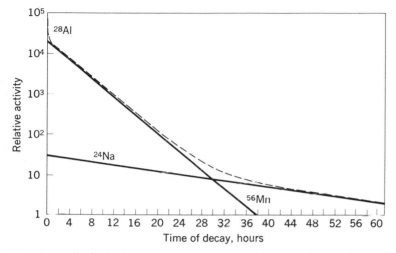

Fig. 16-10. Analysis of manganese-aluminum alloy by neutron activation. (*Analytical Chemistry*.)

aluminum had time to decay and thus not interfere with subsequent measurements.

By reference to the figure, it is seen that the activity (on a logarithmic scale) is a linear function of time after about 40 hr. This straight-line portion extrapolated backward gives the activity at any time due to an isotope with a half-life which can be read from this graph as about 15 hr. This is identified from tables of isotopes as ^{24}Na, with a known half-life of 15.0 hr. Furthermore, the plot shows a linear region extending from 1 to 18 hr which must be due to an isotope of half-life close to 2.5 hr. This is found to be ^{56}Mn (half-life 2.58 hr). The portion of the curve at times shorter than 1 hr is incomplete, as indicated above, and is due to ^{28}Al with a half-life of 2.3 min.

The sensitivity of analysis by means of neutron activation depends upon the intensity of the activating source, upon the ability of the element sought to capture neutrons (called the *neutron-capture cross section*), and upon the half-life of the induced activity. The governing relation[12] is

$$A = N\sigma\phi \left[1 - \exp\left(-\frac{0.693t}{T_{\frac{1}{2}}} \right) \right] \tag{16-4}$$

where A = induced activity at end of period of irradiation, disintegrations sec^{-1}

N = number of atoms present of isotope being activated

σ = neutron-capture cross section, cm^2

ϕ = neutron flux, $cm^{-2} sec^{-1}$

t = duration time of irradiation

$T_{1/2}$ = half-life of product

(t and $T_{1/2}$ must be in the same units). Quantitative analyses are rarely based on calculations from this equation, as sufficiently reliable data are seldom available for σ, ϕ, and $T_{1/2}$; as a further complication ϕ may not be homogeneous and may vary with time. For practical purposes, standard samples are irradiated simultaneously with unknowns, and the analysis carried out by simple comparison.

If a nuclear reactor is available for activation, in favorable cases as little as 10^{-10} g of an element can be detected. For less powerful neutron sources, the method is limited to those elements which have particularly favorable nuclear properties. For example, a neutron source consisting of 25 mg of radium mixed with 250 mg of beryllium, which produces a usable flux of about 100 neutrons per cm^2 per sec, will activate only Rh, Ag, In, Ir, and Dy, but can offer a convenient and accurate method for the determination of these elements, even in trace amounts. A ^{124}Sb-Be source with a flux of 10^3 to 10^4 will activate some 19 elements. A nuclear reactor may have a flux as great as 10^{14}, and will activate nearly all the elements heavier than oxygen, but with greatly differing sensitivities. For some elements (In, Re, Ir, Sm, Eu, Dy, Ho, Lu, V, As, Sb) activation analysis appears to be capable of greater sensitivity than strictly chemical analysis, whereas for others (Fe, Ca, Pb, Bi, Zn, Cd, Na, K, etc.) activation is no better than or distinctly inferior to chemical methods.[13]

Activation may also be produced by bombardment with protons, ^{3}He nuclei, and gamma photons. The experimental difficulties with the first two and the limited applicability of all three have militated against their use.

ISOTOPE DILUTION

This is a technique which is suitable where a compound can be isolated in a pure state but with only a poor yield. A known amount of the same substance containing an active isotope is added to the unknown and thoroughly mixed with it. A sample of the pure sustance is then isolated from the mixture, and its activity is determined. A simple calculation then gives the quantity of the substance in the original material.

Consider a solution which contains W g of a compound which is to be determined. To the solution is added a portion of the same compound which is "tagged" with a radioactive atom; the added portion weighs w g and has an activity of A cpm and a specific activity $S_0 = A/w$. After thorough mixing, g g of this compound is isolated in a pure state and found to have activity B cpm and specific activity $S = B/g$. Now the total amount of activity (assuming loss by decay to be negligible) must be the same after mixing as before, or

$$S_0 w = (W + w)S \qquad (16\text{-}5)$$

from which it follows that

$$W = g\frac{A}{B} - w \qquad (16\text{-}6)$$

If the added material is highly active, then the amount added w can be very small relative to W, and Eq. (16-6) reduces to

$$W = g\frac{A}{B} \qquad (16\text{-}7)$$

Suppose it is required to determine the amount of glycine in a mixture with other amino acids. Glycine can be isolated chemically, but only with a low yield, which makes isotope dilution an appropriate technique. We start by synthesizing or obtaining commercially a sample of glycine which contains an atom of ^{14}C in perhaps one in every million of its molecules. A 0.500-g portion of this active preparation (specific activity, corrected for background, is 25,000 cpm per g) is mixed with the unknown From the mix is obtained 0.200 g of pure glycine with an activity of 1250 counts per 10 min. The background is 100 counts per 5 min. The data may be summarized as follows:

$$w = 0.500 \text{ g}$$
$$S_0 = 25,000 \text{ cpm per g}$$
$$A = wS_0 = 12,500 \text{ cpm}$$
$$g = 0.200 \text{ g}$$
$$B = \frac{1250}{10} - \frac{100}{5} = 105 \text{ cpm}$$

from which

$$W = g\frac{A}{B} - w = 0.200\frac{12,500}{105} - 0.500$$

$$= 23.3 \text{ g}$$

which is the required weight of glycine in the sample. If w were neglected (as in Eq. 16-7) the resulting error in this particular example would have been about 2 percent, whereas the counting error, calculated as detailed in a previous paragraph, is about 3.5 percent.

As a further illustration, let us examine the work of Salyer and Sweet[15] on the electrogravimetric determination of cobalt in steel or other alloys. The reason for using isotope dilution was that cobalt, deposited *anodically* as Co_2O_3, is apt to form a poorly adherent layer; this precludes a conventional gravimetric determination, but loss of some particles of oxide during washing and drying is not objectionable in isotope dilution. Other shortcuts are permissible, such as the substitution of centrifuging for quantitative filtration and washing. A standard curve was prepared by the addition of equal aliquots of ^{60}Co to samples containing various amounts of pure cobalt, then electrodepositing Co_2O_3 in a standardized manner. The unknown was fortified by an aliquot of ^{60}Co immediately upon the dissolution of the sample. Chemical treatment was required in order to remove elements which might interfere with the electrolysis. Cobalt was then deposited, the deposit weighed and counted, and the cobalt in the original sample ascertained by reference to the standard curve. Standard deviations varied from 0.005 to 0.025 percent.

The use of a standard curve in this way tends to eliminate some sources of error, in much the same way that a blank determination does in many types of analysis. However, in many applications the curve is dispensed with, and the answer obtained directly from Eq. (16-6) or (16-7).

RADIOMETRIC ANALYSIS[3]

This term has been applied to analytical procedures in which a radioactive substance is used indirectly to determine the quantity of an inactive substance. An excellent example[9] is the determination of chloride ion by precipitation with radioactive silver, [110]Ag. From the diminution of the activity of a silver nitrate solution following removal of silver chloride precipitated by the unknown sample, the chloride content is easily calculated. A sample computation is given in the original paper. Extremely small quantities of chloride could be determined with greater precision by measuring the acitivity of the precipitate formed.

MÖSSBAUER SPECTROSCOPY[8,18,19]

This term designates a study of the phenomenon of resonance fluorescence of gamma rays. It is comparable to resonance fluorescence in optical regions, but involves *intranuclear* rather than electronic energy levels. An important characteristic of this radiation, under optimum conditions of measurement, is the extreme sharpness of the lines. The resonance gamma ray of [67]Zn, for example, has a width at half-height of only

4.8 $\times$ 10^{-11} eV, but with a photon energy of approximately 93 keV, less than 1 part in 10^{15}. This may be compared with the Zn$K\alpha$ x-ray, which has a half-height width of 4.7 $\times$ 10^{-8} eV for a photon of 1.2 $\times$ 10^{-4} eV, or about 2.5 parts in 1000.

Since the frequency bands are so narrow, extremely slight changes in the energy states of the absorbing nuclei can shift the frequency at which absorption can occur by more than the width of the line of the primary radiation, so that no absorption will take place. The effect of the state of chemical combination on the nuclear levels can be of just this order of magnitude. Such a *chemical shift* can be observed and measured by imposing a translational motion on either emitter or absorber in such a way that the resulting Doppler shift will exactly compensate for the chemical shift. The required motion turns out to be of the order of a few millimeters per second, hence is easily realized in practice.

It seems probable that Mössbauer spectroscopy will take its place along with other physical methods, especially in detailed structural studies. The student is referred to the already extensive literature for a more thorough treatment.

SAFETY PRECAUTIONS

The small amounts of radioactive substances needed for analytical experiments with tracers do not generally present radiation hazards which are difficult to guard against. Beta-active isotopes are safe if kept in ordinary glass or metal containers; when out of such containers, they should be handled with tongs, and the operator should wear rubber or plastic gloves. Gamma emitters may require more extensive shielding, perhaps an inch or so of lead, depending on the photon energy of the specific isotope. Pipetting of these solutions by mouth is never permissible.

A survey meter should always be available so that cleanup of accidental spillage can be checked. Often the hazard to the experimenter is less real than the chance of contaminating the laboratory so that the background count is increased unduly. Most active tracer materials are as safe to use in the laboratory as such more familiar materials as benzene and silver nitrate. Safety precautions are no more exacting, merely of a different type.

Active materials in larger than tracer quantities do, of course, require more elaborate safety measures, descriptions of which are readily available elsewhere.

PROBLEMS

16-1 A mixture is to be assayed for penicillin by the isotope dilution method. A 10.0-mg portion of pure radioactive penicillin which has an activity of 4500 cpm per mg, as measured on a particular counting apparatus, is added to the given specimen. From the mixture it is possible to isolate only 0.35 mg of pure crystalline penicillin. Its activity is determined on the same apparatus to be 390 cpm per mg. (Background corrections have been applied.) What was the penicillin content of the original sample, in grams?

16-2 According to the so-called *reverse isotope dilution* method, a weight w of inactive compound is added to a preparation containing an unknown amount W of an active form of the same compound, which has a known specific activity, S_0. A sample is then isolated in pure form, weighed and counted, just as in normal isotope dilution. Show mathematically that Eq. (16-6) applies equally to this case.

16-3 A method for the simultaneous determination of uranium and thorium in minerals[4] requires (1) measurement of combined U and Th by radioactivity and (2) determination of the Th/U ratio by x-ray emission spectroscopy. The combined radioactivity is expressed as "percent equivalent uranium," namely that amount of uranium in pitchblende necessary to give an equal activity. The Th/U ratio is taken as the ratio of peak heights for the x-ray lines: $ThL\alpha$ and $UL\alpha$. The uranium and thorium contents are given by the relation $x + 0.2xy$ = percent equiv. U, where x is the weight percent of uranium, y is the Th/U weight ratio. For a particular counting apparatus 1 percent equivalent uranium corresponds to 2100 cpm above background. A 1.000-g sample of a monazite sand, when prepared and counted according to the standard procedure, gave 2780 cpm (corrected for background). X-ray examination gave peak heights 72.3 scale divisions for $ThL\alpha$ and 1.58 divisions for $UL\alpha$. Compute the uranium and thorium contents of the sample, in terms of weight percent.

16-4 In a study of solubilities of slightly soluble salts, it is necessary to determine the concentrations of oxalate solutions in the part-per-million range. This analysis is to be carried out radiometrically by precipitation of $^{45}CaC_2O_4$. Calculate the oxalate concentration (in parts per million) in a sample from the following: A standard solution is prepared which is 0.680 F in $CaCl_2$ and has an activity of 20,000 cpm per ml (corrected). To a 100.0-ml sample of trace oxalate solution is added 5.00 ml of the standard calcium solution. No precipitate is visible, beyond a slight turbidity. A few drops of $FeCl_3$ solution is added and the solution made alkaline with ammonia to precipitate $Fe(OH)_3$. The precipitate is collected on a small filter paper by suction, washed once, dried, and counted. The counting apparatus is known by prior experiment to have a 30.0 percent efficiency for the ^{45}Ca beta rays. The time required for a preset count of 6000 is 18.60 min. The background is 150 counts per 5 min. (The efficiency correction need not be applied to the standard solution.)

16-5 A method has been described[6] for the determination of oxidation products of propane by the reverse isotope dilution method. The sample to be oxidized is propane-2-[14]C. Among the products was found a considerable quantity of 2-propanol-[14]C. To the mixture of products was added a measured amount of inactive 2-propanol, and a portion isolated by conventional fractionation. The following data were obtained. (The symbol μc stands for *microcuries*, an alternative unit for activity.)

Quantity of propane-2-[14]C	10 mMoles
Specific activity of propane	72.8 μc/mMole
Inactive propanol added	16 mMoles
Specific activity of propanol	5.8 μc/mMole

Compute the percent of propane which was converted to propanol.

16-6 The beta radiation from an active source is to be measured with a Geiger counter. The maximum uncertainty permitted is ± 1 percent. Counts recorded at the end of successive 5-min periods for sample and background are as follows:

Time, min	0	5	10	15	20	25	30
Background, counts/min	0	127	249	377	502	672	793
Sample, counts/min	0	2155	4297	6451	8602	10,749	12,907

(a) What is the minimum time over which the count must be taken to give the required precision? (b) How long would the background have to be counted? (c) What is the actual corrected count in counts per minute, with precision limits?

REFERENCES

1. "Guide to the Use of Ge(Li) Detectors," Princeton Gamma-Tech, Inc., Princeton, N.J., 1967.
2. Boyd, G. E.: *Anal. Chem.*, **21**:335 (1949).
3. Braun, T., and J. Tölgyessy: *Talanta*, **11**:1277 (1964).
4. Campbell, W. J., and H. F. Carl: *Anal. Chem.*, **27**:1884 (1955).
5. Choppin, G. R.: "Experimental Nuclear Chemistry," Prentice-Hall, Inc., Englewood Cliffs, N.J., 1961.
6. Clingman, Jr., W. H., and H. H. Hammen: *Anal. Chem.*, **32**:323 (1960).
7. Deisler, Jr., P. F., K. W. McHenry, Jr., and R. H. Wilhelm: *Anal. Chem.*, **27**:1366 (1955).
8. Friedlander, G., J. W. Kennedy, and J. M. Miller: "Nuclear and Radiochemistry," 2d ed., John Wiley & Sons, Inc., New York, 1964.
9. Hein, R. E., and R. H. McFarland: *J. Chem. Educ.*, **33**:33 (1956).
10. Jacobs, R. B., L. G. Lewis, and F. J. Piehl: *Anal. Chem.*, **28**: 324 (1956).
11. Kuyper, A. C.: *J. Chem. Educ.*, **36**:128 (1959).
12. Lyon, W. S., Jr., (ed.): "Guide to Activation Analysis," D. Van Nostrand Company, Inc., Princeton, N.J., 1964.
13. Meinke, W. W.: *Science*, **121**:177 (1955); *Anal. Chem.*, **30**:686 (1958).
14. Prussin, S. G., J. A. Harris, and J. M. Hollander: *Anal. Chem.*, **37**:1127 (1965).
15. Salyer, D., and T. R. Sweet: *Anal. Chem.*, **28**:61 (1956); **29**:2 (1957).
16. Smith, V. N., and J. W. Otvos: *Anal. Chem.*, **26**:359 (1954).
17. Taylor, T. I., R. H. Anderson, and W. W. Havens, Jr.: *Science*, **114**:341 (1951).
18. Wertheim, G. K.: *Science*, **144**:253 (1964).
19. Wertheim, G. K.: "Mössbauer Effect: Principles and Applications," Academic Press Inc., New York, 1964.

17

Mass Spectrometry

The mass spectrometer is an instrument which will sort out charged gas molecules (ions) according to their masses. It has no real connection with optical spectroscopy, but the names *mass spectrometer* and *mass spectrograph* were chosen by analogy because the early instruments produced a photographic record which resembled an optical line spectrum.

Several distinct methods of producing mass spectra have been devised. In all methods, ions are produced by the collision of rapidly moving electrons with the molecules of the gas to be analyzed.[5] Collisions between electrons and molecules are almost always more effective at producing positive ions than negative ones; with organic molecules the ratio of positive to negative ions may be of the order of 1000:1, hence our discussion will be almost exclusively restricted to positive ion analysis.

If the sample to be analyzed has an appreciable vapor pressure, it is allowed to enter the instrument by diffusion through a tiny orifice called a *leak*. If it is nonvolatile, a portion must be converted to the vapor state by an electric arc or spark, by laser heating, or by other means.

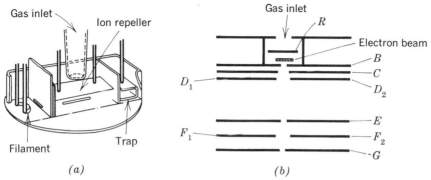

Fig. 17-1 Electron-impact source of ions for mass spectrometry. (a) Ionization chamber; (b) schematic diagram of source, showing the ion-repeller R, draw-out plate C, focusing half-plates D, collimating plates E and G, and beam-centering half-plates F. (*McGraw-Hill Book Company, New York*.)

Figure 17-1 shows a typical electron-impact source. A beam of electrons emitted from the hot tungsten or rhenium filament passes through an orifice into a central region which contains the gaseous sample. Those electrons which do not collide with molecules are caught in an electron trap. Molecules which do not become ionized are removed by a high-vacuum pumping system which is operated continuously.

The ions and those electrons which have been slowed down by collision find themselves in an electric field applied between the *repeller* R (positive), and the *draw-out* plate C (negative). The other plates (B, D, E, F, and G) serve to center the beam of ions and to limit its divergence. The entire assembly is called an *ion gun*.

Up to this point all the types of mass spectrometers are essentially the same. The choice of design for the remainder of the instrument will be concerned with the requirements of (1) the range of masses which must be covered, (2) the resolution, and (3) the sensitivity.

The term *resolution* is unfortunately not always used in the same

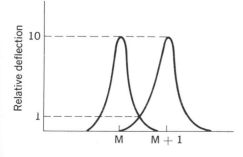

Fig. 17-2 Illustrating the definition of resolution in mass spectrometry.

sense.   We shall follow the most widely used definition: The resolution is defined as the ratio $M/\Delta M$, where ΔM is the difference in mass numbers which will give a valley of 10 percent between peaks of mass numbers M and $(M + \Delta M)$ (Fig. 17-2), where the two peaks are of equal height. Resolution is considered satisfactory if $\Delta M \gg 1$.   Thus if this criterion is met for masses up to 600 and 601, the *unit resolution* is said to be 600.

The resolution is not generally uniform over the whole range of masses which can be detected, becoming poorer at higher mass numbers. For some instruments, unit resolution will be specified as the same as the maximum observable mass and can be expected to be better at lower masses.   In others the unit resolution figure may be much larger than the maximum observable mass, which makes it possible to distinguish species in which the difference in mass is correspondingly less than one mass unit.

We will now consider the several types of mass spectrometers individually.

ELECTROMAGNETIC–FOCUSING SPECTROMETERS[6]

In the best known type of instrument, the beam of ions is led within an evacuated chamber through a powerful magnetic field which forces the

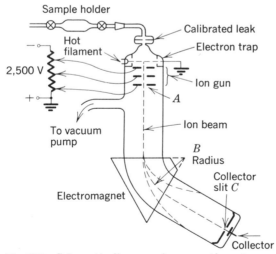

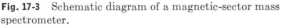

Fig. 17-3 Schematic diagram of a magnetic-sector mass spectrometer.

ions to follow circular trajectories. Figure 17-3 shows an example of this kind of spectrometer.

During their acceleration by the electric field in the ion gun, the ions acquire energy E given by the equation

$$E = eV \tag{17-1}$$

where e represents the charge on the ion and V the applied potential. E must be equal to the kinetic energy of the ions emerging from the gun, so that

$$E = eV = \tfrac{1}{2}mv^2 \tag{17-2}$$

in which m represents the mass of the ion and v its velocity. Algebraic rearrangement gives

$$v = \sqrt{2V\frac{e}{m}} \tag{17-3}$$

which shows that the ions move with velocities determined by their charge-to-mass ratios e/m.

As these ions move into the magnetic field (at normal incidence to the boundary of the field), they are fanned out into a family of circular paths. The magnetic (centripetal) force acting upon the ion is

$$F = Hev \tag{17-4}$$

where H is the magnetic-field strength. This force is equaled by the centrifugal force

$$F = Hev = \frac{mv^2}{r} \tag{17-5}$$

where r is the radius of curvature of the path. Since the ions enter the magnetic field with velocities given by Eq. (17-3), we can write for the radius

$$r = \frac{mv}{He} = \frac{m}{He}\sqrt{2V\frac{e}{m}} = \frac{1}{H}\sqrt{2V\frac{m}{e}} \tag{17-6}$$

The construction of the apparatus (Fig. 17-3) is such that the only ions which can gain access to the collector electrode are those with trajectories of a prescribed radius of curvature. Equation (17-6) shows that for given values of V and H those ions will be collected which have a particular value of the ratio m/e.

Since the ions are produced by the removal of electrons from neutral species, the charge on any ion must be a small integral multiple of the electronic unit charge. In practice the charge is most commonly 1, less frequently 2, and rarely greater than 2 electron units, so that usually the quantity e in Eq. (17-6) may be taken as unity. This means that ions of any desired mass can be collected by making appropriate adjustments of the accelerating voltage V or of the magnetic field H.

There is no convenient way in which the ion beam can be truly collimated, but it is possible to limit its angle of divergence by circular or slit diaphragms. It can be shown that such a divergent beam will be brought to a focus at the diaphragm or mask in front of the collector, provided that the ion source A, the apex of the sector-shaped magnetic field B, and the collector slit C lie along a straight line as indicated in Fig. 17-3.

In the instrument of Fig. 17-3, a 60° sector angle is shown for the magnetic field. This shape is used in several commercial mass spectrometers, but other angles give equally good focusing, and some manufacturers employ 90° or 180° angles. The mass range may be scanned by progressive variation of V with H maintained constant, or the reverse;

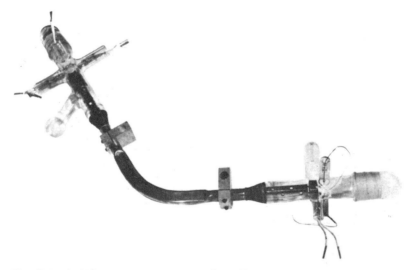

Fig. 17-4 A 60° mass spectrometer tube. The electron-gun and accelerator connections, together with the side arm for connection to the pump, appear at the left, the collector electrode at the lower right. (*General Electric Co., Schenectady, N.Y.*)

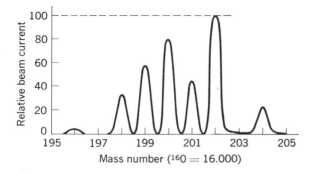

Fig. 17-5 The mass spectrum of mercury vapor, showing isotopes, as determined with a magnetic sector mass spectrometer. (*General Electric Co., Schenectady, N.Y.*)

the two methods are about equally effective.   The signal received by the collector is amplified electronically and made to operate an automatic recorder.   Figure 17-4 shows a 60° spectrometer tube prior to installation.

A typical spectrum, that of mercury vapor, is shown in Fig. 17-5, as determined on a 60° magnetic focusing instrument.   Note that the resolution is easily great enough to show the individual isotopes present. Indeed nearly all the information we have about stable isotopes has been acquired by means of the mass spectrometer.

The resolving power of the instrument described above is fully adequate for a large proportion of analytical applications, but not all. It can distinguish between successive mass numbers as high as 1500 in particularly well-designed instruments, though 500 is a more realistic figure for most.   Applications to organic chemistry almost never require masses above about 500, because molecules as heavy as this cannot be volatilized without decomposition to smaller fragments.

There are, however, important applications which require a much higher resolution, though not necessarily at higher mass numbers.   An example is the need to distinguish between fragments which have equal nominal mass but which contain different elements so that the actual masses differ by a small fraction of a mass unit.

Increased resolving power can be obtained by employing two mass analyzers in series.   The first analyzer is usually electrostatic, the second, magnetic.   The electrostatic analyzer consists of a pair of curved metallic plates with a dc potential across them.   The beam of ions passes between the plates through the radial field V, and is deflected in a circular orbit with radius r given by

$$r = \frac{mv^2}{eV} \qquad\qquad\qquad (17\text{-}7)$$

so that

$$v = \sqrt{Vr\frac{e}{m}} \qquad\qquad\qquad (17\text{-}8)$$

which is seen to be analogous to Eq. (17-3).   Equation (17-7) shows that for a given electric field the ions are dispersed according to their kinetic energies.   If the angle subtended by the cylindrical field is chosen correctly, the emergent ions of each energy content are collimated, i.e., follow parallel paths.   If the ions now pass into a suitably shaped magnetic analyzer, all ions of a particular m/e ratio are brought to a common focus, regardless of their initial kinetic energy.

The most widely employed among several designs of this nature is that originated by Mattauch and Herzog,[4] illustrated in Fig. 17-6.*

Figure 17-7 shows an example of the mass spectra of a mixture of several substances which yield ionic species of the same nominal mass, as observed on a double-focus spectrometer.

CYCLOIDAL FOCUSING[6]

Another geometry which permits double focusing (both electrostatic and magnetic) is achieved by the superposition of two fields in the same region of space.   The ionizing source can then be immersed in the magnetic field and at the boundary of the electric field, with the target immersed in both.   As a result the ions move only within the fields, thus eliminating troublesome edge effects.   Under these conditions, it

---

* A description of a specialized research spectrometer consisting of four units in series, two electrostatic and two magnetic, has been published by White and Forman.[15] The resolution is so great that the usual criterion cannot be applied; the contribution of ions of mass 149 to the signal at 150 is of the order of 1 part in 10^8.

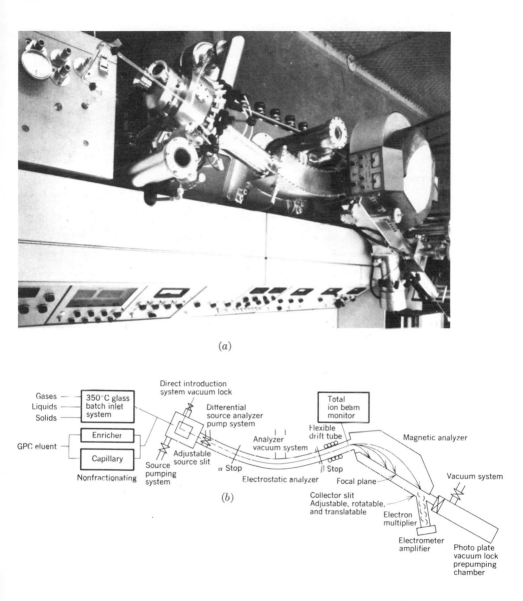

(a)

(b)

Fig. 17-6 A Mattauch-Herzog high-resolution mass spectrometer, (a) photograph, and (b) schematic: the Consolidated 21-110B. The photograph is oriented to bring out the geometrical relation to the schematic. (*Consolidated Electrodynamics Corporation, Pasadena, Calif.*)

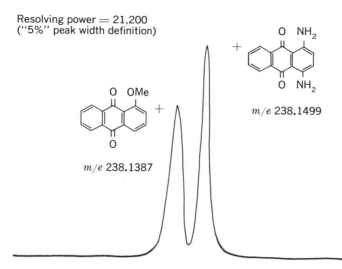

Resolving power = 21,200
("5%" peak width definition)

O OMe

m/e 238.1387

O NH$_2$

O NH$_2$

m/e 238.1499

Fig. 17-7 Mass 238 doublet from the parent peaks of 1-methoxy-anthraquinone and 1,4-diaminoanthraquinone, taken on an EAI-MS9 double-focus mass spectrometer. (*Picker X-Ray Corporation, White Plains, N.Y.*)

can be shown that the ions will follow a *cycloidal* path, as in Fig. 17-8, and those incident at the collector slit will be characterized by the ratio

$$\frac{m}{e} = \frac{kH^2}{V} \tag{17-9}$$

where k is a constant dependent on the geometry of the assembly.

The cycloidal arrangement is doubly focusing, though not in quite the same sense as the Mattauch-Herzog instruments. Both of these types, however, bring ions to a focus regardless of variations in the kinetic energy they may have on leaving the ion gun.

TIME–OF–FLIGHT SPECTROMETERS[6, 16]

The instruments previously described produce a steady ion beam for any given setting of the controls. It is possible, however, to apply the accelerating potential intermittently and to cut up the beam accordingly into pulses. This permits sorting out the ions according to their velocities, which is tantamount to mass sorting, without the need of a magnetic field. An apparatus for doing this, the Bendix time-of-flight spectrometer, is shown schematically in Fig. 17-9. An electron beam

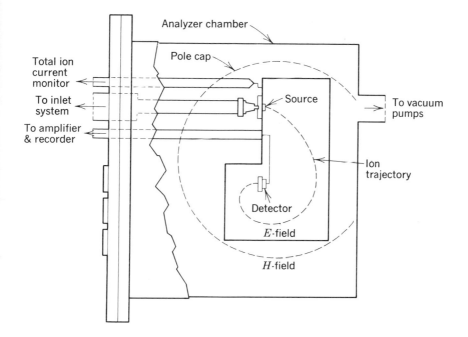

Fig. 17-8 Cycloidal focusing mass spectrometer, schematic. (*Varian Associates, Palo Alto, Calif.*)

ionizes the incoming gas sample as in other spectrometers. An accelerating potential of the order of 100 V is applied to the "ion-focus grid" in the form of a voltage pulse lasting 1 μsec or less and repeating a few thousand times per second. This positive pulse accelerates the ions out through the grid where they are picked up by the field of the "ion-energy grid." All ions receive the same energy; therefore the velocity which they acquire is proportional to $\sqrt{e/m}$ [cf. Eq. (17-3)]. The ions are now allowed to drift at constant velocity in a field-free region (40 cm or so), after which they enter a magnetic electron multiplier. The time of transit through the drift space is $T = k \sqrt{m/e}$, where k is a constant depending on the distance and the parameters of the ion gun. The value of k is not far from unity if T is taken in microseconds, m in atomic mass units, and e in units equal to the charge on an electron. A singly charged nitrogen molecule would accordingly have a flight time of approximately $\sqrt{28} = 5.30$ μsec, whereas for a singly charged oxygen molecule, $T = \sqrt{32} = 5.66$ μsec, and for a xenon ion, $T = \sqrt{132} = 11.50$ μsec.

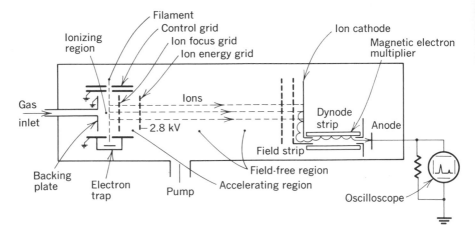

Fig. 17-9 Schematic diagram of the Bendix Time-of-Flight mass spectrometer. (*Bendix Corporation, Cincinnati, Ohio.*)

After the ions pass through the field-free space, they are further accelerated by a negative potential of 1000 V or more on the cathode of the magnetic electron multiplier. The ions bombarding the cathode cause the ejection of electrons, which are constrained by an applied magnetic field to follow curved trajectories, as indicated, to impinge repeatedly on a secondary high-resistance electrode (dynode) where the process is repeated many times. The output from the anode of the multiplier is applied to the vertical deflection plates of a cathode-ray oscilloscope which has its horizontal sweep synchronized with the accelerator pulses. The result is a display of the mass spectrum on the scope tube, as illustrated in Fig. 17-10. The resolution is of the order of 400 by the criterion of Fig. 17-2.

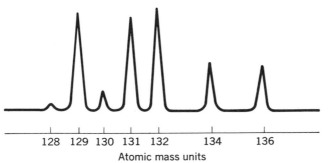

Fig. 17-10 Oscilloscope trace showing the isotopes of xenon, determined with the Bendix spectrometer. (*Bendix Corporation, Cincinnati, Ohio.*)

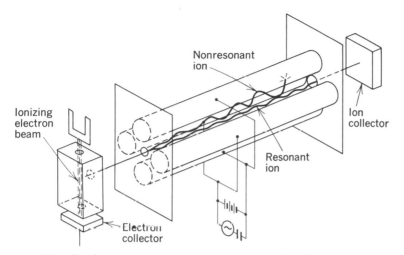

Fig. 17-11 Quadrupole mass spectrometer, schematic. Scanning in this device is accomplished by varying the radio frequency or the magnitude of the rf and dc voltages. (*Research/Development.*)

QUADRUPOLE MASS SPECTROMETERS[1,6]

This is another device in which ions can be resolved according to their m/e ratio without the need of a heavy magnet. It consists of four metallic rods, precisely straight and parallel, so positioned that the ion beam shoots down the center of the array (Fig. 17-11). Diagonally opposite rods are connected together electrically, the two pairs to opposite poles of a dc source and simultaneously to a radio-frequency oscillator.

Neither the dc nor the ac field has a component parallel to the rods; hence neither has any effect on the forward motion of the ions, but lateral motion will be produced by these interacting fields. This can be analyzed in terms of the coordinate system of Fig. 17-12. If

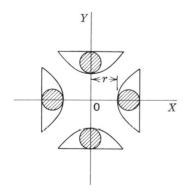

Fig. 17-12 Geometry of the rods in a quadrupole mass spectrometer. The Z axis, along which the ions move, is perpendicular to the plane of the paper.

the rods were of symmetrical hyperbolic cross section, then the potential ϕ at any point (x,y) would be given as a function of time t by the equation

$$\phi = \frac{(V_{dc} + V_0 \cos \omega t)(x^2 - y^2)}{r^2} \tag{17-10}$$

where V_{dc} is the applied direct potential, V_0 the amplitude of the alternating voltage of frequency ω radians per second, and r is the parameter defined in the figure. This relation holds to a good approximation even though the hyperbolic rods are replaced by less expensive cylindrical ones.

The force acting laterally on an ion of unit charge e is obtained by differentiating with respect to x and y:

$$F_x = -e \frac{\partial \phi}{\partial x} = -e \frac{(V_{dc} + V_0 \cos \omega t)2x}{r^2}$$
$$F_y = -e \frac{\partial \phi}{\partial y} = +e \frac{(V_{dc} + V_0 \cos \omega t)2y}{r^2} \tag{17-11}$$

Then, since acceleration can be equated to the ratio of force to mass, we can write as the equations of motion for an ion of mass m and charge e,

$$\frac{d^2x}{dt^2} + \frac{2}{r^2}\left(\frac{e}{m}\right)(V_{dc} + V_0 \cos \omega t)x = 0$$
$$\frac{d^2y}{dt^2} - \frac{2}{r^2}\left(\frac{e}{m}\right)(V_{dc} + V_0 \cos \omega t)y = 0 \tag{17-12}$$

These latter equations indicate that the motions of the ions will have a periodic component of frequency ω, but will also be dependent on the e/m ratio. A further mathematical treatment will show that for $V_{dc}/V_0 < 0.168$ there is only a narrow range of frequencies for which the ionic trajectories are stable (i.e., not divergent) with respect to both the x and y coordinates; outside of this range the ions will collide with one or the other set of rods. Maximum resolution is obtained with the V_{dc}/V_0 ratio approaching as closely as possible the limiting value, 0.168. If the ratio is allowed to become larger than this, then no frequency can be found for which a stable trajectory will result, no matter what the e/m value. Mass selection can be achieved alternatively by keeping the frequency fixed and varying the dc and rf potentials while maintaining their ratio accurately constant. The selected beam of ions is received on an anode plate and amplified.

An example of a quadrupole mass spectrometer is the EAI Quad 300.* This instrument is equipped with analyzer rods 7.5 in. long. It

* Electronic Associates, Inc., Palo Alto, Calif.

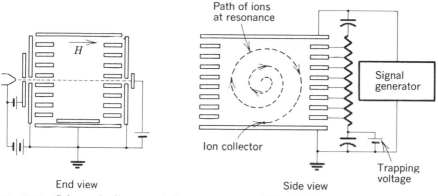

Fig. 17-13 Schematic diagram of the omegatron. (*Physical Review*.)

can cover masses 1 to 500 in three ranges with unit resolution or better; the scanning time is variable from 500 μsec to 30 min. The spectrum is presented either on an oscilloscope or strip-chart recorder. A spectrum taken on this instrument can be found in Fig. 21-13.

RADIO-FREQUENCY SPECTROMETERS[6]

It is possible to separate ions of different masses of radio-frequency fields utilized in several other ways. One of these, developed by Bennett, can be considered the forerunner of the time-of-flight spectrometer, with a high-frequency sine wave rather than a pulse applied to the control grid. Another grid between drift space and detector is fed from the same rf source, with carefully adjusted phase relations, to assist in resolving bunches of ions. This is seldom used today, at least for analytical purposes.

Another system, which finds wider application, is the *ion-resonant* or *cyclotron* spectrometer, often called the *omegatron*.[13] A box-shaped space (Fig. 17-13) is subjected simultaneously to a constant magnetic field and an electrostatic field alternating at radio frequencies. The ions are formed in the center of this space by electron impact. They are then accelerated first one way and then the other by the high-frequency field and given circular paths by the magnetic field. The resultant is an expanding spiral trajectory. Eventually the ion beam is intercepted by a collector electrode. In order that the ions strike the collector, their m/e ratio must be such as to fulfill the following equations. The effect of the magnetic field, as stated in Eq. (17-5), is $He = mv/r$. We can substitute for the linear velocity v its equal ωr, where ω is the angular velocity, or $2\pi fr$, where f (equal to $\omega/2\pi$) is the frequency in hertz. Upon rearrangement, this gives us the working relation,

$$\frac{m}{e} = \frac{H}{2\pi f} \qquad (17\text{-}13)$$

The resolution varies inversely as the mass of the ion and is adequate for distinguishing isotopes.

A simplified omegatron intended for measuring the density of molecular nitrogen in the upper atmosphere via rocket flights has been described by Niemann and Kennedy;[11] their paper includes interesting design details, the selection of operating parameters, and test methods.

COMPARISON OF MASS SPECTROMETERS

Table 17-1 gives a rough comparison of a few mass spectrometers currently on the market. For each class of instrument two examples are included: one highly sophisticated, and one more modest (where such a distinction is appropriate).

Table 17-1   Comparison of mass spectrometers

Type	m/e range	Unit resolution
Simple magnetic sector	1–2000	1500
	2–300	120
Mattauch-Herzog double focus	1–6400	25,000
	1–2000	30,000
Cycloidal double focus	10–2000	1000
	2–230	200
Time-of-flight	1–1200	250
	1–10,000	75
Quadrupole	1–500	500
	1–120	100
Omegatron	1–280	700
	2–50	50

APPLICATIONS

Mass spectrometers are of great usefulness in a number of fields. Small units without scanning capabilities are in wide use as *leak detectors*. The instrumental parameters are set at the factory so that only helium will be detected. The instrument can then be attached to a vacuum system under test, and a fine jet of helium played around any joints or other locations suspected of leaking. One advantage is that several leaks can be located without the necessity of repairing one before searching for the next.

Another application, particularly for spectrometers of limited mass range, is as *residual gas analyzers*. For this purpose, the mass spectrometer amounts to a very sensitive pressure gage which can determine not only the total residual pressure in an "evacuated" chamber, but also the partial pressure of each gas present.

Chemical applications also are varied. The historical search for natural isotopes has already been mentioned. The isotopic variation in natural lead ores, for example, a problem of great significance in geology, can be determined with an error less than 1 percent by examination of the mass spectra of the tetramethyl derivative.

In the study of compounds,[2,12] organic or otherwise, fragmentation is always observed. When a compound in the vapor phase is admitted to the spectrometer, the bombardment by electrons causes the molecules to break apart in all possible ways, giving rise to positive ions of a whole series of masses. For example, n-butane (C_4H_{10}) produces all the ions listed in Table 17-2, and these are observed in the resulting mass spectrum. Since it is characteristic of the mass spectrometer that it cannot distinguish between a singly charged ion of a given mass and a doubly charged ion of twice the mass, the ions producing a record at mass number 29 in Table 17-2, for example, may be in part $C_2H_5^+$ and in part $C_4H_{10}^{++}$.

Table 17-2 Fragmentation pattern of n-butane, relative to mass 58 = 100

Mass	Ions	Intensity	Mass	Ions	Intensity
59	$C_4H_9D^+$, $^{12}C_3\ ^{13}CH_{10}^+$	9	43	$C_3H_7^+$	700
58	$C_4H_{10}^+$	100	42	$C_3H_6^+$	96
57	$C_4H_9^+$	24	41	$C_3H_5^+$	210
56	$C_4H_8^+$	11	40	$C_3H_4^+$	22
55	$C_4H_7^+$	12	39	$C_3H_3^+$	107
54	$C_4H_6^+$	5	38	$C_3H_2^+$	19
53	$C_4H_5^+$	11	37	C_3H^+	10
52	$C_4H_4^+$	6	30	$C_2H_4D^+$, $^{12}C\ ^{13}CH_5^+$	13
51	$C_4H_3^+$	12	29	$C_2H_5^+$, $C_4H_{10}^{++}$	319
50	$C_4H_2^+$	13	28	$C_2H_4^+$	234
49	C_4H^+	6	27	$C_2H_3^+$	277
44	$C_3H_6D^+$, $^{12}C_2\ ^{13}CH_7^+$	33	26	$C_2H_2^+$	50

The presence of various isotopes of each element further complicates the record. Thus the table includes mass number 59, which corresponds to $C_4H_{10}^+$ ions in which either one carbon or one hydrogen is one unit

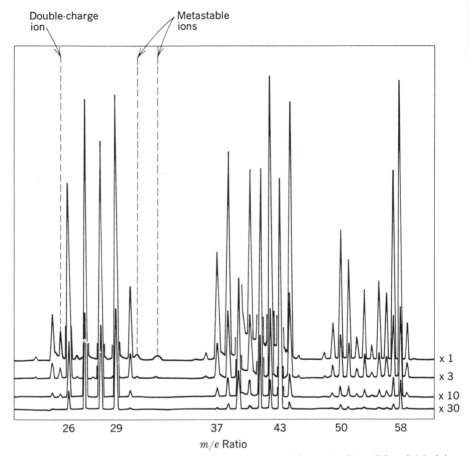

Fig. 17-14 The mass spectrum of *n*-butane, as recorded on the Consolidated Model 21-103 spectrometer. Simultaneous tracings by galvanometers of different sensitivities provide a dynamic range up to 30,000. (*Consolidated Electrodynamics Corporation, Pasadena, Calif.*)

heavier than normal. Ions of masses 44 and 30 are similarly accounted for. The relative abundance of ions of various masses is characteristic of the particular compound under specified conditions of excitation and is known as its *fragmentation pattern*. The abundance figures may be somewhat different if observed on another instrument or with a changed ionizing or accelerating potential.

Figure 17-14 is a reproduction of the actual trace of part of the spectrum of *n*-butane, as recorded on a magnetic sector spectrometer. Simultaneous traces were made with light-beam galvanometers of four different sensitivities, a device which permits a much greater dynamic range than would a single recorder without cumbersome range-changing accessories. Note that the peak for mass 43 (the largest in Table 17-2) is recorded only by the least sensitive galvanometer trace, whereas smaller peaks (e.g., mass 30) require the most sensitive. (The tabulated data were taken from a different instrument than was the trace of Fig. 17-14.)

An experimental difficulty which sometimes arises is the calibration of the mass scale. With compounds of relatively small molecular weight it can often be assumed that the strong peak of largest mass number seen in the spectrum, exceeded only by one or two small "satellite" peaks, will identify the mass number of the compound itself; this is often called the *parent peak*. The satellites are caused by the presence of heavy isotopes. This rule sometimes breaks down when some bond in the molecule is particularly easily severed. Of the spectra illustrated in this chapter, the rule is seen to be followed in *n*-butane (Fig. 17-14) and isobutyl alcohol but not the other butyl alcohols (Fig. 17-15). The rule does not apply to compounds in the higher mass ranges, which always fragment. In such cases, particularly with double-focusing spectrometers, it is often helpful to introduce a small portion of *perfluorokerosine* (PFK) as a mass indentifier. PFK is especially suitable for this because its constituent molecules fragment easily to give a large number of identifiable mass peaks, the fact that fluorine has only a single isotope simplifies the interpretation. PFK thus plays a role comparable to that of the standard iron spectrum in emission spectrography. A list of 73 PFK fragments (described as a "partial" spectrum) has been published by McLafferty.[10]

The fragmentation pattern of a compound reflects the relative stability of groupings of atoms within the molecule and the relative ease with which various bonds can be broken. Hence it is just as valid a tool for both qualitative and quantitative analysis and for the elucidation of covalent structures as an infrared absorption spectrum. The individual peaks are usually sufficiently well resolved that overlapping need not be considered.

Mixtures can be resolved by comparison with spectra of reference compounds determined under the same conditions. In Fig. 17-15 are shown fragmentation patterns for the four isomeric butyl alcohols.[7] The spectrum obtained with a mixture of these isomers is presented in Fig. 17-16. The peak at mass 56 is almost entirely due to the normal alcohol. The peaks at 45, 59, and 74 will serve to measure the amounts of secondary, tertiary, and isobutyl alcohols, in that sequence, but in each case a correction must be made for significant contributions of other isomers. The solution of this problem involves four simultaneous equations, as follows:

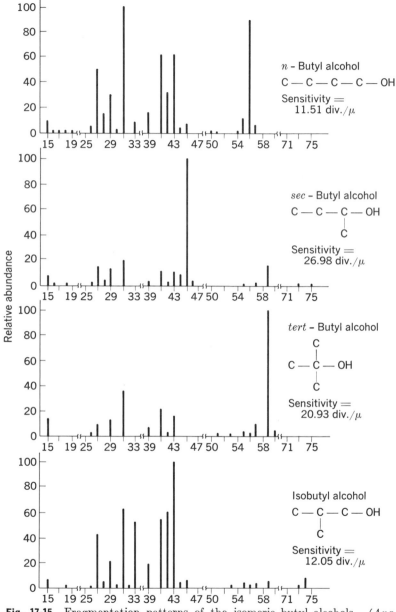

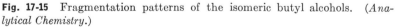

Fig. 17-15 Fragmentation patterns of the isomeric butyl alcohols. (*Analytical Chemistry.*)

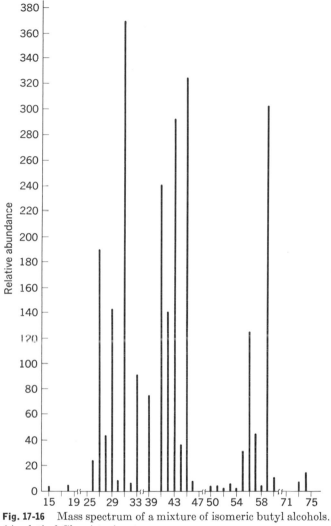

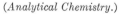

Fig. 17-16   Mass spectrum of a mixture of isomeric butyl alcohols.
(*Analytical Chemistry.*)

Let x_1 be the relative contribution of *n*-butyl alcohol,
  x_2 that of *tert*-butyl alcohol,
  x_3 that of *sec*-butyl alcohol, and
  x_4 that of isobutyl alcohol.

Then (using values from the original tables from which Figs. 17-15 and

17-16 were plotted) we can write

$$M_{56} = 1.267 = 0.9058x_1 + 0.0147x_2 + 0.0102x_3 + 0.0246x_4$$

$$M_{59} = 3.015 = 0.0026x_1 + 1.0000x_2 + 0.1778x_3 + 0.0498x_4$$

$$M_{45} = 3.226 = 0.0659x_1 + 0.0059x_2 + 1.0000x_3 + 0.0503x_4$$

$$M_{74} = 0.148 = 0.0079x_1 + 0.0000x_2 + 0.0029x_3 + 0.0906x_4$$

where the M's are the respective peaks of the mixture. The values of the x's obtained must be divided by the sensitivity values given in Fig. 17-15 for the reference spectra, to give partial pressures in microns $(1\ \mu = 10^{-3}\ \text{torr})$. The results are as follows:

Component	x	Sensitivity, $div./\mu$	Partial pressures, μ	Mole percent Found	Mole percent Known
n-Butyl	$x_1 = 128.71$	11.51	11.18	24.4	24.9
$tert$-Butyl	$x_2 = 239.76$	20.93	11.46	25.0	25.2
sec-Butyl	$x_3 = 305.55$	26.98	11.33	24.8	24.8
Isobutyl	$x_4 = 142.34$	12.05	11.81	25.8	25.1
Total			45.78	100.0	100.0

An accuracy of about ±0.5 percent is obtained for each component. A complete analysis can be performed in about an hour, with the help of a desk calculating machine.

Occasionally a peak is found which does not correspond even approximately to an integral mass number. If it falls halfway between two mass numbers, it is undoubtedly produced by a doubly charged ion with odd mass number. A nonintegral peak which is broader than normal is probably the result of a *metastable ion* which has spontaneously decomposed during its passage through the instrument.[2] Examples of both kinds of nonintegral peaks are shown in Fig. 17-14. Study of metastable peaks is capable of providing additional information about the chemistry of the system.

SOLID SAMPLES

Mass spectra can be obtained for solid samples as well as for gases and volatile liquids, either by direct volatilization if an appreciable vapor pressure exists at a few hundred degrees Celsius, or by pyrolysis to give characteristic fragments. In some instances a high-voltage spark struck between electrodes containing the sample will provide the necessary ions. The spark has the effect of producing ions which have considerable kinetic energy with a Gaussian distribution. This is no disadvantage with spectrometers using Mattauch-Herzog or cycloidal double-focusing geometries, but with other types the accelerating voltage in the ion gun must be higher than is needed with electron impact ionization, so that the velocities of the emerging ions will be nearly homogeneous enough to permit adequate focusing.

TRACER ANALYSIS

Just as radioactive isotopes may be employed as tracers, so may stable isotopes of unusual mass numbers. Compounds can be synthesized with a molecular weight different by one or two or more units from the usual compounds. Then isotope dilution or other tracer procedures can be used, substituting mass-spectrometer measurements for radio-activity data.

A determination frequently required for such studies is the proportion of deuterium in water. In mixtures of H_2O and D_2O, equilibrium between these species and HDO is to be expected, since all these compounds ionize to some extent. Since the heavy isotope is usually present in only a small concentration, the equilibrium $H_2O + D_2O \rightleftharpoons 2HDO$ is displaced toward the right, so that very little D_2O exists as such. In such a mixture, the mass spectrometer shows the presence of ions with masses 17, 18, 19, and 20, as shown in the following tabulation:[14]

		Relative abundance			
Mass	Ions	Natural H_2O	1 percent D_2O	5 percent D_2O	10 percent D_2O
17	OH$^+$ (from H_2O and HDO)	22.6	22.9	23.5	23.3
18	H_2O^+, OD$^+$	100.0	100.0	100.0	100.0
19	HDO$^+$	(0.7)	1.94	8.73	17.73
20	D_2O^+	(0.0)	0.16	0.27	1.01

From these data it is apparent that the ratio of peak heights 19/18 is most sensitive to the heavy-water content. Actually the curve plotted from these and additional data is nearly linear over the range 0.2 to 10.0 percent D_2O. The method by which these data were obtained was not suited to very small amounts of D_2O, such as occur in natural waters.

An excellent example of the isotope dilution method is the elementary analysis of volatile organic compounds.[9] Oxygen, carbon, and nitrogen can be determined on successive samples by combustion and equilibration of the CO_2 or N_2 formed with $^{18}O_2$ in an oxygen analysis, with $^{13}CO_2$ for carbon or with $^{15}NH_3$ for nitrogen. In the case of oxygen, for example, a weighed sample of the organic compound is mixed with a measured quantity of heavy oxygen (a mixture of $^{16}O_2$, $^{16}O^{18}O$, and $^{18}O_2$) in a platinum container. It is then heated electrically to 800°C for an hour. This treatment effects the combustion of the sample and also the catalytic equilibration represented by the reaction $CO_2 + {}^{18}O_2 \rightleftharpoons C^{16}O^{18}O + {}^{16}O^{18}O$. The resulting CO_2 is isolated, freed of uncondensable gases, and examined in the mass spectrometer.

In the determination of nitrogen, the sample is oxidized by hot CuO rather than O_2, to minimize the likelihood of introducing traces of nitrogen. A correction can readily be made for any air which may have leaked into the apparatus by the simple expedient of measuring the peak at mass 40 due to argon. As the ratio of nitrogen to argon in the air is constant, the spurious nitrogen can be computed and subtracted.

The mass spectrographic isotope dilution method is regularly used to determine both total uranium and isotopic distribution in partially spent reactor fuel.[8] Radioactivity methods cannot be applied conveniently because of the intense activity of the fission products which are present.

PROBLEMS

17-1   An organic compound is analyzed for its nitrogen content by isotope dilution.[9] A measured amount of the compound containing ^{15}N in place of ^{14}N is added. After conversion of all the nitrogen to N_2, a mass spectrometer shows the following peak heights:

m/e	28	29	30
Height	978.5	360.6	52.5

Calculate the percent of the nitrogen which is ^{15}N.

17-2   An isotope dilution method for the determination of carbon in submilligram quantities with ^{13}C tracer has been reported by Boos et al.[3] The sample is mixed with a portion of succinic acid, $C_4H_6O_4$, which contains about 30 atom percent ^{13}C. The mixture is oxidized to CO_2 and H_2O, and the resulting CO_2 examined in a mass spectrometer. The ratio of mass 45 to 44 (corrected for the natural isotopic composition of oxygen) is taken as the $^{13}C/^{12}C$ ratio, designated as r. In natural carbon, the abundance of ^{13}C is 1.11 percent, ^{12}C, 98.9 percent, which must be taken into account.

The equations are as follows:

$$^{13}C_8 = W_T \left(\frac{4}{119.3}\right)\left(\frac{r_T}{r_T + 1}\right) + W_S \left(\frac{X_C}{12.01}\right)(0.0111)$$

$$^{12}C_8 = W_T \left(\frac{4}{119.3}\right)\left(\frac{1}{r_T + 1}\right) + W_S \left(\frac{X_C}{12.01}\right)(0.989)$$

$$r_8 = \frac{^{13}C_8}{^{12}C_8}$$

where $^{12}C_8$ and $^{13}C_8$ represent the numbers of milligram-atoms of the respective isotopes present in the mixed sample, r_8 is the observed ratio, W_T and W_S are the weights in milligrams of tracer and sample, r_T is the ratio for pure tracer compound oxidized in the same manner, and X_C is the quantity sought, the weight fraction of carbon in the unknown. (a) Explain the above equations, and from them derive an expression for X_C in terms of W_S, W_T, and r_8. (b) In a particular analysis, 0.156 mg of sample and 0.181 mg of tracer were taken. The ratio r_8 was found to be 0.206. The tracer contained 31.41 percent of its carbon as ^{13}C. Calculate the percent carbon in the sample.

REFERENCES

1. "Quadrupole Residual Gas Analyzer: Theory of Operation," Electronic Associates, Inc., Palo Alto, Calif., 1966.
2. Biemann, K.: "Mass Spectrometry: Organic Chemical Applications," McGraw-Hill Book Company, New York, 1962.
3. Boos, R. N., S. L. Jones, and N. R. Trenner: Anal. Chem., **28**:390 (1956).
4. Duckworth, H. E., and S. N. Ghoshal: High-resolution Mass Spectroscopes, in C. A. McDowell (ed.), "Mass Spectrometry," chap. 7, McGraw-Hill Book Company, New York, 1963.
5. Elliott, R. M.: Ion Sources, in C. A. McDowell (ed.), "Mass Spectrometry," chap. 4, McGraw-Hill Book Company, New York, 1963.
6. Farmer, J. B.: Types of Mass Spectrometers, in C. A. McDowell (ed.), "Mass Spectrometry," chap. 2, McGraw-Hill Book Company, 1963.
7. Gifford, A. P., S. M. Rock, and D. J. Comaford: Anal. Chem., **21**:1026 (1949).
8. Goris, P., W. E. Duffy, and F. H. Tingey: Anal. Chem., **29**:1590 (1957).
9. Grosse, A. V., S. G. Hindin, and A. D. Kirshenbaum: Anal. Chem., **21**:386 (1949).
10. McLafferty, F. W.: Anal. Chem., **28**:306 (1956).
11. Niemann, H. B., and B. C. Kennedy: Rev. Sci. Instr., **37**:722 (1966).
12. Silverstein, R. M., and G. C. Bassler: "Spectrometric Identification of Organic Compounds," 2d ed., John Wiley & Sons, Inc., New York, 1967.
13. Sommer, H., H. A. Thomas, and J. A. Hipple: Phys. Rev., **82**:697 (1951).
14. Thomas, B. W.: Anal. Chem., **22**:1476 (1950).
15. White, F. A., and L. Forman: Rev. Sci. Instr., **38**:355 (1967).
16. Wiley, W. C.: Science, **124**:817 (1956).

18

Magnetic Resonance Spectroscopy

A radically different type of interaction between matter and electro-magnetic forces can be observed by subjecting a sample simultaneously to two magnetic fields, one stationary H, and the other varying at some radio frequency f, 5 MHz or higher. At particular combinations of H and f, energy is absorbed by the sample, and the absorption can be observed as a change in the signal developed by a radio-frequency detector and amplifier.

This energy absorption can be related to the magnetic dipolar nature of spinning nuclei. Quantum theory tells us that nuclei are characterized by a spin quantum number I which can have positive values of $n/2$ (in units of $h/2\pi$; h is Planck's constant), where n can be 0, 1, 2, 3, $\cdot$ $\cdot$ $\cdot$ $\cdot$. If $I = 0$, the nucleus does not spin, and hence cannot be observed by the method here considered; this applies to ^{12}C, ^{16}O, ^{32}S, and others. Maximum sharpness of absorption peaks occurs with nuclei for which $I = \frac{1}{2}$, including among many others ^{1}H, ^{19}F, ^{31}P, ^{13}C, ^{15}N, and ^{29}Si. The first three of these are most easily observed because they constitute substantially 100 percent of the corresponding elements in natural abundance, whereas the others named occur in small proportions.

The spinning nuclei simulate tiny magnets, and so interact with the externally impressed magnetic field $\dot{H}$. It might be supposed that they would all line up with the field like so many compass needles, but instead their rotary motion causes them to precess like a gyroscope in a gravitational field. According to quantum mechanics, there are $2I + 1$ possible orientations, and hence energy levels, which means that the proton, for example, has two such levels. The energy difference between them is given by

$$\Delta E = hf = \frac{\mu H}{I} \tag{18-1}$$

where μ is the magnetic moment of the spinning nucleus. The characteristic frequency f is called the *Larmor frequency*. If now an alternating flux is applied at right angles to the dc field, at frequency f, the nuclei in the lower energy state will absorb the resonant energy, and the absorption can be noted at the output of the detector.

Equation (18-1) can be rewritten in terms of ω, the angular frequency of precession:

$$\frac{\omega}{H} = \frac{2\pi f}{H} = \frac{2\pi \mu}{hI} \tag{18-2}$$

The ratio ω/H is a fundamental constant characteristic of any nuclear species which has a finite value of I. This is called the *gyromagnetic ratio* (sometimes the *magnetogyric ratio*) and given the symbol γ.

At a frequency of 60 MHz (one of the most commonly used values) the energy difference ΔE is less than 10^{-2} cal per mole, which means that the source of radio frequency need not be very powerful, but the detector must be quite sensitive.

Another effect of the imposed alternating field at the Larmor frequency is to cause all the spinning nuclei to precess *in phase*. Thus we have a multitude of nuclear oscillators, which according to electromagnetic theory must radiate energy; since they are in phase with each other, they will act as a *coherent* source. Their radiation can be picked up by another coil in the neighborhood of the sample, positioned with its axis mutually perpendicular to those of the oscillator coil and the fixed field.

Hence two types of NMR spectrometers are possible, the single-coil instrument, in which absorption is measured, and the two-coil variety, which measures resonant radiation. Prior to about 1961 most commercial instruments were of the two-coil type, but now the trend is toward the single-coil arrangement.

INTRODUCTORY EXPERIMENT

Let us consider the single-coil NMR spectrometer of Fig. 18-1, containing a borosilicate glass sample tube filled with distilled water. The rf oscillator is set at 5 MHz, and the magnetic field varied at a constant rate from zero to about 10 kG*, while the output of the detector is monitored. The resulting graph will resemble Fig. 18-2, with resonance maxima corresponding to each isotope of each element present for which $I > 0$. (It is customary to plot NMR spectra as peaks above a base line, no matter whether measured as absorption or as resonant emissions.) The indicated field values correspond to the gamma ratios of the various isotopes at $f = 5$ MHz. The copper peaks result from the wire of the rf coil. Clearly this provides the possibility of effective qualitative analysis. Quantitative measurements are also possible, through integration of the areas beneath the peaks.

An instrument capable of performing an experiment of this type is called a *wide-line* or *low-resolution* NMR spectrometer. Its great potential as an analytical tool has largely been overlooked, no doubt partly because of its relatively high cost. It has found rather limited

* Kilogauss; not to be confused with kg, kilograms.

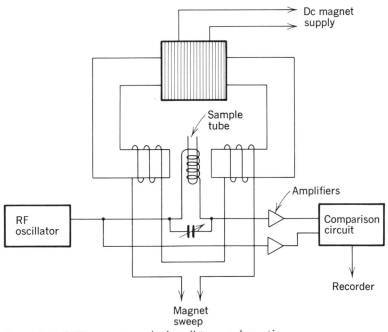

Fig. 18-1 NMR apparatus, single-coil type, schematic.

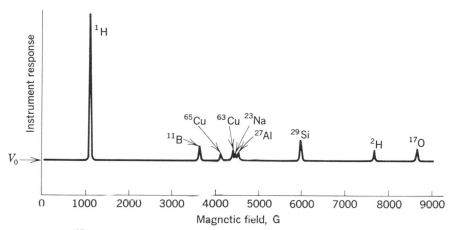

Fig. 18-2 NMR spectrum from a sample of water in a glass container, taken on a Varian low-dispersion instrument with a frequency of 5 MHz. (*Varian Associates, Palo Alto, Calif.*)

applicability in investigations of the physical environment of nuclei, including certain crystal parameters in solid samples.

HIGH-RESOLUTION NMR

In practice it is found that NMR is of maximum chemical utility if restricted to the study of the fine structure in the resonance of a single nuclear species; an instrument for this purpose is called a *high-resolution* NMR spectrometer. The majority of instruments have parameters suited only to detecting the resonance of hydrogen nuclei (protons). This means that both the magnetic field and the frequency can be held constant, except for a necessary variation in one of them by a few parts per million, as we shall see presently. For example, the Varian A-60A* has a magnetic field of 14,092 G, and a frequency of 60 MHz, corresponding to the gyromagnetic ratio which for the proton is 2.675 rad gauss^{-1} sec^{-1}. This instrument features an automatic circuit which maintains constant the H/f ratio, while still permitting a frequency change Δf up to ± 2 kHz. Thus $H/(f + \Delta f)$ may change slightly while H/f is held constant.

In the preceding discussion it has been assumed that the nucleus under consideration is actually subjected to the applied magnetic field with its full-measured intensity. If this were strictly true, then only a single peak would be seen, resulting from resonance with all the protons

* Varian Associates, Palo Alto, Calif.

in the sample, as in a wide-line instrument. This is only a rough approximation to the truth, however. A high-resolution spectrometer can evidence two distinct types of structure in the NMR absorption due to proton resonance, known respectively as the chemical shift and spin-spin interactions.

THE CHEMICAL SHIFT

Every nucleus is surrounded by a cloud of electrons in constant motion. Under the influence of the magnetic field these electrons are caused to circulate in such a sense as to oppose the field. This has the effect of partially shielding the nucleus from feeling the full value of the external field. It follows that either the frequency or the field will have to be changed slightly to bring the shielded nucleus into resonance. It is customary in many instruments to accomplish this by an adjustment of the magnetic field through an auxiliary winding carrying varying direct current which sweeps the field over a narrow span (a few milligauss in a field of 14 kilogauss). The complex electronic circuitry involved (in the Varian A-60A, for example) is such that the value of the added field is converted to its frequency equivalent for presentation to the recorder.

The value of the shift depends on the chemical environment of the proton, since this is the source of variations in shielding by electrons; it is thus called the *chemical shift*. Although the chemical shift is measured as a field or frequency, it is in reality a *ratio* of the necessary change in field to the applied field, or of the necessary change in frequency to the standard frequency, and hence is a dimensionless constant, usually designated by δ, and specified in parts per million.

Since we cannot observe the resonance of a test tube full of protons without any shielding electrons, there is no absolute standard with which to compare shifts. Therefore an arbitrary comparison standard must be adopted. For organic materials, whenever solubilities will permit, carbon tetrachloride (no protons) is used as solvent, and a small quantity of tetramethylsilane (TMS), $(CH_3)_4Si$, is added as internal standard. This material is chosen because not only are all its hydrogen atoms in an identical environment, they are more strongly shielded than the protons in any purely organic compound. The position of TMS on the chemical shift scale is arbitrarily assigned the δ-value of 0. Some authors prefer to give TMS the value 10 and denote the shift by τ where $\tau = 10 - \delta$, as this results in small positive values for nearly all other samples. Greater shielding corresponds to an "upfield" chemical shift. That is, the field must be increased to compensate for the shielding; δ decreases with increased shielding, τ increases. In the older literature (1960 is

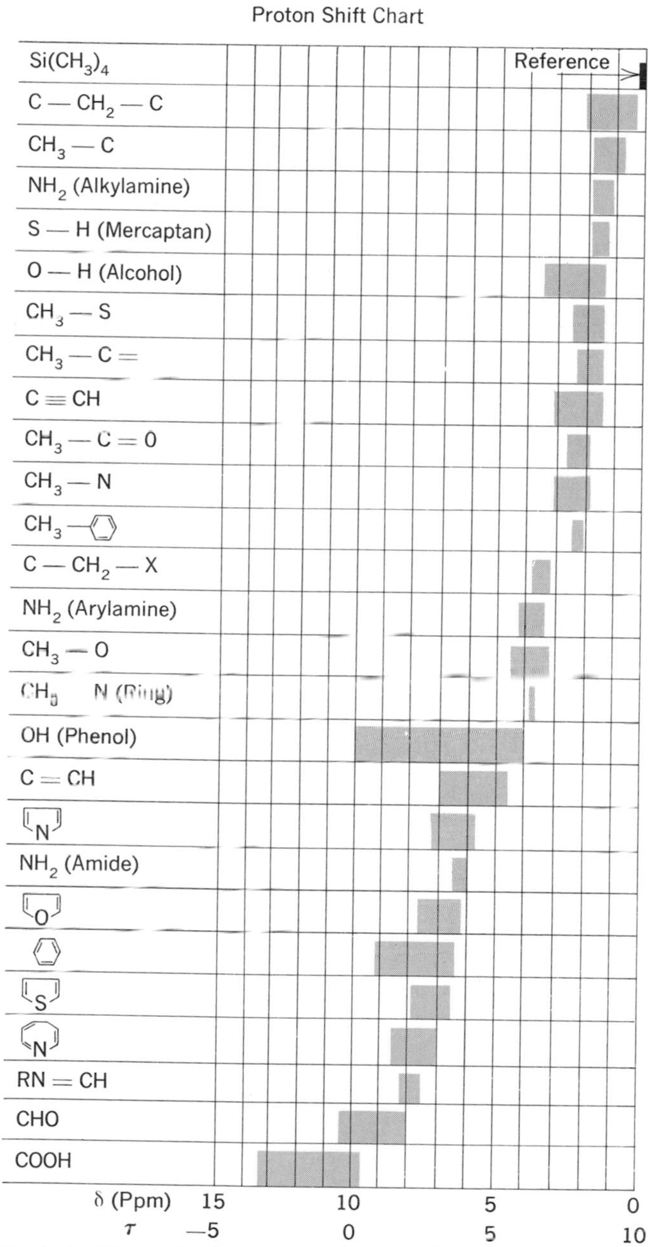

Proton Shift Chart

Fig. 18-3 Chart showing ranges of proton chemical shifts for various molecular environments. (*Chemical & Engineering News.*)

"old" in this field), δ was often referred to the proton resonance in water, benzene, or other substance which has only a single peak.

It is possible to construct a diagram (Fig. 18-3) of approximate ranges of δ or τ for protons in various chemical environments.[1] The exact values depend to a large degree on substituent effects, solvent, concentration, hydrogen bonding, etc., but are reproducible for any given set of conditions.

SPIN–SPIN COUPLING

The second type of structure frequently observed in NMR spectra is due to the interaction of the spin of a proton with that of another proton or protons attached to (usually) an adjoining carbon. The interaction involves the spins of the bonding electrons of all three bonds (H—C, C—C, and C—H), but we need not be concerned with the detailed mechanism. If the protons are in equivalent environments the interaction will not be manifest, but otherwise the maxima at each chemical shift position will be split into a close multiplet.

As an example, consider the spectrum of ethyl iodide in deuterochloroform solution with added TMS (Fig. 18-4). At (a) the spectrum is shown as determined at medium resolution. The areas enclosed beneath the multiplets at δ 3.2 and 1.8 are in the ratio of 2:3; the first is produced by the two methylene protons, the second by the three protons of the methyl group. The maximum at $\delta = 0$ is the TMS reference peak. The methylene resonance is split into a quadruplet in the approximate ratio of 1:3:3:1; these magnitudes are the result of the possible relative orientations of the spins of the three methyl protons: there is an equal probability of their being all lined up *with* the field, which we may denote by $\left[\begin{smallmatrix}\leftarrow\\\leftarrow\end{smallmatrix}\right]$, or *against* it, $\left[\begin{smallmatrix}\rightarrow\\\rightarrow\end{smallmatrix}\right]$, but it is three times as probable that two will face one way and one the other $\left[\begin{smallmatrix}\rightarrow\\\leftarrow\end{smallmatrix}\right]\left[\begin{smallmatrix}\leftarrow\\\rightarrow\end{smallmatrix}\right]\left[\begin{smallmatrix}\leftarrow\\\leftarrow\end{smallmatrix}\right]$, and similarly $\left[\begin{smallmatrix}\leftarrow\\\rightarrow\end{smallmatrix}\right]\left[\begin{smallmatrix}\rightarrow\\\leftarrow\end{smallmatrix}\right]\left[\begin{smallmatrix}\rightarrow\\\rightarrow\end{smallmatrix}\right]$. One direction of orientation will result in slightly greater shielding, the other in slightly less, hence the 1:3:3:1 ratio. By analogous reasoning it can be shown that the two methylene protons should be expected to cause splitting of the methyl resonance in a 1:2:1 ratio.

The spacings between the components of both multiplets are all equal and designated by a *coupling constant* J which has units of frequency. J is typically between 1 and 20 Hz. It will be noted that the intensity ratios 1:3:3:1 and 1:2:1 are not exactly followed. Those peaks of each group which are closer to the other group are larger in proportion; the nearer together the chemical shifts, the more marked is this effect.

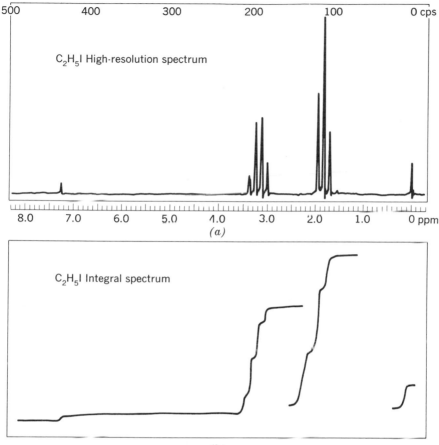

Fig. 18-4 NMR spectra of ethyl iodide dissolved in deuterochloroform ($CDCl_3$); (a) normal spectrum at moderate resolution, (b) integral spectrum. Run on a Varian A-60A spectrometer. (*Courtesy of R. F. Hirsch.*)

At (b) in Fig. 18-4 is shown a trace produced by an integrator which is built into the spectrometer. The step heights are proportional to the areas beneath the corresponding peaks of the NMR spectrum. The integral step corresponding to an entire multiplet gives a quantitative measure of the number of protons contributing to that resonance.

INSTRUMENTATION FOR NMR

The requirements of the design of high-resolution NMR spectrometers are quite severe. The magnetic field must be uniform over a large enough region to cover the area subtended by a sample probe. The diameter of the pole pieces must be at least four times that of the area which is required to be uniform. If the closest peaks to be resolved are separated by unit J-value, in a 60-MHz spectrometer, this field must be homogeneous to within 1 part in 60 million. Even the most carefully machined magnet cannot equal these requirements, so sets of specially shaped auxiliary windings called *shimming coils* are provided and powered with adjustable direct current to counteract any residual inhomogeneity. The field observed by the nuclei under study may still have some lack of uniformity originating in the sample itself and in its container; this is largely averaged out by spinning the sample within the coils by means of a small air turbine.

The radio-frequency excitation is derived from a highly stable crystal-controlled oscillator and fed to the probe coils and the receiver through coaxial cables. Most high-resolution instruments now on the market operate at either 60 or 100 MHz. Varian has one model with a choice of 56 or 60 MHz; 56 MHz is the resonant frequency for fluorine nuclei at the same field which resonates protons at 60 MHz.

There are a number of other electronic components—either optional or essential. An automatic circuit to maintain a constant H/f ratio has been mentioned; it functions by an electric frequency-controlling feedback circuit which "locks onto" the proton resonance of a suitable reference sample in a separate sample holder mounted in the same probe with the analytical sample. Without this automatic control, the performance of the instrument would be dependent on the magnetic surroundings and would be adversely affected by such extraneous events as the operation of an elevator in the vicinity, or even the changing of the location of a steel gas cylinder.

There is usually also an electronic integrator based on an operational amplifier with a capacitance in its feedback, similar to those mentioned in connection with coulometry. This component produces the type of integral trace illustrated in Fig. 18-4(b).

SPIN DECOUPLING

The spin-spin coupling previously described is sometimes of considerable help in the identification of resonances, but in relatively complex molecules it can complicate the spectrum to the point where it becomes impossible to elucidate. Of great assistance in such cases is a *spin decoupler*. This amounts to an auxiliary oscillator which can produce an alternating current at a selectable frequency and impose the corresponding field on the sample with considerable intensity. If this added signal is tuned to the resonant frequency of one set of protons while the contribution of the other set to the spectrum is under observation, it will be found that the multiplet caused by the coupling will collapse, leaving a single sharp peak. In Fig. 18-4(a), for example, if the auxiliary frequency is set at 110 Hz "downfield" (from TMS, that is, $\delta = 1.83$), the quadruplet at $\delta = 3.20$ will change to a singlet at the same location and with the same area. Likewise, if the decoupling field is adjusted to 190 Hz ($\delta = 3.20$), the triplet at $\delta = 1.83$ will collapse to a singlet. (For such a simple spectrum as that illustrated, this is hardly worth the trouble.)

This decoupling is caused by the rapid equilibration of the undesired protons between their two energy states, so that the protons being observed cannot distinguish the separate states and hence cannot be split.

APPLICATIONS OF NMR

The theory of NMR, as already discussed, is sufficient indication of the great utility of the method in the qualitative identification of pure substances. Atlases are available of NMR spectra, comparable to those of optical absorption spectra, for comparison of unknowns with authentic samples already studied. For substances of unknown structure, NMR provides a valuable diagnostic tool. Chemical shifts and spin coupling and decoupling observations are all useful in this connection.

Quantitative analysis is applicable through study of integrator traces. The immediate information obtained is the relative numbers of protons in various molecular environments in the sample. Without the need for high resolution, it is a simple matter to determine the total hydrogen content of a substance, and if it is known to be a pure compound, this will give information with a precision of the order of ± 0.5 percent relative, comparable to the conventional gravimetric combustion method with great saving in time.

Mixtures of compounds can be analyzed in favorable cases with excellent precision. An example is shown in Fig. 18-5. A mixture approximating that which might be found in some types of petroleum was prepared from tetralin, naphthalene, and n-hexane. The aromatic hydrogens showed overlapping multiple resonances in the region marked (a) in the figure. The hydrogens on carbon atoms adjoining an aromatic ring ("alpha to the ring") appeared at (b), and the purely aliphatic hydrogens at (c). (It might be possible to identify the source of some of the multiplicities, but that was unnecessary for the problem in hand.)

Tetralin has four aromatic protons, while naphthalene has eight; only tetralin has alpha protons, of which there are four; finally, tetralin has four protons which are essentially aliphatic, and hexane contributes all 14 of its protons in the same region. A set of three simultaneous

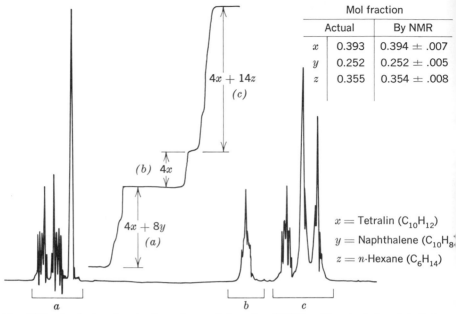

	Mol fraction	
	Actual	By NMR
x	0.393	0.394 ± .007
y	0.252	0.252 ± .005
z	0.355	0.354 ± .008

$x =$ Tetralin ($C_{10}H_{12}$)

$y =$ Naphthalene ($C_{10}H_8$)

$z = n$-Hexane (C_6H_{14})

Fig. 18-5 Analysis of a hydrocarbon mixture by NMR. (See text for details.) (*Varian Associates, Palo Alto, Calif.*)

equations can be solved for the mole fractions of the three components, leading to the data in the table at the upper right of the figure. These are in extremely good agreement with the composition of the mixture as calculated from the weights.

ELECTRON–SPIN RESONANCE (ESR)*

Since electrons always possess a spin, they also have a magnetic moment, and hence the basic magnetic resonance theory applies to electrons as well as to spinning nuclei. The gyromagnetic ratio for an electron turns out to be 1.76×10^7 rad G^{-1} sec^{-1} (as compared with 2.68 for the proton). Since this is an inconvenient number to work with, it is customary to rewrite Eq. (18-1) as

$$E = hf = \frac{\mu H}{I} = g\beta H \qquad (18\text{-}3)$$

where β is the Bohr magneton, a constant with the magnitude 9.2732 ×

---

* This subject is also known as electron magnetic resonance (EMR) and electron paramagnetic resonance (EPR).

10^{-21} erg G^{-1}, and g is called the *splitting factor*.   The quantum number I has the value $\frac{1}{2}$ for the electron, so that it has just two energy states, as does the proton.   The value of g is 2.0023 for free electrons, and varies from this by a few percent for free radicals, transition-metal ions, and other bodies containing unpaired electrons.   If electrons are paired, their magnetic moments effectively cancel each other and they are not observable.

The unpaired electrons have a great tendency to show fine structure in their resonances due to coupling with spinning nuclei in their vicinity. The principles are similar to those involved in the spin coupling of protons, but are often much more complex, frequently to the point where the source of each component of the fine structure cannot be identified. The ethyl free radical, for example, produces an array of four triplets. Spin decoupling can sometimes help.

ESR INSTRUMENTATION

Most of the work in ESR is carried out at a constant frequency in the neighborhood of 9.5 GHz with the corresponding field of about 3400 G. Instruments are available, however, which operate as high as 35 GHz, and others are designed for lower frequencies, in the MHz region.   It is usually considered simpler design to maintain constant frequency, generated by a klystron oscillator, and vary the field to scan the resonance peaks.

The most obvious difference between the instrumentation for ESR and NMR is that because of the high frequencies usually employed, the power is much more effectively conducted from one point to another by rigid waveguides than by flexible coaxial cables.   The sample cell is inserted through an orifice in the waveguide at a point where the magnetic (rather than the electric) vector of the electromagnetic wave is undergoing maximum amplitude fluctuations.   The efficiency of transfer of energy to the sample is generally low (1 to 30 percent), but even so the method is quite sensitive.   Under typical conditions as little as 10^{-10} to 10^{-8} mole percent of free radicals in a 1-g sample can be detected.

It is customary to record ESR spectra in the form of the first derivative of the normal absorption spectrum, as greater resolution is attainable.   Quantitative measurements must then be made on the second integral of the ESR curve.

APPLICATIONS OF ESR

Just as in NMR, a standard reference substance is convenient in ESR. The most widely used is the 1,1-diphenyl-2-picrylhydrazyl free radical:

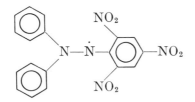

which is a chemically stable material with the splitting factor $g = 2.0036$.
It cannot be used as an *internal* standard with other free radicals, as there
is only slight variation in g-values, and the standard cannot be distinguished from the substance studied; they can be recorded consecutively,
however, with all parameters kept constant. A tiny chip of ruby crystal
cemented permanently to the sample cell has been recommended for a
standard; ruby contains a trace of Cr(III) entrapped in its crystal lattice
and shows a strong resonance ($g = 1.4$).

Free radicals can be studied readily by ESR, even in very low concentrations. Figure 18-6 shows the ESR spectrum of a quinone-hydro-
quinone redox system. It proves conclusively that a semiquinone free
radical anion exists as an intermediate. The five-line pattern is a consequence of the magnetic spin interaction between the odd electron and
the four hydrogens on the ring; statistics show that the intensities should
stand in the ratios $1:4:6:4:1$. Figure 18-7 gives the results of a kinetic
study of the formation and decay of the semiquinone. The ESR instrument was tuned to the point of maximum signal and observed over a

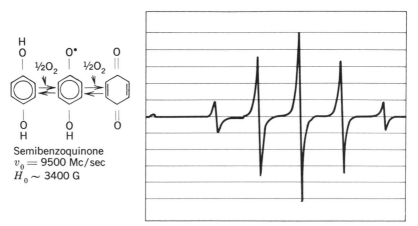

Fig. 18-6 Semiquinone intermediate in the oxidation of hydroquinone. (*Varian
Associates, Palo Alto, Calif.*)

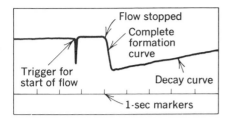

Fig. 18-7 Kinetic study of the formation and decay of semiquinone. (*Varian Associates, Palo Alto, Calif.*)

period of time. The sample cell was designed to permit two solutions (hydroquinone and oxygen, respectively) to mix together and immediately flow through the observation area. The flow rate was faster than the rate of production of the radical, but as soon as the flow was stopped, its production, followed by its further reaction, could be monitored easily. The half-time of the formation reaction under one set of conditions was 0.15 sec.

Cells have been devised which permit the formation of radicals in situ by irradiation with ultraviolet, gamma or x-rays, or by electrolytic redox reactions. The great sensitivity of the method makes possible observations of transient species which could be detected in no other way.

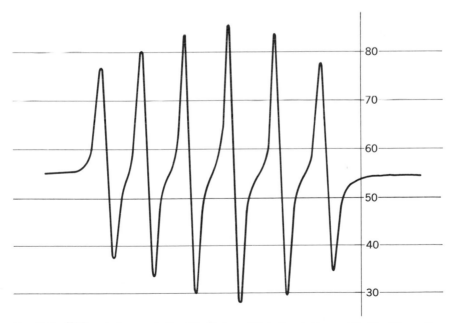

Fig. 18-8 ESR spectrum of the Mn^{++} ion, 10^{-4} F solution. (*Perkin-Elmer Corporation.*)

For example, the fivefold resonance of the ethyl radical becomes visible when ethanol is subjected to x-radiation.

Another important field of application of ESR is in the estimation of trace amounts of paramagnetic ions, particularly in biological work. Figure 18-8 shows the six resonances of the Mn^{++} ion as observed in aqueous solution. The multiplicity is given by $2I + 1$; for Mn, $I = \frac{5}{2}$, which accounts for the six peaks. This ion can be measured to 10^{-6} formal or less, with increased instrumental sensitivity, before becoming lost in the background noise.

PROBLEMS

18-1   Justify the (a), (b), and (c) assignments of Fig. 18-5 by comparison with Fig. 18-3.   Measure as accurately as you can the heights of the three steps in the integral curve, and calculate the amounts of x, y, and z.

18-2   Account for the small peak at $\delta = 7.32$ in Fig. 18-4(a).

18-3   Figure 18-9 shows the NMR spectrogram of a mixture of cyclohexane, toluene, and water.   By comparison with Fig. 18-3, identify the four peaks.   From the integral curve, estimate the composition of the mixture.

18-4   Corsini et al.[2] used ESR evidence to determine which of the following two formulas is correct for the complex obtained between copper and 8-quinolinethiol (C_9H_7NS):

(a) $Cu^{(I)}(C_9H_7NS)(C_9H_6NS^-)$
(b) $Cu^{(II)}(C_9H_6NS^-)_2$

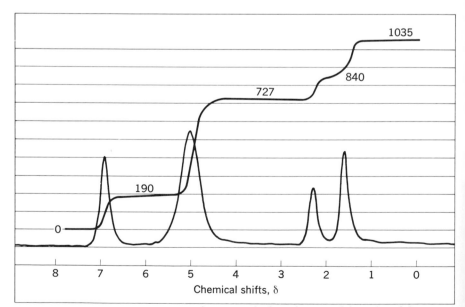

Fig. 18-9   NMR spectrogram at moderate resolution of an emulsion of cyclohexane, toluene, and water.   (*Varian Associates, Palo Alto, Calif.*)

Show that the ESR approach should be expected to distinguish between these structures.

REFERENCES

1. Bovey, F. A.: *Chem. Eng. News,* **43**:98 (1965).
2. Corsini, A., Q. Fernando, and H. Freiser: *Talanta,* **11**:63 (1964).

GENERAL REFERENCES

Chamberlain, N. F.: Nuclear Magnetic Resonance and Electron Paramagnetic Resonance, in I. M. Kolthoff and P. J. Elving (eds.), "Treatise on Analytical Chemistry," pt. I, vol. 4, chap. 39, Interscience Publishers (Division of John Wiley & Sons, Inc.), New York, 1963.

Pople, J. A., W. G. Schneider, and H. J. Bernstein: "High-resolution Nuclear Magnetic Resonance," McGraw-Hill Book Company, New York, 1959.

Roberts, J. D.: "Nuclear Magnetic Resonance: Applications to Organic Chemistry," McGraw-Hill Book Company, New York, 1959.

Silverstein, R. M., and G. C. Bassler: "Spectrometric Identification of Organic Compounds," 2d ed., John Wiley & Sons, Inc., New York, 1967.

19
Thermometric Methods

Many of the analytical methods discussed in other chapters have significant temperature coefficients, but in general their measurement does not itself provide analytical information. In the present chapter we shall consider a number of methods in which some property of the system is measured as a function of the temperature. It will help to clarify the relations between these to list them here for reference (Table 19-1).[8]

Table 19-1 Thermoanalytical methods

Designation	Property measured	Apparatus
Thermogravimetric analysis (TGA)	Change in weight	Thermobalance
Derivative thermogravimetric analysis (DTG)	Rate of change of weight	Thermobalance
Differential thermal analysis (DTA)	Heat evolved or absorbed	DTA apparatus
Calorimetric DTA	Heat evolved or absorbed	Differential calorimeter
Thermometric titration	Change of temperature	Titration calorimeter

There are other possible thermometric methods in addition to those listed, which are less used at present, and will not be discussed in detail. The detection of impurities in nearly pure substances by observation of melting points is an everyday procedure, especially for organic chemists. With some elaboration this method can be rendered quantitative, but is not frequently so employed. Melting-point analysis for other than trace quantities becomes very specific for a given system, depending on the details of the phase diagram involved.

THERMOGRAVIMETRIC ANALYSIS (TGA)

This is a technique whereby the weight of a sample can be followed over a period of time while its temperature is being changed (usually increased at a constant rate). Several examples of thermograms obtained by this method are shown in Fig. 19-1.[3] Curve 1 shows the weight of a precipitate of silver chromate, collected in a filtering crucible. The initial drop in weight represents the loss of excess wash water. Just above 92°C the weight becomes constant and remains so to about 812°C. From there to 945°, oxygen is lost. The loss in weight shows that the decomposition proceeds according to the reaction $2Ag_2CrO_4 \rightarrow 2O_2 + 2Ag + Ag_2Cr_2O_4$. The residue is thus a mixture of silver and silver chromite. It follows that the silver chromate precipitate, if used for a gravimetric chromium analysis, may be dried anywhere in the plateau region between about 100° and 800°C, say at 110°C. Laboratory directions in the older textbooks specified exactly 135°C.

The balance is calibrated, preferably each time it is used, by placing a known weight on the pan to give a reference mark, such as that in the upper right corner of Fig. 19-1.

Curve 2 of the same figure shows a heating curve for mercurous chromate. This compound is stable between about 52° and 256°C and then decomposes according to the equation $Hg_2CrO_4 \rightarrow Hg_2O + CrO_3$. The mercurous oxide is lost by sublimation, and the chromium trioxide remains at constant weight above 671°C. Because of the high atomic weight of mercury, the precipitate of mercurous chromate provides a particularly favorable gravimetric factor for the determination of chromium. It had previously been the practice to ignite the precipitate under a hood and weigh the chromium trioxide. The thermogravimetric study shows this procedure not only to be unnecessary but to result in lowered precision.

Much of the reported work in thermogravimetry has been directed toward the establishment of optimum temperature ranges for the conditioning of precipitates for conventional gravimetric analysis, as the

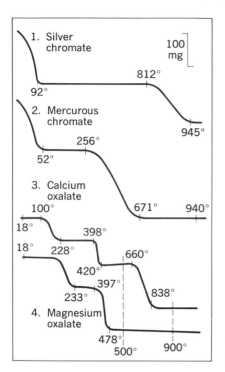

Fig. 19-1 Examples of curves taken from the thermobalance. (*American Elsevier Publishing Company, Inc., New York.*)

preceding examples suggest. The technique, however, has a greater potential than this.

Consider, for example, curves 3 and 4 of Fig. 19-1. A significant difference in behavior is observed between the oxalates of calcium and magnesium which permits their simultaneous determination. Calcium oxalate loses its carbon and excess oxygen in two steps, $CaC_2O_4 \rightarrow CaCO_3 + CO$ and $CaCO_3 \rightarrow CaO + CO_2$, whereas the magnesium compound does not pass through the carbonate stage, $MgC_2O_4 \rightarrow MgO + CO + CO_2$. The ranges of stability are:

Compound	°C		Compound	°C
$CaC_2O_4 \cdot H_2O$	up to 100		$MgC_2O_4 \cdot 2H_2O$	up to 176
CaC_2O_4	226–398		MgC_2O_4	233–397
$CaCO_3$	420–660		MgO	480 and up
CaO	840 and up			

Thus, at 500°C, calcium carbonate and magnesium oxide are stable, whereas at 900°C, both metals exist as the simple oxides. Comparison of the weights of a mixed precipitate at these two temperatures will permit calculation of both the calcium and the magnesium content of the original sample.

Another example is the analysis of copper-silver alloys based on the relative stabilities of the nitrates (Fig. 19-2).[3] $AgNO_3$ is stable up to 473°C, where it starts to lose NO_2 and O_2, leaving a residue of metallic silver above 608°C. $Cu(NO_3)_2$, on the other hand, decomposes in two steps to the oxide CuO, which is the stable form up to at least 950°C. A binary alloy can be analyzed by successive weighings at 400 and 700°C in a short time (perhaps 30 min) with an accuracy of about ±0.3 percent.

The limiting temperatures of the various segments of the thermograms such as those of Figs. 19-1 and 19-2 cannot be considered reproducible without qualification. The thermogravimetric method, as ordinarily carried out, is a dynamic one, and the system is never at equilibrium. Hence the temperatures of distinctive features on the curves are somewhat different as observed on different instruments, or on the same instrument at different rates of temperature scanning, or with different size samples, etc.

THERMOBALANCES

There are more than a dozen manufacturers of thermobalances, and some of them produce several models. The weighing mechanism may be a modification of a single- or double-pan balance, an electronically

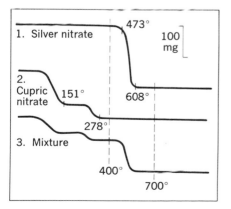

Fig. 19-2 Examples of curves taken from the thermobalance. (*American Elsevier Publishing Company, Inc., New York.*)

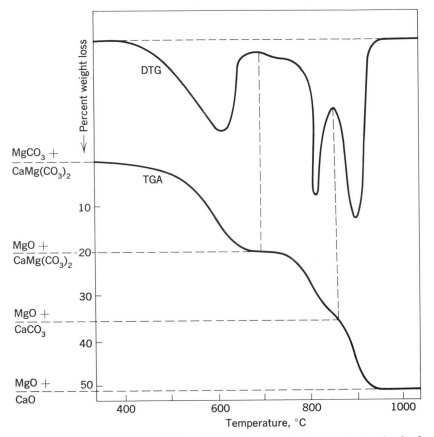

Fig. 19-3 Relation between TGA and DTG curves for the pyrolysis of mixed calcium and magnesium carbonates. (*Fom Paulik et al.*[5])

self-balancing device, a torsion balance, or a simple spring balance. Several models have an electric furnace for heating the sample located beneath the balance, with the crucible suspended within it by a long platinum wire. This design requires stringent precautions against convection effects interfering with the balance. Some designers have preferred to place the furnace above the balance, with the crucible supported at the top of a rod extending upward from the balance beam.

All thermobalances intended for precise work have provision for automatic recording of weight either against time, or, with an X-Y recorder, directly against temperature. If plotted against time, either the temperature must be programmed to increase at a steady rate, or a second pen must be arranged to plot the temperature-time relationship. An example of the latter is seen in Fig. 19-5.

There are also a number of nonrecording thermobalances, though not usually dignified by that name, intended principally for the determination of superficial moisture in bulk materials. The sample is heated, as by an infrared lamp, while on the pan of a balance specially designed to minimize errors produced by air currents. Readings are taken manually until no further change is seen. A precision of a few tenths of a percent water in a 1-g sample can be obtained in a few minutes.

DERIVATIVE THERMOGRAVIMETRIC ANALYSIS (DTG)

There is sometimes an advantage in being able to compare a thermogram with its first derivative, as in Fig. 19-3.[5] The plateau in the thermogram at 700°C is clear enough, but the shoulder at about 870° could not have been pinpointed without the derivative curve.

Several commercial thermobalances are provided with electronic circuits to take the derivative automatically. A two-pen recorder permits a convenient direct comparison of the two curves.

DIFFERENTIAL THERMAL ANALYSIS (DTA)

This is a technique by which phase transitions or chemical reactions can be followed by observation of the heat absorbed or liberated. It is especially suited to studies of structural changes within a solid at elevated temperatures, where few other methods are available.

In a typical apparatus, one set of thermocouple junctions (Fig. 19-4) is inserted into an inert material, such as aluminum oxide, which does not change in any manner through the temperature range to be studied. The other set is placed in the sample under test. With constant heating, any transition or thermally induced reaction in the sample

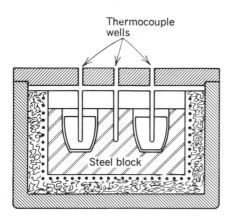

Thermocouple wells

Steel block

Fig. 19-4 Simple arrangement for differential thermal analysis.

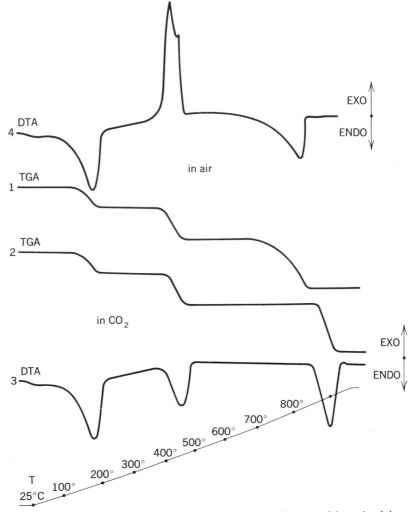

Fig. 19-5 Simultaneous DTA-TGA diagram of the decomposition of calcium oxalate monohydrate in air and in carbon dioxide. (*Plenum Publishing Corp., New York.*)

will be recorded as a peak or dip in an otherwise straight line.   An endothermic process will cause the thermocouple junction in the sample to lag behind the junction in the inert material, and hence develop a voltage; an exothermic event will cause a voltage of opposite sign.   It is customary to plot exotherms upward, endotherms downward, but this convention is not universally followed.

Since the usual mode of operation is to supply heat to the samples, endothermic events are more likely to occur than exothermic, and the latter when they are observed are often caused by secondary processes.

As an example of this approach, consider the curves of Fig. 19-5.[6] Curve 1 is essentially the same as curve 3 of Fig. 19-1; the slight differences result from differing instrumental conditions. Curve 2 represents the same type of experiment, but with the sample in an atmosphere of CO_2 rather than air; as should be expected, no difference is evident until the decomposition of $CaCO_3$, which now requires a higher temperature.

Curve 3 is a differential thermogram (DTA curve) also showing the decomposition of calcium oxalate in an atmosphere of CO_2. It is seen that the three points of weight loss correspond to three endothermic processes: it requires energy to break the bonds in the successive elimination of H_2O, CO, and CO_2. By contrast, the second peak in curve 4, where the atmosphere is air, is sharply exothermic, but corresponding to the same weight loss. The explanation for the difference lies in the exothermic burning of CO in air at the temperature of the furnace.

Comparison of Figs. 19-3 and 19-5 reveals that there is a degree of similarity between the curves obtained in DTG (derivative thermograms) and DTA (differential thermograms). DTG can give information

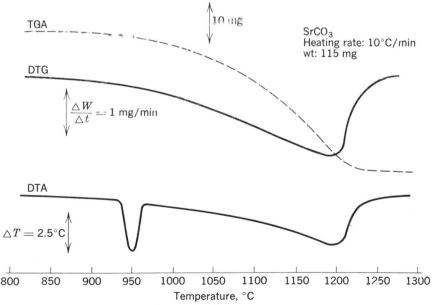

Fig. 19-6 Decomposition of $SrCO_3$ in air, showing the DTA carbonate decomposition endotherm overlapping the endothermic rhombic-hexagonal crystalline transition. (*Mettler Instrument Corporation, Princeton, N.J.*)

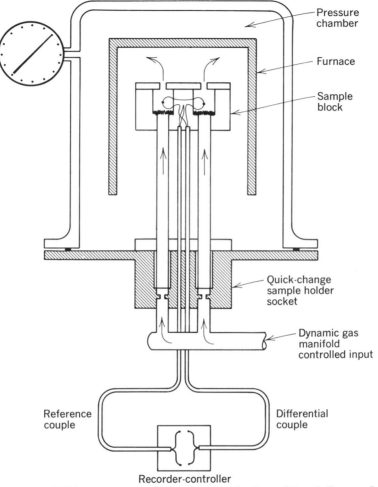

Fig. 19-7 DTA apparatus, schematic. (*R. L. Stone Div. of Tracor, Inc., Austin, Tex.*)

only about weight changes, whereas DTA will reveal changes in energy, regardless of constancy or change in weight. In Fig. 19-5, if DTG curves had been included, they would have been identical up to about 700°C, even though the DTA curves vary greatly. Figure 19-6,[1] on the other hand, shows a situation in which a pronounced endotherm in DTA (at 950°C) fails to show up at all in either TGA or DTG; this is evidence of a crystalline transition, from rhombic to hexagonal modifications of $SrCO_3$, which, of course, does not involve a change in weight.

DTA APPARATUS

There are many companies in the DTA field. Their products vary with respect to such parameters as sample size, temperature range, selectable scanning rates, precision, convenience, and cost. Figure 19-7 shows schematically the essential parts of a typical DTA apparatus, which includes provision for bathing the samples with a controlled atmosphere; the gas can be made to flow *through* the bed of particulate sample, thereby flushing away any gaseous product of decomposition.

In this apparatus, as well as in some thermobalances, provision is made for detecting and analyzing effluent gases, by gas chromatography or other means.

Several companies have combined DTA and TGA capabilities into a single instrument. The record is obtained on a two-pen recorder.

CALORIMETRIC DTA

Conventional DTA as just described is capable of giving good qualitative data about the temperatures and signs of transitions, but it is difficult or impossible to obtain quantitative data—the heat of transition if the purity is known, or the amount of a constituent in a sample if the heat of transition is known. This difficulty arises from uncontrollable and frequently unknown factors such as the specific heat and thermal conductivity of the sample both before and after the transition. The rate of heating, placement of thermocouples, and other instrumental parameters will also affect the areas beneath endotherms or exotherms.

Quantitative results can be achieved by converting the sample compartment in a DTA apparatus to a differential calorimeter.[9] This has been done in three different ways by three different companies. Perkin-Elmer manufactures an instrument called a Differential Scanning Calorimeter (DSC), in which the calorimeter is of the isothermal type.[7] Each sample holder (unknown and reference) is provided with its own resistive heater. When the differential thermocouple starts to register a voltage, an automatic control loop sends just enough power into whichever of the two samples is the cooler, to counteract the trend and keep the two temperatures within a very small fraction of a degree of equality. A second electronic control loop forces the temperature of the reference (hence effectively of both) samples to increase linearly with time. The recorder traces out the electric power which has to be delivered to one or the other sample to maintain isothermal conditions. The resulting thermogram resembles conventional DTA, but the area beneath a peak is now an *exact* measure of the energy supplied to the unknown sample to compensate for an endothermal event, or to the reference material to equal the energy emitted in the unknown when an exothermic event takes place.   Differences in thermal conductivity, heat capacity, etc., are now irrelevant.

Technical Equipment Corporation* has chosen an adiabatic approach in designing their comparable unit, the Deltatherm Dynamic Adiabatic Calorimeter (DAC).[2]   In an adiabatic calorimeter there must be no passage of heat between the sample and its surroundings, so instead of a reference sample, the DAC has a massive copper block, with the sample in a central cavity but thermally isolated from its walls. The block is heated at a constant rate, and the amount of power required to maintain the sample at the same temperature as the surrounding walls is recorded.   This device gives comparable precision in the measurement of heats of transition, but the adiabatic system facilitates the determination of specific heats as well.

The third instrument in this category is the Du Pont DTA apparatus with accessory calorimeter.   This is also adiabatic, but makes use of a reference sample as well.   The reference temperature actuates the X axis of an X-Y recorder, while the temperature differential between the two samples controls the Y input.   The area beneath a DTA peak under these conditions is an accurate measure of the heat of transition, independent of specific heat and other variables; specific heats can also be determined.

THERMOMETRIC TITRATIONS[4,9]

Since practically all chemical reactions are accompanied by a heat effect, it is possible to follow the course of a reaction by observing the heat liberated.   Such a titration can be carried out manually in a small Dewar flask.   The temperature can be read with a thermometer calibrated in tenths or hundredths of degrees or by means of a thermocouple or a resistance thermometer.

Thermometric titration can readily be automated, but apparently the only thermometric titrator manufactured as a complete unit in the United States is the Aminco† Titra-Thermo-Mat.   This instrument makes use of a thermistor detector in a titration beaker surrounded by heat-insulating material.   Solution is added at a constant rate by a syringe pump, and the resistance of the detector monitored on a strip-chart recorder.

A large variety of titrations have been followed with success by the thermometric method.   These include neutralizations of any acid with $pK_a > 10^{-10}$, precipitations, redox reactions, and complex formations.

* Denver, Colo.
† American Instrument Co., Inc., Silver Spring, Md.

Solvents of all sorts can be employed; besides water, work has been reported in hydrogen acetate, carbon tetrachloride, benzene, and nitrobenzene, and in the fused eutectic of lithium and potassium nitrates. Reported precision is usually not poorer than about ± 1 percent standard deviation, sometimes much better.

It is essential that the two reacting solutions do not differ appreciably with respect to extraneous materials that would contribute noticeable heat effects either by reaction with each other or with solvent (heat of dilution).

It has been pointed out by Jordan[4] that thermometric titrimetry constitutes one of a very few methods of titration which is not based solely on consideration of the free energy change ΔG, hence on the equilibrium constant of the reaction. The quantity measured is ΔH, not ΔG, in the familiar thermodynamic equation $\Delta H = \Delta G + T\Delta S$. Hence it is possible that thermometric titrations may give useful results even if ΔG is zero or positive. Two examples will indicate the potential value of this approach.

The neutralization of boric acid follows the reaction $H_3BO_3 + OH^- \rightleftharpoons H_2O + H_2BO_3^-$. Given the first ionization constant $K_a = 5.8 \times 10^{-10}$ (at 25°C), it follows that the standard free energy change of this reaction is -6.5 kcal/mole, corresponding to a neutralization constant $K_n = K_a/K_w = 5.8 \times 10^{-4}$. This may be compared to hydrochloric acid, where the only reaction to be considered is $H^+ + OH^- \rightleftharpoons H_2O$, $K_n = 1/K_w = 1 \times 10^{14}$, and $\Delta G° = -19.2$ kcal/mole. However, it so happens that the entropy term $T\Delta S°$ is -3.7 kcal/mole for boric acid (at 25°C) against $+5.7$ kcal/mole for hydrochloric. These combine to give heats (enthalpies) of neutralization of -10.2 kcal/mole for boric and -13.5 kcal/mole for hydrochloric.

Thus boric acid cannot be titrated successfully by any method which depends on the pH (such as potentiometric or photometric techniques) because the hydrogen-ion activity is determined by the equilibrium constant. But the heats of neutralization of the two acids are comparable in magnitude, and they can be titrated thermometrically with equal ease, as shown in Fig. 19-8.[4]

Another interesting situation arises in the titration of calcium and magnesium with EDTA. The stability constants of the chelates differ by less than two orders of magnitude, so that titration based on an indicator such as Eriochrome Black T can give only the sum of the two ions. However, the entropy of reaction of Mg^{++} with EDTA is twice that for the Ca^{++} reaction. This not only gives a distinct difference in ΔH values, it actually changes the sign: $\Delta H°$ is $+5.5$ kcal/mole for the magnesium reaction and -5.7 kcal/mole for calcium. A titration curve for a mixture of the two is shown in Fig. 19-9.[4]

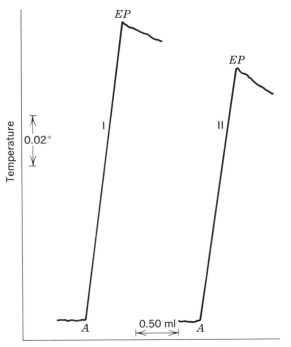

Fig. 19-8 Typical thermometric titration curves in 0.01 *F* aqueous solutions. (I) HCl titrated by NaOH; (II) boric acid titrated by NaOH. *A* indicates the start of the titration, *EP*, the end point. (*Record of Chemical Progress.*)

Thermometric titrimetry can give information in addition to the usual stoichiometry. Figure 19-10 represents a generalized titration curve. Segment AB is a pretitration trace on the recorder, to establish a base line; the titration is started at point B. In the vicinity of C the trace often shows curvature rather than a sharp angle, and this can be related to the equlibrium constant (hence to $\Delta G°$) for the reaction. From C to D, the curve commonly rises as shown, which corresponds to the heat of dilution of the reagent; it may, however, slope downward, indicating either that the dilution is endothermic (which is unusual) or that the reagent is cooler than the contents of the titration vessel. The temperature change which can rightly be ascribed to the reaction is ΔT, from B to the extrapolated intersection of CD with the zero-time ordinate. ΔT is proportional to the total amount of heat liberated

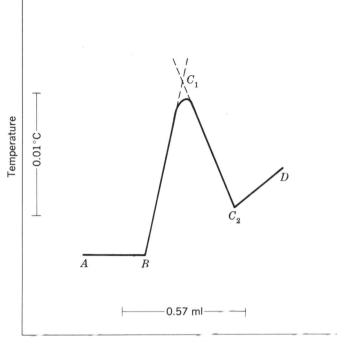

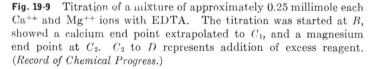

Fig. 19-9 Titration of a mixture of approximately 0.25 millimole each Ca^{++} and Mg^{++} ions with EDTA. The titration was started at B, showed a calcium end point extrapolated to C_1, and a magnesium end point at C_2. C_2 to D represents addition of excess reagent. (*Record of Chemical Progress.*)

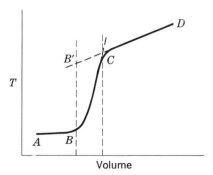

Fig. 19-10 Generalized thermometric and enthalpometric titration curve.

$-N\Delta H$, where N is the number of moles reacting, and ΔH, as usual, the heat *absorbed* per mole. The constant of proportionality is the heat capacity of the vessel and its contents, which we will designate as k in kcal/deg, so that

$$\Delta T = -\frac{N\Delta H}{k} \tag{19-1}$$

The quantity k can be determined by a simple calibration step, for example by a known amount of electrical heating. Then N can be determined if $\Delta H°$ is known, or vice versa, if we make the approximation that in the dilute solutions usually employed, ΔH and $\Delta H°$ do not differ materially. The value of N (or of concentration) can be determined more accurately, of course, by a complete titration. Note that the calorimetric determination of N does not require a standardized titrant. This procedure is called *enthalpy titration*.

PROBLEMS

19-1 A sample of a copper-silver alloy is dissolved in nitric acid in a small crucible, the crucible placed in the thermobalance, and its temperature gradually raised to 750°C. The trace obtained (weight against time) is shown in Fig. 19-11, together with the weights of residue as read from the graph. Calculate the composition of the alloy in terms of percent by weight.

19-2 Devise a procedure, analogous to that in the preceding problem, for the analysis through the chromate precipitates of mixtures of silver and mercurous ions.

19-3 Microcosmic salt, $Na(NH_4)HPO_4 \cdot 4H_2O$, upon heating, first liberates four water molecules, then another water and a molecule of NH_3, to end up as sodium metaphosphate, $NaPO_3$. Sketch curves which you might expect from a study of this substance (a) with a thermobalance, (b) by differential thermal analysis.

19-4 What is the end product of pyrolysis of $SrCO_3$, as shown by Fig. 19-6?

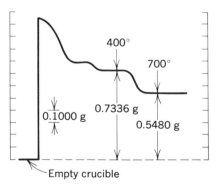

Fig. 19-11 Thermobalance trace of a mixture of copper and silver nitrates.

REFERENCES

1. *Tech. Bull.* T-102, Mettler Instrument Corp., Princeton, N.J.
2. Dosch, E. L.: *Instr. Soc. Am. Conf. Proc. Preprint* no. 2.6-5-64 (1964).
3. Duval, C.: "Inorganic Thermogravimetric Analysis," 2d ed., American Elsevier Publishing Company, Inc., New York, 1963.
4. Jordan, J.: *Record Chem. Progr.*, **19**:193 (1958); *J. Chem. Educ.*, **40**:A5 (1963).
5. Paulik, F., J. Paulik, and L. Erdey: *Hung. Sci. Instr.*, **1**:3 (1964).
6. Vaughan, H. P., and W. G. Wiedemann: An Integrated Vacuum Thermoanalyzer for Simultaneous TGA and DTA, in P. M. Waters (ed.), "Vacuum Microbalance Techniques," vol. 4, Plenum Publishing Corp., New York, 1965.
7. Watson, E. S., M. J. O'Neill, J. Justin, and N. Brenner: *Anal. Chem.*, **36**:1233 (1964).
8. Wendlandt, W. W.: "Thermal Methods of Analysis," Interscience Publishers (Division of John Wiley & Sons, Inc.), New York, 1964.
9. Wilhoit, R. C.: *J. Chem. Educ.*, **44**: A571, A629, A685, A853 (1967).

20

Introduction to
Interphase Separations

Separation is not per se an analytical technique, but it is so commonly required prior to analyses that it is a subject of major concern to analytical chemists. It has a valid place in this book because many of the procedures involve instrumentation which may be highly sophisticated. We will not be concerned with those separation techniques which are primarily suited for preparation or purification.

A large and important class of separation schemes involves the transfer of one or more substances from one phase to another. Hence we can classify separation methods according to the type of phases between which equilibrium takes place (or is approached). There are four such classes: gas-liquid, gas-solid, liquid-liquid, and liquid-solid.

It will be advantageous to establish a general notation to systematize the several analytical situations which may arise.* We will let A and B represent substances to be separated. The two phases involved will be designated by subscripts 1 and 2. If substance A is equilibrated between

* Different authors use variations on the notation presented here, so that one must be alert to avoid confusion.

the two phases, the fraction in phase 1 will be denoted by $p_{(A)}$, the fraction in phase 2 by $q_{(A)}$, so that

$$p_{(A)} + q_{(A)} = 1 \tag{20-1}$$

The ratio between p and q will be designated by K, the *distribution ratio*:

$$K_{(A)} = \frac{p_{(A)}}{q_{(A)}} \tag{20-2}$$

Combining the above equations gives us

$$p_{(A)} = \frac{K_{(A)}}{K_{(A)} + 1}$$
$$q_{(A)} = \frac{1}{K_{(A)} + 1} \tag{20-3}$$

The same distribution ratio can be expressed in terms of the volumes of the two phases V_1 and V_2, and the corresponding concentrations C_1 and C_2:

$$K_{(A)} = \frac{C_{1(A)} V_1}{C_{2(A)} V_2} \tag{20-4}$$

The concentrations in the two phases do not necessarily refer to identical chemical species. For example, in the distribution of acetic acid between water (phase 1) and benzene (phase 2), C_1 will refer to the total of ionized and un-ionized aqueous species, and C_2 will include both monomeric and dimeric forms.

The distribution ratio for a given system, even at constant temperature, may not be strictly constant. This is partly because the distribution equilibrium may be affected by competing equilibria within one or both phases, as suggested above for acetic acid, so that K may well change with total concentration. Also, K cannot be taken as a true thermodynamic constant unless activity coefficients are included, and these are usually not readily determinable.

To separate A and B from each other, it is desirable that the distribution ratios be as widely different as possible. We will define α, the *separation factor*, as

$$\alpha = \frac{K_{(A)}}{K_{(B)}} \tag{20-5}$$

Combining Eqs. (20-4) and (20-5) gives

$$\alpha = \frac{C_{1(A)} C_{2(B)}}{C_{2(A)} C_{1(B)}} \tag{20-6}$$

which is independent of volumes. This ratio should be either much larger or much smaller than unity for best separation. The separation is most effective if $K_{(A)} = 1/K_{(B)}$; this is often not attainable, but can be approached by manipulating the volume ratio.

In practice we are almost invariably concerned with repetitive separations; only rarely will a single equilibrium suffice. The repetition can be continuous, as in chromatography, or stepwise, as in solvent extraction with separatory funnels, but actually the difference between these approaches is a matter of degree rather than kind. If the stages in a stepwise procedure are made sufficiently numerous, the net effect is indistinguishable in theory from a truly continuous method. On the other hand it is sometimes convenient to treat a continuous separation mathematically as though it were stepwise, using the theoretical-plate concept originated in fractional distillation theory.

Experimentally we can distinguish between *countercurrent* systems, in which both phases are replenished with fresh material not containing the sample, and *crosscurrent* systems, in which only one phase is replenished. Elution chromatography comes in the first category; an example of the second is the extraction of a substance by successive contact with fresh portions of an immiscible solvent.

Complete separation by either crosscurrent or countercurrent method is theoretically impossible, but separation can be carried to any required degree, so that high purity can be attained at the expense of yield.

It must be emphasized that the present discussion is general; the mechanism of transfer of the sample between phases may involve ion-exchange, surface adsorption, solubility, volatility, or other phenomena.

CROSSCURRENT SEPARATIONS

Let us assume that a substance A exists initially solely in phase 2. A portion of phase 1 is added and the system equilibrated, following which the fractions of the total quantity of A present in the two phases will be $p_{(A)}$ and $q_{(A)}$. The phases are then separated mechanically, and a new portion of phase 1 added and equilibrated. The fraction remaining in phase 2 is now $q_{(A)} \cdot q_{(A)} = q_{(A)}^2$; the fraction in the combined portions of phase 1 is $(1 - q_{(A)}^2)$. The process can be repeated as many times as necessary, say n times. The fraction remaining in phase 2 and the total fraction extracted become

$$q_{(A)}^n = \frac{1}{(K_{(A)} + 1)^n}$$

$$p_{(A)\text{total}} = 1 - q_{(A)}^n$$

(20-7)

These relations, together with Eq. (20-4), can be used to determine the number of equilibrations necessary to effect a desired degree of separation, or to determine the separation attainable by a given number of equilibrations, provided that K is known.

If two substances, A and B, both initially in phase 2, are to be separated, it follows that the ratio of the fraction of A remaining in phase 2 to the fraction of B also remaining in phase 2 after n stages is

$$\left(\frac{q_{(A)}}{q_{(B)}}\right)^n = \left(\frac{K_{(B)} + 1}{K_{(A)} + 1}\right)^n \tag{20-8}$$

and the ratio of the total amounts passing into phase 1 is

$$\frac{p_{(A)\text{total}}}{p_{(B)\text{total}}} = \frac{1 - q_{(A)}^n}{1 - q_{(B)}^n} \tag{20-9}$$

Stepwise crosscurrent separations are applicable to the extraction of a substance from a liquid solution or from a porous solid by an immiscible liquid, or to extraction from a liquid by ion exchange, when these are operated batchwise. Separation by simple distillation also falls into this category. In each case, equilibrium conditions can be assumed to be attained at every step.

In some cases crosscurrent methods can be applied on a continuous basis, as in the removal of oxygen from a solution by bubbling nitrogen through it, or in various liquid-liquid extractors. The extraction of solubles from ground coffee beans in a percolator is an example, usually of only peripheral interest to analytical chemists.

COUNTERCURRENT SEPARATION

This is a powerful method, which is capable of separating in a reasonable time substances with more nearly identical distribution ratios. Whereas in crosscurrent methods only one of the phases is continually renewed, countercurrent calls for both to be renewed. One can visualize the sample standing still while phase 1 moves by it in one direction, phase 2 in the opposite. Components of the sample, if they differ at all in their K values, will have different tendencies to be pulled one way or the other.

It is seldom convenient, at least on a laboratory scale, to have both phases actually mobile; one is usually stationary while the other moves past it. The sample components move also, but at a slower rate than the mobile phase, as their partial affinity for the fixed phase tends to hold them back.

We will first consider stepwise equilibrations. Let us assume that we are dealing with an indefinitely long row of beakers, test tubes,

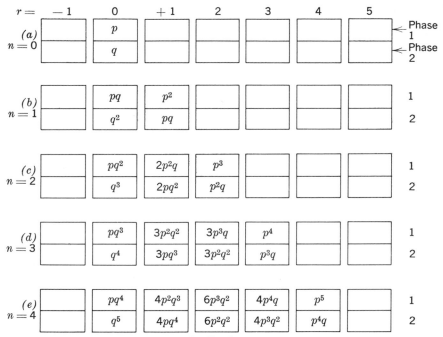

Fig. 20-1   Stepwise countercurrent distribution.

separatory funnels, or other appropriate containers, which we will refer to as "tubes." We will represent these schematically in Fig. 20-1 as rectangles composed of two segments corresponding to the two phases. The tubes are designated by "r numbers." Substance A is present initially only in tube $r = 0$, distributed in the ratio of p to q in accordance with Eq. (20-2). (For the moment we will simplify by omitting the subscript A.) Now let each segment of phase 1 be moved one space to the right with respect to phase 2 (or the equivalent operation, moving phase 2 to the left), as at (b) in the figure. After the transfers, each segment is reequilibrated, with the results shown in terms of p and q; the total amount of A in position $r = 1$ must be p, and when this equilibrates with a fresh portion of phase 2, the ratio in the two phases must be $K = p/q = p^2/qp$. Analogously, in position $r = 0$, the quantity q will distribute itself in the ratio $p/q = pq/q^2$.

If we now repeat the relative motion, we obtain the sequence shown at (c). In this case, the total amount in tube $r = 1$ is the sum of pq (coming from $r = 0$ in phase 1) plus another pq (from $r = 1$ in phase 2), giving $2pq$, which reequilibrates in the ratio of $K = p/q$, or $p(2pq) = 2p^2q$ in phase 1 and $q(2pq) = 2pq^2$ in phase 2.

By similar reasoning, the quantities after further transfers and equilibrations are as shown at (d) and (e) of the same figure. It will be noted that the numerical coefficients are identical to those of the binomial expansion.* The quantities in successive positions of the upper phase (phase 1) are $p(p + q)^n$ and in the lower phase $q(p + q)^n$, where n is the number of transfers, r is the tube number, and where

$$(p + q)^n = \frac{n!}{r!(n - r)!} \, p^r q^{(n-r)} \tag{20-10}$$

This equation will give directly the total fractional amount of the corresponding substance in tube r after n equilibrations. It will be more conveniently utilized if we substitute the value of K in terms of p and q from Eqs. (20-3), to obtain

$$(p + q)^n = \frac{n!}{r!(n - r)!} \frac{K^r}{(K + 1)^n} \tag{20-11}$$

We are now in a position to calculate the relative fractions of two substances, A and B, present in any tube r after n equilibrations. Let us take, for example, substances for which $K_{(A)} = 0.10$ and $K_{(B)} = 12.0$, and calculate the total fraction of each present in tube $r = 2$ after $n = 10$ transfers. The factorial portion of Eq. (20-11) becomes $10!/2!8!$ which equals 45. The equation then tells us that the total fractional amount of A is $(45)(0.10)^2(1.10)^{-10} = 0.174$, and of B is $(45)(12)^2(13)^{-10} = 4.71 \times 10^{-8}$. Thus, given equal quantities of A and B to start with, there will be about 30 million times more A than B in tube 2 after 10 stages.

The figures for the same substances at $r = 8$ after 10 stages show that here there will be about 1 million times more B than A. The fractional amounts of A and B (with the K values specified above) are plotted in Fig. 20-2, as calculated for each value of r, for both $n = 10$ and $n = 5$. Note that for $n = 10$, positions corresponding to $r = 4$, 5, and 6 contain no significant quantity of either A or B, so that 10 transfers and equilibrations represent a waste of time and effort. Five stages $(n = 5)$ would achieve essentially complete separation, since at $r = 2$ there is negligible B and at $r = 3$, negligible A. The two curves are not quite symmetrical because $K_{(A)}$ is taken (intentionally) not exactly the reciprocal of $K_{(B)}$, though not far from it.

A more complex case is illustrated in Fig. 20-3, the attempted separation of three substances with $K_{(A)} = 0.90$, $K_{(B)} = 1.15$, and $K_{(C)} = 12.0$, calculated only for $n = 10$. Separation here is not adequate, particularly for substances A and B, which are not resolved to any

* For example, $(a + b)^4 = a^4 + 4a^3b + 6a^2b^2 + 4ab^3 + b^4$.

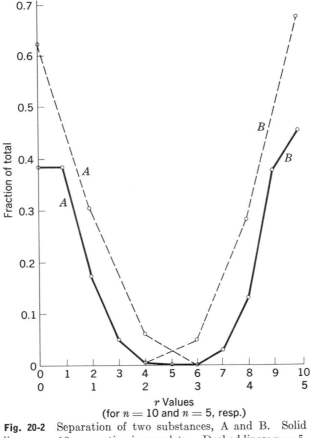

Fig. 20-2 Separation of two substances, A and B. Solid lines: $n = 10$, separation is complete. Dashed lines: $n = 5$, separation is nearly complete. $K_{(A)} = 0.1$; $K_{(B)} = 12$.

useful extent. All three could be separated with a much larger number of stages, but binomial calculations become unwieldy above $n = 20$ or 25.

When the number of equilibrations must be greater than this, calculation can be facilitated by use of the *Gaussian* or "normal" distribution, which can be considered the limiting form of the binomial distribution as n becomes large. The Gaussian distribution can be stated as

$$(p + q)^n = (2\pi npq)^{-\frac{1}{2}} \exp\left(-\frac{r^2}{2npq}\right)$$

$$= (K + 1)(2\pi nK)^{-\frac{1}{2}} \exp\left(-\frac{r^2(K + 1)^2}{2nK}\right) \qquad (20\text{-}12)$$

The standard deviation σ is $(npq)^{1/2}$, wherein it must be noted that n may not always be an integer, since the Gaussian equation is derived on the basis of a continuous function. The advantage of this equation is that tabular data are readily available whereby one can determine the fraction of the enclosed area, hence the fraction of the desired substance, between any selected value of r and r_{max}, the value of r corresponding to the maximum in the distribution curve.

CONTINUOUS COUNTERCURRENT SEPARATION

The most significant analytical methods under this category are the several varieties of *chromatography*. Chromatography involves a mobile phase, either liquid or gas, which passes over the surface of a stationary phase, which may be a solid or may be a liquid immobilized by some method, as by adsorption on the surface of a solid. The sample is inserted at or near the point where contact is first made between the two phases. Its components are then carried along at various rates depending on their relative affinities for the two phases, and if the experiment is successful, are cleanly separated.*

* We are referring to elution chromatography only, and will not discuss the less useful frontal and displacement techniques.

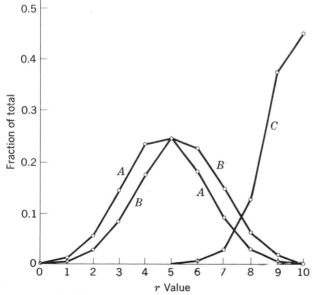

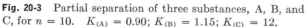

Fig. 20-3 Partial separation of three substances, A, B, and C, for $n = 10$. $K_{(A)} = 0.90$; $K_{(B)} = 1.15$; $K_{(C)} = 12$.

Chromatographic methods are usually classified according to their most obvious features, such as paper chromatography, ion-exchange chromatography, gas-liquid chromatography, etc. Sometimes it is not clear exactly what separation mechanism is operative, and we find non-committal terms such as column chromatography.

Whatever the mechanism, the same general mathematical approach can be taken. In our discussion, it will be convenient to refer to the bed of the stationary phase as the *column*, even though sometimes the "column" will be a sheet of paper.

The sample may be introduced in either phase, as convenient, but always in as small and compact a form as possible, to provide a sharp starting point. As the experiment proceeds, two major effects will be noted: (1) the movement of the *zone* or slug of solute with reference to the column, and (2) the broadening of the zone. Other effects may also be observed, including decreased symmetry of the zone, which may take the form of *tailing*.

The position of the zone is best observed at its maximum; the width is less easily defined. If the shape of the zone is Gaussian (often a good approximation), then its width can be specified in terms of the *standard deviation* σ, which is defined as half the width of the Gaussian curve measured at 0.607 of the maximum height.* The value of 4σ is often taken for the width, since better than 95 percent of the area beneath a Gaussian lies within a width of 4σ, and because this distance is easily measured graphically. The relation between the various quantities can be seen in Fig. 20-4. Note that each peak can be approximated by drawing a triangle with sides tangent at the points of inflection of the Gaussian, which is just the height where σ is defined. This shows that the distance y is equal to 4σ.

It can be shown, by both theory and experiment, that the separation between bands $x_{(A)} - x_{(B)}$ in Fig. 20-4 increases in proportion to the distance traversed, while the width of the band increases as the square root of the distance:

$$\frac{y_{(A)}}{y_{(B)}} = \left(\frac{x_{(A)}}{x_{(B)}}\right)^{\frac{1}{2}} \tag{20-13}$$

From this relation it would seem that by simply increasing the length of the column, any degree of separation could be achieved, since the peaks will draw apart faster than they broaden. This is true, but of somewhat limited usefulness for practical reasons; the longer the column, the longer the time which must be devoted to each experiment. Also "peaks" may broaden to the point where they are hardly detectable.

---

* $0.607 = e^{\frac{1}{2}}$, where e is the base of natural logarithms; this is consistent with the use of the term standard deviation in the previous section.

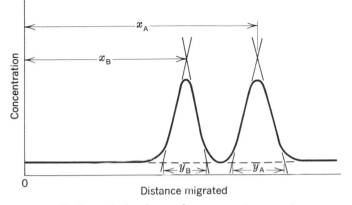

Fig. 20-4 Peak resolution in continuous countercurrent separation.

THEORY OF CHROMATOGRAPHIC MIGRATION

The theory of migration is based on the repeated passage of solute molecules (or comparable entities) back and forth between phases. Any one molecule (on the average) will spend time t_s in the stationary phase and time t_m in the mobile phase, as it passes through a given distance. During the time t_m, it is moving forward with the velocity of the carrier v; during time t_s, it is not moving forward at all. Its motion, then, is stepwise, as it passes into and out of the mobile phase. The relative magnitudes of t_s and t_m will determine how quickly the solute moves along the column. The ratio t_s/t_m is equal to the distribution ratio K previously defined [Eq. (20-2)].

The various factors contributing to the efficiency of separation may be examined conveniently by the HETP (height equivalent to a theoretical plate) approach. A "theoretical plate" is a fictitious concept, which does not correspond to any actual entity in the column. For evaluation purposes it is a very convenient parameter, which is its only *raison d'être*. It is defined as that length of column which will yield an effluent in equilibrium with the mean concentration over that length in the stationary phase. For high efficiency, a large number N of theoretical plates is desirable, and to avoid very long columns, the HETP should be as short as possible.

The HETP relates directly to the width of a peak; in fact it can be shown by statistical considerations to be equal to σ^2/x, where x is the distance traveled, as in Fig. 20-4. It is more convenient to measure y than σ on a recorded chromatogram, so we can use the expression

$$H = \frac{y^2}{16x} \tag{20-14}$$

(where H is the average HETP over the distance x) and the corresponding expression for the number of theoretical plates:

$$N = \frac{x}{H} = 16 \left(\frac{x}{y}\right)^2 \tag{20-15}$$

From the theoretical approach,* it has been shown by van Deemter et al. that the broadening of a peak is the summation of effects from several sources, not completely independent of each other. Expressed in terms of the equivalent HETP, the *van Deemter equation* (in slightly simplified form) is

$$H = A + \frac{B}{v} + Cv \tag{20-16}$$

where A, B, and C are constants for a given system, and v is the rate of flow of the carrier phase.

The A term arises from geometrical effects involving the size and uniformity of solid grains in a packed column, and the existence of numerous parallel channels of various dimensions through which the mobile phase can flow (the *eddy* effect). The B term, which becomes less important as the flow rate v increases, has to do with the longitudinal diffusion of the solute in either or both phases. The last term, Cv, which predominates at higher flow rates, is contributed by transverse diffusion in the mobile phase, as from one channel to another, and by the kinetic effects concerned with the transfer of solute between phases. The C term is sometimes ascribed to the departure of the system from true equilibrium, or to the resistance to mass transfer between phases (which are different ways of saying the same thing).

A fourth term must be added if more precision is required; this involves second-order interactions between the previously mentioned factors.

The van Deemter equation is plotted in Fig. 20-5, to show the qualitative relations between the three terms. There is an optimum flow velocity v_{opt} for the system where H will be a minimum. In practice this is usually determined by trial and error. The optimum velocity will not be the same for different substances in a mixture, and should be selected for the component most difficult to separate. Chromatographers sometimes choose to use a velocity greater than v_{opt} in order to cut the time of analysis, even though sacrificing some resolution.

RESOLUTION

The resolution is a measure of the success in separating zones of similar substances A and B. It is obviously related to the separability α [Eq.

* The mathematical theory is readily available in several reference works.

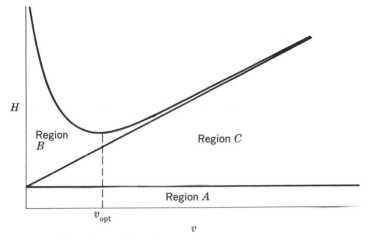

Fig. 20-5 Plot of van Deemter equation.

(20-5)], to the difference between the two peaks of Fig. 20-4, which we can label $\Delta x = x_{(B)} - x_{(A)}$, and to the widths of the peaks, $4\sigma_{(A)}$ and $4\sigma_{(B)}$. Since the peaks are close together, their widths will be essentially equal, and we can drop the subscripts A and B. The resolution can be defined as $\Delta x/4\sigma$, which can be shown to be equal to $N^{1/2}(1 - \alpha)$. To obtain a specified degree of resolution the condition which must be fulfilled is

$$ N^{1/2} = \frac{\Delta x}{4\sigma}\left(\frac{1}{1 - \alpha}\right) \tag{20-17} $$

For example, if two substances with $\alpha = 0.9$ are to be separated with unit resolution (that is, $\Delta x/4\sigma = 1$), $N^{1/2} = 1/(1 - 0.9)$ and $N = 100$ theoretical plates. If $\alpha = 0.99$, 10^4 plates would be required.

Note that the quantities x and σ must be expressed in the same units, but these may be distances measured along the column, retention volumes, elution times, or simply distances taken from a recorder chart.

COMPARISON OF GAS AND LIQUID CHROMATOGRAPHY (GC AND LC)

It is interesting to compare these two varieties of elution chromatography, which both follow the theoretical treatment just outlined. It is obvious that the fields of applicability are different, though many solutes can be handled either way. The specific details of each will be treated in later chapters. We are concerned here with the factors which determine the efficiency and resolution of a column.

The *length* of a column for LC is ordinarily from 10 to 100 cm, while for GC, packed columns are usually longer, perhaps 1 to 40 m; in GC it is possible to use an open capillary column of much greater length, up to the order of 1 km.

The *diameter* of packed columns is comparable for LC and GC, 5 to 10 mm; the long GC capillaries run 0.25 to 0.5 mm.

The HETP for optimum flow rate turns out to be of the same order of magnitude for packed columns, and roughly comparable to the average diameter of the particles of the packing material. Some representative values are given in Table 20-1.

Table 20-1   HETP values for representative chromatographic systems*

System	Solute	HETP, mm
1. Ion-exchange		
Dowex 50–HCl	Na⁺, K⁺	0.35
Dowex 1–formate buffers	Ribonucleotides	0.66
CM-cellulose–phosphate buffers	Corticotrophins	0.67
Amberlite IRC-50–phosphate buffers	Lysozyme	1.5
2. Liquid-liquid partition		
Celite-water-cyclohexane	Phenols	0.27–0.33
Celite-water-methanol-hexane	Steroids	0.75
Celite-water-methanol-cyclohexane-benzene	Aldosterone	0.75
Celite-Cellosolves–aqueous buffers	Insulin	0.2–0.5
Celite-0.1 N HCl–2-butanol	Insulin	0.1–0.5
Sephadex G–25	Uridylic acid	0.1–0.5
3. Vapor-liquid partition		
Nylon capillary–dinonylphthalate	Hydrocarbons	0.5
4. Adsorption		
Silica-ether-ligroin	Nitro compounds	0.85

* Data from C. J. O. R. Morris and P. Morris;[2] references to the original literature are there given.

The *number of theoretical plates*, as determined from the ratio of the column length to the HETP, varies between 100 and 1000 for LC, 1000 and perhaps 50,000 for GC with packed columns. Values up to 10^6 have been attained with capillaries.

The *flow rate* for a gas is considerably greater than for a liquid, for comparable columns, largely because of the viscosity difference (about a hundredfold).

The *speed* of separation is proportional to the value of the van Deemter C term, which may be 10^3 to 10^5 smaller for a gas than a liquid mobile phase. The simplified equation is

$$t_{\text{zone}} = NC(K + 1) \tag{20-18}$$

where t_{zone} = time required for solute to move down length of column

N = number of plates in same length of column

K = distribution ratio [Eq. (20-2)]

This makes GC much faster than LC for substances fairly easy to separate, so that an excessively large value of N will not be required. However, for two solutes which are very similar to each other, another factor must be appended to the equation, which is a rather complex function of the pressure drop across the column, and which favors liquid chromatography for such cases.

The *resolution* $\Delta x/4\sigma$ can also be expressed as $(L/H)^{1/2}(\Delta x/X)$, where X is the mean of x_1 and x_2. So for a column of a given length L, the resolution will change with the inverse square root of the HETP, for which numerical orders of magnitude have been given above.

REFERENCES

1. Giddings, J. C.: Principles and Theory, in "Dynamics of Chromatography," pt. 1, vol. I, Marcel Dekker, Inc., New York, 1965.
2. Morris, C. J. O. R., and P. Morris: "Separation Methods in Biochemistry," Interscience Publishers (Division of John Wiley & Sons, Inc.), New York, 1963.
3. Rogers, L. B.: Principles of Separations, in I. M. Kolthoff and P. J. Elving (eds.), "Treatise on Analytical Chemistry," pt. I, vol. 2, chap. 22, Interscience Publishers (Division of John Wiley & Sons, Inc.), New York, 1961.

21

Gas Chromatography

This technique is beyond doubt the most extensively employed (for analytical purposes) of all the instrumental separation methods, and so merits first consideration. It provides a quick and easy way of determining the number of components in a mixture, the presence of impurities in a substance, and, in many instances, prima facie evidence of the identity of a compound. The only requirement is some degree of stability at the temperature necessary for the production of vapor. Thus a gas chromatograph (GC) is an essential tool to the chemist who is concerned with the synthesis or characterization of covalent compounds of moderate molecular weight.[1,5]

It is probably fair to say that its greatest usefulness is qualitative or semiquantitative, but with careful calibration, quantitative measurements can also be made.

Figure 21-1 shows schematically the essential parts of a gas chromatograph. There is a great deal of latitude between a basic unit which will serve for many identifications, and a highly sophisticated instrument suitable for the varied and stringent requirements of research. We will

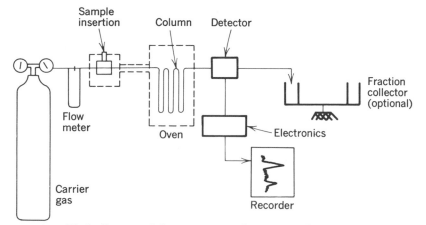

Fig. 21-1 Block diagram of elementary gas chromatograph.

consider first the physicochemical systems—the choice of materials for the fixed and mobile phases appropriate for various kinds of samples—then detailed instrumental features.

THE STATIONARY PHASE

Gas chromatography is divided into two subclasses, according to the nature of the stationary phase. In one (called GSC, for *gas-solid chromatography*) the fixed phase consists of a solid material such as granular silica, alumina, or carbon. The separation process involves adsorption on the solid surface. It is quite limited in applicability, largely because of the tailing caused by nonlinear adsorption isotherms, partly because of the difficulty in reproducing surface conditions, partly because of excessive retention of reactive gases which reduce the available area. Surface catalysis may also play a restricting role. GSC is of chief value in the separation of the permanent gases and low-boiling hydrocarbons.

By far the more important class is GLC, *gas-liquid chromatography*, in which the fixed phase is a nonvolatile liquid held as a thin layer on a solid support. The *solid substrate* generally has no effect on the chromatographic process, and is selected for its ability to hold the liquid film in place. The most common supports are inert porous materials, particularly diatomaceous earths and crushed firebrick. The particle size should be fairly uniform and not too fine. Typical diameter ranges are 60–80 mesh (about 0.25–0.18 mm), 80–100 mesh (0.18–0.15 mm), and 100–120 mesh (0.15–0.13 mm). The smaller the grains, the more pressure is required to force gas through the columns. Occasionally, for special purposes, other packings are selected, such as granular Teflon or glass beads.

In some systems the support introduces complications by partially adsorbing a solute from the liquid phase. The result is sluggish release of the compound to the moving gas, as evidenced by tailing of a peak. This effect can often be reduced by treating the solid with dimethyl-chlorosilane, the same material sometimes applied to glass to make it hydrophobic, a process called *silanizing*.

CAPILLARY COLUMNS

It is possible to eliminate the granular support by using a long capillary of metal, glass, or organic polymer, in which the walls act as support for the stationary liquid phase. Typical dimensions are 0.25 mm inside diameter by 50 m in length, with an HETP of less than 1 mm. The advantages are the ability to handle extremely small samples (<5 μg), and high efficiency in terms of the large number of theoretical plates which can be contained in a given size of oven.

THE LIQUID PHASE

There are hundreds of liquids which have been reported as particularly suited to specific separations. These differ for the most part with respect to their degree of polarity and the temperature range over which they are useful. For the majority of applications a limited number of liquids will suffice. The following list gives the 13 materials which the Perkin-Elmer Corporation has found to provide a versatile selection (in the order of increasing polarity):

Material	*Temp.,* °C
1. Squalane ($C_{30}H_{62}$, branched)	150
2. Apiezon-L grease (A.E.I., Ltd., England)	250–300
3. Didecyl phthalate	165–170
4. Di-(2-ethylhexyl) sebacate	150
5. Methyl silicone oil, low viscosity (DC-200, Dow-Corning)	200
6. Phenyl silicone oil (DC-550, Dow-Corning)	180–220
7. Methyl silicone gum (SE-30, General Electric Co.)	300–350
8. Polyethyleneglycol (Carbowax 1540, Union Carbide)	150
9. Polyalkyleneglycol (Ucon Oil LB-550-X, Union Carbide)	180–200
10. Polyalkyleneglycol (Ucon Oil 50-HB-2000, Union Carbide)	180–200
11. Polyphenylether (OS-138)	200–225
12. Butanediol succinate polyester "BDS"	200–205
13. Diethyleneglycol succinate polyester "DEGS"	205–210

The temperatures quoted are the upper useful limits; they depend on other factors, hence are not to be considered hard limits. Thus some detectors will tolerate larger partial pressure of substrate vapor than will others. Also, it may be permissible to heat a liquid to its upper limit, or even beyond, if it is held there only for a very short time. The lower temperature limit (not listed) depends on such factors as freezing or greatly increased viscosity.

The *polarity* of the liquid phase is not usually specified in terms of dielectric constant, but rather empirically by its ability to separate appropriate compounds under chromatographic conditions. Nonpolar solutes, such as pentane, butane, and propane, can be resolved easily on a nonpolar liquid such as squalane, whereas their peaks fall much closer together on a column of similar dimensions but containing a polar liquid, such as one of the succinates. The converse applies to the separation of polar solutes such as alcohols.

The amount of liquid carried by the solid supporting material is specified in terms of percent *loading* by weight. The usual coating procedure is to dissolve the required amount of the liquid in a volatile solvent, mix it thoroughly with the dried granular solid in an open container, then remove the solvent by evaporation. The solid with its liquid coating has the appearance of a free-flowing sand which can be poured into a long, straight, metal tube (with tapping or vibrating to promote even packing). The tube, fitted with screw-type connectors, is coiled loosely *after* filling.

Capillary columns are usually coated by forcing a small amount of a 10 percent solution of coating material in a volatile solvent through the column.

Another column packing is available which can be considered intermediate between the bare solids of GSC and the coated support of GLC.[11] This is a packing consisting of porous beads of a copolymer of styrene and divinylbenzene. It appears that the components of the sample exchange directly between the gas phase and the porous amorphous beads, the latter acting more like a solvent than an adsorbent. This material gives remarkably clean separations. The maximum permissible temperature is about 250°C.

Any of these columns will usually need further conditioning prior to use. This is accomplished by flushing with nitrogen gas for a few hours at the highest permissible temperature.

CARRIER GAS

By far the most common carrier is helium gas, in spite of its cost. There are two principal reasons for this choice. One of the most useful detectors depends upon the thermal conductivity of the gas, a property which is much greater for hydrogen and helium than for any other gases. Hydrogen has two drawbacks, its fire and explosion hazard, and, more fundamentally, its reactivity toward reducible or unsaturated sample components. The other advantage of helium, also shared by hydrogen, is that because of its low density, greater flow rates can be employed, thus reducing the time required for a separation.

Other gases, such as argon or nitrogen, are required by certain detectors, as will be detailed later.

An interesting possibility is the use of water both as the fixed

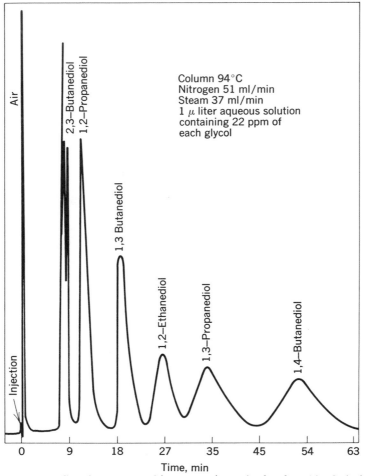

Fig. 21-2 Gas-chromatographic separation of glycols. (*Analytical Chemistry.*)

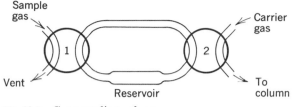

Fig. 21-3   Gas sampling valves.

liquid and as a component (with nitrogen) of the carrier gas.[17]   It is essential that the support be silanized.   This provides one of the best systems for separating very similar polar compounds, such as the homologous alcohols and glycols.   Figure 21-2 shows a chromatogram from such a system and is an excellent example of the simplicity of separations possible with GC.

SAMPLE INJECTION

An outstanding feature of GC is its ability to utilize small samples—from 0.1 to 50 μl of a liquid is usual.   There are three methods of inserting measured samples: by valve, by ampoule, and by syringe.

The valve method is especially convenient for sampling gas streams. Figure 21-3 shows an example, which consists of a pair of identical dual-path stopcocks.   In the position shown, the carrier gas is passed through to the column, which is in a standby condition.   To take a sample, stopcock No. 1 is turned 90°, so that the reservoir (of calibrated volume) is filled with sample gas.   Then No. 1 is returned to its original position, and No. 2 turned through 90°, whereupon the measured quantity of gas is flushed into the column.   Many ingenious modifications of this device have been constructed, to use a single stopcock with multiple openings, or the equivalent valve with a linear sliding motion.

The introduction of samples by ampoule is probably the most precise, but least convenient.   The sample, cooled if necessary, is sealed in a fragile glass ampoule and weighed.   The ampoule is then inserted into a special heated chamber at the head of the column, where it is crushed mechanically while surrounded by flowing carrier gas.   The temperature is such that the sample is vaporized almost instantaneously and swept onto the column.

The syringe technique is the most widely used.   The device employed is essentially the same as the medical hypodermic syringe, and is available in many calibrated sizes which will deliver from 0.1 μl up. The chromatograph is provided with an inlet port closed with a replaceable septum of natural rubber, neoprene, or, especially for high-temperature work, silicone rubber, through which the needle can be inserted. Syringes can be used with gases or with liquids of low viscosity.

In order to obtain narrow peaks it is important to minimize dead spaces in all parts of the apparatus. Also it is essential that the entire sample be swept into the gas stream as nearly instantaneously as possible. To this end the inlet chamber is made small in volume and is heated (for liquid samples) well above the boiling point.

Solids can sometimes be analyzed by GC directly, if they can be melted and handled as liquids, or via pyrolysis. In the latter case, a fragmentation pattern will result, analogous to that discussed in connection with mass spectrometry.

Another path by which nonvolatile organic materials can be studied is by the formation of volatile chemical derivatives. An example of this method is the separation of amino acids as the N-acetyl, n-amyl esters on a Carbowax column.[12] Another useful series of derivatives can be prepared by *silylation*, the insertion of the TMS group, $-Si(CH_3)_3$, or DMS, $-SiH(CH_3)_2$, in place of reactive hydrogen in compounds containing such functional groups as $-OH$, $-COOH$, $-SH$, $-NH_2$, and $=NH$.[20] This was first reported for making volatile derivatives of sugars, but is applicable to many other classes of compounds. Several reagents are available for the formation of these derivatives, including trimethylchlorosilane, $(CH_3)_3SiCl$, hexamethyldisilazane, $(CH_3)_3$-$SiNHSi(CH_3)_3$, N,O-bis-(trimethylsilyl)acetamide (BSA), $(CH_3)SiO$—$C(CH_3)=N$—$Si(CH_3)_3$, and their $-SiH(CH_3)_2$ analogs.

DETECTORS

In principle the measurement of any property which has different values for different gases can be incorporated into a detector for GC, and 15 or more have been described.[19] As pointed out by Halász, detectors can be classed in two major families.[8] In the first family are those which respond to the *concentration* (in mole fraction) of solute in the carrier whereas those in the second family respond to the *flow rate* of the solute (in moles per unit time).

Members of the second family characteristically destroy the sample in the process of detecting it, while those in the first family do not. This can be important, as sometimes it is desirable to collect successive fractions of the solute for further characterization. On the other hand, precise quantitative analysis is more readily implemented with second-family detectors. Due to the fact that the entire sample component is consumed, the integrated area beneath the signal-time (or signal-volume) curve must correspond exactly to the mass m of substance detected. The height of the curve at any point is proportional to the mass flow rate of the sample $v_s = dm/dt$, hence the area beneath the recorded curve is

$$A = \int v_s \, dt = \int \frac{dm}{dt} \, dt = m \tag{21-1}$$

and thus m is obtained directly. Examples of this family are several detectors which depend on combustion of the sample in a flame.

First-family detectors can also give quantitative results, but only by careful control of variables so that unknowns are duplicated by standards, especially with respect to total gas flow rate v (sample plus carrier). The signal from the detector is a measure of the mole fraction x_s of solute, a dimensionless quantity, so the area beneath a peak on the recorded chromatogram is $\int x_s \, dt$. But $x_s = v_s/(v_s + v_c)$, where $v_s + v_c$ is v, the sum of the flow rates of sample and carrier. So we can write

$$A = \int x_s \, dt = \int \frac{v_s}{v_s + v_c} \, dt \tag{21-2}$$

This is proportional to m only if $v = v_c + v_s$ is held constant, which is not a simple matter experimentally (flow regulators exert control over v_c only). Very small samples will have negligible effect on v, so that

$$A = \frac{1}{v} \int v_s \, dt = \frac{1}{v} \int \frac{dm}{dt} \, dt = \frac{m}{v} \tag{21-3}$$

Hence with detectors of this type, if the measurement is to be absolute, the measured area must be multiplied by the flow rate. For determinations relative to a standard, this factor can be included in the overall calibration. First-family detectors include several very useful examples, the thermal-conductivity, argon, helium, and electron-capture detectors, as well as those based on the absorption of radiant energy.

FIRST–FAMILY DETECTORS: THERMAL CONDUCTIVITY

The *thermal-conductivity* (*TC*) *detector* usually consists of a block of metal with two cylindrical cavities machined into it, each cavity provided with a centrally positioned thin-wire filament or thermistor.[14] These resistive elements form two arms of a Wheatstone bridge (R_1 and R_2 in Fig. 21-4). If the bridge is balanced and the total current indicated by the ammeter A is varied by changing R_6, the galvanometer G will continue to show no deflection. An increase in current causes a rise in temperature of both R_1 and R_2, and hence a change in their resistance, upward for a wire, downward for a thermistor; the change is equal for both arms, and the bridge remains balanced. If, however, the gas surrounding one of the resistors is replaced by a different one, the heat developed in the two arms will in general be conducted away through the gases at different

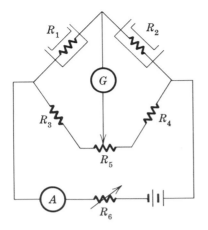

Fig. 21-4 Electric circuit of thermal-conductivity detector. R_1 and R_2 are temperature-sensitive elements, R_3 and R_4 are ratio arms, R_5 is a zero adjustment, R_6 is a current limiter; in practice the galvanometer G is usually replaced by an electronic voltmeter.

rates, the two arms will then be at different temperatures, and the bridge will no longer be balanced. It is assumed, of course, that R_3 and R_4 are identical and preferably of low or zero temperature coefficient. Thus the galvanometer can be made to respond to changes in the thermal conductivity of the gas surrounding R_1 or R_2.

Alternatively, the bridge can consist of four filaments or thermistors, all identical, arranged so that the sample surrounds opposed resistors simultaneously (R_1 and R_4, for example) while the reference gas passes over the other two. This doubles the sensitivity.

The TC bridge is always operated in a differential manner, as indicated schematically in Fig. 21-5. One resistor (or opposed pair) is

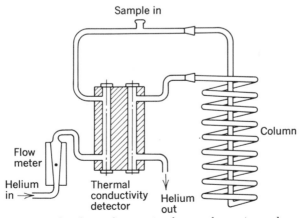

Fig. 21-5 One form of apparatus for gas chromatography. Both column and detector must be enclosed in a thermostated chamber.

exposed to the carrier gas prior to the introduction of the sample, the other to the column effluent. The detector must be held at a temperature at least as high as that of the column to prevent condensation.

It will be seen from Table 21-1 that helium and hydrogen conduct heat considerably better than any other gases. Therefore one of these is the best choice when the detection is by thermal conduction.

Thermal conductance finds wide application in industry apart from

Table 21-1 Some properties of selected gases

Gas	Ionization potential,* eV	Thermal conductivity,† cal sec^{-1}cm^{-1}deg^{-1}	Velocity of sound,‡ m sec^{-1}
He	24.5 (19.6§)	36.0	965
Ne	21.5 (17.6§)	11.6	435
Ar	15.7 (11.5§)	4.25	319
H$_2$	15.6	44.5	1284
N$_2$	15.5	6.24	334
CH$_4$	14.5	8.18	430
CO$_2$	14.4	3.96	259
CO	14.4	5.98	338
Cl$_2$	13.2	2.11	206
SO$_2$	13.1	2.27	213
N$_2$O	12.9	4.13	263
Br$_2$	12.8	1.16	
H$_2$O	12.8	4.25	494 (134°C)
C$_2$H$_6$	12.8	5.12	308 (10°C)
O$_2$	12.5	6.35	316
C$_2$H$_4$	12.2	4.91	317
C$_2$H$_2$	11.6	5.08	
NH$_3$	11.2	5.86	415
NO$_2$	11.0	8.50	
CH$_3$OH	10.9	3.68	335 (97.1°C)
C$_2$H$_5$OH	10.6	3.47	269 (97.1°C)
n-C$_6$H$_{14}$	10.6	3.47	
H$_2$S	10.4	3.68	289
I$_2$	9.7	0.95	
C$_6$H$_6$	9.6	2.56	202 (97.1°C)
NO	9.5	6.20	324 (10°C)

* In C. D. Hodgman (ed.), "Handbook of Chemistry and Physics," 44th ed., pp. 2647–2649, Chemical Rubber Publishing Co., Cleveland, 1962.
† A. P. Hobbs, in L. Meites (ed.), "Handbook of Analytical Chemistry," pp. 4–11 *et seq.*, McGraw-Hill Book Company, New York, 1963. All values are taken at 80°F (27°C).
‡ G. E. Becker, in C. D. Hodgman (ed.), *op. cit.*, pp. 2598–2599. Values for 0°C unless otherwise noted.
§ Electronic excitation energy, eV.

gas chromatography, because the equipment is simple, with no moving parts, and the precision is good.   One of the most extensive applications is in the determination of carbon dioxide in flue gases, which provides a direct indication of furnace efficiency.*

ELECTRON–CAPTURE DETECTORS

For many years TC detectors were used almost exclusively in GC, because of their inherent simplicity and wide applicability.   With the advent of capillary columns, which are limited to smaller samples, and of trace analysis generally, greater sensitivity was required than the TC unit could provide.

One method of detection which can give greater sensitivity is a modification of the ionization chamber long used for radiation detection.[15] The effluent from the chromatographic column is allowed to flow through such a chamber, which is subjected to a constant flux of beta-ray electrons from a permanently installed radioisotope.   A titanium foil containing adsorbed tritium makes the most satisfactory source, though ^{90}Sr (with its daughter ^{90}Y) can also be used.   Both are pure beta sources, which makes for easy shielding against radiation hazard.

Hydrogen is probably the best carrier gas because of its low cross section for electrons, but nitrogen is nearly as good.   The noble gases cannot be used, for reasons which will appear in the following paragraph. The sensitivity for organic solutes is dependent on their large cross section and hence the probability of their being ionized.   The ion current through the chamber will be in the nanoampere region, hence a high-impedance, high-gain amplifier (an electrometer) will be required.

NOBLE-GAS DETECTORS

The outer electrons of helium, neon, and argon, when exposed to a flux of beta particles, are easily promoted from the ground state to an excited metastable level with a half-life of the order of milliseconds.   Collision of a compound gas molecule with a metastable noble gas atom will result in ionization by transfer of the energy of excitation from one species to the other.   This will happen provided only that the excitation energy of the noble gas is greater than the ionization potential of the compound. This results in an extremely sensitive first-family ionization detector of wide applicability.[5]

Table 21-1 includes ionization potentials of a selection of gases, and also indicates the excitation energies of helium, neon, and argon. Either He* or Ne* has sufficient energy to ionize all gases (except He

---

* The thermal-conductivity unit is sometimes called a *katharometer*.   A review of its applications outside of GC is given by Cherry,[3] and a detailed analysis of its chromatographic application has been presented by Lawson and Miller.[14]

and Ne), while Ar* can only ionize those with ionization potentials less than 11.5 eV. The majority of organic compounds (only a few of which are included in the table) ionize at potentials lower than 11.5 eV, while the permanent gases require higher voltages. Argon is more widely employed in this type of detector than helium, because the likely impurities in commercial helium, namely Ne, Ar, H_2, N_2, CH_4, CO_2, H_2O, and O_2, cannot be ionized by active argon, but (except for Ne) can by active helium. Where analyses for these gases are desired, helium can be used if specially prepurified, as by passing through a bed of molecular-sieve material at liquid-nitrogen temperature. Neon is considerably more expensive than helium and has no advantage to offer as a carrier gas.

ULTRASONIC DETECTION

Table 21-1 also lists the velocity of sound in a variety of gases. It will be seen that there is some degree of correlation between this property and thermal conductivity. Hydrogen has the highest value in both, helium the second, and all others much lower. The theory relating sonic velocity to the requirements of a GC detector has been worked out, and the method shown to be capable of good sensitivity and generality.[16] A detector based on this principle has been developed by Micro-Tek.* Figure 21-6 shows schematically how this is designed. Two piezoelectric transducers respectively transmit and receive an ultrasonic wave at a frequency of 6 MHz. The electronic circuitry compares the signal picked up by the receiving transducer with a reference signal from the oscillator. Even a minute change in velocity of the wave will be evidenced as a phase difference, and this can be converted to an output suitable to drive a recorder. The gas volume of the detector is only

* Micro-Tek Division of Tracor, Inc., Austin, Tex.

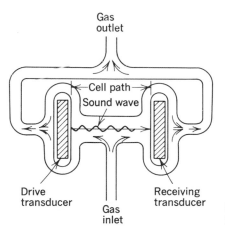

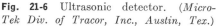

Fig. 21-6 Ultrasonic detector. (*Micro-Tek Div. of Tracor, Inc., Austin, Tex.*)

150 μl. The sensitivity for trace analysis is considerably better than that of the TC detector, but the electronic requirements are more complex.

SECOND-FAMILY DETECTORS: FLAME IONIZATION

Most organic compounds are readily pyrolyzed when introduced into a hydrogen-oxygen flame, and produce ions in the process. The ions can be collected at charged electrodes and the resulting current measured by means of an electrometer amplifier.[5] Figure 21-7 is a diagrammatic sketch of a flame-ionization detector. The carrier gas (usually helium or nitrogen) emerging from the column is mixed with about an equal amount of hydrogen and burned at a metal jet in an atmosphere of air. The jet (or a surrounding ring) is made the negative electrode, and a loop or cylinder of inert metal surrounding the flame is made positive. The sensitivity to organic solutes varies roughly in proportion to the number of carbon atoms. It is generally somewhat less than that of the argon detector. This is, nevertheless, one of the most popular detectors.

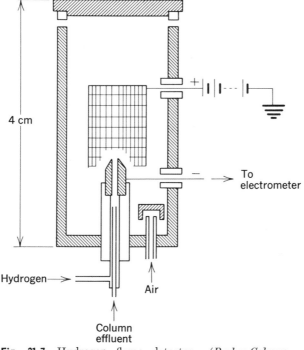

Fig. 21-7 Hydrogen flame detector. (*Barber-Colman Co., Rockford, Ill.*)

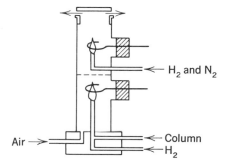

← H$_2$ and N$_2$

Air →

←— Column

←—H$_2$

Fig. 21-8 Schematic of hydrogen flame-ionization detector specific for halogens and phosphorus. (*Analytical Chemistry.*)

DETECTORS FOR SPECIFIC ELEMENTS

The detectors described so far have been nonselective, except in the negative sense that some will not respond to certain gases. There are, however, a few detectors which will signal the presence of specific elements.[2]

Karmen has described a flame-ionization detector which is especially sensitive to the halogen and phosphorus content of organic compounds.[13] His detector utilizes two flames, stacked one above the other (Fig. 21-8). The sample compounds are burned in the lower flame, and the hot products of combustion pass through a screen coated with an alkali halide into the chamber containing the second flame. Each flame is monitored with ionization electrodes. The first detector responds to most organic compounds, whereas the second shows a signal only for those compounds containing halogen or phosphorus. Figure 21-9 shows a comparison of traces from the two detectors.

Subsequently Varian Aerograph, Walnut Creek, Calif., has developed a similar device in which the two flames are combined into one. This amounts to a conventional flame ionization detector but with the burner tip fabricated of a compressed and drilled block of cesium bromide.[10] In this form the detector is several thousand times more sensitive to phosphorus compounds than to other organics, whether containing halogen or not. The reason for this relative sensitivity is not clear. A few picograms of suitable compounds can be detected. The carrier gas is nitrogen. This detector is particularly useful in trace analysis of pesticides, many of which are organophosphorus compounds.

Another specific detector, available from Micro-Tek, is based on the luminous emission from a hydrogen-air flame in the presence of compounds containing sulfur (350 to 450 nm wavelength) or phosphorus (500 to 575 nm).[10] The detector consists of a hydrogen-air burner with a photomultiplier so located as to see only the upper portion of the flame, which does not emit appreciably in the wavelength regions of interest except in the presence of the specified element. Interchangeable optical

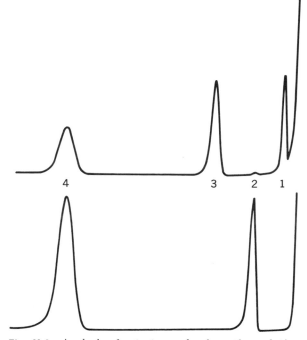

Fig. 21-9 Analysis of a 1-µl sample of an ether solution containing (1) 1 percent acetone by volume, (2) 0.1 percent chloroform, (3) 1.0 percent toluene, and (4) 1.0 percent chlorobenzene. The upper graph is the record of the electric conductivity of the lower flame, the lower graph that of the upper flame of the detector shown in Fig. 21-8. (*Analytical Chemistry.*)

filters enable selection of one or the other of these elements. The sensitivity is comparable to that of the Varian phosphorus detector.

An important feature in all analytical GC is the dead volume of the system, including especially the detector. With a 6 to 8 mm (inside diameter) packed column, the rate of flow of gas passing through is adequate to keep any ordinary detector flushed out. But with a capillary column, the rate may be too slow, so that the contents of the detector become nearly stagnant, thus broadening the peaks which the column has just cleanly separated. It is essential to design the detector to have as little holdup volume as possible, but there are practical limits to this approach. The problem can usually be eliminated by adding an extra supply of pure carrier gas directly to the detector, to keep it flushed out. This is called *scavenging*.

Table 21-2 gives some comparative data regarding the sensitivity

Table 21-2   Comparison of some GC detectors

Detector	Minimum detectable, $g\ sec^{-1}$	Dynamic linear range
Thermal conductivity	10^{-5}	10^4
Ultrasonic velocity	10^{-7}	10^6
Electron capture	10^{-14}	10^3
Argon or helium activation	10^{-12}	10^5
Cross section	10^{-6}	10^4
Flame ionization	10^{-11}	10^7
Phosphorus-sensitive flame ionization	10^{-13}	10^3

and range of a number of detectors.   The figures are illustrative only, and mostly represent average values of comparable items from several manufacturers.

DUAL DETECTION

Since different detectors have different sensitivities for various classes of compounds, added information about samples can often be obtained by the use of two (or even more) detectors simultaneously at the output of the same column.   The tandem flame detector of Fig. 21-8 is an exam-

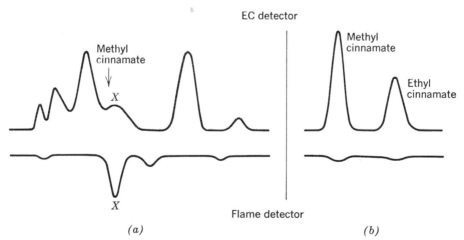

(a)   *(b)*

Fig. 21-10   (a) Dual-detection gas chromatogram of the high-boiling components of peppermint oil.   Peak X was initially thought to be methyl cinnamate (on the basis of the known retention time, indicated by the arrow).   The results in (b) on authentic methyl cinnamate show much greater response on the electron-capture than the flame detector, which rules it out as the source of peak X.   (*Varian Aerograph, Walnut Creek, Calif.*)

ple of this approach, but an unusual one in that the sample is actually destroyed in the first flame.   More commonly, either the first of a tandem pair must be nondestructive, or the two detectors must operate in parallel, with a stream-splitting device to direct part of the gas to one detector, part to the other.   Figure 21-10 shows an example taken from the work of Hartmann et al.,[10] in which dual traces were taken from electron-capture and flame-ionization detectors.   Several peaks can be seen in each trace which are absent or much less pronounced in the other. The relative responses of a number of compounds to these two detectors are given in Table 21-3.   It is evident that a dual record like this can be of great help in the identification of substances.   The fact that a peak occurs in one trace and not in the other can rule some compounds out of consideration, but can never alone prove an identity.

Table 21-3   Approximate ratio of sensitivities: electron capture to flame ionization*

Hexane	10^{-6}	Methyl salicylate	1.2
Carvone	0.01	Diethyl maleate	53
Pulegone	0.01	Diacetyl	53
Menthol	0.1	Ethyl cinnamate	65
Ethyl crotonate	0.9	Benzylideneacetone	65
Benzaldehyde	1.0	Cinnamaldehyde	200
Anisaldehyde	1.0	Carbon tetrachloride	10^6

* Data from Hartmann, Ref. 10.

TEMPERATURE PROGRAMMING[9]

In the separation of a number of compounds of similar type but widely varying volatility, a difficulty arises if the experiment is carried out at constant temperature: the low-boiling components are eluted quickly and bunch together on the record chart, while the less volatile species take much longer and their peaks are much broader and shallower. This can be overcome by increasing the temperature of the entire column at a uniform rate.   The result is that the peaks are more evenly distributed along the chart paper, and are more nearly equal in sharpness. Figure 21-11, taken from the original paper on the subject, illustrates this admirably.[4]

This figure shows an increase in the base line on going to higher temperatures (to the left in the figure), which is characteristic of programmed-temperature GC.   It is due to increased *bleeding* or volatilization of the liquid support as the temperature is raised, which would not be evident if the column were held at one temperature.   In the present

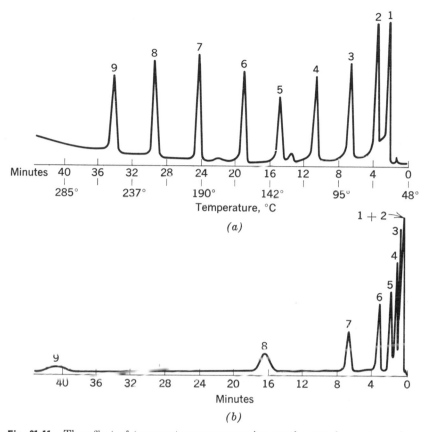

Fig. 21-11 The effect of temperature programming on the gas chromatography of alcohols. (*a*) Programmed temperature chromatogram of (1) methanol, (2) ethanol, (3) 1-propanol, (4) 1-butanol, (5) 1-pentanol, (6) cyclohexanol, (7) 1-octanol, (8) 1-decanol, and (9) 1-dodecanol. (*b*) The constant-temperature chromatogram (165°C) of the same mixture. (*Analytical Chemistry.*)

figure, this effect is not very pronounced, but sometimes it can reduce excessively both sensitivity and accuracy. The effect can be minimized or even eliminated by the use of two parallel columns with identical packing placed together in the same oven. The two columns have identical detectors, or, in the case of the TC detector, the effluent from each column passes through one side of the TC bridge. The carrier gas is split into two streams *before* entering the columns, and the sample is inserted into one side only. The detectors are electrically connected so as to balance out the effect of the bleeding liquid, which is the same in the two columns.

QUALITATIVE ANALYSIS

As suggested above, some degree of qualitative information can be had from observation of the relative sensitivity of various column liquids and various detectors. But this approach will seldom go further than to identify the class to which a compound belongs. For further information, one must turn to observation of *retention times* (or *retention volumes*). This refers to the time which elapses between the injection of the sample and its appearance at the detector. (At constant flow rate, this can equally well be expressed as a volume.) For a given column, flow rate, and temperature, the retention time of a particular gas will be constant, but it is not practicable to transfer such data from one set of conditions to another except by some application of the internal standard principle. Various suggestions have been made as to the best way of doing this,

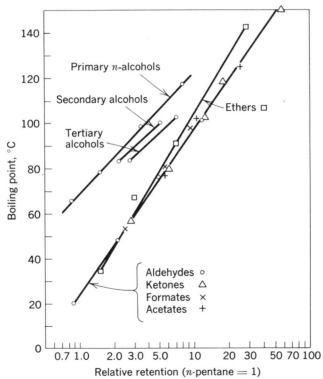

Fig. 21-12 Plot of relative retention of several types of organic compounds as a function of the boiling points. Liquid substrate: Convachlor-12, a high-molecular-weight chlorinated oil. (*Analytical Chemistry.*)

but there is no general agreement at present, perhaps because various schemes are best suited to different situations.

Relative retention times are often specified. In the determination of these values, a standard sample is run through the column before the unknown mixture, and retention times taken as a ratio to the standard. *n*-Pentane is widely used for this purpose, but for work on a polar column at elevated temperature, some other substance, such as methyl palmitate, may be more appropriate. An inherent difficulty is that the relative retention time is often different at different temperatures. An example of the useful relation between relative retention and boiling points of comparable types of compounds is illustrated in Fig. 21-12.[21]

A decided improvement is the *Kováts retention index* system, which presents a uniform scale rather than a single fixed point for comparison.[6] The retention index I was originally defined in terms of volumes, but we will use the equivalent time expression:

$$I = 100 \frac{\log t_x - \log t_{C_z}}{\log t_{C_{z+1}} - \log t_{C_z}} + 100z \tag{21-4}$$

where t_x is the corrected retention time for substance x (that is, the time for x to appear at the detector, less the time required for an unretained gas, frequently air, to pass through), t_{C_z} and $t_{C_{z+1}}$ are the corresponding times for normal hydrocarbons with z and $z + 1$ carbon atoms, respectively. The logs are taken because this function produces a linear scale for successive hydrocarbons. This definition requires that t_x lie between t_{C_z} and $t_{C_{z+1}}$. The index is nearly linear with temperature, at least over short ranges.

SIMULATED DISTILLATION

GC has found a valuable field of application as a replacement for analytical distillation, particularly in the petroleum field. A fractional distillation, as conventionally carried out with a high degree of precision, takes something like 100 hr to accomplish, hence is useless for refinery control purposes. Equally good or even better results can be obtained by dual-column, temperature-programmed GC in only 1 hr.[7] An automatic electronic system continuously integrates the detector signal and prints the accumulated total at intervals of a few seconds. These data, plotted against column temperature, give a curve identical in shape with that produced by the 100-hr distillation. A correction must be made to the temperature scale, if true boiling points are required, because the partial pressure of a component in the carrier gas is not 1 atm, as required in the definition of boiling point. Several manufacturers have distillation simulators in their product lines.

QUANTITATIVE ANALYSIS

It is possible to collect sample components as they are eluted from the column or following passage through any nondestructive detector. The sample collector may take many forms, from a test tube standing in a paper cup full of ice, to elaborate automated and refrigerated sample collectors. The collected substances can be weighed directly or analyzed by any appropriate method. This is not usually done simply for quantitation, but for further study and identification of unknown components.

For known substances, quantitative determinations are generally performed on the recorded chromatogram. If the peaks are sharp and narrow, little error results from simple height measurement. For broader peaks, the included area must be determined. The simplest way is to make use of the triangular approximation of the Gaussian curve, mentioned in the preceding chapter, whereby the area is equated to the height divided by half the base width 2σ (Fig. 20-4).

The recorders provided with many commercial gas chromatographs have built-in mechanical integrators which print directly along the edge of the chart a series of spikes proportional in frequency to the area beneath the curve. Counting the spikes associated with a peak will give a measure of the quantity of corresponding material.

Since integration is an operation required in many fields besides GC, it will be dealt with in some detail in a later chapter.

GC AS A MEMBER OF A TEAM

Gas chromatography can play a valuable role in combination with any other instrumental technique which can accept gaseous or volatile liquid samples and which is compatible in speed. The most important are mass spectrometry and infrared spectrophotometry. As microwave spectrometry becomes more widely employed, a similar marriage with GC will be most appropriate.

In the case of infrared, the ideal procedure would be to record absorption spectra repeatedly over the entire analytical wavelength range in a time short enough that several scans would be covered during the very few minutes that it takes to elute a peak. The difficulty is that common infrared detectors cannot respond fast enough to do this without excessive loss of resolution.* The Beckman IR-102 is designed for this service, and can run a spectrum in as little as four seconds; even with the

---

* An Interrupted Elution GC built in England by Philips (available in the United States through Philips Electronic Instruments, Mount Vernon, N.Y.) is designed to interrupt itself for 15 min following each peak, to give time for the eluted material to be examined in a variety of parallel techniques.[18]

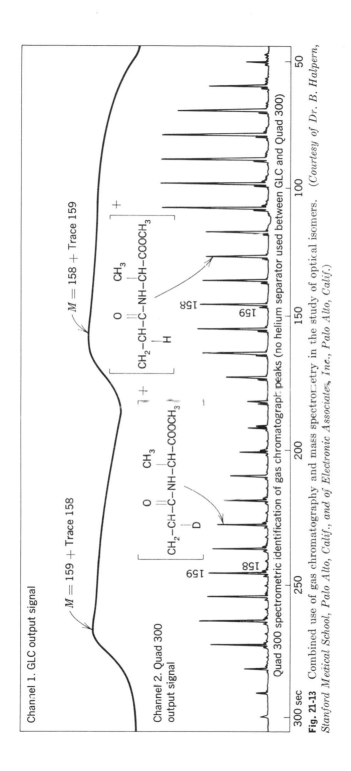

Fig. 21-13 Combined use of gas chromatography and mass spectrometry in the study of optical isomers. *(Courtesy of Dr. B. Halpern, Stanford Medical School, Palo Alto, Calif., and of Electronic Associates, Inc., Palo Alto, Calif.)*

low spectral resolution, identification can often be made. Ultraviolet and visible spectra can be obtained quickly with good resolution, but are not so generally useful for identification.

The first direct combination of GC with mass spectrometry was implemented with the Bendix TOF instrument, which gives a complete mass spectrum almost instantaneously. The quadrupole type is also fully suitable. Figure 21-13 shows an example of the latter, which also illustrates another capability of separation by GC. The sample is a mixture of the two optical isomers of the N-acetyl derivatives of the amino-acid alanine, in which the L-form is labeled with an atom of deuterium. These isomers can be separated on a special column of an optically active nature. The upper trace is the normal signal from the GC detector. From this trace alone there is no way of knowing which peak is due to which compound. The lower trace shows a mass scan repeated at approximately 10-sec intervals, which shows unequivocally that the isomer with the deuterium atom corresponds to the *second* GC peak.

In this particular experiment, as indicated on the chart, no helium separator was used. In many applications such a separator is required because the large excess of carrier gas would interfere with the operation of the mass spectrometer. Helium separators are available which depend on the greater diffusibility of helium compared with the organic species sought. The effluent passes through a jet into a second tube with an open space of a few millimeters between. A large fraction of the helium atoms, but practically none of the heavy molecules, will diffuse out through the space.

PROBLEMS

21-1 It is sometimes advantageous to analyze a hydrocarbon mixture quantitatively by oxidizing the column effluent prior to detection. The water vapor formed can be removed by a trap, and the CO_2 admitted to the TC detector. In a particular experiment, seven components gave peaks with integrated areas as follows:

Peak no.	Compound	Rel. area
1	n-Pentane	2.00
2	n-Hexane	5.72
3	3-Methylhexane	2.21
4	n-Heptane	8.15
5	2,2,4-Trimethylpentane	1.92
6	Toluene	3.16
7	n-Octane	5.05

(a) Point out some advantages and disadvantages of this preoxidation procedure. (b) Should the oxidation step precede or follow passage through the column, and why? (c) Compute the composition of the sample giving rise to the data cited, in terms of mole-percent of total hydrocarbons.

21-2 Show that if the retention time t_R is defined as $t_{air} - t_s$, where t_{air} is the time when an air peak appears (if the column has no affinity for air), and t_s is the amount of time the component spends in the stationary phase, then the ratio t_R/t_{air} is equal to K, the distribution ratio defined in Eq. (20-2).

21-3 A chromatogram shows peaks as follows, in terms of distance from the injection point, measured on the recording paper:

Air	2.2 cm
n-Hexane	8.5 cm
Cyclohexane	14.6 cm
n-Heptane	15.9 cm
Toluene	18.7 cm
n-Octane	31.5 cm

Calculate the Kováts indices for toluene and cyclohexane.

REFERENCES

1. Bennett, C. E., S. Dal Nogare, and T. W. Safranski: Chromatography: Gas, in I. M. Kolthoff and P. J. Elving (eds.), "Treatise on Analytical Chemistry," pt. I, vol. 3, chap. 37, Interscience Publishers (Division of John Wiley & Sons, Inc.), New York, 1961.
2. Brody, S. S., and J. E. Chaney: *J. Gas Chromatog.*, **4**:42 (1966).
3. Cherry, R. H.: Thermal-conductivity Gas Analysis, in D. M. Considine (ed.), "Process Instruments and Controls Handbook," p. 6–186, McGraw-Hill Book Company, New York, 1957.
4. Dal Nogare, S., and C. E. Bennett: *Anal. Chem.*, **30**:1157 (1958).
5. Dal Nogare, S., and R. S. Juvet, Jr.: "Gas-liquid Chromatography," Interscience Publishers (Division of John Wiley & Sons, Inc.), New York, 1962.
6. Ettre, L. S.: *Anal. Chem.*, **36**:(8) 31A (1964).
7. Green, L. E., L. J. Schmauch, and J. C. Worman: *Anal. Chem.*, **36**:1512 (1964).
8. Halász, I.: *Anal. Chem.*, **36**:1428 (1964).
9. Harris, W. E., and H. Habgood: "Programmed Temperature Gas Chromatography," John Wiley & Sons, Inc., New York, 1966.
10. Hartmann, C. H., et al.: *Aerograph Research Notes*, Varian Aerograph, Walnut Creek, Calif., fall issue, 1963; spring issue, 1966.
11. Hollis, O. L.: *Anal. Chem.*, **38**:309 (1966).
12. Johnson, D. E., S. J. Scott, and A. Meister: *Anal. Chem.*, **33**:669 (1961).
13. Karmen, A.: *Anal. Chem.*, **36**:1416 (1964).
14. Lawson, Jr., A. E., and J. M. Miller: *J. Gas Chromatog.*, **4**:273 (1966).
15. Lovelock, J. E., G. R. Shoemake, and A. Zlatkis: *Anal. Chem.*, **36**:1410 (1964).
16. Noble, F. W., K. Abel, and P. W. Cook: *Anal. Chem.*, **36**:1421 (1964).
17. Phifer, L. H., and H. K. Plummer, Jr.: *Anal. Chem.*, **38**:1652 (1966).

18. Scott, C. G.: *Nature,* **209**:1296 (1966).
19. Seligman, R. B., and F. L. Gager, Jr.: Recent Advances in Gas Chromatography Detectors, in C. N. Reilley (ed.), "Advances in Analytical Chemistry and Instrumentation," vol. 1, p. 119, Interscience Publishers (Division of John Wiley & Sons, Inc.), New York, 1960.
20. Sweeley, C. C., R. Bentley, M. Makita, and W. W. Wells: *J. Am. Chem. Soc.,* **85**:2497 (1963); see also W. W. Wells, C. C. Sweeley, and R. Bentley, Gas Chromatography of Carbohydrates, in H. A. Szymanski (ed.), "Biomedical Applications of Gas Chromatography," p. 169, Plenum Publishing Corp., New York, 1964.
21. Tenney, H. M.: *Anal. Chem.,* **30**:2 (1958).

22
Liquid Chromatography

In this chapter we shall consider those forms of chromatography employing a liquid rather than a gas as the moving phase. The stationary phase may be a solid or a liquid on a solid support. The mechanical form may be a packed column, or it may be a thin layer of permeable material. The mechanisms responsible for distribution between phases include surface adsorption, ion exchange, relative solubilities, and steric effects.

COLUMN CHROMATOGRAPHY

For this technique, a vertical glass tube may be employed as container (1 by 30 cm is typical). It is usually constricted at the lower end, and a disk of fritted glass or some equivalent support inserted to hold up the packing material. The packing, in the form of a fine, granular solid, may be any of a large variety of substances if it is to function as adsorbent, or an inert hydrophilic material such as hydrous alumina or silica if it is to support an aqueous stationary liquid, or an ion-exchange resin if the solutes to be separated are ionic. It is essential that the solid material be firmly and evenly packed. This is often most easily accomplished by pouring it into the column as a slurry in a suitable solvent. Back-flushing with solvent is a useful procedure, especially for eliminating air bubbles which have become entrapped in the bed. Once the column is filled, it is important that the bed always be kept covered with solvent, as otherwise air bubbles will enter, which usually will require shutting down operations to perform a back-flush. With care a filled column

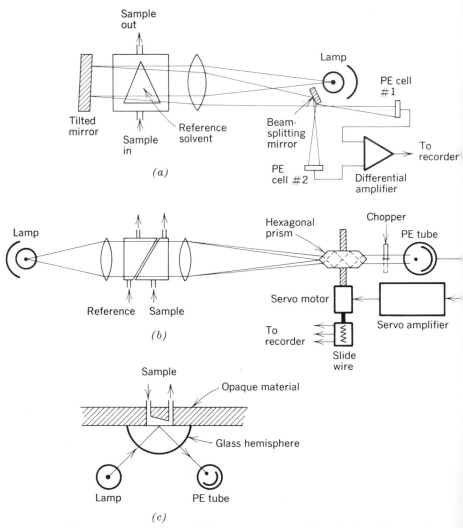

Fig. 22-1 Three forms of differential refractometers: (*a*) Model R-4 (*Waters Associates, Framingham, Mass.*); (*b*) Phoenix (*Phoenix Precision Instrument Co., Philadelphia*); (*c*) design of Johnson et al. (*Analytical Chemistry.*)

can be used repeatedly for a long period of time, until it eventually becomes clogged or coated with some substance which cannot be eluted. The column is operated in a manner similar in principle to that for GC: the components of the sample are swept down the column at varying rates and eluted one by one.

REFRACTOMETRIC DETECTORS

One of the most generally applicable detection systems is the differential refractometer,[9a] in which the refractive index of the eluate is continuously compared with that of the pure solvent.   Figure 22-1 shows the principles of comparison used in three models of refractometer designed for this service.   Those shown at (a) and (b) are both based on the angular displacement of a light beam on passing through two liquid-filled prisms. If the two liquids are identical, the displacement will be nil, but even a change of a few parts per million in the refractive index will make a detectable difference in the angle of the beam.   The Waters* model (at a) records continuously the difference between the signals received by two photocells.   The Phoenix† (b) utilizes a novel optical-null servo system which acts to drive a hexagonal prism into a central position with respect to the light beam; only if the light flux entering on one side of the apex of the prism is equal to that on the other will the two exit beams have equal intensities.   These beams are chopped so that the servo-amplifier will see an ac signal whenever the prism is off-center and will drive the motor to restore balance.

The refractometer cell shown at (c) operates on a different principle. When a light beam strikes an interface at less than the critical angle, the fraction of the incident power which is reflected depends on the refractive indices of both media.[5]   If one of them (the glass) remains constant, the power of the reflected beam will measure the index of the other (the flowing liquid).   The complete instrument utilizes two identical cells like that shown, one for the solvent, one for the eluate.   Two phototubes are connected to read the difference in powers reflected at the two cells.   This design has the advantage that its holdup volume is much smaller, of the order of 0.04 ml, as compared with 1 ml or more for the commercial instruments mentioned.

These differential refractometers have about equivalent sensitivity, of the order of 10^{-6} refractive index unit, which would correspond to a few parts per million of an organic solute in water (this estimate based on handbook values for aqueous maltose at 25°C).

---

* Waters Associates, Framingham, Mass.
† Phoenix Precision Instrument Co., Philadelphia.

OTHER DETECTORS

Karmen has shown that a useful detector for liquid chromatography can be devised by adaptation of well-established GC detectors.[6] He arranged a method whereby the effluent from the column is gradually heated in a stream of nitrogen which is then fed into a flame-ionization detector. The solvent is evaporated in a preheater so that it does not enter the flame. The various solutes present produce nanoampere currents in the ionization chamber. The resulting trace resembles that from GC, but with somewhat more tailing. Good results were obtained with such compounds as volatile esters, slightly volatile steroid esters, and nonvolatile triglycerides. The less volatile substances were pyrolyzed prior to entry into the detector.

Many other detectors can be employed in special cases. Such devices include adaptations of ultraviolet or visible absorptiometers (infrared is likely to be absorbed too strongly by the solvent), fluorimeters, electrolytic conductance detectors, radioactivity[7] and dielectric constant meters. Many times, however, successive fractions are collected for subsequent analysis by conventional methods, and no direct detector is required.

There are several possible variants in liquid chromatography which have no counterparts in GC. One such is the physical separation of peaks or "zones" of solutes while still on the column (in this modification a packing can be used only once). As soon as the components become adequately separated on the column, the liquid phase is allowed to drain out without replenishment, and the bed, resembling wet sand, is cut up with a knife to separate the zones. Formerly this involved extruding the packing from the glass column by pushing from the bottom with a rod; more recently it has been found convenient to substitute thin-walled plastic tubing ("sausage casing") for the glass, and to cut through the whole column without extrusion. The chief difficulty with these procedures is determining where the zones are located. If they are visibly colored, there is no problem. Some solvents will fluoresce under ultraviolet irradiation and can be located by that means. Otherwise it may be necessary to streak or spray the extruded column with a chromogenic reagent to locate the bands of solute.

Another procedure not applicable to GC is *gradient elution analysis*, whereby the eluting solvent is changed with time, either by gradually mixing a second solvent of greater eluting power with the first, less powerful, solvent, or by a gradual change in pH or other property. The result is quite comparable to linear temperature programming in GC.

ADSORPTION CHROMATOGRAPHY

This was the first of the chromatographic techniques to be examined and utilized systematically. It is principally useful for the separation of nonvolatile organic compounds in nonpolar or slightly polar solvents. The detailed theory of adsorption is much too complex to discuss here. The reader is referred to the extensive literature on the subject; see for example Morris and Morris,[12] and the chapter by Snyder.[16]

The retarding forces, which we lump together under the term "adsorption," may be dipole-dipole interactions, van der Waals forces, or hydrogen bonding. A major difficulty with quantitative separations is the nonlinearity of the adsorption isotherms encountered as the concentration of solute is raised. It is desirable that the linear region (i.e., where the amount adsorbed is proportional to concentration) extend as far as possible. In practice it is found that this linear region is longer for weakly adsorbed substances than for those that are strongly adsorbed. This is interpreted as indicative of discrete adsorption sites active with respect to dipole and hydrogen-bond interactions, whereas the weaker van der Waals forces can be effective over a larger area. Snyder has found that a small percentage of water added to a usually anhydrous column becomes firmly attached to the sites of greatest activity, leaving available the somewhat less active ones, which has the effect of increasing markedly the linear region for organic solutes.

It is useful to tabulate adsorbents and solvents in sequences which roughly indicate their relative effect on the adsorption phenomenon (Tables 22-1 and 22-2).[17] These can only be general guides, because the different mechanisms of adsorption will vary in their contributions from one system to another. Hence one should expect some variation

Table 22-1 Adsorbents	Table 22-2 Solvents
(Least active adsorbents at top)	(Adsorption is greatest from those listed first)
Sucrose, starch	Petroleum ether
Inulin	Carbon tetrachloride
Magnesium citrate	Cyclohexane
Talc	Carbon disulfide
Sodium carbonate	Ethyl ether
Potassium carbonate	Acetone
Calcium carbonate	Benzene
Calcium phosphate	Toluene
Magnesium carbonate	Esters of organic acids
Magnesium oxide	Chloroform, methylene chloride, etc.
Calcium oxide	Alcohol
Silica gel	Water (varies with pH and dissolved salts)
Magnesium silicate	Pyridine
Aluminum oxide	Organic acids
Charcoal	
Fuller's earth	

in the sequence of solvents when used in conjunction with one or another adsorbent, and vice versa.

Table 22-3 is presented as an example of the kind of separations which can be achieved. This summarizes a study by Brockmann on para-substituted stilbenes and azobenzenes.[3] The adsorbent was alumina,

Table 22-3   Separation of para-substituted stilbenes and azobenzenes on alumina

R—COOH*†
R—CONH₂
R—OH
R—NH₂
R—COOCH₃, R—N(CH₃)₂
R—NO₂
R—OCH₃
R—H

* R is either C_6H_5—CH=CH—C_6H_4— or C_6H_5—N=N—C_6H_4—.
† The compounds at the top are most strongly adsorbed.

the solvents benzene or carbon tetrachloride. No significant difference was observed between the two series of compounds.

PARTITION CHROMATOGRAPHY

In this category we consider systems where the granular solid serves as a mechanical support for a stationary liquid phase, just as in gas-liquid chromatography, while a second liquid forms the mobile phase. Because of the requirement that the two liquids be immiscible, it follows that they will differ markedly in degree of polarity. Either the more polar or the less polar liquid may be immobilized. Most commonly a polar solvent, such as alcohol or water, is held on a porous support of silica (diatomaceous earth), alumina, or magnesium silicate. A nonpolar solvent can be held on the same materials following silanization which renders them hydrophobic; this is sometimes called *reversed-phase* partition chromatography.

An example of partition chromatography is the separation of organic acids on a column consisting of methanol adsorbed on silica.[18] The sample of mixed acids is added as a solution in petroleum ether, which is the mobile solvent. The zones corresponding to the separated solutes can be made visible by the incorporation of an indicator such as bromcresol green with the methanol phase.

It is possible to calculate a value for K [Eq. (20-2)] from the observed separations. In the majority of systems the values so obtained are identical with distribution ratios determined by the conventional batch method, which, of course, is an excellent confirmation of theory.

A special situation can occur when powdered cellulose is substituted for the silica. Here, apparently, a layer of water is held by hydrogen-bonding so firmly that it is possible to use flowing water as the mobile phase at the same time that bound water is the fixed phase. The fixed phase acts as though it were a concentrated aqueous solution of a carbohydrate. This system has been employed successfully for the separation of soluble carbohydrates.

MOLECULAR SIEVES AND GEL PERMEATION CHROMATOGRAPHY

The term *molecular sieve* refers to a class of natural and synthetic crystalline materials, including the *zeolites*, which are characterized by a high degree of porosity, with all pores the same size. The synthetic varieties* are available in granular form with pore diameters of 0.4, 0.5, 1.0, and 1.3 nm, which are of the same order of magnitude as the dimensions of low-molecular-weight organic molecules. Molecules appreciably smaller than the pore size readily diffuse into the interior of the grains and are adsorbed there very strongly, while molecules larger than the pores obviously cannot enter at all.

These materials can be used in columns, and provide excellent

* Manufactured by the Linde Division of Union Carbide Corp., New York.

Table 22-4 **Separation of components on Linde molecular sieves*†**

Adsorbed on both 4A and 5A‡	Adsorbed on 5A but not on 4A	Adsorbed on 13X but not on 4A or 5A
Ethane, propane	Propane and higher n-paraffins	Branched paraffins
Ethylene, acetylene	Butene and higher n-olefins	Benzene and other aromatics
Methanol, ethanol, n-propanol	n-Butanol and higher n-alcohols	Branched, secondary, and tertiary alcohols
	Cyclopropane	Cyclohexane
Water, ammonia, carbon dioxide, hydrogen sulfide	Freon-12	Carbon tetrachloride, sulfur hexafluoride, boron trifluoride

* From publications of the Linde Division of Union Carbide Corp., via H. C. Mattraw and F. D. Leipziger, in I. M. Kolthoff and P. J. Elving (eds.), "Treatise on Analytical Chemistry," pt. I, vol. 2, p. 1102, Interscience Publishers (Division of John Wiley & Sons, Inc.), New York, 1961.

† The pore sizes of the sieves are as follows: 4A, 0.4 nm (4 Å); 5A, 0.5 nm; 13X, 1.0 nm.

‡ At temperatures below about $-30°C$, appreciable quantities of carbon monoxide, nitrogen, oxygen, and methane are adsorbed on both 4A and 5A.

removal of small molecules from a flowing stream, allowing the larger ones to pass through unhindered.   Table 22-4 lists some materials which are adsorbed by the several sieves.   It is clear that this type of separation can be of great value, but as it is not based on any distribution ratio, the mathematical relations of Chap. 20 do not apply and the method is not truly chromatographic.

An extension of the molecular sieve concept which does obey the laws of chromatography is known as *gel filtration* or *gel-permeation chromatography*.   The stationary phase consists of beads of porous polymeric material.   One of the most widely used of these is a cross-linked dextran (a carbohydrate derivative) which is manufactured and sold under the name of Sephadex by Pharmacia.*   Others are the copolymer of styrene and divinyl benzene mentioned in the preceding chapter, and a variety of polyacrylamide gels.   These materials have much larger pores than the zeolites, and are saturated with solvent (usually, but not always water) prior to use.   The solvent causes the particles to swell considerably, which is one of the attributes of a gel.   The sieve effect is now of such a magnitude that it is more convenient to specify it in terms of the molecular weight of a solute which is just excluded.   Table 22-5 gives such data for several varieties of Sephadex; the other polymers mentioned have comparable grades.   The exclusion limits are rather wide, because the shape of the molecules (globular, linear, folded, etc.) will have considerable effect.

Table 22-5   Exclusion limits for Sephadex*

Type	Limiting molecular weight excluded
G-25	3500–4500
G-50	8000–10,000
G-75	40,000–50,000
G-100	100,000
G-200	200,000

* Data from Pharmacia Fine Chemicals, Inc.

Figure 22-2 shows the results of a separation of oligosaccharides on a 3 by 120 cm column of Sephadex G-25, 100 to 200 mesh (0.15 to 0.074 mm), eluted with distilled water.[4]   The flow rate was 20 ml/hr and the experiment lasted 40 hr.   The molecular weight of the heaviest fraction

* AB Pharmacia, Uppsala, Sweden, and Pharmacia Fine Chemicals, Inc., Piscataway, N.J.

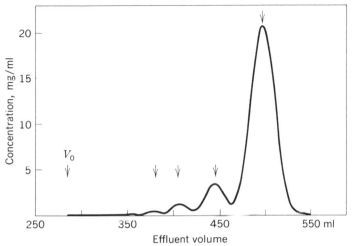

Fig. 22-2 Separation of oligosaccharides in a column of Sephadex G-25, 200 to 400 mesh. Arrows indicate, from left to right, the void volume V_0, isomaltotetraose, isomaltotriose, isomaltose, and glucose. The sugars were identified by paper chromatography. (*Pharmacia Fine Chemicals, Inc., Piscataway, N.J.*)

in this separation, a tetraose, is about 700, much smaller than the exclusion limit for G-25 Sephadex. If a substance of, say 6000 molecular weight were present, it would have passed through the column without retention and produced a peak at the point corresponding to the efflux of one column of water (at V_0 on the graph).

ION-EXCHANGE CHROMATOGRAPHY

Ion-exchange resins consist of beads of highly polymerized, cross-linked, organic materials containing large numbers of acidic or basic groups. Although the resins are insoluble in water, the active groups are hydrophilic and have varying degrees of affinity for ionic solutes. There are four types of resins, which are listed with some illustrative applications in Table 22-6. The useful pH ranges are significant. Below pH 5 the weak acid resins are so slightly dissociated that cation exchange becomes negligible; the converse is true for weakly basic types above pH 9.

Three examples will be described to show the versatility of ion-exchange chromatography. The first is the separation of simple cations on a strongly acid exchanger. For monovalent ions, the relative affinities against water are $Li^+ < H^+ < Na^+ < NH_4^+ < K^+ < Rb^+ < Cs^+ < Ag^+ < Tl^+$ (that is, Li^+ is held least strongly on the resin). A comparable scale for divalent ions is $UO_2^{++} < Mg^{++} < Zn^{++} < Co^{++} < Cu^{++} <$

Table 22-6 Ion-exchange resins for chromatography*

Resin class	Nature of resin	Effective pH range	Chromatographic applications
1. Strongly acidic cation exchange	Sulfonated polystyrene	1–14	Fractionation of cations: inorganic separations; lanthanides; B vitamins; peptides; amino acids
2. Weakly acidic cation exchange	Carboxylic polymethacrylate	5–14	Fractionation of cations; biochemical separations; transition elements; amino acids; organic bases; antibiotics
3. Strongly basic anion exchange	Quaternary ammonium polystyrene	0–12	Fractionation of anions; halogens; alkaloids; vitamin B complexes; fatty acids
4. Weakly basic anion exchange	Polyamine polystyrene or phenol-formaldehyde	0–9	Fractionation of anionic complexes of metals; anions of differing valence; amino acids; vitamins

* From data on Amberlite resins, Rohm & Haas Co., Philadelphia, via Mallinckrodt Chemical Works, St. Louis, Mo.

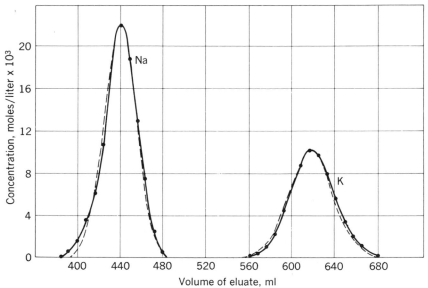

Fig. 22-3 Ion-exchange separation of sodium and potassium on a cation-exchange resin, Dowex 50, eluted with 0.7 F HCl. (*Analytical Chemistry.*)

$Cd^{++} < Ni^{++} < Ca^{++} < Sr^{++} < Pb^{++} < Ba^{++}$. Figure 22-3 shows the complete separation of sodium and potassium.[1] The mixed sample was placed on the top of the column and eluted with 0.7 F HCl. The collected samples were evaporated to dryness to eliminate excess HCl, then redissolved and analyzed by titration of the chloride ion by the Mohr procedure. The error averaged 0.25 mg of alkali halide in samples up to about 350 mg. The dashed curves represent theoretical predictions based on Gaussian distributions. The average HETP was determined to be 0.05 cm.

The next example suggests the possibilities of gradient elution in ion-exchange chromatography (Fig. 22-4), although the figure was the result of a stepwise rather than continuous change in eluant.[8] The sample consisted of a number of transition metal salts; the column contained a strongly basic anion-exchange resin in the chloride form. The column was initially filled with 12 F HCl, and the sample inserted at the top. Elution was carried out by successively more dilute solutions of HCl. The Ni(II) was not retained at all, even in the presence of concentrated HCl, though none of the other metals present moved appreci-

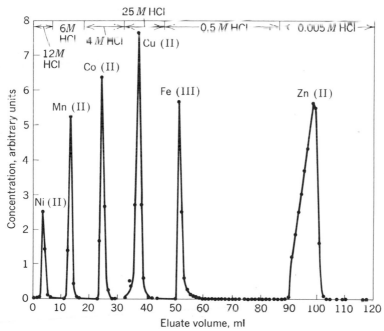

Fig. 22-4 Ion-exchange separation of several transition metals on an anion-exchange resin, Dowex 1, eluted with successively more dilute HCl. (*Journal of the American Chemical Society.*)

ably.   When diluted to 6 F, the acid caused the elution of Mn(II); Co(II) came out at 4 F, Cu(II) at 2.5 F, Fe(III) at 0.5 F, and Zn(II) only at 0.005 F.   This sequence reflects the relative stabilities of the complex chloride anions as well as the varying affinity of the resin and of water for these ions (and for chloride ion).   The chromatography of the lanthanides on a cation-exchange column in the presence of citrate buffers is another outstanding example of a difficult separation involving the interplay of two sets of equilibrium constants.[7]

The third example of ion-exchange chromatography is one which is of great significance in biochemistry: the separation of the amino acids from the hydrolysis of protein.   Since these compounds have both basic and acidic functions, they can be separated on either anion or cation columns, as indicated in Table 22-6.   Many early attempts were made with various resins, and with varying degrees of success, but it was not until the work of Drs. Moore, Spackman, and Stein at the Rockefeller Institute in New York (in 1958) that a practicable separation scheme was achieved.[11]   Their method uses a sulfonated polystyrene resin in a very ingenious semiautomatic system, which includes automatic addition of ninhydrin as color-forming reagent and the measurement of absorbance at two wavelengths, 440 nm which detects proline and hydroxyproline, and 570 nm for all other amino acids and ammonia. Several manufacturers now market apparatus following this basic design.

A number of synthetic inorganic ion-exchangers have been reported and are preferred over organic resins for some purposes.   The reprocessing of nuclear fuels is a significant application, as organic materials are decomposed by the intense radiation.   Some inorganic exchangers are highly specific: ammonium phosphomolybdate shows selective adsorption of Cs^+ over Rb^+ of 28 times, compared with about 1.5 for the most favorable organic resin.

It should be mentioned that ion exchange has important analytical applications other than chromatographic.   These include removal of interfering ions of opposite charge and preliminary increase of concentration of trace quantities of ionic material.   Refer to Samuelson's book for further discussion.[14]

RECYCLING COLUMN CHROMATOGRAPHY

In the separation of substances with close values of the distribution coefficient by the column methods previously described, a very long column may be required.   This is inconvenient to pack and to operate; also the weight of the packing material is apt to crush the resin beads near the bottom, reducing the efficiency markedly.   Porath and Bennich have shown that better results can be obtained by pumping the effluent back onto the same column repeatedly.   It is possible to do this because the zones occupied by the two very similar solutes cover at any moment

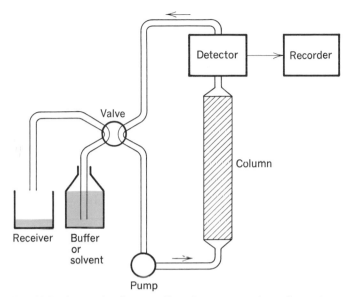

Fig. 22-5 Apparatus for recycling chromatography, schematic.

only a small part of the column, the rest of which is not being utilized. These authors set up a column, a continuous analyzer, sampling valve, and pump in a closed loop (Fig. 22-5). The solution passes upward through the column, opposing gravitation, thus tending to prevent unduly dense consolidation of the packing material.

Figure 22-6 shows results obtained by recycling a biochemical preparation eight times on a Sephadex column.[13] The detector was an ultraviolet absorptiometer at 254 nm. Three cycles could be completed before the advancing first component nearly caught up with the tail of the second component; at this point the valve was opened to simultaneously bleed off component B and admit an equal volume of fresh buffer. A small amount of B remained and was bled off after the sixth cycle. Pure A was pumped out in the eighth cycle. The column height in this experiment was a little less than 1 m; without the recycling feature, a column 8 m high would have been required. This procedure does not shorten the operating time. An apparatus for this service is available from LKB.*

PAPER CHROMATOGRAPHY

As an alternative to the column technique, chromatography can be carried out on a sheet or strip of absorbent paper. Cellulose filter paper

* LKB-Produkter AB, Stockholm, and LKB Instruments, Inc., Washington, D.C.

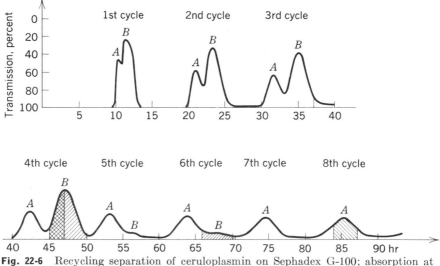

Fig. 22-6 Recycling separation of ceruloplasmin on Sephadex G-100; absorption at 254 nm in a 3-mm cuvet. Hatched areas indicate bleedings of the column. (*Academic Press Inc., New York.*)

is to be preferred for most purposes.   It is so hydrophilic that it normally holds a coating of water (grossly imperceptible) adsorbed from the air. Hence filter paper chromatography is almost always a *partition* phenomenon, an example of liquid-liquid chromatography.

A typical procedure, the analysis of mixed amino acids, can be carried out as follows: A strip of filter paper about 1 by 15 cm in size is laid flat, and a minute drop of the solution to be analyzed is placed in the center, a centimeter or so from one end (Fig. 22-7).   This point is marked with a pencil for future measurement.   The solvent is then evaporated, and the strip suspended from the far end in a tall vessel

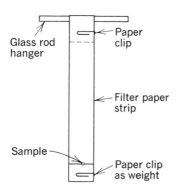

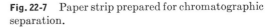

Fig. 22-7   Paper strip prepared for chromatographic separation.

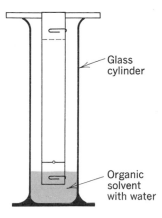

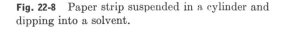

Fig. 22-8 Paper strip suspended in a cylinder and dipping into a solvent.

(Fig. 22-8). The bottom end is dipped into a pool of mixed butanol and water. The top of the cylinder is covered with cardboard so that the hanging strip is bathed in the vapors of both solvents. Upon standing, the liquid climbs up the strip by capillarity, carrying the constituents of the sample along with it at various speeds according to their partition coefficients between the adsorbed water and the interstitial butanol. When the advancing front of the solvent approaches the top, the strip is removed and the point of farthest advance marked upon it. The strip is then dried in air and sprayed with a solution of ninhydrin. A number of spots will be found distributed along the strip; they should be marked in pencil for a permanent record. The individual components can be characterized by their R_f values, where

$$R_f = \frac{\text{distance moved by the component}}{\text{distance moved by the solvent front}}$$

This technique may be extended by the use of large squares of filter paper, perhaps 15 to 30 cm on a side. The sample is applied at a point near one corner and the paper suspended so that one edge dips into the solvent. The constituents of the sample move upward as before. After this step, the paper is dried, turned through 90°, and suspended with the other edge adjacent to the sample corner dipping into a second solvent in which the R_f values are different. This will cause the spots to move across the paper in a direction at right angles to the first motion and hence will produce a two-dimensional chromatogram with correspondingly better separation of a complex mixture (Fig. 22-9).[10] Rectangular glass vessels are convenient for this work, and several sheets of paper may be treated simultaneously.

The preceding describes the *ascending* solvent technique. It is also possible to carry out a paper-chromatographic separation with *descending*

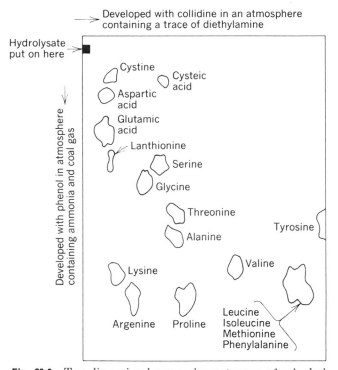

Fig. 22-9 Two-dimensional paper chromatogram of a hydrolysate from wool proteins. A drop of the mixture was placed near the corner as marked; the paper was hung from the long side, and the chromatogram developed with *s*-collidine. The paper was then dried and hung from the short side and the chromatogram further developed with phenol. It was dried again, sprayed with ninhydrin, and heated. The clump of unresolved amino acids in the lower right corner can be separated if benzyl alcohol and butanol are substituted for collidine and phenol, but then the other acids are crowded together. (*Endeavour*.)

solvent.   The paper is hung from a trough of solvent, the whole assembly being mounted in a closed jar in which the air is saturated with solvent vapor to prevent evaporation.

Paper chromatograms can be quantitatively translated into a graph of optical absorbance against position on the paper by measuring the transmission of selected wavelengths directly through the paper.   Several filter photometers have been designed specifically for this application, and in addition paper-holding adaptors are available for several commercial spectrophotometers.

In special instances other physical methods can be utilized to locate and measure the spots, both on paper and on columns. Oscillometry has been used to follow the separation of organic acids. Radioactivity observations with or without added tracers are an important technique. Specially designed counting apparatus is available for this purpose. That shown in Fig. 22-10 is so arranged that the chart drive of a recorder pulls the filter paper strip over the orifice of a windowless flow counter; the activity on the strip is measured and automatically recorded.

An interesting application of two-dimensional chromatography has been described by Blumer, who was interested in analyzing mixtures of porphyrins which contained esters and metal complexes as well as free acids.[2] The esters and complexes were separated first by a carbon tetrachloride–isooctane solvent, which is without effect on the free acids. The unresolved acids were then esterified in situ with diazomethane added dropwise to the spots on the paper. A second development with the same solvent at right angles to the first separated the newly formed esters.

Applications of paper chromatography are legion. Most are of organic and biochemical interest, but inorganic separations are also quite possible, and are receiving increasing attention. In principle it should be possible to devise a complete system of inorganic qualitative analysis based on chromatographic separations, especially by making use of filter paper impregnated with various reagents [15]. In practice, however, this method is best combined with other techniques as it is more advantageous in some separations than in others.

Other kinds of paper are sometimes to be preferred over cellulose.

Fig. 22-10 Automatic recording radioactivity detector for paper strips. (*Picker-Nuclear, White Plains, N.Y.*)

Paper made of fiber glass permits organic materials to be located by a spray of concentrated sulfuric acid, which produces charred spots readily visible on the white surface. Papers are available which are loaded with ion-exchange resins for use with ionic materials.

THIN-LAYER CHROMATOGRAPHY (TLC)

Although paper chromatography could with some justification be classed under TLC, the term is customarily reserved for separations taking place in a coating of powered material adhering to a smooth support such as a glass plate.[9] Any of the solids used in column chromatography can be adapted to TLC, if a suitable binder can be found to ensure adherence to the glass. Separations accordingly may depend on adsorption, partition, ion-exchange, or gel-permeation.

The technique resembles that of paper chromatography with the added requirement of plate preparation. The coating must be as nearly uniform as it is possible to make it, if clean and reproducible separations are to be attained. A uniform particle size is essential. There are a variety of tools on the market to facilitate even spreading of the coating material (as a slurry) onto the plates. Alternatively precoated plates can be purchased.

The sample is spotted at one edge (or end) of the plate and developed by the ascending technique. The plate can then be dried and sprayed with reagent.

An added feature, often useful, is the incorporation into the coating of a fluorescent pigment. Exposure to ultraviolet will produce luminescence everywhere except at those spots where ultraviolet-absorbing solutes are present.

TLC is faster and often more generally applicable than other liquid chromatographic techniques, and so is widely employed as an aid to following a laboratory synthesis, establishing presumptive evidence of purity, etc., just as in GC.

REFERENCES

1. Beukenkamp, J., and W. Rieman, III: *Anal. Chem.*, **22**:582 (1950).
2. Blumer, M.: *Anal. Chem.*, **28**:1640 (1956).
3. Brockmann, H.: *Discussions Faraday Soc.*, **7**:58 (1949).
4. Flodin, P.: "Dextran Gels and Their Applications in Gel Filtration," p. 57, AB Pharmacia, Uppsala, Sweden, 1962.
5. Johnson, H. W., Jr., V. A. Campanile, and H. A. LeFebre: *Anal. Chem.*, **39**:32 (1967).
6. Karmen, A.: *Anal. Chem.*, **38**:286 (1966).
7. Ketelle, B. H., and G. E. Boyd: *J. Am. Chem. Soc.*, **69**:2800 (1947).

8. Kraus, K. A., and G. E. Moore: *J. Am. Chem. Soc.*, **75**:1460 (1953).
9. Maier, R., and H. K. Mangold: Thin-layer Chromatography, in C. N. Reilley (ed.), "Advances in Analytical Chemistry and Instrumentation," vol. 3, p. 369, Interscience Publishers (Division of John Wiley & Sons, Inc.), New York, 1964.
9a. Maley, L. E.: *J. Chem. Educ.*, **45**: A467 (1968).
10. Martin, A. J. P.: *Endeavour*, **6**:21 (1947).
11. Moore, S., D. H. Spackman, and W. H. Stein: *Anal. Chem.*, **30**:1185 (1958); D. H. Spackman, W. H. Stein, and S. Moore, *Anal. Chem.*, **30**:1190 (1958); also summarized in Ref. 12.
12. Morris, C. J. O. R., and P. Morris: "Separation Methods in Biochemistry," Interscience Publishers (Division of John Wiley & Sons, Inc.), New York, 1963.
13. Porath, J., and H. Bennich: *Arch. Biochem. Biophys.*, suppl. **1**:152 (1962).
14. Samuelson, O.: "Ion Exchange Separations in Analytical Chemistry," John Wiley & Sons, Inc., New York, 1963.
15. Schneer-Erdey, A., and T. Tóth: *Talanta*, **11**:907 (1964).
16. Snyder, L. R.: Linear Elution Adsorption Chromatography, in C. N. Reilley (ed.), "Advances in Analytical Chemistry and Instrumentation," vol. 3, p. 251, Interscience Publishers (Division of John Wiley & Sons, Inc.), New York, 1964.
17. Strain, H. H.: "Chromatographic Adsorption Analysis," Interscience Publishers (Division of John Wiley & Sons, Inc.), New York, 1945.
18. Vandenheuvel, F. A., and E. R. Hayes: *Anal. Chem.*, **24**:960 (1952).

23
Solvent Extraction and Related Methods

The separation of substances by selective extraction from one solvent into a second, immiscible with the first, is another method following the general mathematical outline of Chap. 20. In contrast with liquid-liquid chromatography, both liquid phases are in bulk form rather than as adsorbed layers.

Two pure liquids are often applicable, especially for the distribution of somewhat polar covalent compounds as solutes, so that solubility is appreciable in both solvents. In other instances additional solutes in one or both phases will greatly improve the separation achievable. The added substances generally exert their effect through competing equilibria. The pH of an aqueous phase will have a marked influence on the apparent solubility of an acidic or basic substance, hence buffered solutions are of importance.

The formation of metallic complexes with appreciable solubility in nonpolar liquids is likewise important. The complexogen may itself constitute the nonaqueous solvent, as in the instance of the pure liquid acetylacetone, which will extract a number of transition metals from

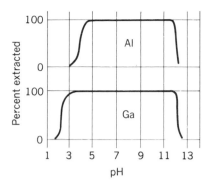

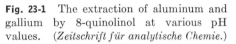

Fig. 23-1 The extraction of aluminum and gallium by 8-quinolinol at various pH values. (*Zeitschrift für analytische Chemie.*)

water at appropriate pH levels to form the complex acetylacetonates. The complexogen may be a solid, hence applied to best advantage in solution. A solution of dithizone in $CHCl_3$ or CCl_4 will extract various transition metals from water, where the nonaqueous solvent alone would be ineffectual. Figure 23-1 shows the pH-dependence of the extraction of two similar elements, aluminum and gallium, into a chloroform solution of 8-quinolinol.[9]

The ligands may be provided in the aqueous phase; an example is the extraction into ether of Fe(III) as the chloride from a solution containing hydrochloric acid.

Table 23-1 gives approximate distribution coefficients for a number

Table 23-1 Selected distribution coefficients

Solute	Solvent 1*	Solvent 2*	K'†
Cl_2	Water	CCl_4	0.10
Br_2	Water	CCl_4	0.044
I_2	Water	CCl_4	0.012
I_2	0.25 F KI	CCl_4	2.1
CdI_2	Water	Ether	5.0
$FeCl_3$	3 F HCl	Ether	5.7
$UO_2(NO_3)_2$	Water	Ether	1.2
CH_3COOH	Water	Benzene	16.0
$CH_2ClCOOH$	Water	Benzene	28.0
$CHCl_2COOH$	Water	Benzene	27.0
CCl_3COOH	Water	Benzene	7.5
Fumaric acid	Water	Ether	0.90
Maleic acid	Water	Ether	9.65
o-Nitroaniline	Water	Benzene	0.016
m-Nitroaniline	Water	Benzene	0.043
p-Nitroaniline	Water	Benzene	0.107

* Each solvent is presaturated with the other.
† K' is the ratio of the concentration of the solute in solvent 1 to that in solvent 2.

of substances which can be extracted from one solvent to another. The *distribution coefficient* K' is related to the distribution *ratio* K of Eq. (20-2) by the ratio of volumes of the two phases:

$$K' = K \frac{V_2}{V_1} \tag{23-1}$$

Combined with Eq. (20-4) this gives

$$K' = \frac{C_1}{C_2} \tag{23-2}$$

where the C's are the equilibrium total concentrations of the substances at equilibrium in the two solvents. Note that the separation factor α, which is of importance in the differential extraction of two substances, is the ratio of either the K or K' values, since the volume ratio cancels.

To a first approximation, the distribution *coefficient* can be taken as equal to the ratio of the solubilities of the substance in the two solvents, as found in reference tables. This convenient relation must be applied with caution, because the conditions in an extraction procedure are seldom identical with those which prevail when solubilities are being determined. Specifically, the two solvents are likely to have finite mutual solubility; a solute would not be expected to have identical solubilities, for example, in pure water and in water saturated with ether.

For quantitative analysis, simple extraction, repeated as many times as necessary, can be used to isolate the desired substance from the impure sample or to remove interfering constituents. Thus uranium can be extracted quantitatively as the tetrapropylammonium trinitrate from a complex mixture of fission products. Ferric chloride can be extracted from an acid solution, but several other chlorides will also pass into the ether. This serves, however, as a convenient way to remove most of the iron from an alloy, in preparation for the analysis of aluminum.

In case only one component of a mixed sample is extractable, then separation can be achieved by repeated extractions with fresh portions of the solvent. Each operation will remove the same fraction of the material present. For example, uranyl nitrate, according to Table 23-1, has a distribution ratio of 1.2, which means that at equilibrium there is 1.2 times as much solute in the water as in the ether (for equal volumes of solvents). If the initial concentration (in water) is 100 g per liter, the quantity remaining after one extraction will be 54.5 g, after two extractions 29.7 g, after three 16.2 g, etc. It will be reduced to 0.1 g after 12 operations.

This process can be accelerated by the use of an apparatus for continuous extraction. One form, designed for removing a nonvolatile solute

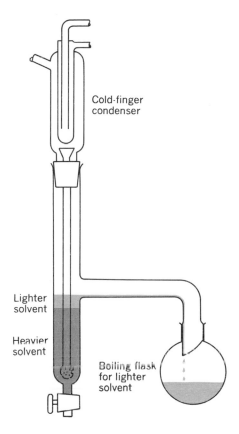

Cold-finger
condenser

Lighter
solvent

Heavier
solvent

Boiling flask
for lighter
solvent

Fig. 23-2 Apparatus for the continuous extraction of a heavy liquid by a lighter one.

from a heavy liquid by extraction with a lighter one, is shown in Fig. 23-2. The heavier liquid, containing the sample, is placed in the lower part of a long vertical tube. A funnel tube with a perforated bulb at the lower end is inserted, also a cold-finger condenser at the top. The lighter solvent is boiled in a flask connected to a side arm. The vapor condenses on the cold finger and is led by the funnel tube to the bottom of the extractor, whence it rises in fine droplets through the heavy solution, extracting solute as it goes, and runs back through the side arm to the boiling flask to be redistilled. Hence the solute collects in the flask, and the original solution is continually in contact with fresh portions of solvent.

COUNTERCURRENT EXTRACTION

More complicated procedures must be employed to separate from each other two or more substances which are simultaneously extracted, but to different extents.

Consider first a mixture of 1000 parts of A with 1000 parts of B, where both A and B are soluble in each of two solvents 1 and 2, with the distribution coefficients

$$K'_A = \frac{C_1}{C_2} = 0.6$$

$$K'_B = \frac{C_1}{C_2} = 1.5$$

This means that A is more soluble in 2 than in 1 and that B is more soluble in 1 than in 2. The distribution between the two solvents after equal volumes are shaken together will be, for A, 375 parts in 1 and 625 parts in 2 and for B, 600 parts in 1 and 400 in 2. After separation each layer is shaken with an equal volume of the opposing solvent, allowed to settle, then separated again, and the process repeated, in a scheme resembling Fig. 20-1, the light solvent from each equilibration combined each time with the heavy solvent from the adjoining vessel. The amount of the solute in each solvent in each tube at each stage can be calculated from the binomial expansion as discussed in Chap. 20.

This method of attack is potentially of great importance in chemistry, as it often permits the separation of substances which are too similar to each other to be resolved readily by other means. For example (see Table 23-1), the several chlorinated acetic acids, as well as fumaric and maleic acids, and the isomeric nitroanilines can be separated quantitatively by the distribution method.

The factor which formerly restricted the application of this technique is the tedium involved in carrying out perhaps several hundred individual extractions with a bank of separatory funnels. The development which made the method practicable was due to Lyman C. Craig and his coworkers at the Rockefeller Institute in New York.[2]

The Craig technique follows exactly the system outlined above, with a special piece of apparatus designed to perform a large number of extractions simultaneously.

One type of apparatus, constructed entirely of glass, consists of a series of identical cells, one of which is shown in Fig. 23-3. Each cell is connected to the two adjoining ones either by a glass seal or by a ground joint. The whole series of cells is mounted on a metal spindle (Fig. 23-4) and can be tilted around a horizontal axis to the three successive positions shown in Fig. 23-3. The equilibration of two liquid phases takes place within each cell as it is oscillated back and forth between positions A and B. After a predetermined number of strokes (5 to 50), the cells are advanced to position C, whereupon the upper liquid drains out through the side arm c into d, which is called a *transfer tube;* the

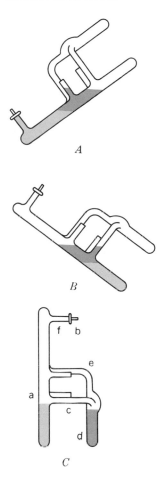

A

B

C

Fig. 23-3 A single stage of the Craig glass counter-current extraction apparatus, shown in three positions. The heavy and light shading represent the heavy and light solvents, respectively.

lower liquid remains in the main tube of the cell beneath the point marked *a*. After the decantation is complete, the cells are returned to position *A*, so that the lighter liquid is fed from *d* through *e* into the next cell in the series. In this manner, each of the two solvents is repeatedly equilibrated with new portions of the other in a continuous counter-current series. Each cell has an opening at *f*, normally closed by a stopper *b*, which is used for filling and emptying. The extractor in Fig. 23-4 has 30 cells; others have been constructed with several hundred cells. The larger assemblies are equipped with an electrical drive which gives the apparatus the preselected number of tilts and separations automatically.

Figure 23-5 shows the kind of results obtainable from a 25-stage apparatus.[8] The curves represent the absorbance, at two wavelengths

Fig. 23-4 Glass apparatus of 30 stages for countercurrent distribution separation. (*H. O. Post Scientific Instrument Co., Middle Village, N.Y.*)

in the ultraviolet, of the contents of each of the tubes after the distribution of a sample between *n*-butanol and 10 percent acetic acid.   The sample consisted of a mixture of substances separated from the urine of patients receiving an antimalarial drug.   This extraction procedure showed unquestionably that the mixture consisted of three components differing in their relative absorbances at the two wavelengths chosen. They could be isolated in a good state of purity from the appropriate tubes of the apparatus.

In Fig. 23-6 is given the distribution pattern for a much more complex mixture.[2]   A synthetic mixture of 300 mg each of 10 amino acids was subjected to fractionation in a 220-cell glass apparatus, between dilute hydrochloric acid and *n*-butanol.   The apparatus was operated automatically until 780 transfers had been made (about 20 hr).   An automatic fraction collector retained separately the portions of light solvent ejected at each decantation.   After the 780 transfers, the effluent samples and the contents of the tubes were analyzed by weighing the hydrochloride residues remaining after evaporating all solvent.   The right-hand part of the figure corresponds to the ejected light solvent fractions, labeled according to the number of the transfer which ejected them.   (No solute can possibly be ejected during the first 220 transfers,

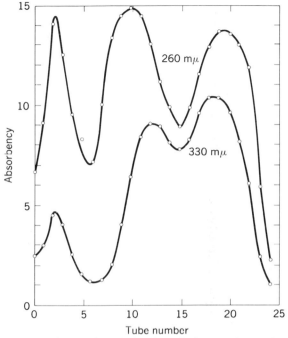

Fig. 23-5 Separation of the degradation products of an antimalarial drug by countercurrent distribution. (*Journal of Organic Chemistry.*)

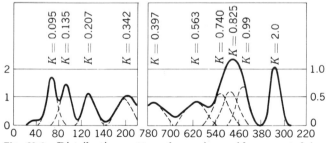

Fig. 23-6 Distribution pattern for amino acids separated in a 220-stage countercurrent apparatus. From left to right the bands represent tryptophane, phenylalanine, leucine, isoleucine, tyrosine, methionine, valine, α-aminobutyric acid, alanine, and glycine. (*Analytical Chemistry.*)

as it is still traveling through the apparatus.)   The left section represents the quantities remaining in the apparatus.   It will be seen that all but three of the amino acids are resolved.   These three could be separated successfully by being returned to the apparatus for continued fractionation.

A number of modifications of the Craig apparatus have been directed toward such features as compactness or ability to handle small volumes. Post and Craig[7] have described a modified tube comparable to that of Fig. 23-3 which will transfer the light solvent in one direction and the heavy solvent simultaneously in the other direction.   Meltzer and his coworkers[6] have constructed an elaborate apparatus for carrying out automatically differential distributions between *three* mutually immiscible solvents, such as heptane, nitromethane, and water.

A major difficulty with solvent extraction is the frequent formation of emulsions which may be difficult to break.   Sometimes a few drops of an antifoaming agent added to the system helps.

BUBBLE SEPARATION METHODS

Recently several separation techniques have been introduced which employ adsorption at the gas-liquid interface of bubbles.[4]   Of these we will discuss briefly the methods known as solvent sublation and foam fractionation.

SOLVENT SUBLATION

In this technique, bubbles of an inert gas such as nitrogen are forced upward through a glass frit into a column containing two immiscible liquids (Fig. 23-7).   The lower solvent, usually aqueous, contains the substances to be separated, along with a small concentration of a surfactant.

As the bubbles move upward through the lower solvent, the surface-active substance will tend to collect at the surface of the bubbles and be carried along into the upper solvent.   Separation can come about in either of two ways.   First, a surface-active component of a mixture can be removed from substances which do not have this property.   Second, an added ionic surfactant will form an ion pair with substances of opposite charge, and thus cause them to be swept into the upper solvent.

Karger et al. have demonstrated both approaches with the dyes methyl orange (MO) and rhodamine B (RB) at pH 10.5, where MO is an anion and RB exists as a zwitterion.[1,3]   RB is easily transported by nitrogen from the basic water to overlying butanol, without the addition of any reagent, which indicates that RB is itself surface-active, at least at this pH.   The MO, however, stays in the aqueous layer unless a

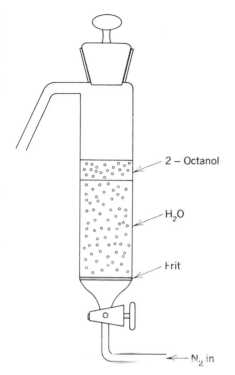

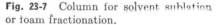

Fig. 23-7 Column for solvent sublation or foam fractionation.

cationic surfactant is added: hexadecyltrimethyl ammonium bromide (HDT) is very effective. The presence of HDT, however, represses the transport of RB, which is interpreted as indicating competition between two surface-active substances (HDT and RB) for adsorption space on the bubble surface.

The mechanism of transport does not depend on any kind of partition equilibrium, hence the mathematics of Chap. 20 does not apply. Indeed, it is quite possible to transport more of a solute from the aqueous to the organic layer than would be predicted by the partition coefficient for the same materials.

Solvent sublation has so recently been brought to the attention of analytical chemists that its full possibilities have yet to be explored.

FOAM FRACTIONATION

This method of removal of specific materials from solution was developed earlier than solvent sublation, but depends on somewhat similar principles.[5] The same apparatus (Fig. 23-6) is applicable, but only a single liquid phase is employed. The flow of nitrogen is controlled so that a foam is produced by any surfactant present at such a rate that it will reach the top of the vertical tube before collapsing.   The selective adsorption of material at the bubble surfaces is similar to that described in the preceding section.

The adsorbate can be increased in amount by refluxing for a period of time (of the order of 15 min).   Then an increase in the rate of gas flow will cause foam to be carried over through a sidearm into a receiver where it is broken either thermally or mechanically.

This method suffers from the disadvantage that an appreciable amount of bulk solvent (with the corresponding concentrations of other solutes) is carried over with the foam.   This is not a significant source of difficulty with solvent sublation.   The lack of a second solvent in the foam method may be considered an advantage in itself.

PROBLEMS

23-1   Copper is extracted by acetylacetone at pH 2 to 5, to the extent of 87.3 percent. How many successive unit extractions would be necessary to remove 99.99 percent of the copper from a copper sulfate solution?

23-2   A volume of 100 ml of acetylacetone containing 70.8 mg of beryllium as the acetylacetonate was shaken for 15 min with 100 ml of acidulated water (pH 3) saturated with acetylacetone.   The aqueous layer was then separated and analyzed gravimetrically.   It was found to contain 1.7 mg of beryllium.   Calculate the distribution coefficient for beryllium.

REFERENCES

1. Caragay, A. B., and B. L. Karger: *Anal. Chem.*, **38**:652 (1966).
2. Craig, L. C., W. Hausmann, E. H. Ahrens, Jr., and E. J. Harfenist: *Anal. Chem.*, **23**:1236 (1951).
3. Karger, B. L., A. B. Caragay, and S. B. Lee: *Separ. Sci.*, **2**:39 (1967).
4. Karger, B. L., R. B. Grieves, R. Lemlich, A. J. Rubin, and F. Sebba: *Separ. Sci.*, **2**:401 (1967).
5. Karger, B. L., R. P. Poncha, and M. M. Miller: *Anal. Chem.*, **38**:764 (1966).
6. Meltzer, H. L., J. Buchler, and Z. Frank: *Anal. Chem.*, **37**:721 (1965).
7. Post, O., and L. C. Craig: *Anal. Chem.*, **35**:641 (1963).
8. Titus, E. O., L. C. Craig, C. Golumbic, H. R. Mighton, I. M. Wempen, and R. C. Elderfield: *J. Org. Chem.*, **13**:39 (1948).
9. Umland, F.: *Z. anal. Chem.*, **190**:186 (1962).

24
Electrical Separation Methods

This chapter deals with separation techniques in which an electric field is utilized to produce or affect the relative motion of charged species in solution. These techniques generally fall into two classes: electrophoresis, where only the electric field causes motion, and electrochromatography, in which the motion of a particle is produced by the resultant of an electric and a gravitational or other nonelectric force. Separation by controlled-potential electrodeposition has been discussed in Chap. 14.

UNSUPPORTED ELECTROPHORESIS

This general method has found its greatest applications in the medical and biochemical fields, in the resolution of mixtures of proteins, nucleic acids, and similar entities. The earliest practical embodiment was in the work of Tiselius at the University of Uppsala, Sweden, first reported in 1937. Tiselius developed a method based on a vertically *moving boundary* between a buffered protein solution and a less dense, supernatant solution composed of the buffer alone. An electric gradient of a few volts per centimeter applied across the boundary causes migration of the protein in the direction dictated by its charge at the given pH, and the boundary moves accordingly.   The position of the boundary is observed by the abrupt change in refractive index.   A complex optical system based on *schlieren photography* is employed to give a result which can be interpreted both qualitatively and quantitatively.

A mathematical treatment shows that the derivative $\partial n/\partial x$, where n is the refractive index and x is distance measured in the direction of migration, follows a Gaussian curve as the boundary is crossed.   Hence in general a mixture of proteins will show, after electrophoresis, a series of Gaussian peaks, more-or-less separated according to the relative mobilities of the various species.   For further theoretical discussion, refer to the reviews by Cann[3] and by Alberty,[1] and for optical details, by Longsworth.[5]

SUPPORTED ELECTROPHORESIS

Electrophoresis can also take place in the pores of a sheet of filter paper or in the medium of a gel.   Figure 24-1 shows the elements of an apparatus for *paper electrophoresis*.   A strip of filter paper tape is laid on a glass table, with the ends dipping into buffer solutions.   A pair of electrodes, connected to the terminals of an adustable dc voltage source, make electric contact through the two solutions.   The strip is moistened with a dilute electrolyte, the sample is placed at the center with a medicine dropper, and a potential of several hundred volts is applied.   The paper strip must be covered in some way to prevent evaporation of the solvent, and, except for the smallest units, must be water cooled, as considerable electric power will be dissipated in the paper.   These two functions can be combined in a cover placed in contact with the paper and provided with cooling coils.

The separation of charged species will depend primarily on their relative mobilities.   In some instances there may be specific interaction between a solute and the paper, which will tend to retard passage along

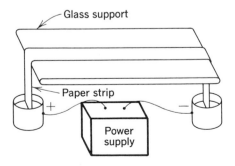

Glass support

Paper strip

+

−

Power supply

Fig. 24-1 Essential features of an apparatus for paper electrophoresis.

the paper, as in chromatography. Note that any chromatographic effects here are acting *parallel* to the electric field.

A typical biochemical application is the analysis of the proteins in human serum, which has been critically discussed by Frijtag Drabbe and Reinhold.[4] They found that with careful attention to details of procedure, including temperature control, preconditioning of the paper, method of applying the sample, and drying, the results were equal in precision to classical methods, and they could be obtained from samples containing as little as 2 or 3 $\times$ 10^{-4} g of total protein, in which less than 5 percent of this amount was one particular protein, determinable with a standard deviation of about 10 percent (i.e., within approximately 1 μg).

This method can also be valuable in the separation of inorganic ions. A supporting electrolyte serves to carry the current. An example is the separation of the ions of Ba and La, and of Ra, Pb, and Bi, carried out in 0.1 F lactic acid with a potential gradient of 3.5 V per cm.[8] In a period of about 24 hr, Ra moved about 100 cm, Ba 90 cm, Pb 50 cm, La 30 cm, and Bi 10 to 15 cm. The locations of the ions were determined by radioautography with natural radioactivity or added tracers. It has been found possible to separate Li and Na from each other and from the other alkali metals in ammonium citrate solution.

A comparable procedure is *gel electrophoresis*. A thin layer of a gel can be cast directly on the bed of an apparatus like that of Fig. 24-1, with filter paper lapping over the ends to act as wicks to make contact with the electrode vessels. The following gels have been found useful: starch gel, agar, agarose, copolymers of polyvinyl chloride and acetate, polyacrylamide, and cellulose acetate.[9]

Porous glass has been employed as a support for electrophoresis.[6]

A disruptive effect which must always be guarded against in electrophoresis carried out in a porous medium is *electroosmosis*. This arises from the charge acquired on the surface of the solid (the *zeta potential*). The charge causes the material to act somewhat like a weak ion-exchanger, attracting oppositely charged ions. There will therefore be a net motion of ions (mostly buffer components) toward the electrode of sign opposite to that of the charge on the solid. This will produce an actual flow of solvent in the same direction, by osmosis, tending to maintain uniform buffer concentration.[7]

Different supports vary in both sign and magnitude of their zeta potentials. Sometimes it is possible to mix two materials in preparing a gel so that the electroosmotic effect of one will be offset by the other. When this cannot be done (as with paper or porous glass), then the effect must be measured and corrected for. Measurement is usually carried out by adding an uncharged species along with the sample and observing how far it moves during the experiment.

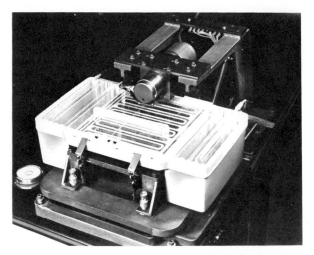

Fig. 24-2 Gel tray for B. & L. Spectrophor-I; photo-multiplier housing above, light source below the tray. (*Bausch & Lomb, Inc., Rochester, N.Y.*)

Figure 24-2 shows the gel-supporting tray and associated parts of the B. & L. Spectrophor-I.* This instrument contains a monochromator (ultraviolet-visible) and photomultiplier, so arranged that the electrophoresis patterns produced simultaneously by as many as eight samples can be scanned sequentially, and the transmittance plotted on a built-in recorder. Figure 24-3 shows a typical trace.

ELECTROCHROMATOGRAPHY

The concurrent application of the two techniques of electrophoresis and chromatography on a paper support has proved very fruitful. The sheet of paper is usually suspended vertically with the solvent (often a buffer) descending from a reservoir at the top. The electric field is applied horizontally via some type of strip contacts along the sides of the paper curtain.

Figure 24-4 shows a commercial unit for continuous separation. The mixture is fed from a capillary to a point near the top of the paper, down which buffer is flowing. A steady state is quickly established, and the components are collected in a series of tubes fed from triangular tabs along the lower edge of the paper. The farthest deviation of a trajectory on the paper from the vertical corresponds to a combination of greatest chromatographic affinity for the stationary phase and greatest electrolytic mobility.

* Bausch & Lomb, Inc., Rochester, N.Y.

The same apparatus can be operated for small samples. A few drops are placed on the paper by pipet and allowed to separate as far as possible without running off the paper, then the voltage is disconnected, the solvent flow stopped, and the paper removed for drying. Subsequent treatment is the same as described previously for paper chromatography.

A critical discussion of design and operational features of paper electrochromatographs has been given by Strain.[10] He considers such features as optimum paper size and shape, and methods of establishing electric contact with the paper while avoiding contamination from electrolysis products. He recommends a tapered sheet of paper, narrower at top than bottom, so that the potential gradient will be greatest near the point where the sample is injected. Figure 24-5 is a drawing of one of Strain's papers showing the effective separation of Ag, Cd, Bi, and Al ions, in a supporting electrolyte of ammonium malonate.

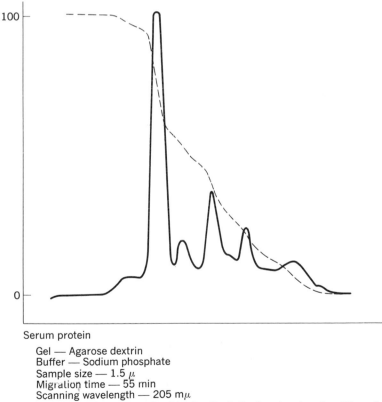

Serum protein

Gel — Agarose dextrin
Buffer — Sodium phosphate
Sample size — 1.5 μ
Migration time — 55 min
Scanning wavelength — 205 mμ

Fig. 24-3 Separation effected with the B. & L. Spectrophor-I. (*Bausch & Lomb, Inc., Rochester, N.Y.*)

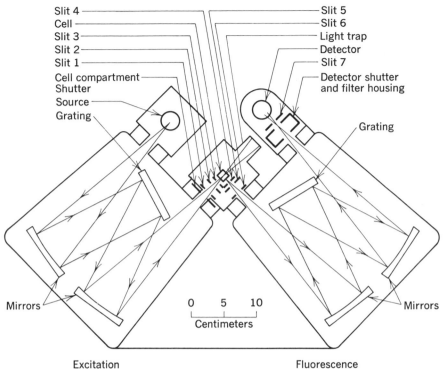

Slit 4 — Slit 5

Cell — Slit 6

Slit 3 — Light trap

Slit 2 — Detector

Slit 1 — Slit 7

Cell compartment — Detector shutter and filter housing

Shutter

Source

Grating — Grating

Mirrors — Mirrors

0   5   10

Centimeters

Excitation   Fluorescence
monochromator   monochromator

Fig. 24-4 Apparatus for continuous-flow electrochromatography. (*Beckman Instruments, Inc., Fullerton, Calif.*)

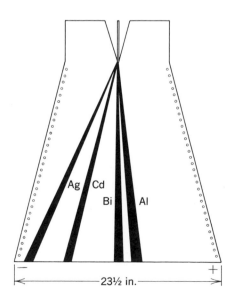

Ag Cd

Bi   Al

—   +

23½ in.

Fig. 24-5 Paper for electrochromatography, cut in the manner described by Strain, and showing the separation of ions of four metals. (*Analytical Chemistry.*)

Many modifications of such apparatus are on the market. Several units which appear at first glance to fall into this category are actually better described as instruments for electrophoresis with transverse flow of solvent. The solvent occupies the entire volume of a thin space (ca. 0.5 mm) between two glass plates and flows in at one side and out at the other, between two membrane-protected electrodes. The flowing liquid may help to avoid the cooling problem and simplify the input device and the collection of samples, but it cannot contribute to the separation per se.

ELECTRODEPOSITION CHROMATOGRAPHY

Blaedel and Strohl have suggested the possibility of combining the principles of controlled-potential electrodeposition at a mercury cathode and liquid-liquid chromatography with mercury as the stationary phase.[2] The distribution coefficient for such a system is

$$K' = \frac{[M]}{[M^{n+}]} \tag{24-1}$$

where [M] is the concentration of a metal dissolved in mercury and $[M^{n+}]$ is the corresponding concentration of the aqueous ion. Combining Eq. (24-1) with the Nernst equation gives (at 25°C)

$$\log K' = \frac{n}{0.059} (E° - E) \tag{24-2}$$

where n, E, and $E°$ have their usual significance. This means that the distribution coefficient can be controlled by control of the applied potential.

The authors used mercury plated on a bed of small bits of platinum wire as their stationary phase and cathode, with a silver–silver chloride reference half-cell. The samples studied contained Tl^+, Pb^{++}, In^{+++}, or Sn^{++}. A flow-through polarographic cell served as chromatographic detector. They found Tl^+ and Pb^{++} to follow the predicted equations. In^{+++} did not, presumably because of the limited solubility of indium in mercury; Sn^{++} did not, although tin is soluble enough, perhaps because of the formation of an intermetallic compound.

This method is described here, not as a finished, ready-to-use procedure, but rather because it indicates the interesting results that can sometimes be obtained by an imaginative extension of and combination of well-established techniques.

REFERENCES

1. Alberty, R. A.: *J. Chem. Educ.*, **25**:426, 619 (1948).
2. Blaedel, W. J., and J. H. Strohl: *Anal. Chem.*, **37**:64 (1965).

3. Cann, J. R.: Electromigration and Electrophoresis, in I. M. Kolthoff and P. J. Elving (eds.), "Treatise on Analytical Chemistry," pt. I, vol. 2, chap. 28, Interscience Publishers (Division of John Wiley & Sons, Inc.), New York, 1961.
4. von Frijtag Drabbe, C. A. J., and J. G. Reinhold: *Anal. Chem.*, **27**:1090 (1955).
5. Longsworth, L. G.: *Ind. Eng. Chem., Anal. Edition*, **18**:219 (1946).
6. MacDonell, H. L.: *Anal. Chem.*, **33**:1554 (1961).
7. Morris, C. J. O. R., and P. Morris: "Separation Methods in Biochemistry," pp. 632ff., Interscience Publishers (Division of John Wiley & Sons, Inc.), New York, 1963.
8. Sato, T. R., W. P. Norris, and H. H. Strain: *Anal. Chem.*, **27**:521 (1955).
9. Scherr, G. H.: *Anal. Chem.*, **34**:777 (1962).
10. Strain, H. H.: *Anal. Chem.*, **30**:228 (1958).

25
General Considerations
in Analysis

We have now completed a survey of some of the most useful analytical methods available to the chemist. This rather imposing array of techniques may well appear confusing. We must give some consideration to the problem of choosing the most appropriate method for any analytical problem which may arise.

Suppose that you, as an analytical chemist, are asked to devise a procedure for the quantitative determination of substance X. Here is a checklist of some of the questions you might ask before undertaking the task:

1. What range of values can be expected?
2. What is the matrix or host material in which the desired substance is found?
3. What impurities are likely to be present, and in approximately what concentration?
4. What degree of precision is required?
5. What degree of accuracy is required?

6. What reference standards are available?
7. Is the analysis to be performed in a laboratory, in a plant location, or in the field?
8. What power sources and other facilities can be utilized?
9. How many samples are expected per day?
10. Is it essential that the answers be obtained quickly? If so, how quickly?
11. To what extent is long-term reliability required (as for continuous unattended operation), and to what extent can it be traded off to lower the cost of equipment?
12. In what physical form is the answer desired (automatic recording, printed or punched tape, written report, telephoned report, etc.)?
13. If special training of personnel is required, can it be arranged?
14. What special and unusual facilities are available which may affect the selection of a method (e.g., an atomic reactor)?

It may happen that a compromise is necessary. High precision may not be compatible with speed, for example. In many instances, personal preference may well prove the deciding factor. Thus colorimetric and polarographic methods may be made to yield about the same accuracy with similarly dilute samples, the time consumed in the two procedures is comparable, and even the cost of apparatus is about the same. The analyst is then free to choose the method with which he is more familiar. Many of the more generally applicable methods of analysis are listed in Table 25-1, with comments designed to help in the selection of a procedure for various kinds of samples.

SENSITIVITY

The sensitivity S of an analytical method or instrument may be defined as the ratio of the change in the response R to the change of the quantity (i.e., concentration) C which is measured:

$$S = \frac{dR}{dC} \text{ or } \frac{\Delta R}{\Delta C} \tag{25-1}$$

It will be instructive to consider a number of functions of the form $R = f(C)$, relating instrument response to a concentration term, which might describe the behavior of some analytical systems. In Fig. 25-1 curves are plotted for four such functions: (1) linear $R = k_1 C$; (2) square-law or power series $R = k_2' C^2 + k_2 C$; (3) reciprocal $R = k_3/C$; (4) logarithmic $R = k_4 \log C$. The derivatives $\partial R/\partial C$ for each of these are also plotted. It is apparent from the figure that the reciprocal and logarithmic functions give very steep curves, hence great sensitivity,

Table 25-1   The comparative applicability of various analytical procedures

Type of sample	Procedure	Application
1. Alloys, ores	a. Spectrography	General; rapid.
	b. Electrodeposition	General; slower; less-expensive apparatus.
	c. Colorimetry	More specific; especially for lesser constituents.
	d. Activation	Specific; less convenient except in special cases.
	e. X-ray absorption	Where sought element and impurities vary widely in atomic weight.
	f. X-ray fluorescence	General; rapid.
2. Traces of metal ions	a Colorimetry ⎫	
	b. Nephelometry ⎪	These are of comparable sensitivity
	c. Fluorimetry ⎬	and accuracy; highly specific.
	d. Polarography ⎭	
	e. Stripping analysis	Specific and highly sensitive.
3. Gaseous mixtures	a. Gas chromatography	General; some specificity.
	b. Gravimetric	Especially for carbon dioxide or water.
	c. Volumetric (Orsat, etc.)	Mixtures; to determine several constituents.
	d. Manometric (Warburg)	Evolution or uptake; small samples.
	e. Infrared absorption	Routine assay for single component.
	f. Mass spectra	General; expensive apparatus.
4. Mixtures (complete separation not required)	a. Infrared spectra ⎫ b. Raman spectra ⎭	Especially for organic compounds.
	c. X-ray diffraction	Crystalline solids.
	d. Isotope dilution	Analysis for single component.
	e. Mass spectra	For simple volatile compounds.
	f. NMR	For liquids.
5. Mixtures (separation procedures)	a. Ion exchange	For ionic materials.
	b. Countercurrent distribution ⎫	Must be partially soluble in each of
	c. Partition chromatography ⎭	two immiscible liquids.
	d. Adsorption chromatography	Chiefly for organic compounds.
	e. Electrodeposition	For metallic cations.

at low concentrations, while the square function gives greater sensitivity at higher concentrations. For the linear function, as expected, the sensitivity as defined above is constant over the whole range.

The slopes in Fig. 25-1 are written as partial derivatives to emphasize the fact that there are likely to be other variables which will affect the sensitivity. An example is spectrophotometric analysis, where the sensitivity can be varied by change of wavelength. The response curve will be of the same form, but the horizontal scale will be compressed or expanded.

It is difficult to generalize about the relative sensitivities of various methods, since in so many instances the sensitivity is widely different from one element or type of compound to another. Morrison[1] has

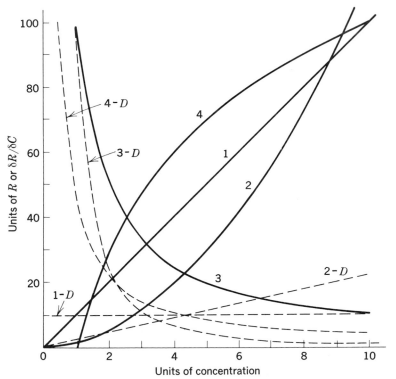

Fig. 25-1 Various functions of concentration and their derivatives.

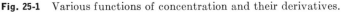

(1) $R = k_1C$	(1-D) $\partial R/\partial C = k_1$	$k_1 = 10$
(2) $R = k_2'C^2 + k_2C$	(2-D) $\partial R/\partial C = 2k_2'C + k_2$	$k_2 = k_2' = 1$
(3) $R = k_3(1/C)$	(3-D) $\partial R/\partial C = -k_3(1/C^2)$*	$k_3 = 100$
(4) $R = k_4 \log C$	(4-D) $\partial R/\partial C = 0.435k_4(1/C)$	$k_4 = 100$

* The negative sign is neglected for convenience in plotting.

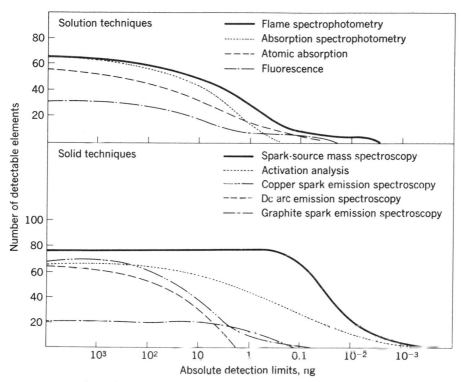

Fig. 25-2 Absolute detection limits of nine analytical methods and the number of elements detectable at each level. (*John Wiley & Sons, Inc., New York.*)

compiled a valuable comparison of the sensitivities of nine analytical methods as applied to all of the elements for which data were available. The methods covered are absorption spectrophotometry, ultraviolet-visible fluorescence, atomic absorption, flame spectrophotometry, neutron activation, spark-source mass spectroscopy, and emission spectrography with a dc arc, a copper spark, and a graphite spark. One finds from this compilation, for example, that the element europium can be detected in a quantity as small as 0.5 pg (picogram = 10^{-12} g) by neutron activation, but only to 1 ng (nanogram = 10^{-9} g) by flame emission, 100 ng by atomic absorption, etc. On the other hand, iron is determinable only to 5 μg by activation and to 3 ng by flame emission.

Figure 25-2 summarizes diagrammatically the limiting concentrations of these nine techniques with respect to the number of elements detectable.

The limiting minimum concentration C_m can be defined in terms of the *signal-to-noise ratio* S/N, where S is the magnitude of the desired signal, and N is the spurious signal called *noise*, resulting from the random error inherent in the system.[2] The required value of C_m is that concentration for which the S/N ratio is given by:

$$\frac{S}{N} = \frac{t\sqrt{2}}{\sqrt{n}} \tag{25-2}$$

where t is the Student-t statistic which can be obtained from handbook tables, and n is the number of *pairs* of readings taken (i.e., one reading for the blank or background, one for the sample). See the reference for the derivation and further significance of this relation.

PRECISION

This quantity is measured inversely by the relative standard deviation s. The smaller the value of s, the better the precision. It is intimately related to the *accuracy*, which is the closeness of agreement between the observed result and the known or "true" value.

The precision of measurement can be improved by repetition with suitable statistical treatment of the data. A procedure which has a

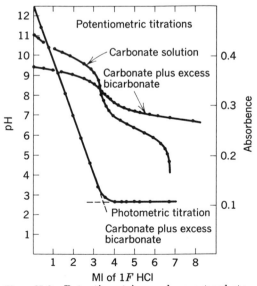

Fig. 25-3 Potentiometric and spectrophotometric titrations of 3.4×10^{-2} F sodium carbonate solution containing 1 F sodium bicarbonate. (*Analytical Chemistry.*)

similar effect is titration. In an instrumental titration one has the opportunity to (and in many situations must) take a whole series of measurements, both before and after the end point; drawing a smooth curve through these points has about the same effect on the overall precision as would be obtained by taking the same number of individual readings on a solution without titrating. (It must be remembered, of course, that the information obtained with and without titration may not be, even ideally, the same, but may pertain to different equilibrium states.)

The precision obtainable from one method as compared with another is often affected by the form of the response curve, apart from the inherent ability of the instrument to detect signals. Figure 25-3 shows two titration curves corresponding to the same reaction, the titration of carbonate by strong acid in the presence of a large concentration of bicarbonate.[3] The potentiometric curve is almost useless, but the photometric titration (at wavelength 235 nm, where the carbonate ion absorbs but bicarbonate does not) shows an excellent end point, obtained by extrapolation of two straight-line segments.

COMPARISON WITH STANDARDS

The majority of the analytical methods which have been discussed involve the comparison of a physical property of the unknown with the corresponding property of a standard or a series of standards containing the same material in known amount. This may be achieved by means of a calibration curve, which is a plot of the magnitude of the physical property against the concentration of the desired constituent (or some simple function of the concentration, such as its logarithm or reciprocal). In some instances the shape of the curve is predicted by theory (Beer's law, the Ilkovič equation, etc.), and it may be more convenient to perform a calculation based on the equation than to employ a calibration curve. This is the case, for example, in the determination of a cation by the measurement of a half-cell potential: the Nernst equation will give the desired information directly, but it actually represents a curve which can be drawn (potential difference against the logarithm of concentration) for the graphical comparison of unknowns with the standard solution from which $E°$ was originally evaluated (Fig. 11-1).

Another general procedure for comparison of unknowns and standards is to bracket the unknown between two suitably close standards, one slightly below and one slightly above it with respect to the quantity measured. This finds application especially in Nessler tubes and other optical comparators where the intensity of colors is matched directly by eye.

In all comparisons it is highly desirable that the standards duplicate the unknown as closely as possible. This principle results in the substantial reduction of systematic errors which have the same effect on all solutions. In some cases the precision can be greatly increased, since the full-scale span of the instrument can be applied to measuring the difference between two rather close magnitudes, rather than the distance of each magnitude from zero. This has been discussed in its application to photometric analysis in Chap. 3, but the principle may be equally valuable in other fields.

Closely related is the type of apparatus in which the comparison between standard and unknown is made directly in a single operation. Examples are the potentiometric concentration cell, the thermal conductivity detector in gas chromatography, and those photometers and spectrophotometers which employ a balanced system of two light beams passing through two samples.

It must be remembered that comparison with standards cannot improve the precision of an analysis, but it may have an effect on the accuracy, which can never be better than the standards. The preparation and preservation of standards for extremely dilute solutions (micromolar to nanomolar) can be quite difficult. The walls of a glass vessel have a tendency to adsorb solute, and may reduce the concentration significantly below the intended value; this can be overcome in favorable cases by the precaution of rinsing out the vessel with some of the solution to be stored.

An important aid in the direction of overall standardization is provided by the extensive series of standard samples made available at nominal cost by the National Bureau of Standards in Washington. Each sample is accompanied by a certificate bearing the concentration of each constituent, from the major elements down to those present in only a few thousandths of a percent. Nearly any variety of analysis in the fields of metals, alloys, and ores can be tested as to accuracy and precision by means of these samples.

STANDARD ADDITION

This is a very generally applicable method of implementing the comparison with a standard. It has been mentioned in the discussion of a few instrumental methods (polarography, for example), but it can readily be adapted to others. A reading is taken on the sample to be analyzed, then a measured quantity of standard is added to the sample with mixing, and the measurement repeated. If the analysis is destructive (as a titration usually is), then the standard must be added to a second aliquot.

In many circumstances this procedure will serve to identify the feature of the record which pertains to the desired material and at the same time give the information needed for a quantitative analysis. The dilution of the sample by addition of the standard must be allowed for or shown to be negligible.

This technique has the great advantage that the standard and unknown are measured under essentially identical conditions. Even if the kind and quantity of other substances present are not known accurately, they can be taken as identical in the two measurements.

PROBLEMS

25-1 Identify several instrumental methods as corresponding to each of the functions plotted in Fig. 25-1, and show equations to verify your choices.

25-2 How would you best determine water quantitatively in each of the following circumstances (outline a procedure where possible):

(a) Water vapor in tanks of compressed H_2 and O_2.
(b) Water dissolved in "pure" chloroform or ether.
(c) Water content of the atmosphere in a sealed space vehicle.
(d) Water collected in the bottom of a large gasoline storage tank.

25-3 Devise an instrumental method for the determination of TEL (tetraethyl lead) and TML (tetramethyl lead) when either or both may be present in a gasoline.

25-4 One of the most difficult separations is between zirconium and hafnium. A result of the "lanthanide contraction" is that Zr and Hf atoms are almost precisely the same size, which, together with their similar electronic structure, makes them chemically nearly identical. Suggest at least two tentative methods by which you might attack their separation. In each method, what data would you look for to evaluate the success of the method? What nonseparative analytical methods might be used to analyze mixtures of these elements?

25-5 In the colorimetric analysis of manganese and chromium in steel, the color of the ferric iron will cause interference unless suitable precautions are taken. It is possible to remove the bulk of the iron by ether extraction in the presence of hydrochloric acid, or by ion exchange following oxidation, or the iron may be complexed with citrate, tartrate, or other reagent to destroy its color. Alternatively, the iron may be allowed to remain in the solution and its absorbance corrected for by taking photometric readings before and after oxidation of the manganese and chromium; or combinations of these methods may be employed. Compare these procedures critically. Why is mercury-cathode electrolysis inapplicable as a separatory procedure?

25-6 In Chap. 16, mention was made of the use of sodium iodide crystals "activated" by the addition of a trace of thallous iodide, as luminescent crystals for the detection of radioactivity. Devise a nondestructive method for determining the amount of thallous iodide in sodium iodide crystals. Estimate the precision.

25-7 In the following table is given the average chemical composition of sea water (in milligram-atoms per kilogram). Note that in certain instances the element or radical may be present in more than one form: thus carbon dioxide may be partly as dissolved gas and partly as carbonate or bicarbonate; several of the metals may be present in more than one oxidation state. Suggest appropriate analytical methods by which sea water may be analyzed for each of these elements or ions. Any procedure by which more than one ion can be determined simultaneously will, of course, be an advantage.

Cl	535.0	Br	0.81	NO_3^-	0.014	Ag	0.0002
Na	454.0	Sr	0.15	Fe	0.0036	NO_2^-	0.0001
SO_4^{--}	27.55	Al	0.07	Mn	0.003	As	0.00004
Mg	52.29	F	0.043	P	0.002	Zn	0.00003
Ca	10.19	Si	0.04	Cu	0.002	Au	2.5×10^{-7}
K	9.6	B	0.037	Ba	0.0015	H	10^{-6}–10^{-7}
CO_2	2.25	Li	0.015	I	0.00035		

REFERENCES

1. Morrison, G. H., and R. K. Skogerboe: General Aspects of Trace Analysis, in G. H. Morrison (ed.), "Trace Analysis: Physical Methods," chap. 1, Interscience Publishers (Division of John Wiley & Sons, Inc.), New York, 1965.
2. St. John, P. A., W. J. McCarthy, and J. D. Winefordner: *Anal. Chem.*, **39**:1495 (1967); T. Coor, *J. Chem. Educ.*, **45**:A533 (1968).
3. Underwood, A. L., and L. H. Howe, III: *Anal. Chem.*, **34**:692 (1962).

26
Electronic Circuitry for Analytical Instruments

The great majority of the instrumental methods of analysis treated in this book require electric circuitry, and this, in all but the simplest cases, implies the need for electronics. So for a full understanding of analytical instruments, their limitations, and what can go wrong with them, some knowledge of electronics is essential. The brief account in this chapter can only be considered a survey. No attempt is made to derive the fundamental mathematical relations; more complete treatment can be found in numerous texts.

The heart of any electronic device is one or more components which act directly on electrons. This includes vacuum and gas-filled tubes in which electrons are liberated by thermionic emission from a hot cathode, by the photoelectric effect, by a glow discharge, or otherwise. It also includes a variety of solid-state components, particularly semiconductor rectifiers, transistors, and photocells. None of these active elements is self-sufficient; they all require more-or-less complex supplementary circuits composed of resistors, capacitors, inductors, meters, etc. In most cases a power supply is also required, which may be a selection of batteries, or may be rectified alternating current. In addition there will frequently be incorporated other components for convenience or safety, such as switches, fuses, and pilot lights.

Electronics first became important to chemists in the 1930s with the introduction of the glass electrode for pH measurement, which made necessary the development of an electronic pH meter. Other electronic instruments followed soon, notably spectrophotometers to displace the inconvenient and often inexact photographic observation of absorption spectra. During and since the Second World War, advancing electronic technology has made obsolete prior instruments in nearly all fields.

In the years since about 1960, another revolution has taken place, in which vacuum tubes have been supplanted in new designs of laboratory instruments by the introduction of transistors. At first, "transistorization" was mostly a process of replacing tubes by transistors with a minimum of circuit changes—something like a literal translation of one language into another. More recently, as the idiomatic capabilities of transistors have been more fully understood, new circuits have been designed to fit existing needs. Present-day instrument engineers rarely choose vacuum or gas tubes except for a few specialized applications.

It now appears likely that discrete transistors will be largely displaced in the future by the integrated solid-state circuitry known as *microelectronics*.

Since so many vacuum-tube instruments are still in use, we must study both tube and transistor circuits, with emphasis on their similarities and differences. Microelectronics, however, needs little attention in our context, as it represents primarily an innovation in methods of fabrication and makes use of the same basic circuits that apply to discrete solid-state devices.

ACTIVE ELECTRONIC COMPONENTS

VACUUM TUBES

The simplest evacuated electronic device of importance is the *diode*, which consists of two electrodes in a glass or metal envelope. One electrode, the *cathode*, is heated electrically, and releases electrons by thermionic emission. The second electrode, the *anode* (or *plate*), collects the electrons, but cannot itself emit electrons to a significant extent. Thus current can pass through a diode in one direction but not the other, and it can serve as a *rectifier*.

Diodes are useful in two main areas. In the larger sizes they are employed as power rectifiers to convert alternating to direct current as a power source for other devices. Smaller diodes are applied to the

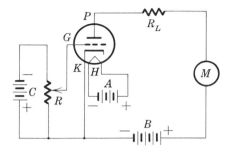

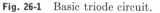

Fig. 26-1 Basic triode circuit.

rectification of ac signals; in this service they are often called *detectors* or *demodulators.*

Vacuum triodes are similar to diodes in that they possess an electron-emitting cathode and an electron-collecting anode, but between these is mounted a *grid,* which consists of an array of fine, parallel wires. This element introduces another degree of freedom, the possibility of regulating the flow of electrons from cathode to anode by control of the potential applied to the grid.   Figure 26-1 shows schematically a test circuit for a triode.   In this diagram K represents the cathode, which is heated to redness by power liberated in a heater H, connected across a battery A. P is the anode (plate), and G the control grid.   The anode is connected to the positive terminal of the battery B, through the load resistor R_L.   The potential of the grid can be varied by means of a voltage divider R placed across another battery C. M is a dc milliammeter.   The flow of electrons across the evacuated space is controlled by the potential of the grid, which is normally maintained slightly negative with respect to the cathode.   The more negative the grid, the fewer electrons are able to reach the plate; conversely, the less negative it is, the more electrons pass.   Electrons from the cathode are not attracted to the negative grid; neither can electrons leave its cold surface.   Hence the grid acts solely as a potential-sensing element and carries no significant current.

The *amplification factor* μ of a tube is defined as the ratio of the change in anode potential to the change in grid potential (of opposite sign) which will maintain the anode current constant.   The ratio lies in the range 10 to 100 for common triodes.

A second grid, called the *screen grid,* may be introduced between the control grid and the plate, to diminish the capacitance between them. Such a tube is known as a *tetrode.*   The added electrode, which is operated at a positive potential, produces a substantial increase in μ, but at the same time causes difficulties with secondary emission.   In all tubes, bombardment of the plate by electrons causes the emission from the plate of more electrons.   In the diode and triode these are not objectionable,

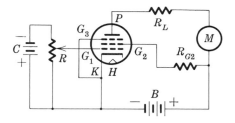

Fig. 26-2 Basic pentode circuit; G_1 is the control grid, G_2 the screen grid, G_3 the suppressor grid (other symbols as in Fig. 26-1).

because the plate is the only positive element and the secondary electrons are drawn back into it. In a tetrode, however, many of the secondary electrons end up on the positively charged screen, which is detrimental to the normal action of the tube and constitutes a serious limitation to its usefulness.

This difficulty can be overcome by inserting a third grid, called the *suppressor*, between screen and plate, forming a *pentode*. The suppressor is normally operated at the same potential as the cathode, and acts to suppress the emission of secondary electrons. This permits increase of the amplification factor up into the thousands. The basic circuit for a pentode is shown in Fig. 26-2.*

The operation of vacuum tubes can be clarified by the so-called *static characteristic curves*, typical examples of which are shown in Fig. 26-3 for a triode, and in Fig. 26-4 for a pentode. Note that the plate current in the pentode is quite insensitive to changes in plate voltage except at low voltages, whereas the corresponding curves are quite steep for the triode. The dependence of plate current on grid voltage is com-

* Note that an electric connection between wires is indicated by a dot at the intersection: ┿. A crossing of wires with no dot + indicates no electrical connection; this was formerly shown by the symbol ⌐.

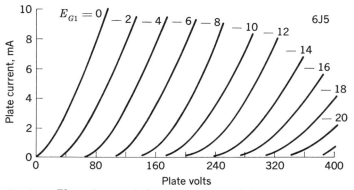

Fig. 26-3 Plate characteristics of the 6J5, a triode.

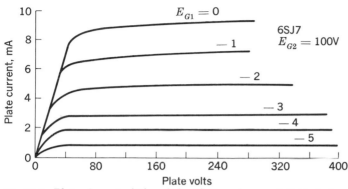

Fig. 26-4   Plate characteristics of the 6SJ7, a sharp-cutoff pentode.

parable in both, as indicated in Fig. 26-5.   It can be seen from the latter figure that the plate current for both the 6J5 triode and 6SJ7 pentode is reduced effectively to zero by a grid voltage of approximately -6 V, which is called the *cutoff* point.   All triodes show such a cutoff, but pentodes are of two classes: the sharp-cutoff (the 6SJ7 is an example) and the remote-cutoff tubes in which a much larger negative grid voltage is required for cutoff, so that ordinarily some plate current always flows; the 6SK7 is an example of this kind of pentode.   Curves like those of Fig. 26-5, relating the output to the input, are called *transfer plots.*

There is, of course, an extremely wide variation with respect to current- and power-handling capacity between tube types, following the variation in physical sizes from subminiature tubes (ca. 20 by 6 mm) up to the giant tubes used in radio transmitting.

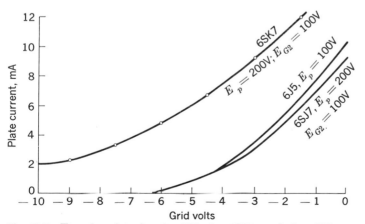

Fig. 26-5   Transfer plots for the 6J5, the 6SJ7, and the 6SK7, a remote-cutoff pentode.

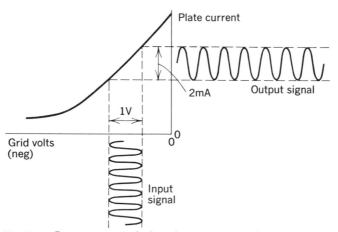

Fig. 26-6 Input-output relations for a vacuum tube.

VACUUM–TUBE AMPLIFICATION

The amplifying action of a triode or pentode results when the signal (shown in Fig. 26-6 as a sine wave) is impressed on the grid so that the total voltage variation is within the nearly linear portion of the characteristic. Let us suppose that the input signal originates in a circuit of moderately high impedance, say 10^5 Ω, and changes by 1 V. The power level of this signal is

$$P = EI = \frac{E^2}{R} = \frac{1^2}{10^5} = 10^{-5}\,\text{W}$$

In the example of Fig. 26-6, which approximately corresponds to the 6SJ7 tube, the variation of the output would be about 2 mA, giving an output power level (for a load resistor of 10^5 Ω)

$$P = EI = RI^2 = (10^5)(2 \times 10^{-3})^2 = 0.4\,\text{W}$$

which shows a *power gain* of $0.4/10^{-5} = 40{,}000$. Frequently more useful is the *voltage gain*. In this example the variation of voltage across the load would be $RI = (10^5)(2 \times 10^{-3}) = 200$, so the voltage gain is $200/1 = 200$.

GAS TUBES

Electronic tubes which are filled with a gas at low pressure have properties which contrast quite strongly with the analogous vacuum types. Particularly important in our context are the glow tube and the thyratron. The *glow tube* is a two-element tube (diode) in which both electrodes are cold (i.e., operate only slightly above ambient temperature).

It is filled with an inert gas, helium, argon, or neon.   Such a tube will conduct only when the applied potential is greater than a certain well-defined threshold value, which is determined in part by the ionization potential of the gas and its pressure, in part by the nature of the electrode surface (its "work function"), the spacing of the electrodes, etc.

The characteristic current-voltage curve for a glow tube is shown in Fig. 26-7.   As the applied potential is increased from zero, no current will flow until a particular voltage is reached.   The starting potential varies somewhat with the degree of incident illumination or ionizing radiation, either of which will help initiate the glow discharge.   As soon as the discharge commences, the potential across the tube drops off slightly and assumes a value which is very nearly constant even though the current may vary over a considerable range.

A triode or tetrode containing gas is called a *thyratron*.   It has

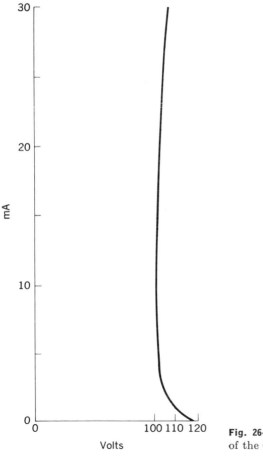

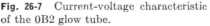

Fig. 26-7 Current-voltage characteristic of the 0B2 glow tube.

properties analogous to an electromechanical relay: it either conducts or it doesn't, it is either on or off, with no intermediate stages. If a thyratron be inserted in the test circuit of Fig. 26-1, no current will pass until the grid potential reaches a certain value, when a gas discharge starts between anode and cathode. While the discharge is present, the grid is ineffective; it can neither increase the tube current nor cut it off. The only way in which the discharge can be stopped is by lowering the plate voltage nearly to zero. An alternating potential is often applied to the plate of a thyratron; the grid in this case can regain control once in every cycle, as the plate current cannot flow during the period when the plate is negative. The power delivered to a load can be controlled easily by regulation of the grid potential.

Thyratrons can be used in place of relays, and have the advantages of no mechanical inertia and no contact points with their tendency to corrode and freeze. Thyratron tetrodes require less driving power than the triodes, and hence are better suited to applications where the signal is derived from a high-impedance source, such as a vacuum photocell.

SEMICONDUCTORS

A semiconductor is a solid substance which is intermediate between metallic conductors on the one hand and nonconducting insulators on the other. It is characterized by a relatively large *negative* temperature coefficient of resistance, whereas the coefficient is *positive* for metals; this provides a convenient criterion for distinguishing the two types of conductors. The most used semiconductors are silicon, germanium, selenium, and a variety of metallic oxides and sulfides.

In a crystal of high-purity germanium, each atom is bound covalently to each of four other atoms (the diamond structure), and since each atom has just four valence electrons, it is fully satisfied by this structure. To make the crystal useful for electronic purposes, it must be "doped," that is, a trace of impurity added. The foreign atoms must be of such nature that they can replace some of the germanium atoms in the crystal lattice. If the impurity is a pentavalent element, such as arsenic or antimony, then each of its atoms possesses an extra valence electron beyond those needed for the covalent lattice bonds. The extra electrons are easily torn loose from their parent atoms by thermal energy, and then are free to wander at random throughout the lattice. The impurity atoms become unipositive ions imbedded in the crystal, and hence immobile. On the other hand, if the impurity is trivalent gallium, indium, or gold, then there will be a deficiency of one electron per atom. The spot where the electron is lacking is spoken of as a *hole*. Occasionally an electron in a normal covalent bond located near one of the impurity atoms will have sufficient thermal energy that its vibrations will bring it so close to the hole that it will escape completely from its previous berth and move into the hole. The result of such a process is that the hole has moved from one spot to another within the lattice. In a piece of germanium doped in this way, the holes appear to wander freely through the crystal in an exactly analogous fashion to the surplus electrons in a piece doped with arsenic or antimony. The mobility of the holes, however, is somewhat less than that of the electrons.

If an electric field is applied to doped germanium, a current will flow which is carried almost exclusively by the excess electrons or holes provided by the impurity.* Germanium in which the current carriers are electrons (negative) is called n-type germanium, whereas that in which the predominant carriers are holes (positive) is p-type.

CRYSTAL DIODES

A crystal diode is a two-terminal semiconductor device which has the ability to pass current with ease in one direction but to block it from flowing the other way. Hence it is analogous to a thermionic diode and is useful as a rectifier.

The crystal diode is made from a small bar of silicon or other semiconductor (called a *wafer* or *chip*), part of which is p type, part n type, as indicated schematically in Fig. 26-8. When it is connected to a source of potential so that the n region is negative and the p region positive, the dominant carriers in both sections tend to move toward the p-n junction. At the junction, electrons from the n side neutralize holes from the p side, and so current flows easily. In the reverse connection, with the n region positive and the p region negative, both holes and electrons are pulled away from the junction, leaving an intermediate region with very few carriers of either type, hence a high resistance.

In Fig. 26-9 is shown the characteristic current-voltage curve for silicon and germanium crystal diodes. For positive (i.e., forward) values of potential, the current follows approximately an exponential function of the voltage. For negative (reverse) potentials, the current is practically zero until (for silicon) a critical value E_z is reached, at which the current increases negatively until limited by the series resistance in the circuit. This critical point, called the *zener* or *breakdown* voltage, corresponds to the potential necessary to tear electrons out of covalent bonds in the crystal, thus creating pairs of positive and negative carriers which are swept toward the respective circuit connections. By varying the manufacturing techniques, silicon diodes can be prepared with zener

* It should be emphasized that germanium or silicon which is doped, even relatively heavily, is still from the *chemical* standpoint extremely pure. A controlled impurity of the order of 1 ppm will usually be adequate to impart the desired electrical properties.

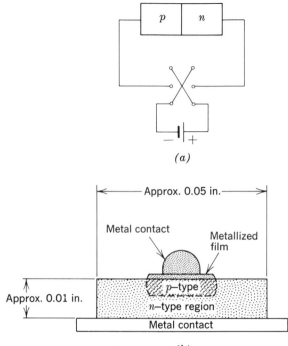

Fig. 26-8   A p-n junction diode: (a) schematic; (b) physical structure of one form; the diode is manufactured by diffusing a p-type impurity into the n-type starting material.   [*Part b, courtesy Education Development Center (formerly Educational Services, Inc.), Newton, Mass.*[2]]

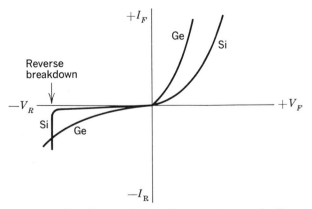

Fig. 26-9   Dc characteristics of germanium and silicon diodes.

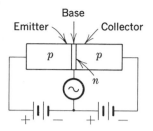

Fig. 26-10 A *p-n-p* junction transistor, schematic.

voltages anywhere between about 2 and 200 V. Germanium diodes have a lower reverse resistance than silicon, and no well-defined zener breakdown potential.

Semiconductor diodes can be used for the same applications as mentioned for vacuum diodes. For service as a rectifier, the zener voltage must be more negative than the greatest inverse voltage to be encountered. We will see later that the zener effect can be useful to provide a fixed-reference voltage. Note the similarity between Fig. 26-7 (gas diode) and the zener portion of Fig. 26-9; both show nearly complete independence of voltage drop and current over a wide range.

JUNCTION TRANSISTORS

The semiconductor analog of the vacuum triode is the transistor, which exists in many modifications, the most prevalent being the junction transistor. This may be visualized with the aid of Fig. 26-10, which shows schematically a wafer made up of two *p*-type segments separated by a thin layer of *n* material. One way of making such a unit is to start with a wafer of relatively high resistance (i.e., lightly doped) *n*-type silicon, deposit bits of indium on each side, then heat it to a high enough temperature to melt the indium (mp 155°C) and to cause it to diffuse into the silicon, converting *n* to *p* type for a region on each surface. The result resembles Fig. 26-11, which shows a very thin portion of the original *n* silicon separating two *p* regions to which electric connections are made through the indium. For transistor action, the junction between the smaller *p* region, called the *emitter*, and the *n* material, called the *base*, is given a small forward bias, while the junction with the larger *p* region, the *collector*, is reverse-biased. Current can flow easily across the forward-biased emitter-base junction, but since the emitter region has more holes as carriers than the base has electrons, the major part of the emitter current is carried by holes. The base region, being both thin and lightly doped, is poor in electrons, so that most of the holes from the emitter diffuse across it, as they have a greater probability of being "collected" by the large collector junction than of combining with an electron. Any current which may be injected into the transistor

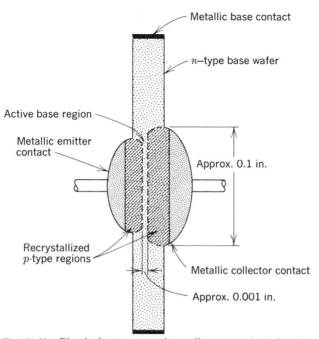

Fig. 26-11 Physical structure of an alloy $p\text{-}n\text{-}p$ junction transistor; the p-type regions are obtained by alloying a metal containing large amounts of p-type impurity into the n-type semiconductor wafer. [*Courtesy Education Development Center (formerly Educational Services, Inc.), Newton, Mass.*[2]]

through the base connection will provide additional electrons to carry a portion of the emitter-base current, and hence will have a considerable effect on the number of holes available to flow into the collector. This is the basis of the amplifying ability of the transistor.

The transistor described above is of the $p\text{-}n\text{-}p$ type, but it is equally possible to fabricate a unit with opposite characteristics, an $n\text{-}p\text{-}n$ transistor. The two can be used in equivalent circuits, except that the polarities of all voltages and currents must be reversed. Figure 26-12 shows the standard symbols for the two types. The arrowhead on the

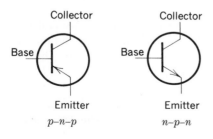

Fig. 26-12 Symbols for $p\text{-}n\text{-}p$ and $n\text{-}p\text{-}n$ transistors.

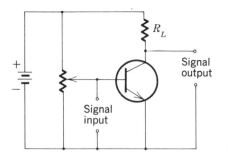

Fig. 26-13 Transistor test circuit.

emitter indicates the direction of easy positive current flow for the emitter-base junction.

A simple circuit to demonstrate the amplification possible with a transistor is given in Fig. 26-13, which may be compared with the analogous tube circuits of Figs. 26-1 and 26-2. It is convenient to make use of a single battery E with a divider to control the dc bias potential for the base. This is possible since the potentials applied to collector and base, relative to the emitter, are always of the same polarity, which is not the case with vacuum tubes. A reasonable value for E is between 5 and 20 V, compared with several hundred volts for tube circuits.

Figure 26-14 gives a family of collector characteristics for a typical silicon n-p-n transistor, the 2N5182, showing the effect of the base current as parameter (compare Figs. 26-3 and 26-4 for vacuum tubes). A transfer plot for this transistor takes the form shown in Fig. 26-15 (compare Fig. 26-6). The current-amplifying ability of a transistor is usually expressed as the ratio of a small change in collector current to the corresponding change in base current (at constant collector voltage) denoted by β.

FIELD–EFFECT TRANSISTORS (FET)

This semiconductor unit works on a principle somewhat different from that of the transistors we have previously discussed. Consider the sketch in Fig. 26-16. This device consists of a bar of n-type silicon, called the *channel*, with connections at both ends, called respectively the *source* and *drain*. The channel is sandwiched between layers of p material (connected together) called the *gate*. In some constructions, the gate is all in one piece, completely surrounding the central bar. Both n- and p-channel FETs are available.

The FET is classified as a *unipolar* transistor, as there is effectively only a single junction, whereas those transistors having two junctions (emitter-base and base-collector) are *bipolar*.

The operation of an FET can be followed by the aid of Fig. 26-17. If no voltage is applied to the gate, then current will flow unhindered

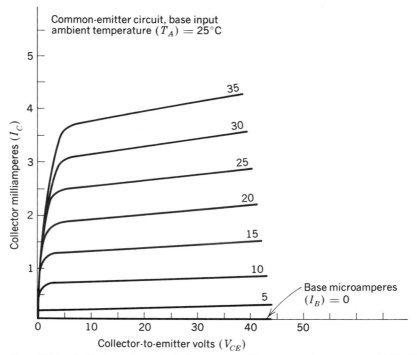

Fig. 26-14 Collector characteristics of a silicon transistor, type 2N5182. (*Radio Corporation of America, Harrison, N.J.*)

through the channel, electrons passing from source to drain (the channel current is carried entirely by majority carriers). In its normal mode, the gate-to-channel junction is reverse-biased. This has the effect of depleting of electrons the area shown in dotted lines, hence increasing the effective resistance of the channel. This arrangement has the added feature that the reverse bias prevents the flow of appreciable current in the gate circuit. The voltage applied to the gate may be several volts, say up to 10 or so. Since no current flows in this circuit, the gate characteristic is given in volts, just as in vacuum tubes, rather than in current units as in bipolar transistors which draw appreciable base currents. The voltage impressed between drain and source may be quite high (requiring some current-limiting component in the circuit) and hybrid operation with vacuum tubes is therefore more practicable than with bipolar transistors.

The characteristic curves for an FET resemble those of Fig. 26-14, if drain current (in milliamperes) is plotted against the drain-source voltage, with the gate-source voltage as parameter. The principal advantage of the FET is its high input impedance, resulting from the reverse-bias

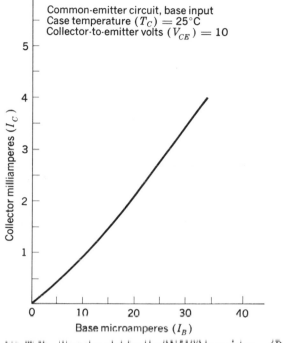

Fig. 26-15 Transfer plot for the 2N5182 transistor. (*Radio Corporation of America, Harrison, N.J.*)

condition. This impedance may be in the range of tens or even hundreds of megohms.

The *insulated-gate FET* (sometimes called MOSFET for metal oxide semiconductor) is a modification of the FET, wherein a thin film of insulating material (silicon dioxide) separates the gate from the channel. This eliminates the rectifying junction, so that the gate can be of either polarity without drawing any current. The electrostatic field between gate and channel is still able to modify the distribution of holes or electrons in the channel and so determine its resistance. The MOSFET has the highest input resistance of any transistor, higher indeed than that of the majority of vacuum tubes.

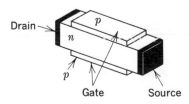

Fig. 26-16 Junction field-effect transistor, schematic.

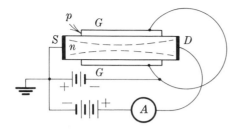

Fig. 26-17 n-Channel field-effect transistor, connections.

There are many other semiconductor devices which find occasional application in laboratory instruments. One of the most widely useful is the *thermistor*, a resistor with a large negative temperature coefficient, made of the sintered or fused oxides of transition metals.[1] It provides a sensitive temperature detector, though with rather less reproducibility than a resistance thermometer made of a metal such as platinum. It can also serve in temperature-compensator circuits to counteract the positive coefficients of other components or to lower the bias at the base of a transistor if the temperature rises, thereby stabilizing the gain of the transistor.

ELECTRONIC CIRCUITS

The majority of modern laboratory instruments are powered from the 115-V 60-Hz ac lines. Heater power for tubes is obtained directly from a step-down transformer, except in very sensitive circuits, where batteries are to be preferred. Direct current is obtained by rectification, either with vacuum diodes or with silicon rectifiers. Silicon units are more convenient for the low-to-moderate currents and voltages needed for instruments and hence are rapidly replacing vacuum diodes.

There are a number of possible connections for the rectifiers, each of which has merit for particular types of application. The simplest is the *half-wave* rectifier, Fig. 26-18. (In this and subsequent diagrams, the symbol for a semiconductor rectifier $-\triangleright\!\!\mid-$ will be employed with the understanding that it can be replaced by a vacuum diode.) The transformer T has two functions: it isolates the circuit from the ac supply line,

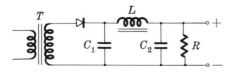

Fig. 26-18 Half-wave rectifier with filter.

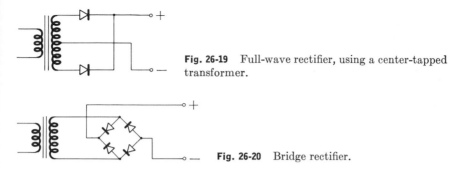

Fig. 26-19 Full-wave rectifier, using a center-tapped transformer.

Fig. 26-20 Bridge rectifier.

and it permits a selection of the voltage level, as may be required.   The C_1-L-C_2 network constitutes a *filter* to reduce the residual ripple to a low value.   With the type of filter shown, the voltage of the dc output will be somewhat greater than the ac voltage at the transformer secondary.   If C_1 is omitted, the output will be lower.   For low-current requirements, the choke L may be replaced by a resistor.   The resistor R, called a *bleeder*, ensures the discharge of the capacitors when the supply is turned off; it is particularly important in high-voltage units.

The filter requirements are much less stringent with a *full-wave* rectifier, Fig. 26-19.   The output voltage approximates *half* that of the transformer secondary.   This is one of the most popular configurations, especially with twin vacuum rectifiers containing two diodes with a common cathode in one envelope.

The *bridge rectifier* of Fig. 26-20 is often used with semiconductors, but seldom with vacuum tubes.   It provides full-wave rectification just as does the preceding circuit but with the advantage that the output voltage is essentially that of the entire secondary, not half of it.

Figure 26-21 shows one form of *voltage-doubler* rectifier which gives an output about twice the transformer voltage.   The two capacitors each charge to the secondary voltage on opposite half cycles of the alternating current but in such a sense that their voltages add together for the output.

High voltages at low currents (a few kilovolts at less than 1 mA) are required for ionization chambers, counters, photomultiplier tubes, and cathode-ray tube applications.   Such requirements can be met by means of a half-wave rectifier with a simple RC filter, but the transformer may

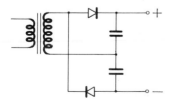

Fig. 26-21 Voltage-doubling rectifier.

be unduly bulky and expensive.   The same result may be achieved with an oscillator operating at perhaps a few hundred kilohertz.   The output can be stepped up in voltage by an air-core rf transformer, followed by a half-wave rectifier or even a voltage doubler to give very high dc potentials.   Filtering requirements are easily met because of the high ripple frequency.   Modern trend is toward the use of an oscillator employing two power transistors, which with transformer and associated parts can be enclosed in a single shield can.   This unit can be powered either from a single battery or from a line rectifier.

VOLTAGE REGULATION

In the rectifier power supplies just described, the output voltage and current are both subject to variation resulting from any change in load resistance or in ac line voltage.   The effects of line-voltage fluctuations can usually be reduced to the point where they are negligible by the insertion of a *constant-voltage transformer* between the instrument and the line.   The effects of changes of load are not so easily eliminated, as they depend on the effective internal resistance of the rectifier-filter combination.   A choke-input filter gives a more stable voltage than does a capacitor-input, while a voltage-doubler shows less constancy.   Two types of circuit for the regulation of voltage will be discussed; they are effective against variations either from line-voltage or load changes.

The first of these makes use of the constant-voltage characteristic of a glow-discharge tube or a zener diode (Figs. 26-7 and 26-9).   The diode is connected across the load, but with a series resistor between it and the rectifier, as in Fig. 26-22.*   The sum of the currents through the load and through the diode must remain constant, even if the load resistance should change, so that the potential drop across the series resistor R will be constant, to meet the requirements of a constant voltage across the diode.   Thus any increase of load current will be at the expense of diode current, and the voltage will not change.

* A black dot within the symbol of a tube (diode, triode, etc.) indicates that it is gas-filled.

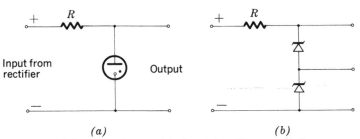

(a)   *(b)*

Fig. 26-22   Voltage regulators: (*a*) glow tube, (*b*) zener diodes.

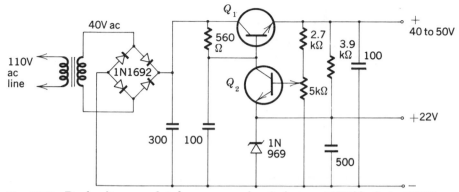

Fig. 26-23 Dual-voltage-regulated power supply; Q_1 is type 2N2108, Q_2 is 2N697; capacitors in μfd.

Zeners are available in various voltage and power ratings, whereas glow tubes are restricted to a few specific voltages and to currents less than 30 or 40 mA.

The regulation which can be achieved by this simple circuit is of the order of a 2 percent change in voltage for a load change from the maximum design current to zero current. This is adequate for some purposes, but quite inadequate for others.

Better regulation can be obtained by means of a more complex circuit, such as that in Fig. 26-23. Here a high-current transistor Q_1 is introduced in series with the load. Any change in either load current or line voltage will produce a change in the voltage supplied by the rectifier bridge to the collector of Q_1, but the zener voltage is continually compared with a portion of the load voltage by transistor Q_2, which supplies just enough base current to Q_1 to maintain a constant overall output. This circuit will keep the output voltage within about 0.5 percent for a change of load from 0 to 100 mA, and within 2 percent up to 400 mA. Other regulators can be designed which will give several orders of magnitude better constancy than this.

Regulator circuits generally respond in a time interval small compared with the period of the alternating source, so only a minimum of conventional filtering is required.

CURRENT REGULATION

A source of constant current is required for coulometric applications, and also as excitation for some types of light sources, such as the hydrogen discharge lamp. One approach which is applicable to small currents in electrochemistry lies in the use of a constant high-voltage source with a large dropping resistor in series with the electrolysis cell (Fig. 26-24).

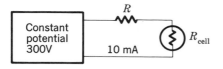

Fig. 26-24   Constant-current source.

Suppose 10 mA is desired and the drop across the cell is of the order of 1 V.   Then for a 300-V supply the series resistor R must be $299/0.01 \cong 30$ kΩ.   If the cell resistance should change by 100 percent, i.e., from 100 to 200 Ω, the current would only change by the factor $302/301$ or about 0.3 percent.

A more versatile device, which also eliminates any possible high-voltage hazard, consists of a regulating circuit comparable to Fig. 26-23, wherein the voltage supplied to the sensing transistor is the potential drop across a precision resistor in series with the load.

In the case of a power supply for a light source, it is desirable to initiate control by an auxiliary photocell which monitors the lamp.

AC AMPLIFIERS

The circuits previously shown are intended to help explain the properties of triodes, pentodes, and transistors.   More detailed schematics are necessary to represent operable amplifiers.   Figure 26-25 shows such a diagram for a pentode arranged to amplify small alternating signals. The signal, assumed to be in the audio-frequency range, is introduced through capacitor C_1 onto the control grid of the tube.   The grid is connected to ground through a resistance R_1 so that any electrons which may be intercepted by the grid can leak off to ground.   The amplified

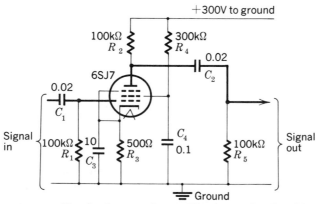

Fig. 26-25   Circuit of a pentode amplifier for ac signals.   The path taken by the signal is denoted by heavy lines; capacitances are given in μfd.

signal which appears at the plate is taken off through the *blocking capacitor* C_2 to the output (which might be the grid of another tube).   The purpose of C_2 (and similarly C_1) is to prevent the dc plate potential of one tube from exerting any influence on the grid of the next.   The required negative dc potential for the grid is obtained from the voltage drop across R_3, the *cathode resistor*, which carries the entire tube current.   The cathode is operated a few volts positive (dc), but the *bypass capacitor* C_3 ensures that the cathode is at ground potential so far as the signal is concerned.   The screen grid is maintained at a positive dc potential by R_4, but any signal that it may pick up is shunted to ground by C_4.   This circuit is said to be *capacitively coupled*, in reference to the fact that the signal is led in and out through capacitors.   Transformer coupling is also possible: the signal is passed from one stage to another inductively, while dc blocking action results from the lack of a direct connection between primary and secondary windings.

A comparable amplifier using a *p-n-p* transistor is shown in Fig. 26-26.   The signal is fed through a capacitor to the transistor base, where it serves to control the current flowing from emitter to collector. The purpose of the 100-K to 10-K voltage divider is to establish a suitable bias potential on the base.   Similar stages can be connected in series, just as with tube amplifiers.

The voltage gain, defined as the ratio of the output to input signal voltages, is about the same for these two amplifiers, about 165.   Notice that in both instances, the output signal is inverted with respect to the input: a rise in input produces a decrease in output.

CATHODE FOLLOWERS AND EMITTER FOLLOWERS

The output of an amplifier can be taken across the cathode or emitter resistor instead of at the anode or collector.   This is shown in Fig. 26-27. It can be shown that the voltage gain with this arrangement is just slightly less than unity—0.98 to 0.999.   This slight loss in voltage is more than offset by the gain in *current* which is available.   This results

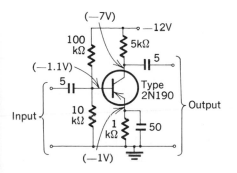

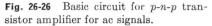

Fig. 26-26 Basic circuit for *p-n-p* transistor amplifier for ac signals.

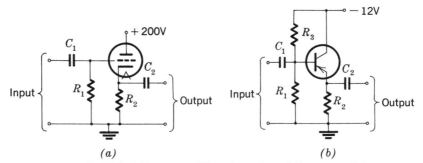

Fig. 26-27   (*a*) Cathode-follower amplifier; (*b*) emitter-follower amplifier.

from the fact that R_2 can be very much smaller than R_1, and since the potentials across the two are essentially equal, much greater power is available at the output than in the input circuit.   Thus we have a power amplifier but not a voltage amplifier.   The name comes from the fact that the signal voltage on the cathode or emitter *follows* exactly whatever variations are presented to the grid or base.   The follower is utilized as an *impedance converter* to couple a high-impedance circuit to one of low impedance.   The input impedance is high enough to receive signals from sources of moderately high impedance, though not usually high enough for measurements with the glass electrode or with ionization chambers.

ELECTROMETERS

When it is necessary to measure extremely small currents (in pico-amperes), such as those passed by some types of ionization gages and GC detectors, or potentials arising in circuits of very high resistance (in gigohms), as in glass electrodes, an amplifier of unusually high input impedance is required.   Any instrument capable of such service is called an *electrometer*, no matter what kind of components it contains.

There are three basic kinds of electrometers in use today, based on (1) special vacuum tubes, (2) vibrating capacitors, and (3) field-effect transistors.   Representative of special electrometer tubes is the type 5886, a subminiature tetrode which is constructed with particular attention to excellence of insulation between elements.   It is operated at a low anode potential so that the electrons will not acquire sufficient energy to eject secondary electrons nor to ionize residual gas molecules with which they may collide.   The tube is shielded from light so that no electrons will be emitted from the grid by the photoelectric effect.   The quiescent anode current will not exceed the microampere range, but this is sufficient to permit sizable current and power gains.

The vibrating-capacitor electrometer achieves high input impedance by converting the dc signal to a proportional ac signal.   This is accomplished by means of an air-dielectric capacitor C_v (Fig. 26-28), in which one plate is made to vibrate at some frequency, preferably not a multiple or submultiple of the line frequency. The signal current charges C_v at a constant rate, but as C_v is continually changing, the potential developed across it by the constant charging current will also vary at the frequency of vibration, and so a corresponding ac signal will be passed by capacitor C to the amplifier. This arrangement results in a very high input impedance.

The high input resistance of FET and especially MOSFET devices has already been mentioned. These can provide the input circuits for excellent electrometers.

DC AMPLIFIERS

An amplifier built around a single tube or transistor will amplify signals at any frequency from 0 Hz (that is, direct-current) up to some maximum which varies with the type of device and circuit. Special difficulties arise with dc signals when multiple stages are required. For one thing, the plate of a tube is usually at a high positive (quiescent) potential, and hence cannot be connected directly to the grid of the next stage, which must be slightly negative. Another difficulty is concerned with *drift*. Any gradual change in the first stage, such as might be produced by a change in ambient temperature or by ageing of some component, will be amplified by the successive stages and appear at the output indistinguishable from true signal. Neither of these problems occurs in ac amplifiers because the capacitive (or inductive) coupling between stages prevents passage of dc potentials.

One possible approach to dc amplification lies in the use of a regulated source of voltage with a bleeder resistor provided with a large number of taps. It is then possible to select potentials from the bleeder so that each of a series of tubes has the correct relative voltages for proper operation. This approach has been widely employed in the past

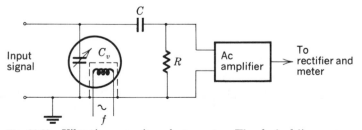

Fig. 26-28 Vibrating-capacitor electrometer. The dashed line represents an electrostatic shield. The frequency f should not be a multiple or submultiple of line frequency.

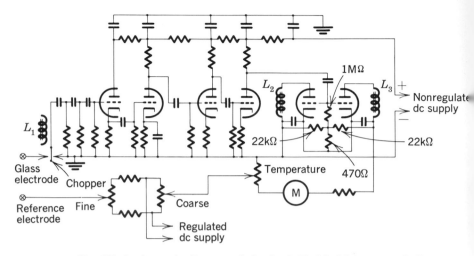

Fig. 26-29 Simplified schematic diagram of the L. & N. Model 7401 pH Indicator. An input-switching arrangement has been omitted for clarity. L_1, L_2, and L_3 are all secondaries on the same 60-Hz transformer. (*Leeds & Northrup Co., North Wales, Pa.*)

(the Beckman Model H-2 pH meter is an excellent example), but is now outmoded, partly because it made no provision for the elimination of drift problems.

A highly successful method of eliminating drift is by *chopping* or *modulating* the signal, which has the effect of converting the dc signal to an equal ac signal. An ac amplifier, which is not subject to drift, may then be utilized. The vibrating-capacitor electrometer presents an example of this principle.* Less expensive, but nevertheless capable of excellent performance, is a type of chopper in which a vibrating metallic reed touches first one electric contact, then another, in synchronism with the ac line.

Figure 26-29 shows an example of chopper amplification in the Leeds & Northrup Model 7401 pH Indicator. The input to the first tube is connected (through an RC filter) alternately to ground and to the glass electrode. If the potential of the electrode is either above or below ground, then an ac signal is produced. This signal is passed through a cathode follower for impedance conversion, then through three conventional stages of amplification to a double-triode *phase detector*. The twin plates of the phase detector are fed from two secondary windings on a

---

* Note the similarity between this chopping of a dc electric signal and the chopping of a beam of radiation by means of a mechanical shutter, as is done in most recording spectrophotometers.

60-Hz power transformer phased so that one plate is positive when the other is negative. Consider the situation at an instant when the left-hand plate is positive and the right-hand plate negative: only the left section can conduct. If the signal arriving from the amplifier at this instant is at the negative peak of its cycle, the current through the left triode section will be reduced; if positive, it will be increased. Since the phase of the signal reverses at the same frequency as the plate potentials, the effect of a given signal on the right-hand triode will be just the opposite to that on the left. The indicating meter and the RC network at the cathodes have a sufficiently long time-constant that the meter shows a constant deflection which is proportional to the dc input signal, and the direction of the deflection indicates the polarity of the input. Controls for standardization and temperature compensation are located in the line leading to the reference electrode. The chopper amplifier eliminates the need for an electronically regulated rectifier to supply voltage to the tubes, but a low-current regulated supply is provided for the calibrating potential.

There are several alternative ways in which a chopper can be connected to an ac amplifier, each with its own advantages. Two of these are shown in Fig. 26-30.

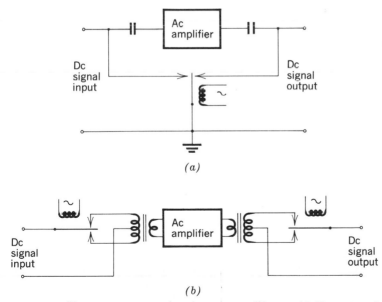

(a)

(b)

Fig. 26-30 Two arrangements for chopper amplifiers. (a) Economy of parts, more prone to noise pickup; (b) more parts, but permits isolation of output from input and gives better discrimination against noise.

OPERATIONAL AMPLIFIERS

We will now consider a class of dc amplifiers especially designed to perform mathematical operations (hence the name) on signals presented to them.   Appropriate external connections establish conditions such that the output voltage will be (1) the algebraic sum of two or more input voltages, (2) the product of an input voltage multiplied by a constant factor, (3) the time integral of the input, or (4) the derivative of the input with respect to time.   Other mathematical functions, such as taking logarithms or antilogarithms, squaring, or multiplying or dividing one variable quantity by another, can be implemented by the use of nonlinear components along with the amplifiers.

To be suitable for such applications an amplifier must have the following attributes: (1) it must have a large negative gain, at least -10^4, with many commercial units well beyond -10^6; the negative sign signifies inversion—the output is of opposite sign to the input; (2) it must have a large input impedance, not less than $10^5 \, \Omega$, often up to 10^{12} or even higher; (3) it must be capable of being nulled, that is, to give zero output for zero input; and (4) it must have only minimal drift.

Most operational amplifiers have two input terminals, only one of which produces an inversion of sign.   The terminals are conventionally marked $+$ and $-$, as in Fig. 26-31.   These designations do not mean that the terminals are to be connected only to potentials of the indicated sign, as would be the case with similar marks on a voltmeter, but rather that the one marked $-$ gives sign inversion and the other does not.   In those operational amplifiers which have only a single input, it is always the noninverting one which is omitted.   In case the noninverting input

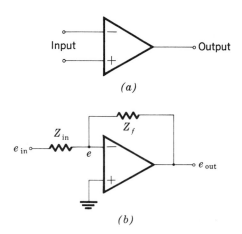

(a)

(b)

Fig. 26-31 Operational amplifiers.   (a) General symbol; (b) connected as an inverting amplifier; $e_{\text{out}} = -(Z_f/Z_{\text{in}})e_{\text{in}}$, provided the internal gain $A \gg 1$; the triangular symbol implies that a suitable power supply is provided; all potentials are referenced to ground.

is not required in a particular application, it should be grounded to avoid instability.

The basic connections are shown in Fig. 26-31(b). Most circuits using operational amplifiers depend on negative feedback: a connection is made through a suitable impedance Z_f from the output to the inverting input. If the signal to be sensed by the amplifier is a voltage, then it must be applied through an impedance Z_{in}. Since the input to the amplifier proper draws negligible current (it might well be the grid of an electrometer tube) the current flowing in Z_{in}, namely $(e_{in} - e)/Z_{in}$, must be equal to that in the feedback loop. But the current in the feedback loop, given by $(e - e_{out})/Z_f$, can come only from the output of the amplifier. Therefore, when an input signal is applied, the amplifier must adjust itself so that the feedback and input currents are precisely equal, or

$$\frac{e_{in} - e}{Z_{in}} = \frac{e - e_{out}}{Z_f} \tag{26-1}$$

which leads to

$$e_{out} = \frac{e(Z_f + Z_{in}) - e_{in}Z_f}{Z_{in}} \tag{26-2}$$

This relation can be greatly simplified by taking into consideration the high inherent gain of the amplifier (often called its *open-loop gain*). This means that potential e at the summing junction* must be very small compared to e_{out}. If the gain is 10^6, then an output of 10 V means that the input to the amplifier will be at a potential only 10 μV removed from ground. This is so close to ground that the summing junction is commonly said to be at *virtual ground*. Hence in Eq. (26-2), the term involving e can be neglected, giving

$$e_{out} = -e_{in}\left(\frac{Z_f}{Z_{in}}\right) \tag{26-3}$$

which is the basic working equation of an operational amplifier. In practice the ratio Z_f/Z_{in} is seldom made greater than 100 nor less than 0.01.

If Z_f and Z_{in} are purely resistive, they can be replaced by corresponding R's, and Eq. (26-3) shows that the output voltage will be the negative of the input multiplied by a constant, adjustable between 0.01 and 100.

Several inputs can be connected simultaneously to the summing

---

* The summing junction is a designation given to the inverting input connection of an operational amplifier for reasons that will appear shortly.

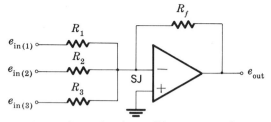

Fig. 26-32 Operational amplifier connected as a summer. SJ denotes the *summing junction*. For $A \gg 1$,

$$e_{\text{out}} = -R_f \left[\frac{e_{\text{in}(1)}}{R_1} + \frac{e_{\text{in}(2)}}{R_2} + \frac{e_{\text{in}(3)}}{R_3} \right]$$

junction, as in Fig. 26-32, in which case the output becomes the negative sum of the inputs, each multiplied by the appropriate ratio. It is because of this important property that the inverting input is called the *summing junction*. Any of these multiple inputs can be given a negative signal, resulting in subtraction.

To perform integration, the feedback element must be a capacitor, as in Fig. 26-33. Since negligible current can enter the amplifier proper, the input and feedback currents must be equal, and we can write

$$\frac{e_{\text{in}}}{R_{\text{in}}} = -C \frac{de_{\text{out}}}{dt} \tag{26-4}$$

which is equivalent to

$$e_{\text{out}} = -\frac{1}{R_{\text{in}}C} \int_0^t e_{\text{in}} \, dt \tag{26-5}$$

To differentiate, the positions of R and C must be interchanged, and we obtain

$$e_{\text{out}} = -R_f C \frac{de_{\text{in}}}{dt} \tag{26-6}$$

This simple circuit is little used in practice because it overemphasizes the effect of random noise at the input, and may be unstable.

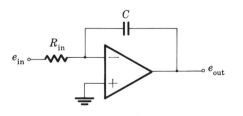

Fig. 26-33 Operational amplifier connected as an integrator. For $A \gg 1$,

$$e_{\text{out}} = -\frac{1}{R_{\text{in}}C} \int_0^t e_{\text{in}} \, dt$$

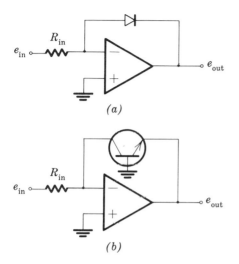

(a)

(b)

Fig. 26-34 Operational amplifier connected to give a logarithmic function. Feedback is (a) a silicon diode, (b) a silicon transistor with grounded base, both shown oriented for positive e_{in}. For $A \gg 1$,

$$e_{out} = k \log e_{in} - k'$$

in which k and k' are numerical constants.

Another important function is the logarithm, which can be obtained by utilizing the exponential characteristic of a forward-biased p-n junction. Figure 26-34 shows the circuit. The equation is derivable, as before, by setting equal the currents through the input connection and the feedback. The current-voltage relation of a p-n diode is approximated by

$$\log I = k_1 V \qquad \text{or} \qquad I = k_2 \text{ antilog } V \tag{26-7}$$

Hence we can write

$$\frac{e_{in}}{R_{in}} = k_2 \text{ antilog } e_{out}$$

which gives for e_{out}

$$e_{out} = k_3 \log e_{in} - k_4 \tag{26-8}$$

Experiment shows that a silicon diode is more satisfactory for this purpose than one of germanium, but that a silicon transistor with grounded base [Fig. 26-34(b)] is better yet; it can give a linear-log relation over at least four logarithmic decades.

Antilogarithms can be taken by interchanged connections, as in Fig. 26-35. It is possible to carry out multiplication or division of variables by taking their logarithms, adding or subtracting, then extracting the antilogarithm. It is essential that all the functional transistors (or diodes) be held at the same temperature or compensated for temperature changes.

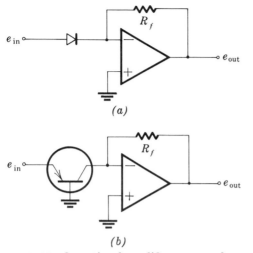

Fig. 26-35 Operational amplifier connected to give an exponential (or antilogarithmic) function. Input impedance is provided (*a*) by a silicon diode, (*b*) by a silicon transistor, both shown for positive values of e_{in}. For $A \gg 1$,

$$e_{out} = k \exp\left(-k'e_{in}\right)$$

in which k and k' are numerical constants.

TRANSDUCER APPLICATIONS OF OPERATIONAL AMPLIFIERS

Transducers by which chemical information is converted into electric signals can be classified according to the electrical quantity which represents the signal. There are only three major categories in which transducers can be considered to act, respectively, as (1) a variable resistor, (2) a source of potential, or (3) a source of current.

The class of *resistive transducers* includes photoconductive cells, vacuum (or gas-filled) phototubes, photomultipliers, thermistors, metallic resistance thermometers, and cells for electrolytic conductivity (most operational amplifiers will work at audio frequencies as well as at dc).

In principle, any of these can be used either as Z_{in} or Z_f in the circuit of Fig. 26-31. If e_{in} is replaced by a constant potential E_{in}, then observation of e_{out} will allow the unambiguous determination of the resistance of the transducer. If the quantity desired is actually *conductance*, then the transducer can conveniently be placed in the input so that the reciprocal will be obtained directly. Difficulty may be encountered if the resistance to be measured is larger than about 100 MΩ, as may occur with a photoelectric cell in near darkness.

If a *ratio* of two resistances is desired, as might be the case in some dual-beam photometers, then it may be possible to use them as Z_{in} and

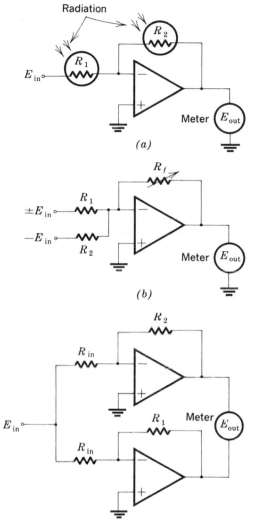

Fig. 26-36 Resistance measurement with operational amplifiers. (a) The ratio of two resistances (shown as light-sensitive): $E_{out} = -(R_2/R_1)E_{in}$; (b) the sum or difference of resistances; the two E_{in}'s must have the same absolute value, then adjustment of R_f to make $E_{out} = E_{in}$ will give $R_f = (R_2 \mp R_1)$; (c) provided the two resistors marked R_{in} are equal, then $E_{out} = (E_{in}/R_{in})(R_1 - R_2)$.

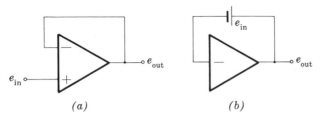

(a)   *(b)*

Fig. 26-37   Potential measurement with an operational amplifier.   *(a)* "Potential follower."   *(b)* Comparable circuit for single-input amplifier.   In either case, always provided $A \gg 1$, $e_{out} = e_{in}$.

Z_f with a single amplifier, Fig. 26-36(a).   If their *difference* is required, rather than their ratio, the circuits of Fig. 26-36(b) or (c) can be utilized. Circuit (b) requires two standard voltages of equal value but opposite sign, which may not always be available.   Circuit (c) needs only a single reference voltage, but requires two amplifiers; it has an added feature which may be a disadvantage, that the output meter cannot be grounded.

Transducers which produce *potentials* include the many electrode combinations for potentiometry and chronopotentiometry, photovoltaic cells, and thermocouples.   The general circuit of Fig. 26-31(b) is not suitable for potential measurement, as it requires current to flow from the source.   Instead, either of the circuits of Fig. 26-37 is to be preferred. Circuit (a) is better in that the potential to be measured can be referenced to ground, but it requires a differential amplifier.   The arrangement at (b) for single-input amplifiers will give equally valid results, but the potential being measured must be floating, i.e., not connected to ground, even though one end is at a *virtual* ground.

Currents produced by transducers are of importance in voltammetry, polarography, amperometry, and also with photovoltaic cells and with ionization chambers and flame-ionization GC detectors.   These current sources can be connected directly to the summing junction of the amplifier, as in Fig. 26-38.   In order to keep this junction at virtual ground, the amplifier forces an equal current through the feedback resistor, thereby building up a potential which is the output e_{out}.

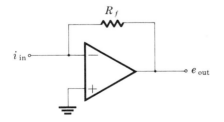

Fig. 26-38   Current measurement with an operational amplifier. $e_{out} = -R_f i_{in}$

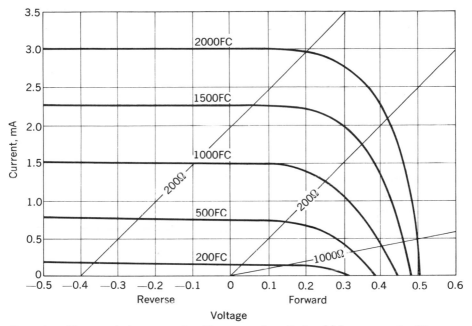

Fig. 26-39 Characteristic curves of a silicon *p-n* photodiode which can operate either as a photovoltaic or a photoconductive device. (*Solar Systems, Inc., Skokie, Ill.*)

There are a few transducers which can be operated in alternative ways to produce desired mathematical functions and degrees of sensitivity. An example is a silicon *p-n*-junction photocell. Figure 26-39 shows a family of current-voltage curves for such a unit. This device can be operated as a photovoltaic cell at zero current, using one of the circuits of Fig. 26-37, in which mode it will give an approximation to a straight line when the logarithm of illumination (in such units as foot-candles or lux) is plotted against volts. It can also be given a reverse bias of a few tenths of a volt in the circuit of Fig. 26-40, which corresponds

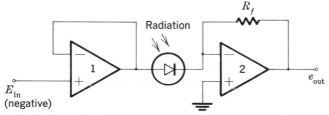

Fig. 26-40 Photosensitive *p-n*-junction diode operated at constant reverse bias. Amplifier 1 can be omitted if the voltage E_{in} is sufficiently well regulated.

to a vertical operating line to the left of the center in Fig. 26-39; the output will be a very closely linear function of illumination. If the cell is connected directly to a low-resistance meter, its response will follow an intermediate curve; for example if the resistance of the meter is 200 Ω, the response curve will include the intersections shown in Fig. 26-39 for a 200-Ω "load line" passing through the origin. A combination of negative bias and low-resistance meter may give linear results, as indicated by the 200-Ω load line starting at 0.4 V reverse bias.

OTHER APPLICATIONS OF OPERATIONAL AMPLIFIERS

In addition to their functions in handling the output of transducers, operational amplifiers have wide areas of usefulness in auxiliary positions in connection with analytical systems.

As indicated in Fig. 26-33, an amplifier with capacitive feedback acts as an integrator. Such a circuit is often used as a coulometer in electrochemical studies. If the current to be integrated is small, it can be led directly to the summing junction, omitting R_{in}, so that the governing relation is simply $e_{out} = -C^{-1} \int_0^t i_{in}\, dt$. For larger currents i_s, e_{in} must be taken as the voltage drop across a series resistor R_s, and the overall relation is

$$e_{out} = -\frac{R_s}{R_{in}C} \int_0^t i_s\, dt \tag{26-9}$$

An example of this application will be discussed later.

An integrator provided with a constant input potential will produce a ramp output, that is, $e_{out} = -(RC)^{-1} \int_0^t dt = (RC)^{-1}(t - t_0)$, such that the voltage increases linearly with time, which makes a very convenient polarizing unit for recorded voltammetry and polarography.

An operational amplifier can serve as an adjustable constant-voltage source for comparison potentiometry, so that, for example, the defined standard potential of a Weston cell can be utilized without the possibility of damage due to the passage of current. Figure 26-37(a) or (b) would be suitable.

Constant-current sources can be assembled by following Fig. 26-31(b), where e_{in} and Z_{in} are held constant; the current through Z_f, which becomes the load, is determined by the ratio e_{in}/Z_{in}, and is independent of Z_f. The difficulty with this is that the load cannot be grounded, which is sometimes important. This restriction can be avoided by using either two amplifiers or two reference potential sources, as in Fig. 26-41.

It must be emphasized that in practice other considerations, such

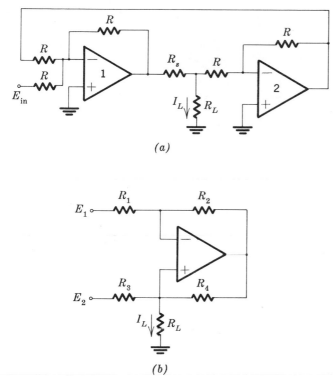

Fig. 26-41 Constant-current sources with grounded load. (a) All resistors marked R are equal; $I_L = E_{in}/R_s$; two amplifiers are required, but only one voltage source. (b) The "Howland" circuit: only one amplifier is needed, but two voltage sources; if $R_1/R_2 = R_3/R_4$, then $I_L = (E_2 - E_1)/R_3$. In both circuits, note that the load current I_L is independent of load resistance R_L.

as voltage and current offset and temperature effects, must not be overlooked for successful application of operational amplifiers.

ANALOG COMPUTERS

An electronic analog computer consists of an assemblage of operational amplifiers together with the necessary power supplies and varying quantities of supporting equipment. Commonly all input and output connections are brought out to a logical array of jacks on a panel. A selection of capacitors and both fixed and variable resistors provided with mating jack plugs is made available, along with numerous patch cords. Most such computers have built-in meters with which to establish input potentials and measure outputs, but the data are usually taken from the computer by recorder or oscilloscope.

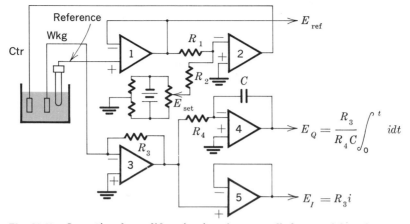

Fig. 26-42   Operational amplifier circuitry for controlled-potential coulometry. The reference potential, working current, and coulombs passed can be monitored simultaneously.

The original purpose of analog computers was the solution of algebraic or differential equations, especially in an engineering context. For such use, sometimes a great many amplifiers (hundreds) are required. For purposes of chemical instrumentation, few applications require more than about 10 amplifiers. Any of the operational amplifier circuits discussed in the preceding pages can be implemented on a small computer, often with greatly increased convenience compared with the use of individual discrete amplifiers.

Figure 26-42 is presented as an example of an analytical instrument utilizing five operational amplifiers.[3] This could easily be assembled with patch cords on the panel of a general-purpose analog computer. In controlled-potential coulometry, three electrical quantities are of interest: the potential of the working electrode against a reference, the current passing between working and counter electrodes, and the number of coulombs required to carry out a chemical process. All of these quantities can be controlled or observed with the analog equipment shown.

Notice first that the working electrode is connected directly to a summing junction (of amplifier 3), and hence is always effectively at ground potential. Therefore, to maintain a desired voltage between the working and reference electrodes, we must establish the reference electrode at a point away from ground. This is accomplished by amplifier 2, as the potentials of the reference electrode and of a manually settable voltage source (E_{set}) are summed at its summing junction. For convenience, R_1 may be made equal to R_2, so that because of the virtual ground restriction one may be sure that the reference electrode potential is equal to E_{set}, but of opposite sign. Amplifier 2 delivers as much current to the counter electrode as it needs to maintain the desired condition. Amplifier 1, connected as a voltage follower, is solely to prevent any current being passed through the reference electrode.

The current flowing through the working electrode must also successively pass through R_3 and R_4, where it charges capacitor C of the integrator (amplifier 4) which measures coulombs. If there is likelihood of the integrator going off-scale, then provision can be made to discharge the capacitor when it reaches a particular voltage level (as 10.00 V) and to count on a mechanical register the number of times such discharging is required. The voltage follower (amplifier 5), being connected between R_3 and R_4, gives a measure of the electrolysis current at any moment.

SPECIAL DEVICES

OSCILLATORS

A source of alternating current (other than line frequency) is often needed in laboratory instruments and is most conveniently obtained electronically. In principle any amplifier can be made into an oscillator by providing a positive feedback path with a suitable frequency characteristic. This means returning part of the output to the input in such a phase that an increase in output causes an increase in input, which in turn causes a decrease in output, and hence in input, and so on ad infinitum.

Figure 26-43 shows one (of many) circuits that will do this: the Hartley oscillator, in both triode and transistor versions. The feedback from the output circuit to the input takes place through the mutual inductance of the two portions of the inductor L. The frequency of

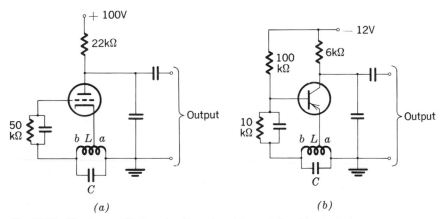

Fig. 26-43 Hartley oscillator circuits using (a) a triode, (b) a transistor.

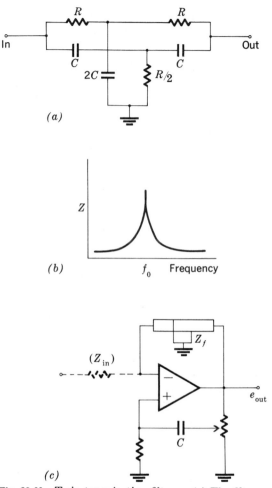

Fig. 26-44 Twin-tee rejection filter.   (a) The filter circuit for which the characteristic frequency is $f_0 = 1/(2\pi RC)$ Hz.   (b) Impedance as a function of the impressed frequency.   (c) An operational amplifier as an oscillator, with twin-tee feedback.

oscillation is $f = (2\pi)^{-1}(LC)^{-\frac{1}{2}}$ Hz, and is most conveniently altered by change of capacitance.

Another way in which frequency can be fixed is by a *twin-tee filter* [Fig. 26-44(a)]. This network will pass all frequencies *except* that given by the formula $f = (2\pi RC)^{-1}$ Hz.   A plot of impedance as a function of frequency shows a very sharp peak [at (b) in the figure]; the sharpness depends on how accurately the resistors and capacitors are matched.

If this network is connected as the negative feedback impedance of an amplifier (which may or may not be an operational amplifier as shown), then the net gain, $e_{out}/e_{in} = -Z_f/Z_{in}$, will be low for all frequencies other than the characteristic frequency of the filter. We may suppose that an unshielded summing junction will pick up enough noise at random frequencies to constitute an input (shown dotted in the figure), and this will cause oscillations to start at the filter frequency, since this is *not* fed back negatively and hence is amplified by the full open-loop gain of the amplifier. To sustain oscillation, a small, untuned, positive feedback may be injected at the noninverting input, as shown. This oscillator is especially convenient for service at one or a few fixed frequencies.

THE "FLIP–FLOP" CIRCUIT

An interesting circuit can be formed by symmetrically interconnecting two triodes or transistors as shown (for the triode case) in Fig. 26-45. In order to make the discussion quantitative, we will assume the tubes both to be type 6J5, so that the curves of Fig. 26-3 apply.

Suppose that initially tube V_1 is conducting and V_2 is cut off. If the plate current in V_1 is 4.5 mA, the drop in the 33-kΩ plate resistor is 150 V, so that the actual potential on the plate is 250 V. The drop in the 22-kΩ cathode resistor is 99 V. A study of Fig. 26-3 will show that for the stated plate current (4.5 mA) and plate-cathode voltage (250 − 99 ≅ 150 V), the grid potential must be −5 V, which value is provided

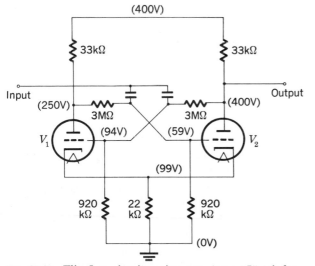

Fig. 26-45 Flip-flop circuit, using two type 6J5 triodes. Voltages are given in parentheses.

by the 3-MΩ–920-kΩ voltage divider between ground and the plate of
the other tube V_2. Note that the plate potential of V_2 is 400 V, since
the tube is cut off and there is no voltage drop across its plate resistor.
Analogous voltage divider action between the plate of V_1 and ground
fixes the grid of V_2 at 59 V above ground, or 40 V negative with respect
to the cathode, so that in spite of the high voltage on its plate, V_2 remains
cut off. All resistance values in Fig. 26-45 are symmetrical, so that the
voltages indicated for V_1 could equally well be interchanged with those
for V_2, whereupon V_1 would cut off and V_2 conduct. Thus this is a
bistable device.

If now a positive pulse is impressed through the capacitors onto
the two grids, there will be no primary effect on V_1, which is already
conducting, but the grid of V_2 will be driven more positive than it was,
which will result in a reduction of the potential of its plate, which in turn
drives the grid of V_1 negative. So the net result of a positive pulse is to
transfer the conduction from one tube to the other; a second positive
pulse returns it to its first condition. This is called a flip-flop action.
Negative pulses have the same effect as positive ones, though the
mechanism is slightly different.

The utility of this circuit arises from the fact that the output, taken
from the plate of either tube, shows a single pulse for every *two* input
pulses. From this point of view, the circuit is often called a *scale-of-two*.
Any number of these units can be placed in series, resulting in overall
scale factors of 2, 4, 8, 16, ... The application of such scalers in
nuclear measurements is described in Chap. 16.

A simple modification of this circuit, including a feedback connection, permits it to trigger itself and run freely without the need for input
pulses. In this configuration, it is called a *multivibrator* and is a convenient source of square waves. The frequency is determined by the
RC time constants in the cross-coupling segments.

SERVOMECHANISMS

An instrument servo system in its most common form consists of a small
motor, usually a two-phase ac type, which is controlled in speed and
direction of rotation by an amplifier. The system is provided with some
sort of *feedback* so that the turning of the motor produces a signal which
is automatically compared with a standard or reference signal. The
difference, called the *error signal,* is returned to the amplifier to control
the motor.

A servo may be included, for example, in a recording spectrophotometer to control the slit width. The beam of radiation which
passes through the reference cuvet (solvent) serves as the signal to
operate the slit-control motor in such a manner as to maintain the

transmitted energy constant as the spectrum is scanned. For another application, see Fig. 14-3.

An important servo application is to the self-balancing potentiometric recorder; this will be described in the next section.

AUTOMATIC RECORDERS

Electric recorders fall into two general classes: deflection and null instruments. The deflection meters, which may be galvanometers, voltmeters, ammeters, wattmeters, etc., are generally less complex in design and can be much the faster in response, following variations of input up to perhaps 100 Hz. However, their accuracy is generally inferior to null types, and they are restricted to narrow recording paper, sometimes with curvilinear coordinates.

A major drawback to deflection instruments is the loading effect on the system being measured. It requires a considerable amount of power to operate the deflection system, and this must be provided by the circuit connected to the meter. Thus a deflecting millivoltmeter draws a significant current, and a deflecting milliammeter causes a significant voltage drop in its circuit.

Deflecting recorders which are especially designed for high speed of response are known as recording oscillographs. They are particularly useful in engineering measurements on dynamic system, vibration studies, and the like, and are also applied to biological and clinical studies, including encephalography and electrocardiography. The principal application in chemistry is in mass spectrometry, particularly high resolution (see Fig. 17-14).

Null recorders may be typified by the self-balancing potentiometer shown schematically in Fig. 26-46. The unknown potential E_x is connected in series opposition with a potential E_p taken from a slide-wire potentiometer. A vibrating chopper is so connected that it throws the

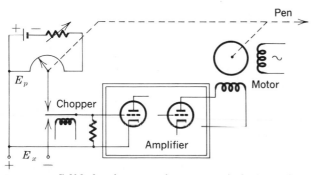

Fig. 26-46 Self-balancing potentiometer, typical schematic.

two potentials alternately onto the input of an ac amplifier, at the frequency of the power line.   The output of the amplifier energizes one winding of a two-phase motor, the other winding of which is connected to the power line.   The motor controls the moving contact on the potentiometer slide wire.   At the point of balance, $E_x = E_p$, and the amplifier receives no 60-Hz signal from the chopper, hence supplies no 60-Hz output, and the motor is idle.   If E_x becomes larger than E_p, a proportionate signal is observed and amplified, and the motor turns in such a direction as to increase E_p to rebalance the circuit.   If E_x becomes smaller than E_p, a similar action takes place, but the output of the amplifier is shifted in phase so that the motor turns in the opposite direction to attain balance.   The motor shaft is mechanically linked to the recording pen, causing it to move a distance proportional to the angular displacement of the sliding contact.   This is therefore a measure of the unknown potential E_x.

The working current through the slide wire must be adjusted so that the deflection of the pen agrees with the calibration of scale and paper. This is generally accomplished by means of a standard cell and rheostat.

REFERENCES

1. Boucher, E. A.: *J. Chem. Educ.*, **44**:A935 (1967).
2. Gray, P. E., D. DeWitt, and A. R. Boothroyd: "SEEC Notes I; PEM: Physical Electronics and Circuit Models of Transistors," copyright 1962 by Education Development Center (formerly Educational Services, Inc.), published by John Wiley & Sons, Inc., New York.
3. Propst, R. C.: "A Multipurpose Instrument for Electrochemical Studies," A.E.C. Research and Development Report DP-903 (1964).

GENERAL REFERENCES, WITH COMMENTS

The following four books are introductions to electronics intended for chemists and other scientists not making electronics their career:

Benedict, R. R.: "Electronics for Scientists and Engineers," Prentice-Hall, Inc., Englewood Cliffs, N.J., 1967.
Brophy, J. J.: "Basic Electronics for Scientists," McGraw-Hill Book Company, New York, 1966.
Malmstadt, H. V., C. G. Enke, and E. C. Toren, Jr.: "Electronics for Scientists," W. A. Benjamin, Inc., New York, 1962.   (Many laboratory experiments are included.)
Phillips, L. F.: "Electronics for Experimenters in Chemistry, Physics, and Biology," John Wiley & Sons, Inc., New York, 1966.   (Available as a paperback.)

The next two, representative of a large selection, are texts on general electronics:

Angelo, E. J., Jr.: "Electronic Circuits, A Unified Treatment of Vacuum Tubes and Transistors," 2d ed., McGraw-Hill Book Company, New York, 1964.

Lurch, E. N.: "Fundamentals of Electronics," John Wiley & Sons, Inc., New York, 1960.

Somewhat more specialized:

Sevin, L. J., Jr.: "Field-effect Transistors," copyright by Texas Instruments, Inc., 1965; published by McGraw-Hill Book Company, New York.
Walston, J. A. and J. R. Miller (eds.): "Transistor Circuit Design," copyright by Texas Instruments, Inc., 1963; published by McGraw-Hill Book Company, New York.

The most complete and extensive of the manufacturers' manuals, with respect to semiconductor theory and circuitry:

Cleary, J. F. (ed.): "G. E. Transistor Manual," 7th ed., General Electric Co., Syracuse, N.Y., 1964.

Laboratory Experiments

In the procedures which follow, emphasis is placed on the analytical principles involved, the preparation of the sample, adjustment of such conditions as pH, concentration, etc., and the treatment of results. For the most part, where specialized apparatus is required, the details of operation are not included. It is assumed that the instructor will either demonstrate the use of such apparatus or provide copies of the manufacturer's instructions. If the modular apparatus from A. R. F. Products, Inc., Raton, New Mexico, is to be employed, directions for its use should be obtained from the manual written for the purpose.*

There are certain features common to many different instrumental assemblies, which should be discussed before laboratory work is undertaken.

INDICATING METERS

These are usually moving-coil (d'Arsonval) instruments, and must be treated with care. Small meters are equipped with jeweled bearings

* G. W. Ewing, "Analytical Instrumentation, a Laboratory Guide for Chemical Analysis," Plenum Publishing Corp., New York, 1966.

or taut suspensions. The former, in which electric connection to the coil is made through a pair of fine, coiled springs resembling the hair spring of a watch, are more susceptible to damage by mechanical shock than the latter type. Taut-suspension meters for panel mounting use the same metallic wires or ribbons for mechanical support of the coil and for electric connection. The most sensitive galvanometers, employing a quartz fiber for suspension, are extremely delicate; they are not encountered in the analytical laboratory so much as formerly.

When a sensitive meter is to be moved or stored, the coil should always be short-circuited by connecting its terminals with a wire. This reduces markedly the likelihood of damage from rapid motion of the coil.

Nearly every meter is provided with some mechanical device for adjusting the indicator to the zero of the scale when no current is flowing. In panel meters, the mechanical zero is usually adjusted with a screwdriver, and is so constructed that turning the adjustor too far cannot damage the meter. In some suspension types, a knurled knob at the top serves the same purpose, and turning it too far may break the suspension.

Meters are usually designed so that an overload of 150 to 200 percent will cause no damage. Some are provided with built-in fuses for protection.

Recording meters are described in Chap. 26.

OPTICAL PARTS

Lenses, prisms, cuvets, and other items with optical surfaces exposed to the atmosphere or to chemical solutions require frequent cleaning, which must be done with care to avoid the possibility of scratching the surfaces. Fingerprints (which should be avoided in the first place), films of ammonium chloride, and similar nongritty substances can usually be removed by gentle wiping with lens tissue moistened slightly if necessary by breathing on the glass surface (*not* on sodium chloride surfaces!). Good quality facial tissue, if entirely free of grit, is satisfactory in place of lens paper.

When washing is necessary, use a solution of detergent followed by rinsing with tap water and distilled water. Neither chromic acid cleaning solution nor aqua regia should be employed unless absolutely necessary, and not without the instructor's specific approval. Diffraction gratings and front-surface mirrors should be cleaned only with the lightest strokes of a clean camel's-hair brush; if this is not adequate, an expert must do the job.

If any optical surface becomes fouled with cement or other massive substance, an expert should be consulted. The same is true if it is suspected that an optical system is out of alignment.

MECHANISMS AND MOTORS

Caution must be exercised in oiling moving parts. The usual tendency when any mechanism shows signs of sticking is to oil it. This service should be left to the instructor as a guard against excessive oiling. Too much oil can be nearly as deleterious as too little, and it is more difficult to remedy.

UNKNOWNS

Unknowns may be solids, liquids, or occasionally gases. The dissolution of solid samples sometimes causes trouble. Specific directions for dissolving ores, steels, etc., can be found in many texts on quantitative analysis, and so are not detailed here.

In some experiments, it will be suggested that duplicate determinations be made with different apparatus. The purpose is to provide a basis for comparison of the advantages and disadvantages of comparable instruments.

WATER

In nearly every instance where distilled water is specified, demineralized (deionized) water may be used instead.

REPORTS

All numerical data taken during a laboratory session should be entered directly in a permanent notebook, together with adequate descriptive matter for the later identification of the numerical entries and notes to bring back to mind each step performed, particularly any deviations from, or additions to, the printed instructions. *Always date everything.* A complete report, written later, is to be submitted to the instructor for approval and grading. The final report should follow a definite outline; the following is suggested:

1. Name (and, in parentheses, name of partner, if any).
2. Date performed, and date of writing report.
3. Number and title of experiment.
4. Objective of the experiment.
5. Theory. (A brief exposition; give literature references for any papers or books consulted, other than the text.)
6. Procedure. (Outline form; it is not profitable to copy laboratory directions from the book, but any changes therefrom should be described and justified.)
7. Apparatus. (Identify individual instruments when possible, preferably by giving serial numbers.)

8. Data. (Pertinent entries should be copied from the laboratory notebook; any erroneous entries or abortive attempts need not be included.)
9. Sample calculation. (The detailed calculations, particularly if lengthy, need not be given for all data, but should be presented once, with only the results from the remaining determinations.)
10. Graphs should be neatly and carefully constructed on good-quality graph paper. Particular care should be taken in choice of scales. It is desirable to have the curves cover nearly the whole paper, but this goal must often be modified for the sake of a convenient scale. One unit may be taken for every 1, 2, 5, or 10 divisions of the paper for convenient plotting of a variable with random values.
11. Results. (Each result should be calculated separately, followed by any appropriate averaging procedure. Be careful to indicate correctly the number of significant figures, and, where possible, the standard deviation.)
12. Discussion of results. (Comparison of the results obtained by different methods or upon different instruments should be included here, and comparison with known "true" values, if available. Any comments on the method or suggestions for improvement are in order. This section should tie in with the prior statement of the objective of the experiment, indicating how that objective has been achieved.)
13. Any questions included in the directions should be answered at the conclusion of the report.

EXPERIMENT 1 ABSORPTIOMETRY: A COMPARISON OF METHODS

REFERENCE: Text, Chap. 3.

PLAN: Various means of determining and plotting photometric data will be studied with a series of copper sulfate solutions of graded concentrations. These solutions will be measured in a low-dispersion spectrophotometer or a filter photometer with red and green filters. The colors observed will be those of the hydrated and the ammoniated cupric ion. The results will be plotted in several ways to bring out the advantages of each.

PROCEDURE: (1) Weigh out 5.0 g of granular $CuSO_4 \cdot 5H_2O$ (rough balance), dissolve, and make up to 100 ml in a volumetric flask with

0.05 F H_2SO_4.   Label this solution A.   Set up a pair of 50-ml burets, one containing solution A, the other 0.05 F H_2SO_4.

(2) Into a series of five 50-ml beakers place the following solutions:
   (a) 10 ml A
   (b) 8 ml A + 2 ml acid
   (c) 6 ml A + 4 ml acid
   (d) 4 ml A + 6 ml acid
   (e) 2 ml A + 8 ml acid

(3)   Withdraw 10 ml of A from the buret into a 100-ml volumetric flask, and dilute to the mark with acid; designate this solution as B. Empty the remaining A from the buret, rinse it with a little B, and fill it with B.

(4) Into five more 50-ml beakers place serial dilutions of B with 0.05 F H_2SO_4, following the same series as in step (2), substituting B for A. Label these f, g, h, i, and j.

(5) Repeat the dilution procedure of step (3) to give a solution C which is one-tenth the concentration of B.

(6) Repeat step (4), substituting C for B.   Label the five additional beakers k, l, m, n, and o.

(7) Make the initial setting of the photometer at 620 nm or with a red filter, with water in the cuvet.   In succession, insert portions of the 15 copper solutions a through o, returning each to the appropriate beaker after measurement.   Note that not all of the 15 solutions will give useful readings.

(8) Repeat (7) at 500 nm or with a green filter.

(9) Repeat (7) in the red, but with the initial adjustment made with solution B in the cuvet.   (Only solutions a through e need be measured.)

(10) To each of the cuvets or beakers add exactly 2 ml of concentrated aqueous ammonia.   If the precipitate which forms does not completely redissolve in some of the more concentrated samples, add more ammonia dropwise until solution is complete.   Repeat (7) with the ammoniacal solutions.

(11) Repeat (9) with the ammoniacal solutions, the initial adjustment made with solution f.

(12) Repeat (10) with the initial adjustment made with solution k. Measure solutions g through j.

(13) Plot photometric readings as percent T against relative concentration for each series of data (i.e., each combination of water or ammonia, red or green filter, water or copper-containing reference solution). This is Graph 1.

(14) Plot absorbance against concentration for each series (Graph 2).

(15) Plot percent T against the logarithm of concentration (Ringbom plot) for each series. Semilogarithmic graph paper may be used if desired.

QUESTIONS: (1) Over what concentration range would each of the methods show its greatest precision?
(2) Over what concentration regions is Beer's law obeyed for each method?
(3) Which method would you prefer for a sample containing about (a) 20 mg Cu per ml, (b) 2.0 mg Cu per ml, and (c) 0.2 mg Cu per ml?

EXPERIMENT 2 ABSORPTIOMETRIC DETERMINATION OF CHLOROFORM

REFERENCES: (1) Text, Chap. 3.
(2) M. Mantel, M. Molco, and M. Stiller, *Anal. Chem.*, **35**:1737 (1963).

PLAN: The solubility of chloroform in water will be determined at a series of temperatures by means of the Fujiwara reaction, the production of an absorbing species* by reaction with pyridine in the presence of sodium hydroxide. It is suggested that various groups of students select different temperatures and combine their results for a more extensive solubility curve.

MATERIALS PROVIDED: (1) Standard solution containing 1 g $CHCl_3$ per liter of water
(2) 40 percent (w/w) aqueous sodium hydroxide
(3) Pyridine

PROCEDURE: (1) Mix well several 1-ml portions of chloroform with 15 to 20 ml distilled water in erlenmeyer flasks in thermostat baths at various selected temperatures. After equilibration, transfer 0.5 ml of the aqueous layer to a 1-liter volumetric flask, fill to the mark with water, and mix well. (If the equilibration was above room temperature, the

* Speculation as to the nature of the reaction and the chromophoric species will be found in Ref. 2.

large flask should contain some room-temperature water before transfer of the solution. Note that as tap water is free of chloroform, it can be used for this dilution.)

(2) Make up several dilutions of the standard chloroform solution to the proper concentration range (0.2 to 5 ppm). A blank with no chloroform should be carried through the procedure.

(3) Pipet 5 ml of the unknown and of each standard into glass-stoppered erlenmeyer flasks, add 5 ml pyridine and 10 ml 40 percent sodium hydroxide to each, and mix well. Heat on a water bath at 70°C for 15 min with occasional swirling. Cool and transfer each to a separatory funnel, and allow the phases to separate. Transfer the colored pyridine layer to a 10-ml volumetric flask, and dilute to the mark with water.

(4) Read the absorbance against water at 366 nm, and prepare a calibration curve from the standards. If the blank shows appreciable absorbance, the other samples may be measured against a blank in the reference cuvet, or correction may be made numerically.

(5) Report your results in terms of the solubility of chloroform in water at the appropriate temperatures.

[This experiment was suggested by R. F. Hirsch.]

EXPERIMENT 3 COLORIMETRIC DETERMINATION OF CHROMIUM AND MANGANESE IN STEEL

REFERENCES: (1) Text, Chap. 3.
(2) J. J. Lingane and J. W. Collat, *Anal. Chem.*, **22**:166 (1950).

PLAN: The simultaneous determination of two constituents of a sample and the use of absorptivities in the calculation of results will be demonstrated.

MATERIALS PROVIDED: (1) Potassium persulfate crystals ($K_2S_2O_8$)
(2) Potassium periodate crystals (KIO_4)
(3) Silver nitrate solution, 0.5 F, in dropping bottle
(4) Mixed acid (H_2SO_4-H_3PO_4 or $HClO_4$-H_3PO_4, etc.)

PROCEDURE: (1) Weigh out a 1-g sample of steel, and dissolve it in mixed acid. Dilute (CAUTION) to about 150 ml, and heat to dissolve all salts.

Cool, transfer to a 250-ml volumetric flask, and fill to the mark with water.

(2) Pipet a 25-ml aliquot of the sample solution (filtered or centrifuged if not clear) into a 250-ml erlenmeyer flask. Add 10 ml of mixed acid, together with about four drops of 0.5 F silver nitrate (an oxidation catalyst) and 50 ml water. Add 5 g potassium persulfate and swirl the flask to dissolve most of the salt, then heat to boiling. Keep at the boiling point for 5 min, cool slightly, and add 0.5 g potassium periodate. Heat again and keep at the boiling point for another 5 min. Cool, transfer to a 100-ml volumetric flask, and dilute to the mark with water.

(3) Transfer a portion to a 1-cm glass cuvet and determine the absorbance at 440 and 545 nm in a spectrophotometer against a water reference.

(4) Calculate the percent of chromium and manganese in the sample by means of the simultaneous equations in Ref. 2.

(5) For greatest precision, the absorptivities of the permanganate and dichromate ions at each of the two wavelengths should be determined in advance on the same spectrophotometer. Corrections should be made for absorption due to ions of vanadium, cobalt, nickel, and iron, if present in considerable amounts (see Ref. 2). The instructor will provide information about these constituents for purposes of making the corrections.

EXPERIMENT 4 IDENTIFICATION OF A COMPLEX

REFERENCES: (1) Text, Chap. 3.
(2) S. P. Mushran, O. Prakash, and J. N. Awasthi, *Anal. Chem.*, **39**:1307 (1967).

PLAN: The formula of the absorbing complex between tetravalent vanadium and pyrocatechol violet (PCV), pyrocatecholsulfonphthalein, will be established by the methods of Yoe and Jones and of Job.

MATERIALS PROVIDED: (1) Vanadyl sulfate, $VOSO_4$, $2.00 \times 10^{-3}\ F$.
(2) PCV, $2.00 \times 10^{-3}\ F$.

PROCEDURE: (1) Prepare a series of dilutions of both $VOSO_4$ and PCV, in the following concentrations: 2.00, 1.75, 1.50, 1.25, 1.00, 0.75, 0.50, and $0.25 \times 10^{-4}\ F$. Adjust each solution to pH 4.2 by addition of HCl or NaOH as required.

(2) Determine the absorbances of both the 2.00×10^{-4} F solutions at 450 and 600 nm, in 1-cm cuvets, with a Beckman DU or equivalent spectrophotometer.

(3) For Job's method, mix together portions of the two substances to give a total concentration of 2×10^{-4} F in the several combinations suggested by the series of solutions which you have prepared, and determine the absorbance of each at both 450 and 600 nm.

(4) For study by the Yoe-Jones method, mix successive amounts of dye solution to equal concentrations (1.00×10^{-4} F) of $VOSO_4$, measuring the absorbance of each at each of the two wavelengths.

(5) Prepare plots for each method at each wavelength, and deduce the values of n and m in the formula of the complex, $V_n^{(IV)}(PCV)_m$. Comment critically on the precision and usefulness of the four graphs.

SUGGESTIONS FOR FURTHER WORK: (1) Determine the range of concentrations of $VOSO_4$ over which PCV would provide a satisfactory analytical reagent. Ringbom plots should help in this determination.
(2) Study the extent of interference in the analysis of vanadium by the presence of other transition metals.

EXPERIMENT 5   ULTRAVIOLET SPECTRUM OF AN ORGANIC COMPOUND

REFERENCES: (1) Text, Chap. 3.
(2) Any atlas of ultraviolet spectra.

PLAN: An aromatic compound is to be identified by means of its ultraviolet absorption spectrum. The student will be issued as an unknown a pure substance taken from the following list of compounds:

Benzoic acid	Acenaphthene
p-Benzoquinone	1-Naphthol
Picric acid	2-Naphthol
Diphenylacetylene	Fluorene
Biphenyl	Phenanthrene
Tryptophane	Anthracene
Naphthalene	Anthraquinone
Formaldehyde-2,4-dinitrophenylhydrazone	

The absorption spectrum of the unknown will be determined with an ultraviolet spectrophotometer and identified by comparison with authentic spectra.

MATERIALS PROVIDED: In addition to the unknown sample, only a spectro-grade solvent is required.   Water, ethanol, and a hydrocarbon such as cyclohexane or isooctane (2,2,4-trimethylpentane), will cover all the possibilities.

PROCEDURE: (1) Weigh out 3.0 mg of the unknown on a semimicro balance.   Place it in a 100-ml volumetric flask, dissolve in a small amount of solvent, and fill to the mark.   (A preliminary experiment to determine which solvent is appropriate should be made.)   Call this solution A.

(2) Transfer 10 ml of solution A to another 100-ml volumetric flask and fill to the mark with solvent.   This is solution B.

(3) Fill one cuvet with pure solvent, a second with solution A, and a third with solution B.   Place all three in the holder for the spectrophotometer; in case it will accept only two, use solvent and A first, followed by solvent and B.   CAUTION: These silica cuvets are very expensive, and should be handled with extreme care, especially when being rinsed.

(4) Determine the absorbance of both solutions against the solvent, throughout the ultraviolet, from 400 nm down as far as the spectrophotometer will go, usually to about 210 nm.   If a recording instrument is employed, set the scanning rate at approximately 15 or 20 nm per min.   On a manual instrument, the absorbance must be determined point by point.   The intervals between readings should be varied according to the steepness of the curve.   Where the curve is not steep, intervals of 5 nm will suffice, but where it varies rapidly, and in the immediate vicinity of maxima, readings should be made every 1 nm.

(5) Plot the absorbance against wavelength for both A and B solutions, preferably on the same sheet of graph paper; this, of course, is already done for you if you used a recording spectrophotometer.   Inspect the published curves for the compounds listed, in order to identify your unknown.

(6) After the unknown is identified, it becomes possible to calculate the molar absorptivity ϵ.   Determine this value for each maximum in the published curve, and tabulate, together with your own values.

(7) Comment on the agreement between your curves and the published ones, with respect to the locations of maxima and minima on the wavelength scale and in the ϵ values.

SUGGESTIONS FOR FURTHER WORK: (1) Identify a substance supplied by the instructor which is *not* on the above list.

(2) Choose an aromatic amine or N-heterocyclic compound, and determine its ultraviolet spectra in (a) 0.1 F HCl, (b) neutral alcohol, and (c) 0.1 F NaOH. Interpret the spectra in terms of the chromophores to be expected in each medium.

EXPERIMENT 6   ABSORPTIOMETRIC ANALYSIS OF APC TABLETS

REFERENCES: (1) Text, Chaps. 3 and 23.
(2) M. Jones and R. L. Thatcher, *Anal. Chem.*, **23**:957 (1951).

PLAN: The APC tablet is a common pharmaceutical preparation consisting of a mixture of aspirin, phenacetin, and caffeine. Each of these substances has characteristic absorption in the ultraviolet, the principal maxima lying at 277 nm for aspirin, 275 nm for caffeine, and 250 nm for phenacetin (all in chloroform solution).

The method of analysis calls for partition of the dissolved sample between chloroform and 4 percent aqueous sodium bicarbonate; the aspirin alone passes into the aqueous layer. The phenacetin and caffeine are analyzed simultaneously in chloroform. The aspirin solution is acidified, extracted back into chloroform, and determined spectrophotometrically.

MATERIALS PROVIDED: (1) Spectro-grade chloroform; used chloroform solutions should be returned to a bottle designated for the purpose, for recovery of solvent.
(2) Sodium bicarbonate, 4 percent, to which has been added a few drops per liter of concentrated hydrochloric acid.

PROCEDURE: (1) Weigh accurately one tablet. (A common size contains 3.5 grains aspirin, 2.5 grains phenacetin, and 0.5 grain caffeine, plus perhaps a small amount of starch or other inert material as a binder; 1 grain is approximately 65 mg.) Crush the tablet in a beaker with 80 ml of chloroform, then transfer it to a 250-ml separatory funnel, rinsing all particles in with a little more chloroform. Extract the chloroform solution with two 40-ml portions of chilled 4 percent sodium bicarbonate and then with one 20-ml portion of water. Wash the combined aqueous extracts with three 25-ml portions of chloroform, and add to the original chloroform solution. (Before proceding, acidify the bicarbonate solution as directed in step 2, to prevent hydrolysis of the aspirin.) Filter the chloroform solution through a paper previously wetted with chloroform (to remove traces of water) into a 250-ml volumetric flask, dilute to the mark with chloroform, and then dilute further a 2-ml aliquot to 100 ml with chloroform.

(2) Acidify the bicarbonate solution, still in the separatory funnel, with 25 ml of 1 F sulfuric acid. The acid must be added slowly in small portions. Mix well only after most of the carbon dioxide evolution has ceased. The pH at this point should be 1 to 2 (pH test paper). Extract the acidified solution with eight 25-ml portions of chloroform, and filter through a chloroform-wet paper into a 250-ml volumetric flask. Dilute to volume, and then dilute further a 10-ml portion to 100 ml with chloroform.

(3) Prepare standard solutions in chloroform, containing respectively about 75 mg of aspirin, 20 mg of phenacetin, and 20 mg of caffeine per liter.

(4) Measure the absorbances of standard and unknown aspirin solutions at 277 nm in 1-cm silica cuvets. Correct for optical inequalities in the cuvets by interchanging the blank and solution in each case and averaging the results.

(5) With similar precautions, measure the absorbances of standards and unknown containing phenacetin and caffeine at both 250 and 275 nm.

(6) Calculate the quantity of aspirin by direct application of Beer's law, and of phenacetin and caffeine by means of simultaneous equations.

(7) Report the results in terms of milligrams of each constituent per tablet and also as a percentage of the total weight.

SUGGESTIONS FOR FURTHER WORK: Try the feasibility of substituting for chloroform a solvent with better ultraviolet transmission, either carrying out the extractions with the new solvent, or with chloroform followed by transfer to the new solvent for photometry.

EXPERIMENT 7 STUDY OF AN INDICATOR

REFERENCE: Text, Chap. 3.

PLAN: The ionization constant of an indicator is to be determined by means of its absorption spectra at a series of pH values, as determined with a photoelectric spectrophotometer. Thymol blue is chosen for this study because it changes color twice: below pH 1.2 it is red, from 2.8 to 8.0, yellow, and above 9.6, blue. (Any other acid-base indicator can be substituted, with adjustment of the pH values to be observed.)

Spectra of thymol blue will be plotted at each of the following pH values: 1.0, 1.2, 1.6, 2.0, 2.4, 2.8, 3.2, 4.0, 7.0, 7.6, 8.0, 8.4, 8.8, 9.2, 9.6, and 10.0. Each student will have time to determine only a few of these spectra during one laboratory session. It is suggested that the instructor assign the pH values among the students so that each will plot three or four spectra, and each spectrum will be checked by a number of students. All satisfactory curves will be plotted on a single sheet to be discussed at a later class meeting. For indicator theory, see any text on quantitative analysis.

MATERIALS PROVIDED: (1) Glycine, 1.0 F

(2) HCl, 0.1 F

(3) NaOH, 0.1 F

(4) A 0.5 percent aqueous solution of thymol blue, prepared by grinding 0.5 g of the dye in a mortar with 21.5 ml of 0.05 F NaOH, and diluting with water to 100 ml

PROCEDURE: (1) Prepare buffers at the assigned pH values by the following method: a portion of glycine solution is placed in a beaker in contact with the glass and references electrodes of a pH meter. Then acid or base is added slowly, with stirring, until the meter indicates the desired pH. The resulting solution has a fully adequate buffer capacity and accuracy for the purposes of this experiment.

(2) Combine by pipet exactly 2.00 ml of indicator solution with 25 ml of an assigned buffer, and mix thoroughly. Prepare other solutions in the same way for the other buffers.

(3) Fill one cuvet with water; place portions of the colored solutions in other matched cuvets. Be certain that the outer optical faces of the cuvets are clean and dry and that no air bubbles adhere to the inner walls.

(4) Determine the absorbances of each of the colored solutions, referred to water, at 10-nm intervals over the range 400 to 700 nm. If at any point the absorbances becomes so great that accurate readings cannot be made (A greater than about 0.9), it will be necessary to dilute the sample further for the region in question. The dilution must be performed quantitatively and with more of the same buffer.

(5) Plot all your curves on a single sheet of graph paper, as absorbance against wavelength, and determine the two pK values.

(6) Include in your report the structural formulas of the indicator in each of its forms.

SUGGESTIONS FOR FURTHER WORK: (1) The pK value of a colorless weak acid or base can be determined by similar studies in the ultraviolet. See references cited in Chap. 3 of the text.
(2) Devise and execute a similar experiment for study of a redox indicator.

EXPERIMENT 8 INFRARED ANALYSIS

REFERENCES: (1) Text, Chap. 5.
(2) Any atlas of infrared spectra.

PLAN: Commercial xylene will be analyzed quantitatively for its isomers and common impurities by comparison of its infrared spectrum with the spectra of the pure compounds. The concentrations of the three isomers will then be determined by application of the base-line technique to unknown and standards.

MATERIALS PROVIDED: (1) Pure samples of the isomeric xylenes
(2) Commercial xylene
(3) Cyclohexane, spectro-grade

PROCEDURE: (1) Obtain the infrared spectrum of commercial xylene in a 0.025-mm cell with sodium chloride windows. Select a scanning rate of approximately 1 μ per min, and a slit width set to vary automatically from 20 to 770 μ for the rock-salt region (2 to 15 μ). A salt plate equivalent to the thickness of the combined windows of the sample cell should be placed in the reference beam to compensate for absorption and reflection losses.

(2) Compare the recorded spectrum with the appropriate spectra in the atlas. List the compounds identified, with rough estimates of their quantities (as major, minor, or trace constituents).

(3) Make up a solution of 10 ml commercial xylene and 40 ml spectro-grade cyclohexane.

(4) Prepare a standard solution containing 1 ml o-xylene, 15 ml m-xylene, and 6 ml p-xylene made up to 100 ml with cyclohexane.

(5) Obtain the infrared spectra of solutions 3 and 4 in the same 0.05-mm sodium chloride cell over the range 12 to 15 μ. A scanning rate of 0.4 μ

per min and a full-scale pen response of 10 sec for a 10-in. chart are satisfactory.

(6) Apply the base-line technique to calculate the absorbances of the bands at 12.6 μ (p-xylene), 13.5 μ (o-xylene), and 14.5 μ (m-xylene) for both spectra. The points between which to construct the base lines are 12.2 and 14.0 μ for the o- and p-xylenes and 14.05 and 14.75 μ for m-xylene. From these data estimate the approximate composition of the commercial xylene.

(7) Prepare another standard solution such that the unknown will be bracketed between it and solution 4. Obtain its spectrum in the same cell used previously. Apply the base-line procedure as before.

(8) Plot a Beer's law curve (absorbance against concentration) for each of the three wavelengths, from the data for the two standards.

(9) Determine the exact concentrations of the isomers in the unknown by means of the working curves from step 8. Report your results in terms of volume percent in the original sample.

EXPERIMENT 9   TURBIDIMETRIC DETERMINATION OF SULFATE

REFERENCE: Text, Chap. 6.

PLAN: The sulfate content of natural waters is to be determined through the turbidity produced by treatment with barium chloride.

MATERIALS PROVIDED: (1) Conditioning solution: dissolve 120 g sodium chloride in about 400 ml distilled water, add 10 ml concentrated hydrochloric acid and 500 ml glycerol, and dilute to 1 liter
(2) $BaCl_2 \cdot 2H_2O$ crystals, 30 to 40 mesh
(3) Standard K_2SO_4 solution containing 50 ppm of sulfate ion (0.0905 g K_2SO_4 per liter)

PROCEDURE: (1) Make ready a series of eight 100-ml beakers, numbered consecutively. In No. 1 place 50 ml distilled water (as a blank); in No. 2, 50 ml of unknown (by pipet); and in No. 3, 25 ml of unknown and 25 ml of distilled water. If the unknown shows any turbidity, it should be filtered through a dense paper.

(2) Into a 50-ml volumetric flask, pipet 2 ml of the standard sulfate solution, dilute to the mark with water, mix thoroughly, and pour into

beaker No. 4, without rinsing. Repeat, taking the following volumes of the standard: 5, 10, 25, and 50 ml.

(3) To each of the eight beakers add 10 ml of the conditioning solution.

(4) To each beaker in turn add 0.3 g of barium chloride crystals, stir for 1 min, let stand 4 min, stir 15 sec, and then determine the turbidance in a photoelectric turbidimeter or nephelometer, or preferably both. The instructor should be consulted about the method of operation.

(5) Construct a calibration curve from the instrument readings for the knowns, after correction for the blank.

(6) Report the concentration of sulfate in the unknown in parts per million. If both turbidimeter and nephelometer were used, compare their precision.

SUGGESTIONS FOR FURTHER WORK: Run a spectrum of one of the samples in a recording spectrophotometer, and comment on the wavelength dependence of the turbidance.

EXPERIMENT 10 EMISSION SPECTROGRAPHY: QUALITATIVE ANALYSIS

REFERENCES: (1) Text, Chap. 7.
(2) W. R. Brode, "Chemical Spectroscopy," 2d ed., John Wiley & Sons, Inc., New York, 1943.
(3) W. C. Pierce, O. R. Torres, and W. W. Marshall, *Ind. Eng. Chem., Anal. Ed.,* **11**:191 (1939).

PLAN: The object of this experiment is to make a qualitative analysis of a nonferrous alloy by means of a photographic spectrograph. It is assumed that the instrument has already been correctly focused and adjusted and is ready for use.

Before starting work, the student should become familiar with the darkroom facilities and the photographic procedure to be followed. Two developing trays will be required and also a large container of running water. A contact printer or printing frame should be available as well as a suitable safelight.

MATERIALS PROVIDED: (1) Rod-shaped electrodes (two each) of iron (or mild steel), copper, zinc, and of an unknown brass or bronze.
(2) Photographic developer: MQ or universal developer powders to be dissolved as required according to the manufacturer's directions.

(3) Photographic fixer: Prepared from ammonium thiosulfate, which is faster acting than the sodium salt. The solution should contain acid hardener.

(4) Photographic plates or films and contact printing paper: Eastman Spectroscopic Plates, Type 103-F or 103-L, are preferred. Satisfactory results can sometimes be obtained with Eastman's Contrast Process Ortho or other sheet film, which comes in 5- by 7-in. and larger sizes and can be cut to fit the spectrograph. The film should be backed with a piece of cardboard to equal the thickness of the standard glass plates. Some spectrographs are equipped to handle 35-mm film directly. The plates or film must be handled only in complete darkness or under a type of safelight known to be satisfactory with the particular material at hand.

PROCEDURE: (1) After becoming familiar with the construction and operation of the spectrograph, load the plate holder with plate or film, making certain that the sensitized side is placed forward (i.e., face down in the open holder). Return the holder to the spectrograph, and set it at the uppermost position (No. 1).

(2) Place iron electrodes in the arc stand, and strike a dc arc between them (see instructor). CAUTION: *Do not look directly at the arc unless wearing dark glasses.* Adjust the position of the arc stand and the quartz lens to converge the light to a spot of about 1-cm diameter covering the spectrograph slit. Set the slit opening at 0.01 mm. Withdraw the safety slide which covers the plate, and make an exposure of 10 sec. Now lower the plate to the next position (No. 2) and expose for 5 sec, and then to the No. 3 position and expose 1 sec. Replace the safety slide, and turn off the current to the arc. Remove the iron electrodes, and insert copper ones. Strike the arc again, and make 10-, 5-, and 1-sec exposures in the next three positions on the same plate.

(3) Take the plateholder to the darkroom, bathe the plate or film 4 min in the developer, then rinse it briefly in water, and place it in the fixer for another 2 min. While the plate is in the fixer, load a fresh one into the holder. After the plate is removed from the fixer, it should be examined critically by transmitted light in order that the student may choose the most suitable exposure times to use for the next plate. These may, of course, be different for iron and for copper. It may be assumed that the exposure for copper will be about right for zinc, brass, and other similar metals.

(4) The spectra to be recorded on the second plate should be decided upon in consultation with the instructor. For a brass or bronze unknown,

a suitable series would be iron, copper, unknown, zinc, iron. This plate is to be studied in detail, so it should be washed in running water at least 5 min to ensure permanence. It may then be bathed in alcohol for not over 3 min and dried in warm air. If speed is not essential, the alcohol may be omitted.

(5) A convenient method of identifying the lines is described in Ref. 3. In addition, the wavelength tables given in the appendix of Ref. 2 will be useful. As many lines as possible should be identified in the spectrum of the unknown, either by direct comparison with the flanking spectra on the plate itself or by projection onto a chart, as in Ref. 3, or by other means suggested by the instructor. A dispersion curve for the spectrograph will be useful. It gives the approximate wavelengths of lines located at measured distances from the end of the plate. The prominent doublet of copper at wavelengths 3247.5 and 3274.0 Å can readily be identified on the plate and located on the curve to establish the scale of distances.

(6) Report elements certainly present and those tentatively identified in your unknown. A single line ascribed to a particular element should not be accepted as proof of its presence without the supporting evidence of other lines.

(7) A contact print of each plate should be made a part of the report.

EXPERIMENT 11 WATER ANALYSIS BY FLAME PHOTOMETRY

REFERENCES: (1) Text, Chap. 8.
(2) P. W. West, P. Folse, and D. Montgomery, *Anal. Chem.*, **22**:667 (1950).
(3) J. A. Dean and J. H. Lady, *Anal. Chem.*, **27**:1533 (1955).

PLAN: Standard flame emission curves for sodium, potassium, magnesium, and calcium will be prepared. A rough analysis of an unknown water will be made on the basis of these curves. A synthetic sample will then be prepared to duplicate the unknown as nearly as possible. This sample will be analyzed in order to establish correction factors to be applied to the unknown.

MATERIALS PROVIDED: (1) Standard solutions of NaCl, KCl, $MgSO_4$, and $CaSO_4$, each containing 100 ppm of the metal. These solutions should be stored in polyethylene bottles to prevent contamination with sodium from glass.

(2) Distilled water (demineralized water is usually not satisfactory for this purpose).

(3) Tank oxygen with reducing valve and gage.

PROCEDURE: (1) Prepare serial dilutions of each of the four standards, to contain 75, 50, 25, 10, 5, and 1 ppm of the metallic element. All glassware used in this and subsequent steps should be scrupulously cleaned, preferably with chromic acid cleaning solution followed by tap water and distilled water.

(2) Adjust the wavelength scale of the flame photometer to the appropriate setting for the sodium D line, 589 nm. Place some of the 100-ppm sodium solution in the aspirator. Start the flame, after a demonstration by the instructor. (Natural gas and oxygen give a suitable flame; acetylene is not required, but if it must be used, read and heed the caution note included in Experiment 12.) By following the manufacturer's instructions, obtain a photometric reading. The slit should be open only wide enough to bring the reading near the upper end of the photometric scale. The wavelength control should be adjusted to produce the maximum response, even though it turns out to be slightly different from the specified value (this is to correct for any inaccuracy of wavelength calibration). Repeat with each of the other sodium solutions and with distilled water as a blank, leaving the slit width and wavelength controls unchanged. After each use, distilled water should be aspirated to rinse out the liquid passages in the burner assembly.

(3) Repeat step 2 with the potassium solutions at wavelength 767 nm. The slit width will need to be readjusted.

(4) Repeat step 2 with the calcium solutions at wavelength 556 nm.

(5) Repeat with the magnesium solutions, wavelength 371 nm.

(6) Fill the cleaned aspirator with tap water (the unknown), and obtain photometric readings at each of the four wavelengths, with the slit set at the same widths used for the standards.

(7) Plot on a single sheet the four calibration curves from steps 2 to 5. From these curves and the data of step 6 determine the apparent composition of the tap water.

(8) By mixing measured volumes of the four standards, prepare a synthetic sample of known concentration equal to the apparent concentrations in the unknown. In case the tap water is known to contain considerable quantities of other elements (such as iron or manganese),

it may be desirable to include appropriate amounts in the synthetic sample.

(9) Examine the synthetic sample in the flame photometer in the same manner as the unknown.

(10) Determine the apparent concentrations in the synthetic sample by reference to the calibration curves. Any discrepancies between the apparent and known concentrations represent the result of interferences between ions, and should be applied as corrections to the unknown.

(11) Report the concentrations of the four metals in the unknown in parts per million.

NOTES: (1) If the tap water contains a significant amount of bicarbonate, it should be removed by boiling before analysis, to avoid interference.
(2) The radiation buffer technique described in the reference can be used if desired, but if greatest precision is not required, the added steps will probably not be necessary.
(3) It is not expected that the precision of the magnesium analysis will be as high as the other three, but it should be fair if a hot flame, such as oxygen-hydrogen, is used.

SUGGESTIONS FOR FURTHER WORK: By following the method of Dean and Lady (Ref. 3), analyze a sample of limestone or a nonferrous alloy for its iron content.

EXPERIMENT 12 ATOMIC ABSORPTION SPECTROPHOTOMETRY

REFERENCES: (1) Text, Chap. 8.
(2) "Analytical Methods for Atomic Absorption Spectrophotometry," pp. Cr-1, Ni-1, Perkin-Elmer Corporation, Norwalk, Conn., 1966.

PLAN: An unknown solution will be supplied which contains both chromium and nickel. This permits the use of a single multielement hollow-cathode lamp. The analytical chromium line is at 358 nm, but is hard to find because the lamp gives other lines nearby which are not absorbed in the flame; the way to find the proper wavelength will be outlined under "Procedure."

Nickel has two usable lines, 341 nm which is easy to find and gives good precision but relatively low sensitivity, and 232 nm which is more sensitive but harder to find, and gives a nonlinear response at high concentrations.

MATERIALS PROVIDED: (1) Stock solution of a Cr(III) salt [KCr(SO$_4$)$_2$·-12H$_2$O, formula weight 500, is convenient], containing 1000 ppm of chromium
(2) Stock solution of a Ni(II) salt [such as Ni(NH$_4$)$_2$(SO$_4$)$_2$·6H$_2$O, formula weight 395], containing 1000 ppm of nickel
(3) Acetylene and compressed air in cylinders, with reducing valves and pressure gages

CAUTION: (1) Acetylene-air mixtures can be dangerously explosive; do not attempt to ignite the flame until it has been demonstrated; *never* leave the photometer unattended with the flame burning; do not disable the safeguards provided by the manufacturer.
(2) Since heavy metals will be aspirated into the flame, toxic vapors may be formed, making adequate ventilation essential.

PROCEDURE: (1) Dilute the unknown solution quantitatively to 100 ml with water.

(2) Under the supervision of the instructor, ignite and adjust the flame, then find the 358 nm chromium line. This is best accomplished by turning the monochromator control slowly back and forth in the vicinity of the desired wavelength, while alternately aspirating water and a chromium-containing solution, until the point of greatest sensitivity is located.

(3) Without moving the monochromator control, set the meter at zero with distilled water, and at full scale with a dilution of the standard chromium solution containing 20 ppm of the element. Then aspirate the unknown and record the meter reading.

(4) Reset the monochromator at the 232 nm nickel line, and proceed as with the chromium determination.

(5) Repeat the nickel determination at 341 nm.

(6) Calculate and report the concentrations of the two elements in your sample, in parts per million.

SUGGESTIONS FOR FURTHER WORK: (1) Make up a solution containing 150 ppm of nickel, set the full-scale reading at 341 nm with it, then turn the burner sideways and observe the reading. Try this same solution at 232 nm, with the burner in both positions.
(2) Nickel interferes somewhat with the chromium determination. Repeat the chromium experiment with large amounts of nickel present (500 to 2000 ppm) and comment.

[This experiment was suggested by J. M. Fitzgerald.]

EXPERIMENT 13 POTENTIOMETRIC ACID–BASE TITRATION

REFERENCES: (1) Text, Chap. 12.
(2) F. J. C. Rossotti and H. Rossotti, *J. Chem. Educ.*, **42**:375 (1965).

PLAN: The titrations of phosphoric acid and of sodium phosphate will be studied as examples of potentiometric neutralization reactions.   A laboratory pH meter, either line- or battery-operated, equipped with glass and reference electrodes, will be used to follow the titrations.

MATERIALS PROVIDED: (1) H_3PO_4, 0.2 F
(2) Na_3PO_4, 0.2 F
(3) NaOH, 0.5 F
(4) HCl, 0.5 F
(5) Universal indicator, in dropping bottle
(6) Standard buffer, pH 7, for pH meter

PROCEDURE: (1) Standardize the pH meter with pH 7 buffer according to the manufacturer's (or instructor's) directions.   Be sure that the control is set at its standby position whenever the electrodes are not dipping into a solution.

(2) Pipet 10 ml of phosphoric acid into a 250-ml beaker; add about 50 ml water and a few drops of universal indicator.

(3) Bring the solution into position around the electrodes, and, after making sure that the stirrer cannot hit the electrodes, start it operating. Measure and record the pH.   Now titrate with sodium hydroxide added from a 10-ml buret.   After each addition, record the pH and the buret reading.   The intervals may be 0.5 ml in the regions where the pH is changing slowly, but 0.05 ml near the expected end points.   Record also the pH's corresponding to any color changes in the indicator.   Continue titrating until the pH ceases to change appreciably (about pH 11).

(4) Plot the pH against the volume of reagent.   On the same sheet, plot a differential titration curve, from points obtained by measuring the slope of the first curve from point to point.

(5) Similarly titrate a portion of trisodium phosphate by HCl, and plot.

(6) Plot all applicable data by the method of Gran (Ref. 2) and comment on the results.

SUGGESTIONS FOR FURTHER WORK: Titrate a sample of glycinium chloride, $(H_3NCH_2COOH)^+Cl^-$, by NaOH, and sodium glycinate, $Na^+(H_2NCH_2-$

COO)⁻, by HCl, and interpret the curves. Note the relation of this system to the glycine buffers utilized in Experiment 7.

EXPERIMENT 14 IODOMETRIC REDOX REACTION

REFERENCE: Text, Chaps. 12 and 13.

PLAN: A typical redox reaction will be followed by potentiometric and biamperometric titrations with several different electrode systems. A laboratory potentiometer with galvanometer, or an electronic voltmeter or pH meter calibrated in millivolts, can be used with platinum, tungsten, and calomel (or other reference) electrodes. A magnetic or equivalent stirrer is essential. As the titrations are to be recorded, a constant-delivery pump or syringe buret, or lacking those, a constant-head (mariotte) bottle, is required. The recorder can be of the direct deflection type, or a servo (potentiometric) type; in the latter case, it may well be sensitive enough that the electrodes can be connected directly without the need of the electronic voltmeter.

MATERIALS PROVIDED: (1) Crystalline KIO_3
(2) Aqueous KI, about 20 percent
(3) Aqueous $Na_2S_2O_3$, 2 N, containing 0.1 percent Na_2CO_3 as a preservative
(4) H_2SO_4, 1.0 F
(5) Sodium carbonate, anhydrous
(6) Ammonium acetate

PROCEDURE: (1) Weigh out on an analytical balance 0.35 to 0.40 g of potassium iodate, place it in a 250-ml volumetric flask, dissolve, dilute to the mark with water, and mix well.

(2) In a 250-ml beaker, place 10 ml of 20-percent potassium iodide solution and 50 ml of 1.0 F sulfuric acid. Add 2.0 g solid sodium carbonate, cover with a watch glass, and swirl until solution is complete. Introduce by pipet into the stirred solution a 25-ml aliquot of the iodate solution. The brown color of KI_3 will immediately appear. Add about 5 g of ammonium acetate, and insert platinum and calomel electrodes. Start the recorder and stirrer, and titrate with thiosulfate delivered by a constant-flow device. Continue until the equivalence point is exceeded by about 100 percent.

(3) Repeat the titration with tungsten and calomel electrodes.

(4) Repeat with tungsten and platinum electrodes.

(5) Repeat with two platinum electrodes in a biamperometric connection (the electrodes in series with a source of about 30 to 40 mV and a resistor, with the recorder connected to measure the voltage drop across the resistor; the value of the resistor depends on the characteristics of the recorder).

(6) Run a blank titration with whichever set of electrodes is preferred, using the same procedure except for the omission of iodate.

(7) Compare critically the several titration curves you have obtained, and from them calculate the normality of the thiosulfate solution; the potassium iodate can be considered to be a primary standard. Due allowance must be made for the reagent blank.

(8) On the graph from step 4, plot point-by-point the difference between the platinum-calomel and tungsten-calomel curves, and compare with the observed platinum-tungsten curve.

(9) In your report, comment on the convenience and the accuracy of the several titration methods.

EXPERIMENT 15 POLAROGRAPHY: A STUDY OF VARIABLES

REFERENCE: Text, Chap. 13.

PLAN: The electrochemical reaction to be studied will be the cathodic reduction of divalent lead in 0.1 F potassium chloride. The supporting electrolyte will be examined, first alone, then with added lead.

The care of the capillary is especially important. See page 279 for essential precautions. Used solutions can be discarded in the usual way, but precautions must be taken against spilling any mercury in the sink or elsewhere. A container will be supplied for used mercury.

The experiment can be performed with either a manual or a recording polarograph. Refer to the manufacturer's instructions for the instrument at hand.

MATERIALS PROVIDED: (1) Supporting electrolyte stock solution, 2.5 F KCl
(2) Stock solution of $PbCl_2$, $3 \times 10^{-2}\ F$, made up in 0.1 F KCl
(3) Triton X-100, 0.2 percent solution, in a dropping bottle
(4) Tank of nitrogen gas, with reducing valve

PROCEDURE: (1) Prepare 500 ml of 0.1 F KCl by dilution of the stock solution (high accuracy is not required).   This will be referred to in the following as "the electrolyte."

(2) Take about 10 ml of the electrolyte in an erlenmeyer flask and shake to ensure saturation with air.   Pour it into a polarograph cell provided with a calomel reference electrode, and insert the dropping mercury capillary, *with the mercury already dropping.*

(3) Adjust the sensitivity control of the polarograph to 10 μA full-scale, the mercury height to give a drop time between 3 and 5 sec, and record or plot manually the polarogram from 0 to -2.0 V.   The oxygen maxima should appear (see Fig. 13-10).

(4) Set the potential at about -1.5 V and bubble nitrogen through the solution.   Stop the bubbling every minute (oftener if a fritted gas disperser is in use), and after the solution has become quiet, determine the diffusion current.   Continue until no further appreciable diminution occurs.   Now run another complete polarogram on the deoxygenated solution.   It should resemble curve c in Fig. 13-10; if it does not, reducible impurities are probably present, and the instructor should be consulted for advice.

(5) Prepare a solution containing $3.0 \times 10^{-3} F$ Pb(II) in 0.1 F KCl, and bubble nitrogen for the length of time found necessary in the preceding step.   Set the sensitivity at 20 or 25 μA full-scale, and run a polarogram from 0 to -1.0 V.   A wave should appear due to the reduction of lead, but with an initial sharp maximum.   Add 1 drop of Triton X-100 solution, bubble nitrogen just long enough to provide mixing, then run the portion of the polarogram covering the region where the maximum appeared.   Repeat with one more drop of suppressor each time until the maximum has disappeared, then run the polarogram from 0 to -1.0 V.

(6) Prepare a solution containing $2.0 \times 10^{-3} F$ lead in the same electrolyte, with the optimum amount of Triton X-100, deoxygenate, and run the 0 to -1.0 V polarogram.

(7) Repeat for $1.0 \times 10^{-3} F$ lead ion.

(8) Plot the values of the diffusion current (corrected for residual current) corresponding to the point 0.15 V more negative than the half-wave potential, against concentration, for the three solutions, 5, 6, and 7.

(9) With any one of the previous solutions in the cell, and the potential set at the point 0.15 V more negative than the half-wave potential, catch about 20 drops of mercury in a small dipper fashioned from glass tubing, noting the time required.   Dry the mercury carefully and weigh it.

Determine the values of m (mg/sec) and t (sec/drop), and from them evaluate the quantity $m^{2/3}t^{1/6}$.

(10) Compute the diffusion-current constant, $\bar{\imath}_d/(Cm^{2/3}t^{1/6})$, for each value of C. Comment on its constancy.

(11) Compute the diffusion coefficient D by application of the Ilkovič equation.

(12) Plot the quantity $\log \bar{\imath}/(\bar{\imath}_d - \bar{\imath})$ against the potential, and from the curve determine both n and $E_{1/2}$.

EXPERIMENT 16 POLAROGRAPHIC DETERMINATION OF COPPER AND CADMIUM

REFERENCE: Text, Chap. 13.

PLAN: Both Cu(II) and Cd(II) form ammonia complexes in NH_3/NH_4Cl buffer, and both can be reduced at a dropping mercury cathode. However, copper shows two one-electron steps, whereas cadmium, having no stable univalent state, is reduced in a single two-electron step. In this experiment, these elements will be polarographed separately, then in combination, and a mixed unknown will be evaluated.

Note the precautions concerning care of capillary and disposal of mercury in Experiment 15.

MATERIALS PROVIDED: (1) Electrolyte stock solution, containing NH_3 and NH_4Cl, each 2.5 F
(2) Standard solution of a Cu(II) salt, $1.0 \times 10^{-2} F$
(3) Standard solution of a Cd(II) salt, $1.0 \times 10^{-2} F$
(4) Sodium sulfite, crystals

PROCEDURE: (1) Prepare 500 ml of supporting electrolyte 0.1 F in each component, by dilution of the stock. Add about 1 g Na_2SO_3 to reduce dissolved oxygen. Run a trial polarogram on this solution (0 to -1.0 V). If the residual current is too large, consult the instructor; it may be that preelectrolysis between a mercury cathode and a platinum anode is called for.

(2) Make up $10^{-3} F$ standard solutions of copper and of cadmium in your electrolyte, by dilution from the stock solutions provided. Place some of the copper solution in the cell, and insert the DME. Set the potential

at -1.0 V, and adjust the sensitivity to keep the recorder on scale.   Then turn back to zero volts, and run a polarogram from 0 to -1.0 V.

(3) Similarly run a polarogram on the cadmium solution.

(4) Mix together equal quantities of the copper and cadmium solutions, and run another polarogram.

(5) Obtain an unknown solution containing both metals, dilute it as the instructor directs with your electrolyte, and run its polarogram.

(6) Evaluate your unknown by comparison with the standards.   It may be desirable to run one or more additional polarograms on standards mixed in different proportions, simulating the unknown.   Do not omit correction for residual current, unless you can show that it is indeed negligible.

EXPERIMENT 17 AMPEROMETRIC TITRATION OF LEAD BY DICHROMATE

REFERENCES: (1) Text, Chap. 13.
(2) I. M. Kolthoff and Y. D. Pan, *J. Am. Chem. Soc.*, **61**:3402 (1939); **62**:3332 (1940).

PLAN: Lead is to be determined by titration with potassium dichromate. The solution contains potassium nitrate (0.1 F) as supporting electrolyte and an acetate buffer to hold the pH at 4.2.   The reference electrode is the SCE, the indicator the DME.   The dropping-mercury cathode is held either at 0.0 or -1.0 V.   The effect of changing concentration will be observed.

A manual polarograph, or more simply a microammeter (10^{-7} A, full-scale) provided with a variable shunt and a source of 1.0-V potential, is all that is needed.   A recording polarograph can, of course, be used, but is much more sophisticated than needed for titrimetry.

MATERIALS PROVIDED: (1) Supporting electrolyte, consisting of 11.0 g KNO_3, 10 ml $HC_2H_3O_2$ (glacial), and 5.0 g $NaC_2H_3O_2$ per liter
(2) Lead nitrate solution, 0.10 F [33 g $Pb(NO_3)_2$ per liter]
(3) $K_2Cr_2O_7$, primary standard crystals
(4) Triton X-100, 0.2 percent, in dropper bottle

PROCEDURE: (1) Make up 500 ml of approximately 0.05 F potassium dichromate solution, its value known to four significant figures.   Dilute 10.00 ml of this to 100 ml as a second titrating solution.

(2) Dilute 10.00 ml of the stock lead solution to 100 ml with supporting electrolyte. Of this again dilute 10.00 ml to 100 ml with the same electrolyte.

(3) Transfer 25 ml of the 0.01 F lead nitrate to the polarograph cell, add two drops of Triton X-100, and connect the electrodes directly to the galvanometer and shunt (or set the polarograph at zero volts). Titrate with the 0.05 F potassium dichromate from a 10-ml buret. Adjust the shunt as required to keep the meter on scale, always keeping a record of the shunt setting.

(4) Repeat with the more dilute solutions.

(5) Repeat (3) with the cathode at -1.0 V versus SCE.

(6) Repeat (4) at -1.0 V.

(7) Titrate 20 ml of the 0.005 F dichromate to which has been added 20 ml of supporting electrolyte by 0.01 F lead at zero volts.

(8) Repeat (7) at -1.0 V.

(9) Correct all current readings by the dilution factor $(V + v)/V$, where V is the original volume and v the added volume of reagent.

(10) Plot the curves of corrected current vs. volume for all titrations of dichromate by lead on one sheet of graph paper, and lead by dichromate on another. Estimate the precision with which the concentration of an unknown lead solution could be determined by each of these titration procedures.

NOTE: Your cleanup will be made easier if you recall that $PbCrO_4$ is soluble in dilute HNO_3.

EXPERIMENT 18 SEPARATION OF METALS AT A MERCURY CATHODE; FLUORIMETRY

REFERENCES: (1) Text, Chaps. 4 and 14.
(2) "1950 Book of ASTM Methods for Chemical Analysis of Metals," p. 431, American Society for Testing and Materials, Philadelphia, 1950.
(3) A. Weissler and C. E. White, *Ind. Eng. Chem., Anal. Ed.*, **18**:530 (1946).
(4) D. C. Freeman, Jr. and C. E. White, *J. Am. Chem. Soc.*, **78**:2678 (1956).

PLAN: A sample of zinc-base die-casting alloy, which contains small amounts of aluminum, copper, magnesium, iron, etc., is dissolved in acid, and an aliquot is electrolyzed at a mercury cathode to deposit metals less active than aluminum. The aluminum is then determined by a fluorimetric method.

MATERIALS PROVIDED: (1) $K_4Fe(CN)_6$, 5 percent, in dropper bottle
(2) Ammonium acetate, 10 percent
(3) Pontachrome Blue Black RM, 0.1 percent in alcohol
(4) Aluminum standard solution (1 ml = 0.01 mg Al). [Dissolve 0.1760 g $KAl(SO_4)_2 \cdot 12H_2O$ in 1 liter.]
(5) Quinine sulfate stock solution (0.025 g in 20 ml of 0.05 F H_2SO_4, diluted to 1 liter with water)
(6) Methyl red indicator, in dropper bottle

PROCEDURE: (1) Weigh out a 0.1-g sample of the alloy into a 400-ml beaker. Cover it with 25 ml water; then add 5 ml concentrated H_2SO_4. When the sample is completely dissolved (except for specks of copper), transfer the solution quantitatively to the mercury-cathode cell which already contains the requisite amount of mercury. Add about 200 ml of water. The solution at this point should be about 1 N in acid.

(2) Electrolyze at about 1 to 1.5 A for an hour with stirring (or overnight without stirring). (The fluorescence standards and equipment can be prepared while the electrolysis is proceeding.) After this period has elapsed, withdraw 1 ml of the solution and add to it a few drops of $K_4Fe(CN)_6$ solution. If no precipitate or turbidity appears, the deposition of zinc may be considered complete.

(3) After the completion of the electrolysis, withdraw the mercury with the current still turned on. There must be sufficient electrolyte so that the anode remains covered until all the mercury is removed. Now disconnect the power, and draw off the electrolyte into a 500-ml volumetric flask, rinsing down the cell, anode, and stirrer with a stream of water from a wash bottle. Fill to the mark with water, and mix.

(4) Prepare six 100-ml volumetric flasks, numbered serially. Into the first two, transfer 5.00 and 10.00 ml, respectively, of the electrolyzed solution from step 3. Into the next three, pipet respectively 1, 2, and 4 ml of the standard aluminum solution. Flask No. 6 is a blank.

(5) To each flask, add 10 ml of 10 percent ammonium acetate, 50 ml of water, and 10 drops of methyl red. By adding 1 F acetic acid or ammonium hydroxide, adjust the acidity of each to the neutral point for methyl red. It is important that all the solutions be at exactly the same shade. Add 3.0 ml of Pontachrome Blue Black RM. Fill to the mark with water, mix well, and allow all samples to stand at least 45 min, preferably 1 hr.

(6) Determine the intensity of fluorescence of each sample under excitation with ultraviolet light from a mercury arc filtered through an ultraviolet-transmitting glass. The fluorescent light is red, and a red filter should accordingly be placed before the photocell. The method of operating the fluorimeter, standardizing against quinine, etc., will be described by the instructor.

(7) Prepare a calibration curve, percent fluorescence against aluminum concentration. Report the result as percent aluminum in the alloy.

NOTE: Certain other dyes can be used in place of Pontachrome Blue Black RM, particularly Alizarin Garnet R. The latter gives a yellow fluorescence, and is somewhat more sensitive to the presence of traces of iron. See Ref. 4.

EXPERIMENT 19   CONTROLLED-CATHODE ELECTRODEPOSITION

REFERENCES: (1) Text, Chap. 14.
(2) J. J. Lingane, "Electroanalytical Chemistry," 2d ed., Chap. 15, Interscience Publishers (Division of John Wiley & Sons, Inc.), New York, 1958.

PLAN: Copper, lead, and tin will be separated electrolytically at a controlled-potential cathode. Quantitative results will be obtained by gravimetry.

MATERIALS PROVIDED: (1) $CuSO_4 \cdot 5H_2O$, granular
(2) $Pb(NO_3)_2$, crystals
(3) $SnCl_2$, crystals
(4) Sodium tartrate
(5) Succinic acid
(6) Hydrazinium dichloride
(7) Sulfamic acid
(8) Gelatin powder

PROCEDURE: (1) Prepare a test solution containing 0.4 g $CuSO_4 \cdot 5H_2O$, 0.1 g $Pb(NO_3)_2$, 0.1 g $SnCl_2 \cdot 2H_2O$, 10 g sodium tartrate, 2 g succinic acid, and 2 g hydrazinium dichloride. (The metallic salts should be weighed with analytical accuracy.) Dilute to 200 ml, and adjust the pH to 5.5 ± 0.4 with NH_3 or HCl.

(2) Assemble a potentiostat comparable to those shown in Figs. 14-2 and 14-3. Weigh carefully the large platinum cathode. Mount the

three electrodes in the cell provided, and make certain that the stirring device can operate without interference.

(3) Transfer the entire test solution to the electrolysis cell, and start the stirring. Set the voltage at 0.35 V, and electrolyze until the blue color of copper disappears. To test for completeness of deposition of copper, withdraw a few drops of solution with a medicine dropper onto a porcelain spot plate, and mix with a few drops of concentrated ammonia; if the dark blue of the tetrammine does not appear, the deposition can be considered complete. Remove the cathode with great care, with the voltage still applied, by lowering the beaker, while rinsing down the electrodes with a stream of water from a wash bottle. Dry and weigh the cathode.

(4) For the determination of lead, add to the solution in the cell 1 ml of 0.2 percent gelatin solution. Adjust the pH to 5.2 ± 0.2, and electrolyze at 0.65 V until the current becomes constant (at about 10 mA). Completeness of deposition of lead may be tested by a spot test with dichromate. Remove the cathode as before and weigh it. CAUTION: Do not deposit lead or tin directly on platinum, as the electrode may be damaged; if no copper is present in the solution, the electrode should be lightly copper-plated.

(5) Add to the solution from which copper and lead have been removed 20 ml concentrated HCl, and electrolyze at 0.70 V. After the production of gas has nearly ceased, add 1 g sulfamic acid and 2 g hydrazinium dichloride, and continue electrolysis until the current levels off (at about 50 mA). A spot test for tin is the yellow precipitate of SnI_2 produced by KI solution. Weigh the cathode as before.

(6) Report your results for the three elements, in terms of quantity taken compared with quantity recovered, and also *total* quantity taken and recovered. Account for any discrepancies. The cathode may be stripped of all three deposits in 1:1 nitric acid (fume hood!) or by reverse electrolysis.

EXPERIMENT 20 COULOMETRIC TITRATION OF ARSENIC BY BROMINE

REFERENCES: (1) Text, Chaps. 13 and 14.
(2) J. J. Lingane, *J. Am. Chem. Soc.*, **67**:1916 (1945).

PLAN: Arsenic is to be determined by oxidative titration, using electrolytically generated bromine as reagent, and the biamperometric method for detection of the equivalence point. The reaction is carried out in a 250-ml beaker mounted upon a magnetic stirrer. Inserted into the beaker are four platinum electrodes (see figure). Two of these, E_3 and E_4, form part of the indicator system. The other two serve as the generator electrodes for the coulometric titration. The cathode E_1 is isolated from the open solution by a glass tube with a fritted tip. The generator electrodes are connected to a source of constant current (see Chap. 26) and a milliammeter I.

The indicator electrodes are connected to a source of about 50 mV, as measured with voltmeter V. A recording potentiometer with a resistor R_1 of about 200 Ω connected across its terminals is inserted in the detection circuit. (A recording polarograph is well suited to serve as a combined polarizer and detector.) If a recorder is not available, a meter and shunt may be substituted, and a seconds timer will be needed, as shown.

The reaction to be studied is the oxidation of arsenite by bromine:

$$AsO_2^- + Br_2 + H_2O \rightarrow AsO_3^- + 2Br^- + 2H^+$$

The bromine is generated by the electrolysis of aqueous bromide:

$$2Br^- + 2H_2O \rightarrow H_2 + Br_2 + 2OH^-$$

The hydrogen gas and OH^- ions liberated at the cathode are prevented by the porous barrier from being stirred into the solution. The anodically generated bromine reacts with the arsenite and accumulates in the

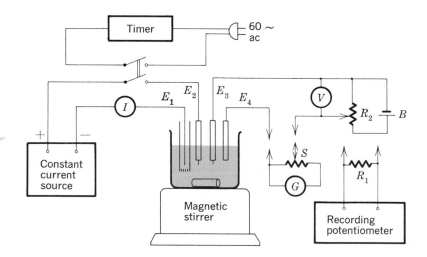

solution only after all the arsenite is oxidized, i.e., after the equivalence point. The indicator electrodes pass current only when both Br_2 and Br^- are present, and the deflection of the recorder will be proportional to the Br_2 concentration, since Br^- is present in large excess.

It is essential to eliminate from the reaction vessel any extraneous material which can react with bromine; this includes rubber, polyethylene, and especially the hydroxide ion.

MATERIALS PROVIDED: (1) Sodium metarsenite, $NaAsO_2$, crystals
(2) 0.2 F HCl
(3) KBr crystals

PROCEDURE: (1) Weigh out two samples of reagent grade $NaAsO_2$ as standards, and two portions of the unknown arsenite. Each portion should contain the equivalent of 30 to 40 mg of As_2O_3. Dissolve each in 0.2 F HCl and make up to 200 ml.

(2) Dissolve approximately 5 g KBr in 75 ml of 0.2 F HCl in a 250-ml beaker. Insert the four electrodes and connect the apparatus as in the figure. Turn on the stirrer and the switch in the generator circuit, and electrolyze for several minutes at about 5 mA. This *pretitration* serves to oxidize any oxidizable materials in the solution. At this point the recorder should show a deflection, indicating the presence of free bromine. Add dropwise a dilute solution of arsenite until the recorder ceases to deflect (at maximum sensitivity). Then electrolyze again until the recorder shows half-scale or more deflection.

(3) Transfer a 10-ml aliquot of one of the arsenite samples into the beaker of acidified bromide. Electrolyze until the recorder again deflects.

(4) Repeat (3) with each of the other arsenite solutions, each added to the same beaker without any need to empty it (if it becomes too full, some of its contents may be removed).

(5) On the chart paper, extrapolate each deflection back to the zero level for measurement. Measure the electrolysis times for standards and unknowns. From the values for the standards, plot a calibration curve, electrolysis time against weight of sample.

(6) Calculate the result in terms of percent As_2O_3 equivalent in the unknown by two methods: (*a*) from the calibration curve, and (*b*) from coulometric data. Compare the results, and comment on the precision to be expected in each method.

EXPERIMENT 21 CONDUCTOMETRIC TITRATIONS

REFERENCE: Text, Chap. 15.

PLAN: A number of representative titrations will be followed conductometrically. Either a Wheatstone bridge or an operational amplifier circuit similar to that of Fig. 26-36(a), where R_1 is the conductance cell, R_2 a stepwise-adjustable resistor, may be used.

MATERIALS PROVIDED:
(1) Hydrochloric acid, 0.05 F and 0.25 F
(2) Acetic acid, 0.05 F
(3) NaOH solution, 0.25 F
(4) Zinc chloride, $ZnCl_2$, crystals
(5) Aluminum sulfate, $Al_2(SO_4)_3 \cdot 18H_2O$, crystals
(6) Platinizing solution, containing about 4 g K_2PtCl_6 and 0.02 g $Pb(C_2H_3O_2)_2$ in 100 ml water

PROCEDURE: (1) Platinization of electrodes: Immerse the platinum electrodes in some of the platinizing solution, and electrolyze with about 25 mA direct current until one electrode is covered with a dark gray or black coating; then reverse the polarity and coat the other electrode. Remove the electrodes, and, without allowing them to dry, place them in dilute sulfuric acid and electrolyze, first with one polarity, then the other (or with alternating current) for about 10 min, to remove traces of chlorine. CAUTION: Do *not* discard the platinizing solution, which is expensive and can be used repeatedly. Following platinization, the electrodes should be kept wet with distilled water when not in use; one platinizing treatment should last for weeks or longer.

(2) Place 25 ml of 0.05 F HCl in a 200-ml beaker with a stirring bar. Insert the conductance electrodes and measure the conductance (or resistance). Add 1-ml increments of 0.25 F NaOH from a 10-ml buret, measuring the conductance each time, until all 10 ml have been added.

(3) Repeat (2) with acetic acid.

(4) Weigh out duplicate samples, approximately 100 mg each, of $ZnCl_2$ into titration beakers, add 5 ml of 0.05 F HCl, and titrate as above, with 0.25 F NaOH. Calculate in advance (roughly) where end points would be expected, so they will not be missed.

(5) Weigh out duplicate samples of about 375 mg of $Al_2(SO_4)_3 \cdot 18H_2O$ into 200-ml beakers, dissolve in water, and make up to approximately

100 ml.   Titrate with 0.25 F NaOH, using all 10 ml of base.   Now, without emptying the beaker, titrate with 0.25 F HCl from a 10- or 25-ml buret until about 12 or 15 ml have been added.

(6) Plot conventional titration curves ($1/R$ against volume added) for procedures 2, 3, and 4.

(7) For procedure 5, plot the NaOH titration as usual, then the HCl titration from right to left on the same coordinates.

(8) Explain each titration curve, with the aid of diagrams comparable to Figs. 15-6(b) and 15-7(b).   In the aluminum titration, an intermediate compound has been reported of the type $Al_2(SO_4)_3 \cdot nAl_2O_3$.   Can you determine a value for n from your curves?

EXPERIMENT 22   THERMOMETRIC TITRATIONS

REFERENCES: (1) Text, Chap. 19.
(2) J. Jordan and T. G. Alleman, *Anal. Chem.*, **29**:9 (1957).

PLAN: Several titrations will be followed thermometrically.   If no ready-made apparatus is available, an operational amplifier circuit like that of Fig. 26-36(a) is recommended; R_1 and R_2 will be a pair of thermistors, R_1 in a comparison vessel, R_2 in the titration vessel.   The two vessels should be identical, and placed close together so that any changes in ambient temperature will affect them equally.   They should contain roughly the same volume of water, so that their heat capacities will be about the same (this is not critical).   It is highly desirable that the titration be carried out automatically with a constant-delivery device (see Experiment 14) and a recorder.

MATERIALS PROVIDED: (1) Hydrochloric acid, 0.05 F
(2) Acetic acid, 0.05 F
(3) Sodium hydroxide solution, 0.25 F
(4) EDTA, tetrasodium salt, 0.01 F
(5) Calcium chloride solution, 1 F
(6) Magnesium chloride solution, 1 F

PROCEDURE: (1) Place 25 ml 0.05 F HCl in the titration flask.   Start the stirring, and allow it to run for a few minutes until a straight line, not too far from horizontal, is established.   Then start the flow of titrant (about 5 ml/min), which is 0.25 F NaOH, and continue until the second change of slope is well established. The elapsed time between the two intersections represents the required amount of titrant, which can be evaluated by knowledge of the flow rate.

(2) Repeat (1) with 0.05 F acetic acid.

(3) Place 25 ml 0.01 F EDTA in the vessel and titrate with 1 F $CaCl_2$, at a flow rate about 1 ml/min.

(4) Repeat (3) with $MgCl_2$.

(5) Repeat (3) with an equimolar mixture of $CaCl_2$ and $MgCl_2$.

(6) Estimate the precision of each titration, and account for the shapes of the various curves.

EXPERIMENT 23 GAS CHROMATOGRAPHY

REFERENCE: Text, Chap. 21.

PLAN: In order to demonstrate the selectivity of a liquid substrate in gas chromatography, a polar liquid phase will be employed to separate a mixture of acetone, methanol, cyclohexane, and n-butyl acetate. A gas chromatograph should be selected which is provided with a thermal conductivity detector, and a column consisting of 30- to 60-mesh Celite or equivalent, coated with Carbowax, Ucon oil, or similar material. A hypodermic syringe will be needed for injection.

MATERIALS PROVIDED: (1) Acetone
(2) Methanol
(3) Cyclohexane
(4) n-Butyl acetate
(5) A mixture of equal volumes of the above
(6) Helium tank, with reducing valve and gage

PROCEDURE: (1) Start the chromatograph operating under the following conditions:

Helium flow	50 ml/min
Column temperature	85°C
Vaporizer temperature	125°C
Chart speed	0.5 in./min or 1 cm/min

(2) Suck into the syringe 0.001 ml of methanol, followed by a small amount of air. Inject this sample into the instrument and mark the

chart for zero time. An air maximum will appear on the chart almost immediately, then, after a few minutes time, the maximum corresponding to methanol. Repeat several times, adjusting the response sensitivity until the methanol maximum is about half-scale (it doesn't matter if the air peak goes off-scale).

(3) Repeat (2) for each of the other single liquids.

(4) Repeat for the mixture. Adjust the sensitivity until all four maxima are conveniently on scale. Identify the several maxima by comparison of retention times, conveniently measured from the air peak, with those of the pure compounds.

SUGGESTIONS FOR FURTHER WORK: (1) Prepare a series of known dilutions of one liquid in another, for example, cyclohexane in methanol, and determine the degree of linearity of peak height (or integrated area) with concentration. Determine the minimum detectable amount of each compound.

(2) Determine the HETP of the column. (See Fig. 20-4.)

(3) If another column is available, carry out similar experiments with it, to compare the ability of various substrates to separate the same compounds. (Note that a change of column often requires a delay to permit attainment of thermal equilibrium.)

(4) Attempt to separate the isomers in commercial xylene on a general-purpose column, such as silicone oil. Compare your results with the infrared analysis of Experiment 8.

EXPERIMENT 24 ION-EXCHANGE SEPARATIONS

REFERENCES: (1) Text, Chaps. 20 and 22.
(2) C. V. Banks, J. A. Thompson, and J. W. O'Laughlin, *Anal. Chem.*, **30**:1792 (1958).
(3) J. S. Fritz and M. J. Richard, *Anal. Chim. Acta*, **20**:164 (1959).
(4) F. W. E. Strelow and C. J. C. Bothma, *Anal. Chem.*, **39**:595 (1967).

PLAN: The lanthanides and actinides have very similar properties. They frequently occur together in minerals such as monazite. The lanthanides are among the predominant products of uranium or plutonium fission. Hence fractionation of mixtures of these elements is necessary to the study of their chemistry. The only generally applicable method of separation is ion exchange.

The procedure to be followed in the present experiment makes use of the sulfate complexes formed by thorium but not by the lanthanides. The latter are represented by samarium.   The thorium is adsorbed onto an anion-exchange resin in the sulfate form, while samarium passes through the column unhindered.   The thorium is eluted from the column by 0.4 F sulfuric acid (Ref. 4).

The concentrations of both metals in the eluted fractions are measured by the absorption of light by their complexes with Arsenazo I:

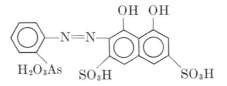

This reagent is sensitive, but not very selective, and therefore can be used only with a separated mixture (Refs. 2 and 3).

MATERIALS PROVIDED: (1) A strongly basic anion-exchange resin, such as Dowex 1X8 or Amberlite CG-400, 100 to 200 mesh, in the sulfate form
(2) Sulfuric acid, 0.40 F and 0.025 F solutions
(3) Buffer, 0.5 F in triethanolamine and 0.25 F in nitric acid
(4) Arsenazo I solution, 0.15 percent (aqueous)
(5) Test solution, 1 to 2 mF Sm(III) and Th(IV) in 0.025 F sulfuric acid

PROCEDURE: (1) Prepare an ion exchange bed by first loosely plugging the lower end of a chromatography column with glass wool, then pouring in a slurry of 10 ml of resin in 0.025 F sulfuric acid.   After the resin settles, the excess acid may be drawn off, but the liquid level should never fall below the top of the bed.

(2) Pipet 1.00 ml of the test solution carefully onto the top of the resin bed.   Draw off enough liquid through the stopcock at the bottom to lower the level just to the top of the resin bed, then add 2.0 ml of 0.025 F sulfuric acid.   Let the liquid flow from the stopcock into a 50-ml volumetric flask, replacing the supply at the top with another 2.0 ml of the same acid.   Continue this process, catching 2.0-ml portions in successive (numbered) 50-ml volumetric flasks.   As each flask is removed, add to it 2.0 ml Arsenazo solution and 15 ml buffer, dilute to the mark with water, and mix.   Measure the absorbance at 595 nm against a reagent blank in a Spectronic-20 or equivalent photometer.   When the absorbance of successive portions drops to near zero, it can be concluded that all the samarium has been eluted.

(3) Now change to 0.40 F sulfuric acid, and continue as before. Note that a different reagent blank is now needed for the photometry.

(4) When all the thorium has been eluted, the resin may be discarded, and the glassware cleaned.

(5) Plot the absorbance against the volume eluted. Mark on the graph the regions corresponding to different acid concentrations in the eluant.

SUGGESTIONS FOR FURTHER WORK: Ref. 4 shows that many other metals can be separated on this column with various concentrations of sulfuric acid. Try the same experiment with the addition of another one or two elements. It may be necessary to change analytical techniques; uranium, for example, is better determined by fluorimetry or polarography. Atomic absorption would be appropriate for many elements.

[This experiment was suggested by R. F. Hirsch.]

EXPERIMENT 25 DISTRIBUTION RATIOS DETERMINED BY RADIOACTIVE TRACER

REFERENCES: (1) Text, Chaps. 16 and 20.
(2) E. R. Tompkins and S. W. Mayer, *J. Am. Chem. Soc.*, **69**:2859 (1947).
(3) G. E. Moore and K. A. Kraus, *J. Am. Chem. Soc.*, **74**:843 (1952).
(4) K. A. Kraus and F. Nelson, *Proc. Intern. Conf. Peaceful Uses At. Energy*, **7**:113 (1956).

PLAN: The procedure to be followed exploits the use of a radioactive tracer to determine the distribution ratio K of cobalt(II) chloride anionic complexes on an anion-exchange resin, as a function of changing HCl concentration.

Before starting work, study the precautions and safety procedures concerned with radioactivity. Approved laboratory coats and disposable plastic gloves must be worn during all preparative steps. A registered film badge should be worn at all times.

MATERIALS PROVIDED: (1) Concentrated HCl.
(2) $CoCl_2$ stock solution, 10^{-4} to 10^{-3} F, in concentrated HCl; this solution contains ^{60}Co tracer.
(3) A strongly basic anion-exchange resin, air-dried.

PROCEDURE: (1) Weigh out five 2.00-g portions of the dry ion-exchange resin, and place in numbered erlenmeyer flasks.   The flasks should be placed in the hood area where tracer work is handled, ready for use later.

(2) Transfer five 5.00-ml portions of the radioactive $CoCl_2$ solution to 100-ml volumetric flasks.   To the first flask, add distilled water to the mark.   To the others add, respectively, 20, 45, 70, and 95 ml concentrated HCl, and fill to the mark with water.

(3) Take 20.00-ml aliquots of the five solutions into the five erlenmeyer flasks containing resin.   Stopper them and agitate gently.   Repeat agitation five times at 5-min intervals.   Release pressure that may build up in the flasks by opening cautiously after each agitation.

(4) After the resin has settled, transfer 5.00 ml of the supernatant liquid from each flask to a test tube of the size that fits the well of the scintillation counter.   Stopper both flasks and tubes.

(5) At this point the experimenter (or one member of the team) should remove his safety gloves (which may be contaminated) and clean the outer surfaces of the test tubes with first moist, then dry, tissues, checking the tissues with the monitor until they are free of radioactivity.

(6) Take the tubes to the counting station and count each in the scintillation well counter for 10 min or 2500 counts, whichever comes first. Determine the background count for a period of 5 min.

(7) Pipet 5.00-ml portions of the solutions prepared in step (2) into test tubes, wipe the tubes as above, and count them.

(8) Cleanup: All solutions go into the designated containers for liquid radioactive wastes.   The used resins go into a similarly designated receptacle for solids, along with contaminated tissues, plastic gloves, etc.   Rinse each piece of glassware with water, detergent, and water again, into the waste container; any subsequent rinsings may go down the sink.   Monitor the work area, the floor, your shoes, etc., with the portable monitor.   In case of any contamination on a nonabsorbent surface, wipe with absorbent paper, checking each piece of paper with the monitor, until all is clean.   In the event of any contamination that you cannot eliminate, consult the instructor.   Turn in your film badge.

(9) Calculate the distribution ratio K for each concentration of HCl by the equation [derived from Eq. (20-4)]:

$$K = \frac{(A_i - A_f)V}{A_f w}$$

where A_i = initial activity of the solution (corrected for background)

A_f = activity after equilibrating with the resin (corrected for background)

V = volume of solution counted

w = dry weight of the resin used

(The instructor will provide the true dry weight as a fraction of the air-dried weight.)   Plot log K against the HCl concentration.   Explain the significance of the curve.

[This experiment was suggested by R. F. Hirsch.]

Appendix

Standard electrode potentials*

Electrode	Reaction	$E°\ V$ vs. NHE
Li+, Li	$Li^+ + e^- \rightleftharpoons Li$	-3.045
K+, K	$K^+ + e^- \rightleftharpoons K$	-2.925
Ba++, Ba	$Ba^{++} + 2e^- \rightleftharpoons Ba$	-2.90
Sr++, Sr	$Sr^{++} + 2e^- \rightleftharpoons Sr$	-2.89
Ca++, Ca	$Ca^{++} + 2e^- \rightleftharpoons Ca$	-2.87
Na+, Na	$Na^+ + e^- \rightleftharpoons Na$	-2.714
La³+, La	$La^{3+} + 3e^- \rightleftharpoons La$	-2.52
Ce³+, Ce	$Ce^{3+} + 3e^- \rightleftharpoons Ce$	-2.48
Mg++, Mg	$Mg^{++} + 2e^- \rightleftharpoons Mg$	-2.37
AlF₆³⁻, Al	$AlF_6{}^{3-} + 3e^- \rightleftharpoons Al + 6F^-$	-2.07
Pu³+, Pu	$Pu^{3+} + 3e^- \rightleftharpoons Pu$	-2.07
Th⁴+, Th	$Th^{4+} + 4e^- \rightleftharpoons Th$	-1.90
Np³+, Np	$Np^{3+} + 3e^- \rightleftharpoons Np$	-1.86
Be++, Be	$Be^{++} + 2e^- \rightleftharpoons Be$	-1.85
U³+, U	$U^{3+} + 3e^- \rightleftharpoons U$	-1.80
Al³+, Al	$Al^{3+} + 3e^- \rightleftharpoons Al$	-1.66
Ti++, Ti	$Ti^{++} + 2e^- \rightleftharpoons Ti$	-1.63
V++, V	$V^{++} + 2e^- \rightleftharpoons V$	-1.18
Mn++, Mn	$Mn^{++} + 2e^- \rightleftharpoons Mn$	-1.18
TiO++, Ti	$TiO^{++} + 2H^+ + 4e^- \rightleftharpoons Ti + H_2O$	-0.89
Zn++, Zn	$Zn^{++} + 2e^- \rightleftharpoons Zn$	-0.763
TlI, Tl	$TlI + e^- \rightleftharpoons Tl + I^-$	-0.753
Cr³+, Cr	$Cr^{3+} + 3e^- \rightleftharpoons Cr$	-0.74
TlBr, Tl	$TlBr + e^- \rightleftharpoons Tl + Br^-$	-0.658
U⁴+, U³+, Pt	$U^{4+} + e^- \rightleftharpoons U^{3+}$	-0.61
TlCl, Tl	$TlCl + e^- \rightleftharpoons Tl + Cl^-$	-0.557
Ga³+, Ga	$Ga^{3+} + 3e^- \rightleftharpoons Ga$	-0.53
Fe++, Fe	$Fe^{++} + 2e^- \rightleftharpoons Fe$	-0.440
Cr³+, Cr++, Pt	$Cr^{3+} + e^- \rightleftharpoons Cr^{++}$	-0.41
Cd++, Cd	$Cd^{++} + 2e^- \rightleftharpoons Cd$	-0.403
Ti³+, Ti++, Pt	$Ti^{3+} + e^- \rightleftharpoons Ti^{++}$	-0.37
PbI₂, Pb	$PbI_2 + 2e^- \rightleftharpoons Pb + 2I^-$	-0.365
PbSO₄, Pb	$PbSO_4 + 2e^- \rightleftharpoons Pb + SO_4{}^{--}$	-0.356
Tl+, Tl	$Tl^+ + e^- \rightleftharpoons Tl$	-0.336
PbBr₂, Pb	$PbBr_2 + 2e^- \rightleftharpoons Pb + 2Br^-$	-0.280
Co++, Co	$Co^{++} + 2e^- \rightleftharpoons Co$	-0.277
PbCl₂, Pb	$PbCl_2 + 2e^- \rightleftharpoons Pb + 2Cl^-$	-0.268
V³+, V++, Pt	$V^{3+} + e^- \rightleftharpoons V^{++}$	-0.255

Standard electrode potentials (continued)

Electrode	Reaction	$E°$, V vs. NHE
Ni^{++}, Ni	$Ni^{++} + 2e^- \rightleftharpoons Ni$	-0.250
Mo^{3+}, Mo	$Mo^{3+} + 3e^- \rightleftharpoons Mo$	-0.2
CuI, Cu	$CuI + e^- \rightleftharpoons Cu + I^-$	-0.185
AgI, Ag	$AgI + e^- \rightleftharpoons Ag + I^-$	-0.151
Sn^{++}, Sn	$Sn^{++} + 2e^- \rightleftharpoons Sn$	-0.136
Pb^{++}, Pb	$Pb^{++} + 2e^- \rightleftharpoons Pb$	-0.126
HgI_4^{--}, Hg	$HgI_4^{--} + 2e^- \rightleftharpoons Hg + 4I^-$	-0.04
H^+, H_2	$2H^+ + 2e^- \rightleftharpoons H_2$	0.000
CuBr, Cu	$CuBr + e^- \rightleftharpoons Cu + Br^-$	$+0.033$
UO_2^{++}, UO_2^+, Pt	$UO_2^{++} + e^- \rightleftharpoons UO_2^+$	$+0.05$
AgBr, Ag	$AgBr + e^- \rightleftharpoons Ag + Br^-$	$+0.095$
TiO^{++}, Ti^{3+}, Pt	$TiO^{++} + 2H^+ + e^- \rightleftharpoons Ti^{3+} + H_2O$	$+0.1$
CuCl, Cu	$CuCl + e^- \rightleftharpoons Cu + Cl^-$	$+0.137$
Hg_2Br_2, Hg	$Hg_2Br_2 + 2e^- \rightleftharpoons 2Hg + 2Br^-$	$+0.140$
Sn^{4+}, Sn^{++}, Pt	$Sn^{4+} + 2e^- \rightleftharpoons Sn^{++}$	$+0.15$
Cu^{++}, Cu^+, Pt	$Cu^{++} + e^- \rightleftharpoons Cu^+$	$+0.153$
$HgBr_4^{--}$, Hg	$HgBr_4^{--} + 2e^- \rightleftharpoons Hg + 4Br^-$	$+0.21$
AgCl, Ag	$AgCl + e^- \rightleftharpoons Ag + Cl^-$	$+0.222$
Hg_2Cl_2, Hg	$Hg_2Cl_2 + 2e^- \rightleftharpoons 2Hg + 2Cl^-$	$+0.268$
UO_2^{++}, U^{4+}, Pt	$UO_2^{++} + 4H^+ + 2e^- \rightleftharpoons U^{4+} + H_2O$	$+0.334$
Cu^{++}, Cu	$Cu^{++} + 2e^- \rightleftharpoons Cu$	$+0.337$
$Fe(CN)_6^{3-}$, $Fe(CN)_6^{4-}$, Pt	$Fe(CN)_6^{3-} + e^- \rightleftharpoons Fe(CN)_6^{4-}$	$+0.36$
VO^{++}, V^{3+}, Pt	$VO^{++} + 2H^+ + e^- \rightleftharpoons V^{3+} + H_2O$	$+0.361$
Ag_2CrO_4, Ag	$Ag_2CrO_4 + 2e^- \rightleftharpoons 2Ag + CrO_4^{--}$	$+0.446$
Cu^+, Cu	$Cu^+ + e^- \rightleftharpoons Cu$	$+0.521$
I_2, I^-	$I_2 + 2e^- \rightleftharpoons 2I^-$	$+0.536$
$AgC_2H_3O_2$, Ag	$AgC_2H_3O_2 + e^- \rightleftharpoons Ag + C_2H_3O_2^-$	$+0.643$
Ag_2SO_4, Ag	$Ag_2SO_4 + 2e^- \rightleftharpoons Ag + SO_4^{--}$	$+0.653$
Fe^{3+}, Fe^{++}, Pt	$Fe^{3+} + e^- \rightleftharpoons Fe^{++}$	$+0.771$
Hg_2^{++}, Hg	$Hg_2^{++} + 2e^- \rightleftharpoons 2Hg$	$+0.789$
Ag^+, Ag	$Ag^+ + e^- \rightleftharpoons Ag$	$+0.799$
Hg^{++}, Hg_2^{++}, Pt	$2Hg^{++} + 2e^- \rightleftharpoons Hg_2^{++}$	$+0.920$
Br_2, Br^-	$Br_2(liq) + 2e^- \rightleftharpoons 2Br^-$	$+1.065$
Pt^{++}, Pt	$Pt^{++} + 2e^- \rightleftharpoons Pt$	$+1.2$
O_2, H_2O	$O_2 + 4H^+ + 4e^- \rightleftharpoons 2H_2O$	$+1.229$
Tl^{3+}, Tl^+, Pt	$Tl^{3+} + 2e^- \rightleftharpoons Tl^+$	$+1.25$
$Cr_2O_7^{--}$, Cr^{3+}, Pt	$Cr_2O_7^{--} + 14H^+ + 6e^- \rightleftharpoons 2Cr^{3+} + 7H_2O$	$+1.33$
Cl_2, Cl^-	$Cl_2 + 2e^- \rightleftharpoons 2Cl^-$	$+1.360$
Au^{3+}, Au	$Au^{3+} + 3e^- \rightleftharpoons Au$	$+1.50$
MnO_4^-, Mn^{++}, Pt	$MnO_4^- + 8H^+ + 5e^- \rightleftharpoons Mn^{++} + 4H_2O$	$+1.51$
Ce^{4+}, Ce^{3+}, Pt	$Ce^{4+} + e^- \rightleftharpoons Ce^{3+}$	$+1.61$
Au^+, Au	$Au^+ + e^- \rightleftharpoons Au$	$+1.68$
Co^{3+}, Co^{++}, Pt	$Co^{3+} + e^- \rightleftharpoons Co^{++}$	$+1.82$
F_2, F^-	$F_2 + 2e^- \rightleftharpoons 2F^-$	$+2.65$

* From W. M. Latimer, "Oxidation States of the Elements and Their Potentials in Aqueous Solution," 2d ed., Prentice-Hall, Inc., Englewood Cliffs, N.J., 1952.

Name Index

Subject Index